BIOMECHANICS: THEORY AND APPLICATIONS SERIES

BIOMECHANICS: PRINCIPLES, TRENDS AND APPLICATIONS

Biomechanics: Theory and Applications Series

Biomechanics: Principles, Trends and Applications
Jerrod H. Levy (Editor)
2010. ISBN: 978-1-60741-394-3

BIOMECHANICS: THEORY AND APPLICATIONS SERIES

BIOMECHANICS: PRINCIPLES, TRENDS AND APPLICATIONS

JERROD H. LEVY
EDITOR

Nova Science Publishers, Inc.
New York

LIBRARY OF CONGRESS CATALOGING-IN-PUBLICATION DATA

Available upon request.

ISBN: 978-1-60741-394-3

Published by Nova Science Publishers, Inc. ✦ New York

CONTENTS

PREFACE

Biomechanics is the application of mechanical principles (statics, strength of materials and stress analysis to the solution of biological problems of living organisms. This includes bioengineering, the research and analysis of the mechanics of living organisms and the application of engineering principles to and from biological systems. This research and analysis can be carried forth on multiple levels, from the molecular, wherein biomaterials such as collagen and elastin are considered, all the way up to the tissue and organ level. This new and important book gathers the latest research from around the globe in the study of this dynamic field with a focus on issues such as; art biomechanics, understanding corneal biomechanics, biomechanical remodeling of the diabetic gastrointestinal tract, biomechanics in children with cerebral palsy, cell biomechanics for orthopaedic implants, and others.

Chapter 1 - While biomechanics has achieved successes in many fields involving locomotion, motor learning, skill acquisition, technique optimization, injury prevention, physical therapy and rehabilitation, one area has heretofore been scarcely represented in the literature – Arts Biomechanics. Biomechanics clearly has significant potential for application in the performance arts, such as music and dance, since skills needed for these activities are visibly related to the human musculoskeletal and nervous systems. In such areas, Arts Biomechanics should begin by focusing on skill analyses and acquisition necessary for the performance of the artistic act. Subsequently it should engage in a deeper discourse that explores the relationship between these and the desired aesthetic outcome. Less apparently, biomechanics may also enhance the analysis and comprehension of other arts, such as painting, where gesture is often embedded in the artwork by means of symbolism, tradition, the process of art creation, or as an inherent product of the existential nature of humanity.

Chapter 2 - The Ocular Biomechanics Group was established in 2002 with one clear target; to develop a virtual reality model of the human eye that can be used effectively and reliably to predict ocular response to surgery, injury and disease. This ambitious, and seemingly illusive, target helped plan our activities over the last 6 years and will still be focusing our efforts as the authors strive to create the necessary knowledge using experimental methods, build the predictive tools using programming and analysis means, and validate the findings in both the laboratory and the clinic. This chapter presents an overview of our biomechanical studies from laboratory material characterisation to finite element numerical simulation. The chapter describes what has been achieved and points at the remaining gaps in our knowledge. It explains that while much remains unknown in ocular behaviour, the authors are now in a good position to use available knowledge to progress predictive modelling and use it in actual applications such as improving the accuracy of tonometry techniques, planning of refractive surgeries and design of contact lenses. The discussion focuses on the cornea, although scleral biomechanics receive some mention. The chapter also refers to microstructural, biomechanical and topographic studies conducted by other research groups. Coverage of these studies has been necessary to provide a more complete image of current understanding of corneal biomechanics.

Chapter 3 - Osseointegrated implants are actually replacements for natural teeth, and, like natural teeth, they are exposed to various forces. The success of osseointegration is based on the clinical outcome; clinicians must ensure that the stresses that the superstructure, implant, and surrounding bone are subjected to are within the tolerable limits of the various components. Structural compatibility is the optimum adaptation to the mechanical behavior of the host tissues. Therefore, structure compatibility refers to the mechanical properties of the implant material, such as elastic modulus, strength, implant design and optimal load transmission (minimum interfacial strain mismatch) at the implant/tissue interface, which is the key to the successful functioning of the implant device.

This chapter reviews some of the reaction, properties and characteristics of the bone and explains how the bone-implant interface will react under loading condition. The chapter also includes characteristics, properties and other important information about the implant biomaterials and implant coating.

Chapter 4 - Gastrointestinal tract sensory-motor abnormalities are common in patients with diabetes mellitus with symptoms arising from the whole GI tract. Common complaints include dysphasia, early satiety, reflux, constipation, abdominal pain, nausea, vomiting, and diarrhea. The pathogenesis of GI symptoms in diabetes mellitus is complex in nature, multi-factorial (motor dysfunction, autonomic neuropathy, glycemic control, psychological factors, etc.) and is not well understood. Histologically, many studies have demonstrated prominent proliferation of different GI wall layers during diabetes. During the past several years, several studies demonstrated that experimental diabetes induces GI morphological and biomechanical remodeling. Following the development of diabetes, the GI wall becomes thicker and the stiffness of the GI wall increases in a time-dependent manner. It is well known that mechanosensitive nerve endings exist in the GI tract where they serve a critical role for tissue homeostasis and symptom generation. Mechanoreceptor-like structures such as intraganglionic laminar nerve endings and intramuscular arrays have been identified. The changes of stress and strain in the GI wall will alter the biomechanical environment of the mechanosensitive nerve endings, therefore, the structure as well as the tension, stress and strain distribution in the GI wall is important for the sensory and motor function. Biomechanical remodeling of diabetic GI tract including alterations of residual strain and increase in wall stiffness will alter the tension and stress distribution in the vicinity of the mechanosensitive afferents with consequences for perception and motility of the GI tract.

Chapter 5 - The gastrointestinal (GI) tract is functionally subjected to dimensional changes. Hence, biomechanical properties such as the stress-strain relationships are of particularly importance. These properties vary along the normal GI tract and remodel in response to growth, aging and disease. The biomechanical properties are crucial for GI motor function because peristaltic motion that propels the food through the GI tract is a result of interaction of the passive and active tissue forces and the hydrodynamic forces in the food bolus and remodeling of the mechanical properties reflects the changes in the tissue structure that determine a specific motor dysfunction. Therefore, biomechanical data on the GI wall are important to understand the pathogenesis to the GI motor-sensory function and dysfunction. Moreover, biomechanical studies of the GI tract pave the way for further mathematical and computational modelling. Biomechanical studies of the GI tract will advance our understanding of GI physiological function, diseases such as dyspepsia and visceral pain, and GI dysfunction due to systemic diseases. Furthermore, integrated GI simulation models will be beneficial for medical education and for evaluation of the efficacy and safety of new drugs on GI function.

Chapter 6 - The force produced by skeletal muscle is controlled by the electrical signals being sent from motor neurons to muscle fibers. These electrical signals, which are known as action potentials, can be recorded as they travel along the muscle cell membrane and are referred to as an electromyogram (EMG) signal. It has been more than a century since the first recording of a voluntary EMG signal was reported, and today it has become a classic technique for evaluating and recording the activation of skeletal muscles during human movement. In recent years, the advent and development of transcranial magnetic stimulation has re-invigorated EMG research, and it is now possible to safely and painlessly evoke EMG signals directly from the motor cortex of conscious humans. This chapter reviews the recording and measurement issues associated with EMG and its respective applications. Particular attention is paid to its role in understanding the neuromechanics of human movement.

Chapter 7 - Children with cerebral palsy (CP) lack the higher-level motor skills present in age-matched typically developing (TD) children. The development of postural control is critical to the acquisition of increasingly complex motor skills as well as to the production of coordinated motor behavior, such as locomotion. This chapter examines recent developments in understanding the abnormal postural control in children with CP and assessments of the effectiveness of rehabilitation techniques using biomechanics measurements. The authors show that the delayed and impaired development of postural control in children with CP is not only due to the immaturity of central nervous system but also abnormal postural alignment and muscle force production.

Children with CP often have difficulty maintaining stability when facing unexpected threat to balance. The authors present studies comparing reactive balance control in children with spastic diplegic cerebral palsy (SDCP) and TD children using support surface perturbation and show that a number of factors contribute to poor balance control in children with SDCP. 1) There was a temporal disorganization of joint torque activation. 2) There was a spatial disorganization of the joint torque profiles. 3) Children with SDCP also showed slower speed to reach peak torque value. In addition, the authors show that when TD children were asked to mimic crouched stance as that seen in children with SDCP, they exhibited abnormal postural control as well, indicating that musculoskeletal constraints are also contributors to the atypical postural muscle response patterns seen in children with SDCP. These findings suggest that the neuromuscular response patterns of some children with SDCP may be appropriate strategies for their musculoskeletal constraints secondary to deficits in the neural system.

In this chapter the authors also discuss the culmination of our findings in relation to clinical applications in the management of musculoskeletal impairments to improve postural control in children with SDCP, and the significance of using biomechanical measures to show a direct relationship between the impairments of the musculoskeletal system and reactive postural control as well as possible coping strategies used by children with SDCP. The authors examine the current biomechanical research used to ascertain the effectiveness of two therapeutic interventions purported to affect the musculoskeletal system for the improvement of function in children with SDCP, ankle foot orthoses (AFO) and strength training.

Finally, the authors examine the gaps in current clinical research when assessing the effectiveness of interventions to reduce the musculoskeletal impairments constraining static, reactive, and dynamic balance control in children with SDCP.

Chapter 8 - The knowledge of corneal biomechanical properties of cornea has gained importance in recent years. Investigators have been trying to find easy and practical ways to establish these biomechanical properties but to date have had to rely on corneal thickness measures to give an idea of corneal biomechanics. This review explores what is known about the biomechanical properties of the human cornea.

An overview of corneal thickness measurements, its impact on measurement of intraocular pressure and its importance in various disease states is discussed. The recent advent of the Ocular Response Analyser, an in-vivo measure of ocular hysteresis and corneal resistance factor and the pulse waveform associated with this will be discussed. The importance of this machine with respect to corneal biomechanics will be presented.

Chapter 9 - This chapter addresses a fundamental aspect of musical performance in string players: how the physical "geography" of the instrument and bow, and the anthropometric dimensions of the player interact to produce the stereotypic motor behavior.

Several factors determine movement, but, unlike most biomechanical tasks, the determining outcome here is acoustic. Both upper extremities are involved in the performance but in very different ways.

The left arm governs the contact position of the fingers on the string, and hence the pitch of a note. The spatial relations of the instrument to the body, the contact point of the finger on the string, the length of the fingerboard, and the dimensions of the arm determine a unique posture, and thus the muscle activation patterns, for the left arm for each individual player. The left arm has quite different postures in relation to the body and the instrument for cello players in comparison to violinists.

The bowing (right) arm draws the bow across the string. Its travel velocity is the principal determinant of loudness, but its distance from the bridge, the contact position along the bow length, bow pressure, and its angle of attack on the string leave a noticeable effect on the tone, or timbre. Bowing movements are essentially determined by the flexion and extension of the elbow, with subtle motions of the shoulder and wrist to keep the bow moving in a straight line perpendicular to the string. Given a fixed spatial relation between the body of the performer and the instrument, the posture of the arm for a given bow/string contact point is uniquely determined.

In addition, the force of gravity plays a role in the control of movement. The authors show, however, that it affects cellists and violinists in very different ways. Left arm movements of cellists are more affected by gravity than those of violinists; whereas gravity affects the right arm more in violinists.

In this chapter the authors focus on a series of specialized topics: 1) control of the left arm during shifting movements; 2) fine control of the left arm during corrections of intonation errors; 3) coordination between the upper arm and forearm; and 4) coordination between the left and right arms.

Chapter 10 - There exist several invasive and noninvasive methods to measure the contact hip stress but due to their complexity only few have so far been tested in clinical trials with large numbers of participating subjects. Consequently, the use of contact hip stress measurements in orthopaedic clinical practice is still in its experimental phase.

Biomechanical studies of human hips based on the analysis of 2-D pelvic radiographs have turned out to be a reasonable compromise between the measurement accuracy and the feasibility in clinical setting. Clinical studies have shown significantly higher values of hip stress in adult dysplastic hips when compared to normal hips. It has been found that the cumulative hip stress independently predicts the WOMAC score after 29-years of follow up in dysplastic hips and does so better than morphological radiographic parameters of hip dysplasia or the resultant hip force alone. The preoperative value of the contact hip stress and the magnitude of its operative correction have been found predictors of the long term success of the Bernese periacetabular osteotomy. Elevated shear stress in femoral neck, but not elevated hip contact stress, has been found to be a risk factor for slipping of the capital femoral epiphysis. A statistically significant correlation between the contact hip stress and the age at the total arthroplasty has been shown in a group of hips with idiopathic hip osteoarthritis.

Through advances in 3-D imaging with MRI and CAT scan, visualisation of the femoral head coverage and pelvic muscle attachment points has improved considerably. However, the need to supplement the morphological hip status with biomechanical analysis remains. The current trend is to combine the kinetic gait measurements of the resultant hip force with 3-D imaging of the hip weight-bearing surface in order to better estimate the contact hip stress for a given activity/body position. The added value of such measurements over 2-D pelvic radiograph analysis has not been established yet in clinical trials.

Chapter 11 - Resistance exercise has long been used to promote musculoskeletal health with the application of training regimens for the clinical treatment and prevention of low back pain growing in popularity over the last couple of decades. A variety of exercise modes have been utilized in an attempt to stimulate and promote increases in muscle function of the lumbar extensors. This chapter examines the current state of knowledge regarding the application of external pelvic fixation during trunk extension exercise and its importance on the concomitant increase of functional outcomes such as muscle strength, muscle activation patterns and compensatory muscle growth.

Chapter 12 - Most revisions of total joint replacements are due to implant loosening, which is mainly caused by wear particles ("wear disease") and inadequate primary implant stability. The optimised integration of cementless total hip and knee endoprostheses into the bone stock is the most adequate approach to achieve secondary implant stability and to prevent implant loosening. Secondary stability is characterized by bone ingrowth of the implant and decreases the amount of relative implant motion between the implant and bone stock. It has also been suggested that prostheses which are fully occupied by bone cells are less susceptible to infection. The economic impact of implant loosening is immense, hence orthopaedic implant manufactures refine their products continuously.

Many technical developments have improved the survival rate of endoprosthetic implants. Modern materials and surface modifications such as coatings help to reduce wear rates, promote cell ongrowth or prevent infections. The cell adhesion of bone cells onto implant surfaces has not been thoroughly investigated so far. However, different methods to measure cell adhesion have been described. Some workgroups investigate short-term adhesion or proliferation of bone cells on implant materials *in-vitro*, but little is known about the long-term adhesion. Proliferation or short-term adhesion cannot predict how strong the bonding between bone and implant will be. In most cases, cost intensive animal studies have to be performed in order to gain expressive data. Hence, it is important to assess the bone cell adhesion forces in an adequate experimental setup *in- vitro*.

The exploration of bone cell adhesion on surfaces of orthopaedic implants encourages the development of bio-compatible, bio-active and anti-infectious surfaces. The authors have developed a test device, based on the spinning disc principle, which allows quantitative measurements of osteoblastic cells on implant surfaces. First results show differences in adhesion forces depending on the substrate. In future assessments different bio-active and anti-infectious surface modifications will be analyzed regarding bone cell adhesion prior to animal studies.

Chapter 13 – The aim of this chapter was to establish and compare the patterns expressed on the fast model of motor learning of children and adults executing a fast and accurate task.

The acquisition of a new motor skill follows two distinct stages with continued practice: first, there is an early, fast learning stage in which performance improves rapidly within a single training session; later, there is a slower learning stage within the time period of several sessions of practice. Motor learning is characterized by a specific set of changes in performance parameters. These changes occur gradually in the course of a learning period. While the decreases or increases in these parameters have been documented in a variety of tasks, it remains to be determined whether the time of fast learning is different for children and adults. Therefore the main aim of this study was to establish if there are differences in reaction time, average and maximal velocity, trajectory, and accuracy as well as the variability of these parameters during motor learning. The tasks involved 5 series with 20 repetitions in each.

Chapter 14 – Fit is one of the most critical factors affecting footwear comfort. Blistering, chafing, bunions and pain may be the result of poor fitting shoes. In long run, it may cause the foot skeleton deformity.

In order to find out the proper fitting of footwear, it involves getting to know the size of feet, shoes, and the subjective perception for the shoes selection. Traditional method in measuring the feet size is to measure the length and width of the feet which can be obtained easily by tape measure and devices like Brannock. However these are considered to be insufficient for good footwear fitting. Furthermore, researchers were also encountering problem in quantifying fit as it is rather subjective which was also suggested to be affected by shoe wearing experience such as tightness and looseness of the shoes. Therefore researchers are exploring new method in measuring footwear fit, both objectively and subjectively.

Chapter 15 – Usually, biomechanical models used for human motion analysis are oversimplified, especially for clinical analyses (Helen Hayes model). The calculated net joint forces and torques are sensitive to the input data: segment kinematics and body segment inertial parameters. It is therefore necessary to improve these input data using new methods and models adapted to the population and movement of interest. The general problem is divided into three parts: (i) minimization of soft tissue artefacts, (ii) joint centre location and (iii) identification of the personalized body segment parameters.

In: Biomechanics: Principles, Trends and Applications
Editor: Jerrod H. Levy, pp. 1-55
ISBN: 978-1-60741-394-3

Chapter 1

ARTS BIOMECHANICS – AN INFANT SCIENCE: ITS CHALLENGES AND FUTURE

Gongbing Shan* and Peter Visentin
[1]Department of Kinesiology,
[2]Department of Music,
University of Lethbridge,
4401 University Drive, Lethbridge, Alberta. Canada, T1K 3M4

1. OVERVIEW

While biomechanics has achieved successes in many fields involving locomotion, motor learning, skill acquisition, technique optimization, injury prevention, physical therapy and rehabilitation, one area has heretofore been scarcely represented in the literature – Arts Biomechanics. Biomechanics clearly has significant potential for application in the performance arts, such as music and dance, since skills needed for these activities are visibly related to the human musculoskeletal and nervous systems. In such areas, Arts Biomechanics should begin by focusing on skill analyses and acquisition necessary for the performance of the artistic act. Subsequently it should engage in a deeper discourse that explores the relationship between these and the desired aesthetic outcome. Less apparently, biomechanics may also enhance the analysis and comprehension of other arts, such as painting, where gesture is often embedded in the artwork by means of symbolism, tradition, the process of art creation, or as an inherent product of the existential nature of humanity.

There are many challenges facing the integration of the Sciences with the Arts. On a fundamental level, the principles and goals of one often seem at odds with the other. In reality, neither science nor art is antithetical to the other.

> "The most beautiful experience we can have is the mysterious. It is the fundamental emotion that stands at the cradle of true art and true science. Whoever does not know it and can no longer wonder, no longer marvel, is as good as dead, and his eyes are dimmed. It was

* Ph: (403) 329-2683, e-mail: g.shan@uleth.ca

> the experience of mystery -- even if mixed with fear -- that engendered religion. A knowledge of the existence of something we cannot penetrate, our perceptions of the profoundest reason and the most radiant beauty, which only in their most primitive forms are accessible to our minds...I am satisfied with the mystery of life's eternity and with a knowledge, a sense, of the marvelous structure of existence -- as well as the humble attempt to understand even a tiny portion of the Reason that manifests itself in nature."(Einstein, 1931)

However, the self-perceptions of artists and scientists may be problematic. Scientists take reductionist and reasoned approaches to the world: there is a position/argument; it is structured in logical steps; a topic sentence names and proves each idea in the discourse; and, the discourse is aimed at a specific result. For the scientist, these give feelings of comfort, power, and/or control over the phenomenon. Conversely, the artist is less interested in the factual and desires to convey the emotional and sensual; reductionist, topical arguments are considered antithetical to the creation of good art; too precisely formulated a conceptual structure is perceived as negating mystery and limiting artistic possibilities; and, the result is a product of the moment – an ever shifting target in time-based performance art. Repeatability is only a desirable quality in that it can be a measure of skill level. Hence it shows little in terms of creative ability. The artist wants power and control over others' perceptions of the phenomenon; each individual audient's experience of the art responding to and thus validating the artwork and, by extension, the artist. Perhaps it is the way in which scientists and artists perceive their own roles that a binary view pervades their respective disciplines – something is or isn't. Given this seeming dichotomy, the question arises "Why bother with Arts Biomechanics?"

The issue that provides the main inertia to expand this nascent field of biomechanics comes from artists. It is a medical one. Epidemic rates of debilitating injury (48-76%) occur in performing arts such as music and dance (Brown, 1997; Fry, 1986, 1987, 1988; Fry, Ross, and Rutherford, 1988; Hagglund, 1996; Hartsell and Tata, 1991; Lockwood, 1988; Middlestadt and Fischbein, 1989; Zaza, 1992, 1998). Unfortunately, there is currently little quantitative research examining the aetiology of performance injuries. Existing strategies to address injuries are largely qualitative and "experience" based. And they are normally employed only after injury has occurred. Even as artists are becoming increasingly aware regarding their physical needs, career longevity and injury downtime are becoming pervading arts industrial issues. Only recently have researchers begun to explore scientific approaches, such as human performance engineering, to trace causal factors related to human bone, muscle, and nervous systems injuries in the performing arts (Chesky, Kondraske, Henoch, Hipple, and Rubin, 2002; W. J. Dawson, 2003, 2007; Solomon and Solomon, 2004). Finally, artists themselves are slowly turning toward science to provide preventative answers and not merely remedial ones. From a phenomenological point of view, the performing arts share many characteristics, including health risks, in common with other skill-oriented activities (Chesky et al., 2002; Lehmann and Davidson, 2002; Wilson, 1986). Commonalities between athletic and artistic performance seem obvious, particularly in the area of motor skill analysis, acquisition during skill learning and performance. Like athletes, elite musicians practice, with many hours of repetition, to perfect complex motor control sequences. However, music students are seldom introduced to basic principles of movement science and physiology that underpin that activity. This is largely because the focus of music teaching is artistic and outcome driven rather than process oriented – this is a difference between practicing and training. While some teachers are knowledgeable regarding efficient use of the body for the benefit of good performance, many are not. In many parts of the world, classical musicians today are taught in a manner virtually indistinguishable from that used 50 years ago. Nowhere in the world could the same be said for elite sport.

As such, it seems logical to build on the successes of Sports Biomechanics in the service of the Arts. Motor learning, skill acquisition and learning while minimizing injuries constitute the main research focuses of Sports Biomechanics (Ballreich and Baumann, 1996). However, there is a fundamental ethical difference between these two fields. Whereas Sports Biomechanics typically directs is energies to achieving specific, goal-driven, quantifiable results that are valued for their repeatability (e.g. faster, higher, stronger, etc.), Arts Biomechanics must be satisfied with guiding the process without appearing to identify an absolute goal – for reasons of artistic and creative freedom. The emphasis of each is on training, but Arts Biomechanics needs to contribute to a demystification of the learning and skill acquisition processes, so that artists can realize their full potential *and* survive their chosen vocation.

> "My early love affair with dance gradually had been replaced by a struggle to become what I could not be. I had selected teachers who felt it was their responsibility to tell me over and over what was wrong, rather than helping me generate the knowledge that would give me tools to make things right" (Evans, 2003)

Clearly, the need to consider artistic values as well as scientific ones creates particular challenges for the field of Arts Biomechanics. Facing these challenges is, in our opinion, best served by a multidisciplinary approach – one where research does not simply adopt a scientific or artistic practice, but engages in discussion that ultimately transcends the viewpoint of each discipline. Meaningful and relevant research results will be those where science informs artistry rather than attempting to modify it. Currently, music and dance are the two dominant areas of biomechanics research in the arts. For this reason, discourse below will primarily focus on these disciplines. The main effort of this chapter is to summarize the state of Arts Biomechanics in the following areas: 1) skill analysis, acquisition and pedagogy, 2) injury risk identification, quantification, prevention and compensation strategies, and 3) Innovative uses of the tools of movement science in the analysis and creation of art. Further, this chapter will provide discussion that identifies some of the challenges facing Art Biomechanics, elaborate on its potential and identify some future directions.

2. Performance Skill Analysis and Acquisition

2.1. Historical Overview

Documentation pertaining to instrumental performance and dance has a long history. In terms of modern classical performance traditions, documentation that incorporates information on mechanics and motor skills begins as early as the 16th century for instrumental performance and the 17th century for dance (Hilton, 1997; Kolneder, 1993). Its very existence can be considered evidence of a general desire to improve overall quality of music/dance performance, both stylistically and technically. Most of the documents are pedagogical in nature and they are clearly not targeted at the mature or virtuoso performer. For example, in his seminal treatise on learning to play the keyboard, Carl Phillip Emanuel Bach clearly states his intent.

> "...keyboard instruction could be improved in certain respects to the end that the truly good which is lacking in so much music, but particularly keyboard music, might thereby become more widespread. The most accomplished performers, those whose playing might prove instructive, are not to be found in such numbers as might perhaps be imagined." (Bach, 1759)

In terms of classical dance, the beginnings of ballet and classical notation are associated with the Court of Louis XIV of France (1639-1714) (Hilton, 1997). Iconography, dance manuals and descriptions of the desired aesthetic provide some of the first rudimentary (bio)mechanics information for analysis and acquisition of dance skills (Figure 1 and 2).

Figure 1. Choreography of a Minuet providing rudimentary instruction on placing the feet. The original source is from an English dance manual by Kellam Tomlinson, and documents the influence of French culture in the 18th century.

Figure 2. A choreography of steps from "Balet de Neuf Danseurs" by Feuillet. Paris, 1700.

Collectively, such sources contain fundamental instruction on music/dance and their contemporary aesthetic goals. They include postural descriptions, instruction pertaining to skill acquisition such as: fingerings in basic positions, embouchure (for wind instruments) and bowing techniques (for strings), and they describe the phenomenology of physically interacting with a musical instrument (how to hold it, the consequences of certain muscular tensions, tablatures revealing mechanical insights, etc.). Choreographed movement in dance also create a gestural map for both the performer and dancer, since the music must maintain aesthetic of bodiliness required in dancing (Bach, 1759; Hilton, 1997; Mozart, 1756; Quantz, 1752; Tromlitz, 1791). Clearly, these point to an awareness of the mechanics of the human-tool interface, whether that interaction is with an instrument or a physical space.

Further, almost on a century-by-century basis, one can observe a progression to more and more systematic approaches toward artistic performance. Documentation pertaining to violin performance can be considered exemplary in this regard. Earliest documentation, from the 16th century, was descriptive, providing relatively little instruction on technique. The 17th Century included some of the earliest tutors. They contained fundamental postural descriptions and instruction on music basics: string tunings, fingerings in basic positions, elementary bowing techniques. Included are some instructions on bow direction, accents, and attempts to describe articulations and expression. Tablatures (musical fingering charts) revealed insights into pedagogy. Information on style and technique may also be gathered from music that incorporated performance indications by the composer. The 18th Century rise of a middle class and a corresponding increased interest in education lead to maturation of instruction methods and pedagogy. More methods and music were published. Contents became more detailed with broader variety of musical examples and more instructions how to coordinate techniques in order to play expressively. Some compositions dealt with specific technical difficulties or material for specific skill acquisition.

The 19th Century showed evidence of a paradigm shift (Baillot, 1835). There was a move from general education and a "collective consciousness" to a focus on individual training and virtuosity. The period marked the beginnings of music pedagogy as a science.

Figure 3. Position of the right arm and wrist held closer to the body when the violinist plays sitting down. The neck of the violin is then held slightly lower. Position of the hand, wrist and fourth finger in extensions (top-right). Forced and improper position of the hand, wrist and fourth finger in extensions (bottom-right).

Documentation used a new tone emphasizing the development of technique as a means to achieve the highest degrees of artistry. The focus was on technical training (Figure 3). During the period, a gradual increase in notion of finding the "best" way to practice can be observed. Some writers advocated a "mechanical approach" to the instrument with attempts to apply scientific methods to describe these mechanics. Others began to approach the subject of performance psychology. Regarding anatomy and physiology, methods to this point may be summarized as rudimentary. They were experience/practitioner-based with superficial descriptions of postures or (bio)mechanics (Baillot, 1835). Attempts to include an understanding of anatomy and physiology were a development of late 19th and early 20th centuries.

Over the course of the 20th century, there was a continued development of systematic teaching methods. Also, there was an increasing trend toward learning from the standpoint of different disciplines: physiology, psychology, (bio)mechanics. Some training exercises were developed (at times by individuals who may have had little to no "practical" experience playing music – e.g. medical doctors, physiology professors) that pushed mechanics to conceptual and physiological limits. As can be seen in Figures 4-6, the body was typically treated as a machine which could be trained, or programmed, through repeated and structured movement variations (Hodgson, 1958).

Most 20th-century writings that attempted to document teaching methods of "successful" practitioner/pedagogues failed to deal with fundamental psyco-physiological learning, biomechanics, or neural control except in superficial terms (G. B. Shan, Visentin, Wooldridge, Wang, and Connolly, 2007). For example, after mentioning postural or internal motor control factors in general terms, most typically continue with experience-based description of techniques and their application in selected passages of repertoire. Thus, they are written in such a way that only those who have experience with the phenomenon and the aural tradition being described can sympathize with the descriptions.

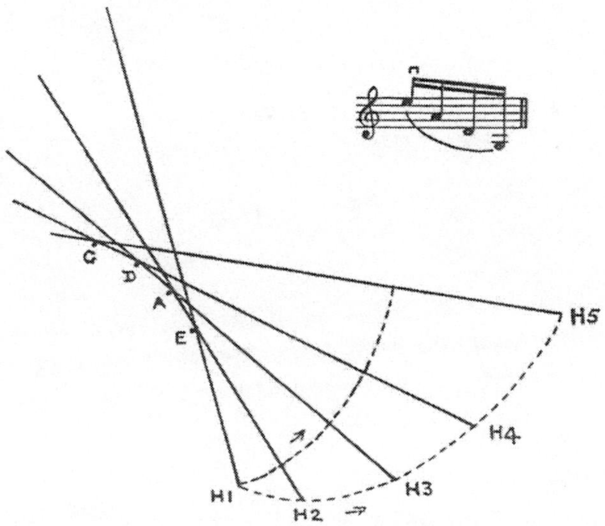

Figure 4. Illustration considering the curve which results from the identified ordering of notes, while using a down bow. The two dotted lines give variations in the track followed by the hand.

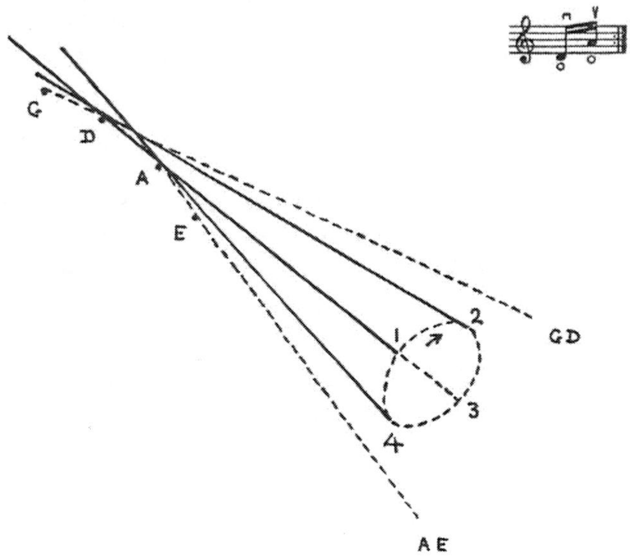

Figure 5. The figure shows a possible elliptical path (marked 1, 2, 3, 4), traversed by the heel of the bow during the playing the identified two notes.

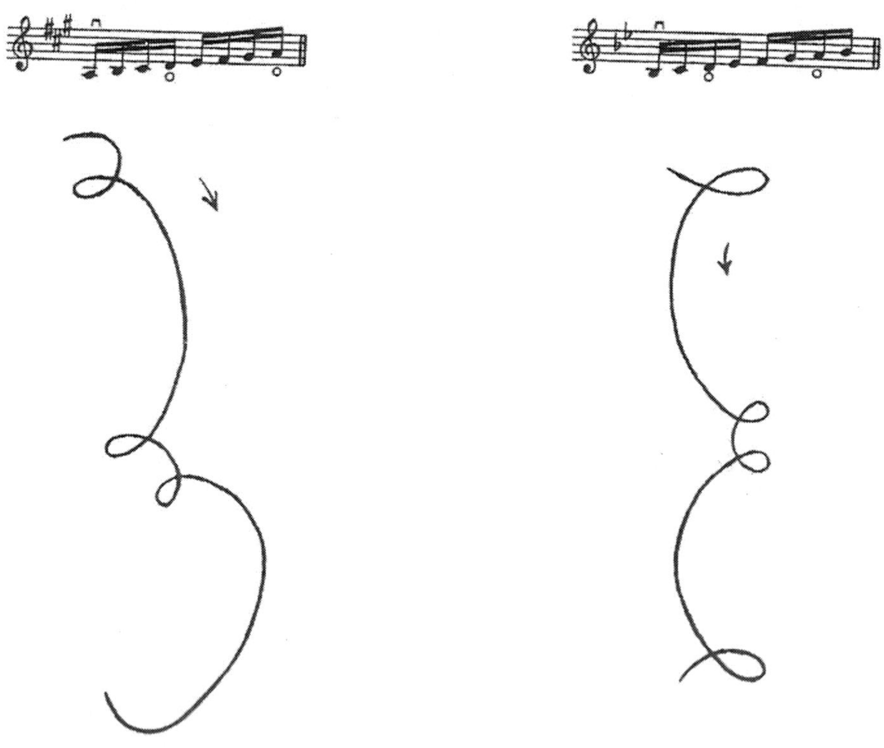

Figure 6. Small loops occurring in the bow-hand that result from string crossings for the identified scales. String crossings occur on an "up-bow" for the figure on the left and on a "down-bow" for the one on the right.

Clearly, in the 21st century, there can be a role for the field of biomechanics in terms of the pedagogical process. Demystification of the mechanics of fundamental skills, and direction regarding the best way to learn them, is in the best interests of these disciplines – especially if it can lower injury rates and/or improve the physical health and well-being of performing artists. Currently, Arts Biomechanics in its infancy as a field of study.

2.2. Challenges Faced by Arts Biomechanics

Notwithstanding efforts to address important artistic concepts through written documentation, the performance arts has and continues to perpetuate its traditions mainly through an apprenticeship learning model. Advanced teaching/learning typically occurs in a one-on-one environment and arts education relies heavily on aural traditions, with pedagogical frameworks by and large limited by conventions and aesthetic values passed from teacher to pupil. Participants in such a model typically take great pride in the provenance of their traditions – often validating their approach by tracing their artistic lineage from teacher to teacher to one or another of the legendary "greats" of the past. Such a mind-set frequently inspires a mystical and, in some cases, a near-religious reverence of the teacher, artistic traditions, concepts, and acquired skills. Nearly everyone who is involved in the performing arts can, through personal experience, cite anecdotal evidence confirming this assertion.

In such a system, and since the performative arts are heavily influenced by the need for an audience, *performance* and *pedagogy* are driven by external validation. Frequently, a teacher's quality is judged by the successes of his/her "prize" students. As artistic success or failure is often determined by micro-increments of "talent", this can lead to an emphasis on immediacy of skill acquisition, sometimes at the expense of sound learning practices (Visentin, Shan, and Wasiak, 2008). According to Ruggiero Ricci, one of the 20th century's premier violinists:

> "Teachers are ruining a lot of kids. Parents love to say 'my kid plays the Mendelssohn Concerto' so they take their child away from a teacher who knows better and gives him to one who will push him. … they are the real criminals in this business" (Schwartz, 1983).

Notwithstanding this somewhat arcane approach, numerous examples of success within the model reinforce the perceived legitimacy of its practices.

However, there are significant problems with such simplistic attitudes and rationales. First, in order to be scientifically defensible, the "talent" must either be an innate characteristic of individuals, somehow defined by their genetic makeup, OR it must be a byproduct learned behaviors and environmental circumstance. At any given point of time, it is relatively easy to observe levels of skill development and ease of skill acquisition (talent) for individuals within a group of learners. In one scientific study, subjects were sorted into two groups based on expert assessments of their "talent" (Levitin, 2006). After following the groups for several years, results proved discrimination on the basis of "talent" to be a false determinant of success over the long term. Learning/practice was the ultimate cause of achievement, and not innate ability. Hence, a focus on the learning process needs to dominate arts pedagogy.

Further, success as defined by the performance is a measure which oft-times fails to concern itself with the long-term health and development of the artist. Since the medical literature shows overwhelming evidence of vocational disaster over the "long-haul" due to technique-related injuries, there has been an increased interest in improving pedagogical practice in the performing arts. Past studies have shown that 48% - 76% of professional musicians need to stop performing for extended periods due to occupational injury (Brown, 1997; Fry, 1986, 1987, 1988; Fry et al., 1988; Hagglund, 1996; Hartsell and Tata, 1991; Lockwood, 1988; Middlestadt and Fischbein, 1989; Zaza, 1992, 1998). As a result, old platitudes such as "the way to Carnegie Hall is practice, practice, practice" have evolved into "the way to Carnegie Hall is practice, *rest*, practice" (Kaplinski, 2003). Further, traditional practices, such expecting students to engage in long hours of repetitive work often with little consideration for its efficiency or biological implications, are now recognized as part of the problem, even if there is a reluctance to changing the mind-set. The performing arts appear to be ready for more scientific collaborations.

Arts Biomechanics can play an important role in challenging the inertia of tradition that exists in the performing arts. By supplying quantitative rather than empirical evidence for improving pedagogy, this sub-discipline of applied biomechanics can improve understanding of physiological phenomena, motor control, the motor learning process and internal loading implications for the body during practice and performance. And, with care, it can do so without being perceived as a threat to the artistic end-product. Ultimately, this is the most significant practical challenge for Arts Biomechanics.

Unfortunately, there is currently little quantitative research in Arts Biomechanics. One reason for this may be that both the performing arts and biomechanics are highly specialized studies each with their own cultures. These cultures may have language barriers (the same words can have different disciplinary meanings, and hence loading), goal differences (personal and subjective experience is desired for one and quantitative observation of a process that leads to specific goals for the other), and/or conflicts between their needs (the arts require flexible outcomes that accommodate changing tastes/ideas and the sciences desire outcomes that are essentially quantifyable/immutable). Further, from the standpoint of arts pedagogy, because of anthropometrical variation between teacher and student, it can be the tendency for an arts teacher to focus on stylistic outcomes, which in turn obliges the student to find their own motor control means to that particular end. Arts Biomechanics can inform the teacher/pupil interface, supplying common ground so that a teacher can adjust his/her understanding to a student's individual anthropometry, and so that the student can better comprehend the teacher's instruction. Biomechanics has great potential to provide linkages between science and the arts. Clearly, collaborative work is needed in order to bridge disciplinary divergences. In short, Arts Biomechanics needs to be an integrative force, a nexus for communication that is not merely coordinative of differing views but accounts for both perceptual experience and scientific quantification at all stages of the work. *This is the most significant conceptual challenge in the application of biomechanics to the arts.*

Finally, the overview of this chapter began by identifying some philosophical challenges for Arts Biomechanics. Individuals within the arts and the sciences have a vested interest in maintaining the fundamental precepts and traditions which define the relationship between discipline and practitioner. For the artist, quantification and transparency represent threats to individuality and artistry. For the scientist, subjectivity is something that merely has not yet been objectified. *Changing attitudes to foster recognition of the potentially symbiotic relationship and finding convergences that can link these ethics are the most significant philosophical challenges for Arts Biomechanics.*

In any motor skill endeavor, skill acquisition can be categorized into three primary stages: cognitive, associative and automotive (Magill, 2001). Clearly, given the cultural characteristic of the performing arts and applied biomechanics disciplines, the greatest potential for collaboration between the two seems to lie in addressing the cognitive and associative stages of the learning process. Such an approach does not challenge artistic values, and it builds on the strengths of biomechanics as a means to understand and further motor skill development. Additionally, it has potential to effectively address injury prevention through quantitative research examining the aetiology of performance injuries. Since most existing strategies to address injuries are largely "experience" based and they are normally employed only after injury has occurred, it appears logical to begin with an examination of pedagogy's contribution to this problem. Arts Biomechanics has an important role in this regard (Visentin et al., 2008). To this end, Arts Biomechanics as a field of study has begun with the following two questions in mind: "can we identify, describe, and *quantify* qualities that lead to artistic success while minimizing the potential of performance injury?" and "how do we communicate these qualities broadly to the appropriate audience?"

2.3. Arts Pedagogy; What Can We Learn from Biomechanics Applications?

The use of biomechanics-based motor learning in sports pedagogy has proven highly successful in improving human performance (Ballreich and Kuhlow, 1980; Magill, 2001). Its focus has usually also been on enhancing physiological limits of athletes. Superficially, sport and the performing arts may appear to have more or less in common, depending on the art. For example, the athletic aspects of dance are obvious whereas those of music performance often are not. All rely upon the training of human motor skills and the development of gross and fine motor control. But, in music and dance the focus is not always overwhelmingly on pushing physiological limits as it is in sports. Rather it is about controlling the process during the performative act. If we continue with a sports analogy, in this regard arts performance can be likened more to acrobatic kinds of sports than to those which rely mainly upon speed and power. In arts performance, fine motor control seems to be a necessarily dominant concern. Fundamentally, both sports and arts performance involve motor learning and the sublimation of neural control to effect a desired physical outcome. In music, although success is judged by an "aural" outcome, it is a product of physical gestures. In dance, the result is a visual one that is obviously guided by gesture. Since sports biomechanics is at an advanced stage of application, it seems sensible to adopt and modify its already proven successful methodologies to the context of the performing arts.

The perspective it can provide may contribute to a general demystification of the teaching/learning process in the arts with a corresponding improvement in overall pedagogical success as measured by both learning outcomes and injury prevention. When adapting a sport-based motor-learning model to the arts, one must consider its multidisciplinary character.

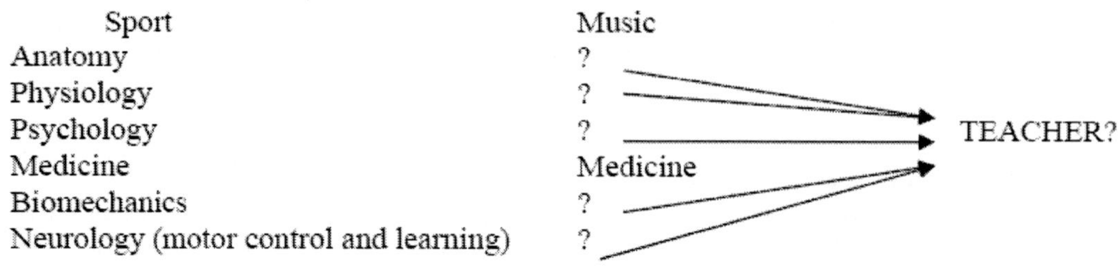

Figure 7. A comparison of the resources used in the pedagogical processes of sport and music.

In sports, intervention of related disciplines to influence pedagogy and learning is a mature science; a team of specialists guides the learning/training. Interdisciplinary research is commonly understood to hold significant potential for improving performance. In performing arts education, responsibility rests primarily, if not solely, with the teacher who may only have minimal training in some of the multiple required areas of competency (Figure 7). Even in the area of arts medicine, prevention is an underutilized strategy for artists. Mostly, medical attention is only sought for injury remediation.

The consequences of poor health and wellness practices for performers can be devastating. They include pain, injury, long-term disability, and even career-ending pathologies. With regard to the intervention of related disciplines, arts pedagogy is in a state of development similar to that of sports pedagogy about fifty years ago. Learned behaviours in the arts are dominated by the vision of the end result. In this regard it is fair to draw from sport history (fifty years ago) as accurately representing and predicting the needs in the arts. One sports anecdote that universally receives sympathy, acceptance and affirmation when presented to the performing arts community involves Tom Simpson, British cyclist and world professional road racing champion in 1965. His mind-set can be summarized by his two most famous quotes. An advocate of performance enhancing drugs, he is noted as saying "if ten will kill me, give me nine". And his final words as he lay dying from cardiac arrest at the top a mountain during a race were "put me back on my bike" (Fotheringham, 2002). Arts practitioners empathize with this "pushing the limits" attitude.

Accordingly, the first aims of those who investigate Arts Biomechanics have been to begin exploring the utility of methodologies and technologies already successfully employed in mature fields of kinesiology and sports medicine to examine arts performance. The bases for most of these explorations have been grounded in two fundamental practices of kinesiology: motor control and motor learning. Not only do these provide a methodological means to analyze movement, cognition and the learning process, they provide an improved basis for discourse through an established commonality of idiom and terminology. Simply, they provide a means for us to all speak the same language.

A Motor-Learning Model

Two foundational theories provide important perspective on skill acquisition in the performing arts. The first, Ballreich's model for motor learning identifies three requisite elements for success: (a) accurate characterization of the skill to be acquired, (b) skill transfer between dissimilar bio-systems, and (c) skill acquisition without injury (Ballreich and Baumann, 1996).

1. *Characterization of the skill.* Characterizing the skill to be acquired involves scientific analysis to identify and describe motor control patterns (i.e. the interaction and coordination of the neural and musculoskeletal systems). Methodologies from kinesiology such as motion analysis are ideally suited to this purpose. Motor control patterns common among expert performers can be generalized, establishing "markers" which can be used as references in the pedagogical process. Using this approach, characteristics of high-level performers can be contrasted with those observed in learners. Such a quantification of fundamental skills begins to address the most practical challenge in the arts – a clear understanding of motor control patterns needed to communicate the desired outcome.
2. *Skill transfer between dissimilar bio-systems.* For a successful motor learning model, fundamental skills must be understood in such a manner that the teacher can transfer the principles to the student regardless of physical differences. The most significant pedagogical weakness in "experience-based" teaching approaches is the fact that not everyone is the same anthropometric shape. Simply, the teacher's experience does not *really* apply. Clear characterization of skills and marker-based learning can facilitate skill-transfer between teacher and student by providing a general model of the specific kinematic gestures vital to skill acquisition. However, for successful skill transfer, "universal" characterizations need to adjust for the "particular", compensating for anthropometric differences among learners. For this to occur, a sufficient body of Arts Biomechanics research is required to inform teachers' practical experience.
3. *Learning without injury.* By directing attention to specific motions or neural control patterns that novices must assimilate into their technique, skill transfer can occur efficiently and effectively; reducing the assimilation time needed to acquire each skill or technique, and thus reduce the risk of Repetitive Stress Injuries.

The second theory addresses skill development at the different stages of the learning process. Relatively simple, the Fitts and Posner model has been proven effective in the discipline of kinesiology (Magill, 2001). It identifies three general stages: cognitive, associative and autonomous.

1. *Cognitive.* The beginner focuses on cognitively oriented problems. Performance is highly variable, showing a lack of consistency from one attempt to the next. Often, learners are confused about what they need to do in order to improve.

2. *Associative.* Cognitive activity changes because the learner has begun to associate multiple references in order to guide the outcome. Through training and repetition learners begin to make fewer gross errors as they refine and combine the multiple skills acquired in the cognitive stage of the process.

3. *Autonomous.* Here the skill becomes almost automatic or habitual. Learners can perform with repeated degrees of accuracy, without conscious thought.

	Cognitive	Associative	Autonomous → Practice time
Goals of each stage	Clarity of description	External and internal processes consciously related	Moves from conscious to subconscious
	Skill transfer	Coordination of multiple fundamentals	Desired artistic outcome drives process
	Injury prevention	Injury prevention	Injury prevention

Figure 8. The stages of motor learning adapted from models of Ballreich and Magill.

Both of these models address different aspects of motor learning. One deals with biomechanical factors influencing the process of motor learning and the other concentrates on the process of skill development (Ballreich and Baumann, 1996; Magill, 2001). Collectively, these two models have important implications for the development of pedagogical practices in the arts. Figure 8 is a combinatorial model, representing Ballreich's model for motor learning on the vertical axis and Fitts and Posner's skill development model on the horizontal one (Visentin et al., 2008). In our opinion, rationalizing this model for arts pedagogy requires careful consideration and a nuanced understanding of disciplinary discourse within the arts. One way in which application of the model may differ from a sports one lies in the emphasis that must be placed at each stage of the learning process. Given arts pedagogy's tendency toward teleology, new ideas need significant evidence (e.g. improved performance outcomes) before they will be adopted. This requires: 1) clarity of definition, 2) convincing methodology, 3) consistent means of situational comparison, and, 4) solutions that do not compromise the artist's perceptions of the end product. Such a rationalization holds significant potential for improving arts pedagogy.

This has clear ramifications for the respective roles and responsibilities of both teacher and pupil. In the cognitive stage, the onus is on the teacher to clearly and accurately characterize the skills to be acquired using instructional strategies such as modeling, clear explanations, analogies, and a vocabulary that accurately describes the phenomena. In the associative stage, teacher and student work more collaboratively toward the "particular". Compensation based on individual anthropometrics needs to be melded with aesthetic goals during this stage, as external (e.g. biofeedback instructions from the teacher) and internal processes (e.g. student's modifications to motor control according to the teacher's instructions) are consciously assimilated. Multiple skills must be coordinated together at this stage. In the autonomous stage of the process "naturalization" occurs as skill control moves from the conscious to the sub-conscious. Here, the onus for successful learning is mostly the student's responsibility and the teacher's role becomes more focused on remediation. As skill control becomes automatic, performers are freed to direct their attentions more toward aesthetic, creative and artistic outcomes, which requires both forward planning and flexible response to the circumstances of the moment. The quality of instruction and practice as well as the amount of practice are important factors during all these developmental stages.

Arts Biomechanics has the potential to positively affect the performance arts through influencing the pedagogical process. It builds upon already proven scientific models of motor learning, adapting these to acknowledge artistic process and values. In the first two stages of the learning process, it can directly influence the quality of teaching/learning that occurs. This can accelerate the skill acquisition process and minimize the necessity of trial and error practice. This will make the process a more enjoyable experience for the learner. This is a motivational prerequisite since it frees a performer to transfer his/her attention from motor control to artistic representation, indirectly improving artists' perceptions about the third stage of the process.

Further, because arts performance involves long-term training, the dangers of vocationally related injury due to poor pedagogical practice is very high. In this regard Arts Biomechanics can highlight injury prevention as a fundamental ethic at each stage of the motor development process. It can provide realistic and practical pedagogical strategies that encourage changes in attitude and behavior. The benefits of this will be improvements to artists' health, wellness and career longevity as well as enhanced self-perception of their artistry.

2.4. The Tools of Arts Biomechanics Research

The scientific quantification of motor skills in arts performance is about decade old (G. B. Shan and Visentin, 2003; Visentin and Shan, 2003). Prior to this, qualitative and some quasi-quantitative image and video-based research exists in several formats; drawings, photographic images, photographic time or multiple exposures, and video (Hilton, 1997; Hodgson, 1958; Moyse, 1973; Palac, 1992; Tulchinsky and Riolo, 1994a). Existing research has primarily been concerned with skill description as it relates to musical values/interpretation and rudimentary analyses. Most of it can be found in medical, pedagogical, and in cognitive/behavioral neuroscience (e.g. Music Perception) journals. As such, biomechanics is mostly used as an indirect or supporting means to examine disciplinary questions in these areas. As a primary area of study, Arts Biomechanics is still young. Clearly, the technologies and methods of biomechanics are an attraction to researchers from many disciplines, but the questions these researchers propose would likely be different than those a biomechanist would proffer. In Arts biomechanics, the technology must be adapted to the context of the arts. But, more importantly, the "right" research questions need to be asked; a condition that requires collaborative methodology. Since performing artists are intimately involved with their art, objectivity can best be provided by an external observer. However, since the artistic product is inherently dependent on the relationship between artist and performance, analytic objectivity has functional limitations.

A review of the most recent literature reveals that research groups looking at the performing arts are becoming more multidisciplinary. The reason for this is simple. Given the complexity of the questions being asked, it is unrealistic to expect a single discipline to provide the required artistic understanding, scientific expertise, and technological know-how. Arts biomechanics as a new area of study can directly contribute in significant ways to the latter two of these needs. Because it influences the health and wellness of performers, it indirectly influences artistic outcomes as well. Biomechanics brings multiple tools and scientific methodologies to bear on research which can be easily adapted to arts applications. Currently, three technologies have dominated arts performance analysis: 3-D motion capture (MoCap), force measurement, and electromyography (EMG). These technologies give rise to additional tools, such as biomechanical modeling and multidimensional signal analysis of simultaneous measurements (Visentin and Shan, 2004). The levels of sophistication currently being demanded in human movement and performance research requires synchronized information from multiple technologies in order to address more and more complex questions.

Motion Capture

3-D motion capture can be used to track movement of any object. It does so by using multiple cameras to view and triangulate the positions of markers that are placed on the item (Figure 9). Some motion capture systems use passive reflective markers and others use active optical markers (diodes). No matter the system, motion capture relies on multiple cameras being able to see each marker since, for complex movements selected points can easily become occluded by the measured object's orientation. At least two cameras must be able to see each marker at all times for accurate reconstruction of XYZ coordinate data.

Using sophisticated software, precision motion capture measurements can be assembled in 3-D computer space, supplying kinematics (positional data, distances, velocities, and accelerations) of the movement. For human subjects, markers need to identify functional anatomical points, such as joints (Lundberg, 1996). Typically a special garment that effectively eliminates the relative movement of clothing and skin, but without restricting motion, is worn by subjects; markers are appropriately positioned on this garment. For example, upper body movement can be quantitatively characterized using markers placed on trunk and joints as follows: the Sternal end of Clavicle, Xiphoid process of Sternum, 7th cervical and 10th thorax vertibra, for the trunk; right and left Acromion, lateral Epicondyle, and Styloid processes of Radius and Ulna for the arms, and; 3rd metacarpal and finger joints for the hands. Head orientation can be determined by placing markers on the head. Frequently additional reference markers are used to determine segmental rotation. If any equipment is used during the activity, its movements can be captured by placing markers on the items. In all instances at least three markers are needed for each measured body segment and each piece of equipment in order to determine both positional translation and rotation.

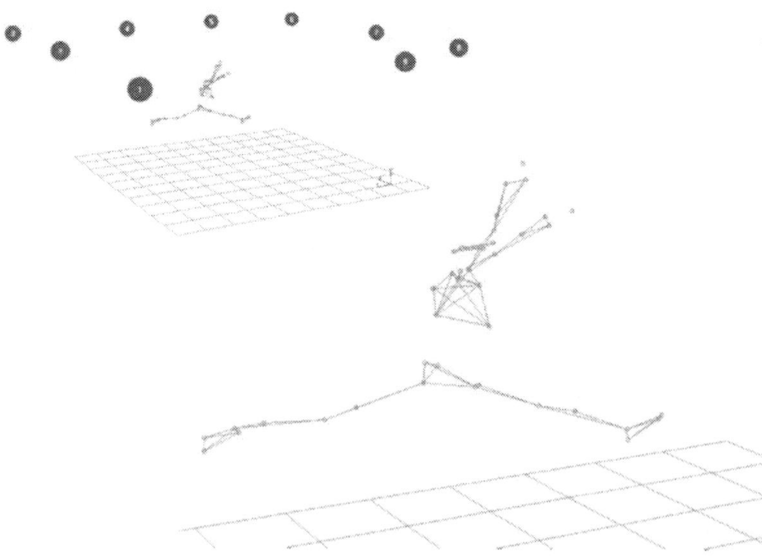

Figure 9. 3-D motion caption of Grande Jeté (Ballet): the set-up of the 9 cameras and a sample frame (G. Shan, 2005).

Biomechanical Modeling

From changes in position over time, computer analysis takes simple kinematic data and translates it into skeletal and muscular movement using the fundamental precepts of physics and anatomy. This is biomechanical modeling. In such biomechanical modeling, inertial characteristics of the body are estimated using anthropometric "norms" found through statistical studies (Shan and Bohn, 2003). Effectively, biomechanical modeling can serve as a platform for data analysis, providing the means to integrate information collected multiple measuring systems such as motion capture, force measurements and EMG.

Task-specific biomechanical models are used for each activity. Modeling techniques such as dynamic and inverse dynamic analysis allow researchers to establish joint coordination (motor control patterns), internal loads (moments and forces) and muscle working patterns (loading types). This information is used to study interactions between humans and their tools (e.g. a violinist and his bow or a golfer and her club), the interface of the body with its surroundings (e.g. dancing) muscle group timing/coordination (e.g. multi-joint tasks such as piano playing), and neural control (Timing article-peter). For example, studies by Shan and Visentin (2003 and 2004) analyzing violin performance used 30 reflective markers to track professional violinists' left and right-arm skills. Cameras were placed in a circle around the subjects, creating a capture volume of approximately 3m × 3m × 2.5m. Of the thirty markers, 24 were placed on the upper body of each subject and 3 were positioned on each the violin and bow. Marker trajectories were used to reconstruct a biomechanical model of violin performance. (Figure 10).

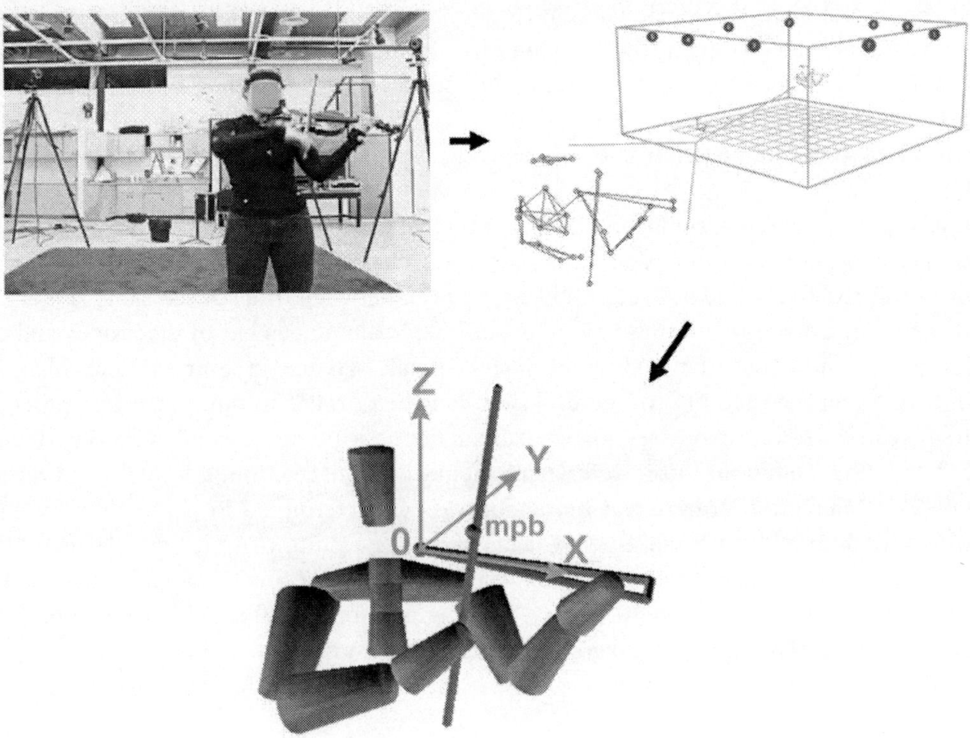

Figure 10. The set-up of 3D motion capture system and the 10-segment biomechanical model of violin playing.

Electromyography

Electromyography is a seductive technology because it provides researchers easy access to data about the physiological processes that cause muscles to generate force, produce movement and engage in the countless functions that allow humans to interact with their world. By placing electrodes on the surface of the skin directly above the longitudinal and medial-lateral centre of a muscle, minute electrical activity can be detected during muscle contractions. These seizure potentials can be directly correlated with the intensity and duration of muscle activity. As such, EMG supplies researchers with a means to determine the timing of neural control and to fractionate the response process of coordination among muscle groups.

EMGs main attraction is that we can to some extent examine what happens to muscles inside the body using non-invasive means. In absolute terms, seizure potentials measured by EMG are very small and an amplifier must be used to enhance electrical signals, which typically register under 10 millivolts. Further, the amplitude of the signal is dependent on the physical characteristics of the subject (e.g. skin conductivity, thickness of fat layers between muscle and skin, training-dependent neural control, etc.). So, comparison of raw data among subjects is senseless. Most significantly EMG signals are stochastic (random) in nature. These would seem to be significant challenges to EMGs utility. But, processing provides a means to overcome such barriers.

In order to interpret EMG data, which is a relative (subject dependent) rather than an absolute value (subject independent), a variety of processing protocols exist. Enveloping is the one most commonly found in the literature (Winter, 1984; Nandedkar and Sanders, 1990; Shiavi, Frigo and Pedotti, 1998). The enveloping process involves full-wave rectification of raw EMG signals and low-pass filtering (Figure 11). Such filtering improves EMG sensitivity as a diagnostic tool by keeping dominant components of the signal and eliminating unstable ones. According to the literature, a 6 Hz filter Fast Fourier Transformation (FFT) low-pass filter is typical for human activities (Arsenault, Winter and Marteniuk, 1986; Shiavi et al.1998).

Enveloping provides a stable signal for researchers. However, it must be noted that such signals are *non-quantitative* measures of neural-musclar activity. Signal intensity cannot be used as a comparative value between subjects, or even between trials of the same subject held on different days, since it is impossible to exactly replicate placement of electrodes and other conditions that influence the quality of EMG signals. Hence in order to provide *quasi-quantification* and possibility of comparison between trials, a normalization process is required. Normalization expresses all measurements as a percentage of a chosen reference value. For EMG, commonly used references include: 1) the maximum signal level during a trial, or 2) a Maximum Voluntary Contraction (MVC) determined in a separate trial under high-load conditions. Using the first of these provides results of questionable quantitative utility since the "reference" may vary from trial to trial. The second approach establishes the physiological limits of the muscle under static contraction, a value which will not change from trial to trial. This allows researchers to examine muscle load intensity in percentage terms of the subject's MVC (or physiological limits). Such a value can provide a reliable means of comparison among different subjects and trials.

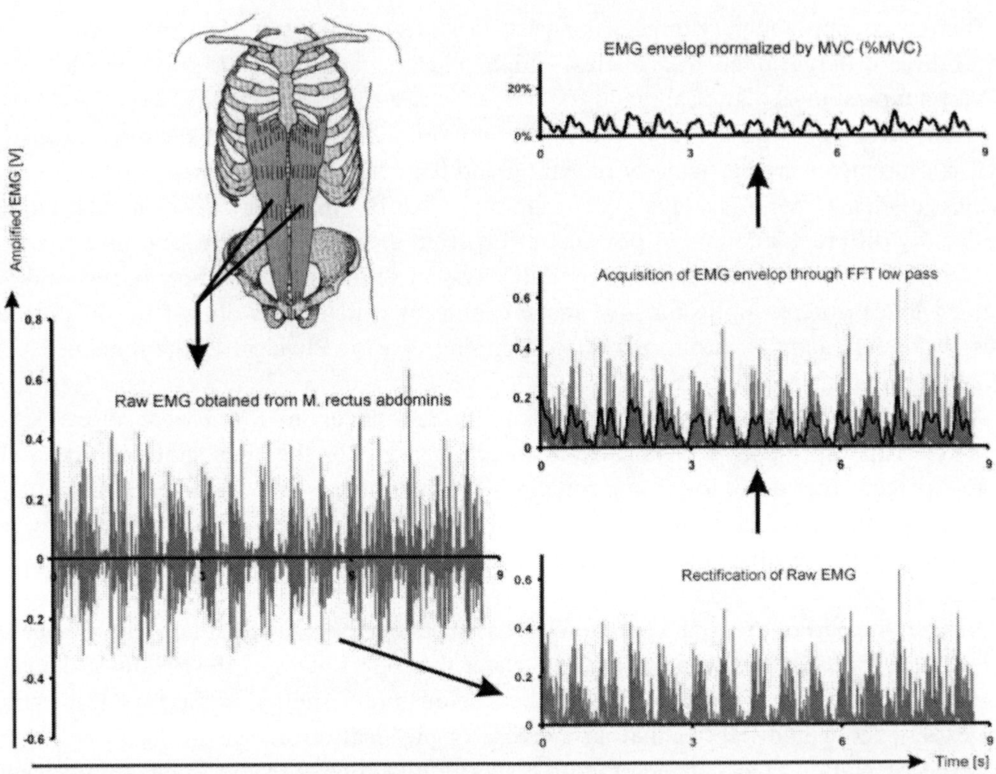

Figure 11. EMG of rectus abdominis during voluntary breathing – the process establishing enveloped and normalized EMG for quasi-quantification.

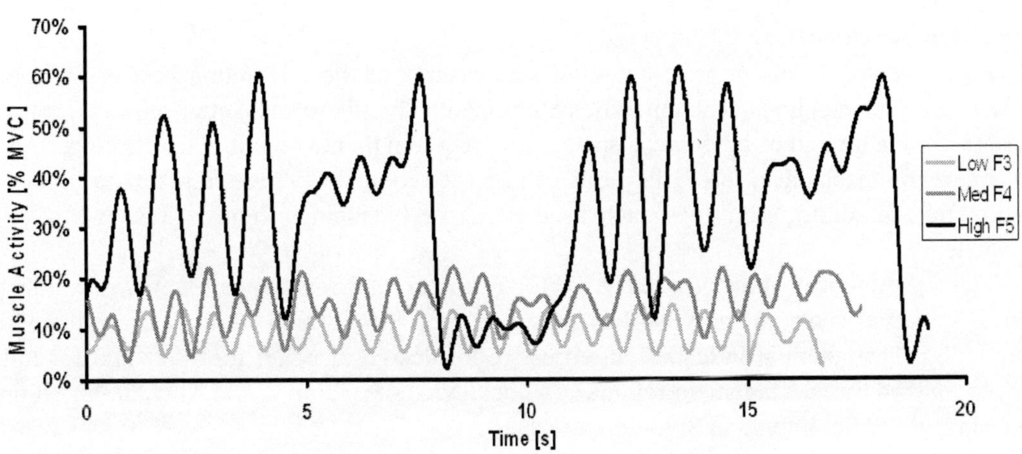

Figure 12. Rectus abdominis activity at different pitches.

Figure 12 is a composite of three different trials measuring the rectus abdominis activity of a professional opera singer (mezzo-soprano). The rectus abdominis MVC was determined in a separate trial under high loading conditions. The protocol involved two sets of singing four short onsets (intonation of a pitch) followed by a sustained tone produced on the same pitch.

There was approximately three seconds of rest between the sets. The same protocol was used at three different pitch frequencies – high (black), medium (dark grey) and low (light grey) for a typical mezzo's vocal range.

In terms of vocal production, analysis of Figure 12 supplies rich details. Normalized EMG measurements are particularly revealing and the consistency/repeatability between sets provides evidence of high levels of training. Clearly, different pitch levels required significantly different amounts of physical effort from the singer. At low and medium pitch, there was almost no noticeable difference in physical effort between short onsets and long sustained tones. Interestingly, for low and medium pitch there was almost no difference in rectus abdominis activity during periods of singing or rest. Physical effort remained under 20% of the subject's MVC.

At high pitch, the story changes significantly. Peak levels of EMG were above 60% of the MVC. This represents a physically demanding task. For the four short onsets, a clear pattern existed – first onset lower and remaining three higher – one that may be explained in terms of the individual's technique. Notably, the period of rest for the high frequency trial, showed lower values than the periods of rest for the other two "easier" ones (low and medium).

Further, the part of the trial where a tone was sustained began with an onset of about 40% of MVC and gradually increased to peak levels of the short onsets as the breath of the singer was expended. These two facts suggest a subconscious pre-planning on the part of the singer, since experience would tell her that an increase of physical effort was going to be required during the generation of a sustained tone. This has implications for vocal pedagogy, since the singer was not perceptibly aware that such patterns were developing. Clearly, in spite of its quantitative limitations, EMG can provide significant information in terms of skill analysis and evaluation of training.

Force Measurements

Force measurements provide quantitative measures of the interaction between a human body and its surroundings. They may be determined using platforms, small sensors, arrays of sensors, or in a myriad of other ways specifically required by the conditions of testing.

In arts biomechanics, force platforms can be used to quantify ground reaction forces and weight transfer during activities such as dancing or balancing. Figure 13 shows a ballet pirouette.

The yellow arrow shows the force vector of the ground reaction force in relation to the dancer's inertial centre of gravity. Not only can such information be used for pedagogical purposes, it has potential industrial significance for design of ballet footwear and for injury prevention and risk management regimes. (Chockley, 2008; Shippen and May, 2008; Shippen and May, 2009; Solomon and Solomon, 2004)

Small sensors in a variety of configurations can be used to answer task-specific research questions. Examples in arts biomechanics includes installation in keyboards (Parlitz, Peschel, and Altenmuller, 1998), in mouthpieces of wind instruments (Devroop and Chesky, 2002), and on places where performers fingers or thumb come into contact with an instrument (Nemoto and Arino, 2007). The data gathered gives information about finger pain, efficiency and effectiveness of physical effort and the effects of long-term training.

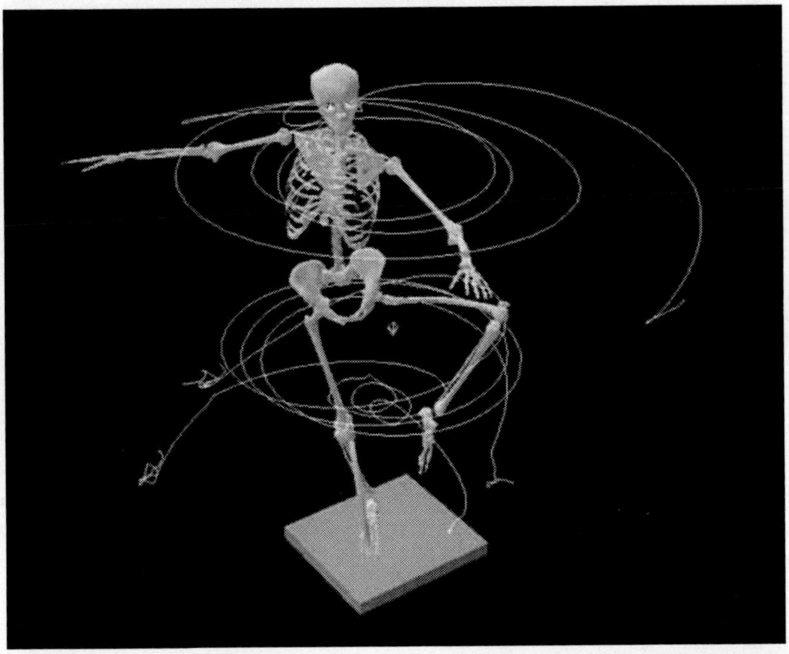

Figure 13. Measurement of ground reaction force using platform during dancing (Shippen, 2009).

Synchronized Measurements

Together, data collection systems such as the ones described supply information on how individuals interact with their environments and tools. The data delineates cognitive and neuromuscular processes used while preparing to produce and during the execution of an action. Synchronized systems – those where data from multiple systems are measured and time-coded simultaneously – are the best means to analyze and understand the complexities of human performance because they provide multiple perspectives of the same phenomenon. The real challenge lies in find meaningful ways to integrate these various perspectives into a holistic view. From an analytic standpoint, it is problematic that inputs can include many different time variables (energy, force, linear and angular positions, speeds, accelerations, bio-signals), each of which may be measured in different units, sampling frequencies and scales. Thus, for meaningful quantitative research, a central analytical tool needs to serve as a nexus for comparing and correlating the multiple signals that are generated. Biomechanical modeling provides just this means (Figure 14).

In summary, synchronized systems coupled with biomechanical modelling reveal more anatomical and biomechanical perspectives than measurements alone. 3D motion capture system supplies the spatial position, velocity, and acceleration of limbs and/or joints. EMG captures muscle activity, which can be utilized to fractionate the response process and get intensity and timing of neural control as well as the coordination between and among muscle groups. Force measurements reveal the interactions between subject and environment. Video other means provide a traditional external view for motion analysis reference. Biomechanical modeling links the coordination of the joints with neural controls (muscle activities). In total, synchronized systems present a more global technique to measure, analyze and understand any human motor skills.

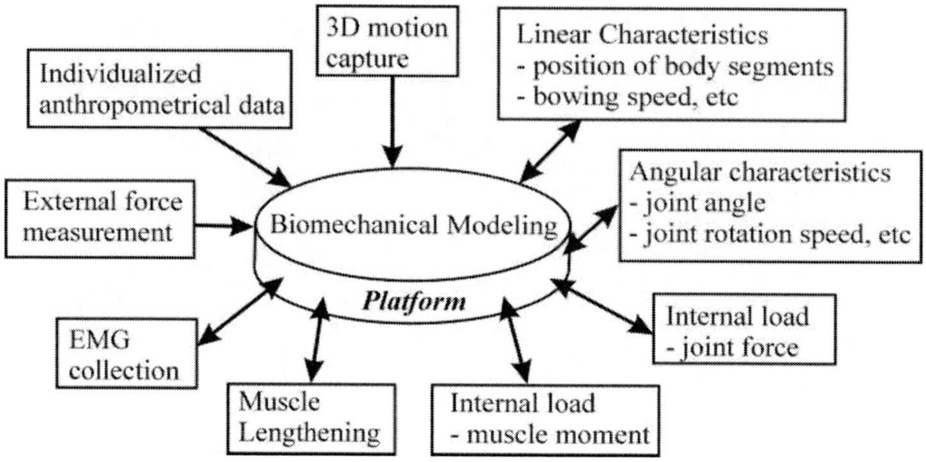

Figure 14. Biomechanical modeling as a platform for integration. Modeling supplies the means to connect and compare multiple inputs. Single-headed arrows represent inputs that are measured. Double arrow heads represent those that both influence *and* are influenced by the model (Visentin and Shan, 2004).

2.5. Samples of Pedagogic Applications

Section 2.3 began with a discussion of the principal challenges faced by Arts Biomechanics as a field of study and ended by identifying two questions that Arts Biomechanics can begin to answer. The first of these, "Can we identify, describe, and quantify qualities that lead to artistic success while minimizing the potential of performance injury?" directly addresses pedagogy. Its goal of reducing the potential of performance injuries will ultimately be unsuccessful unless it provides information for improving pedagogy. Both teachers and learners are primarily interested in much more primitive questions, such as "How can I teach/learn skills faster?" Concerns for injuries typically come low on both students' and teachers' priority lists until they have actually occurred. By engaging in research that addresses both effectiveness and efficiencies in performance, Arts Biomechanics can begin to influence attitudes. If performing artists believe there is a faster/better way to do something, they will listen. But, research in Arts Biomechanics will have to provide practical and convincing scientific evidence, clear language and definitions, and a means to re-evaluate and improve pedagogy in ways that do not compromise artists' perceptions of the end product – their art.

Unfortunately there are relatively few examples of quantitative research in the literature. One example, related to violin, gives a glimpse of the potential of Arts Biomechanics applications in pedagogy (Visentin et al., 2008). In the study the authors set out to determine whether or not it was possible to accurately and objectively characterize and generalize some biomechanical skills related to violin performance. Both smooth and bounced bowing techniques were examined at various speeds and on all four strings of the violin. The goal was to accurately identify postural and kinematic landmarks, kinetic "feel", and motor control patterns (i.e. ranges of motion of single joints and dynamic interactions among joints) that might characterize the movements of expert performers. From a pedagogical standpoint, consistencies observed among expert subjects of differing anthropometric proportions can provide "markers" that should be easily employable in teaching.

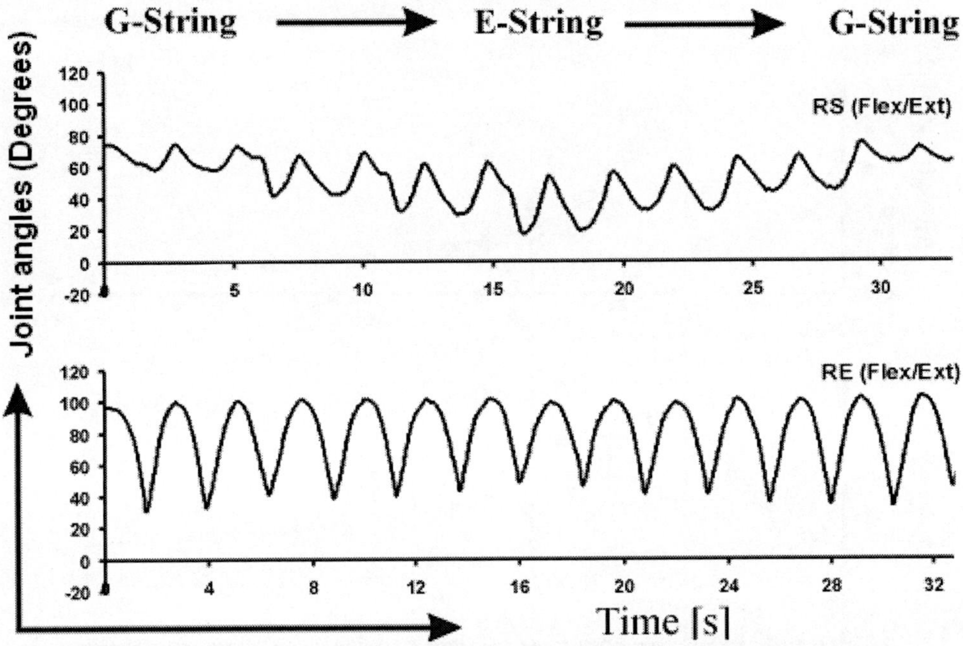

Figure 15. Right elbow and shoulder flexion/extension. Ranges of motion (ROMs) are measured by peak to trough differentials (Visentin, Shan and Wasiack, 2008).

The results of the study showed that, for the two identified bowing techniques, finer motor control movements, such as those found in the wrists, were highly variable among subjects. However, shoulder and elbow movements, those joints that dominated gross motor control, provided generalizable patterns. Using these, several pedagogical markers were identified and quantitatively described. These markers provide information about posture, positions, and time-based dynamic interactions among joints during performance.

Figure 15 shows both that there were absolute changes in the amplitudes of the joints measured and that the control pattern changes of the two joints appeared to be complimentary – one ROM gets bigger as the other gets smaller. From G string to E string overall shoulder flexion/extension angle decreased and shoulder flexion became more active and the ROM increased. The elbow acted in exactly the opposite manner. This is one example of establishing a compound "marker" for a multi-joint task; one that can be used during the teaching/learning process.

Additionally, through biomechanical modeling the study identified tempo-dependent phases of motor control. Three discrete control pattern phases were found for both smooth and bounced bowing (G. B. Shan et al., 2007; Visentin et al., 2008). Forward dynamic analysis revealed each phase to have distinct combination of kinematic factors. The first phase was marked by increased bow acceleration and speed (indicators of exertion), while travel distance of the bow decreased (reaction showing a compensation). The second phase was characterized by constant bow acceleration, revealing workload to remain steady, while distance and speed decreased (showing efficiency gains). And the third, revealing an approach to physical limits of the player, showed bow acceleration, speed and distance to all gradually approach their respective plateaus (Figure 16).

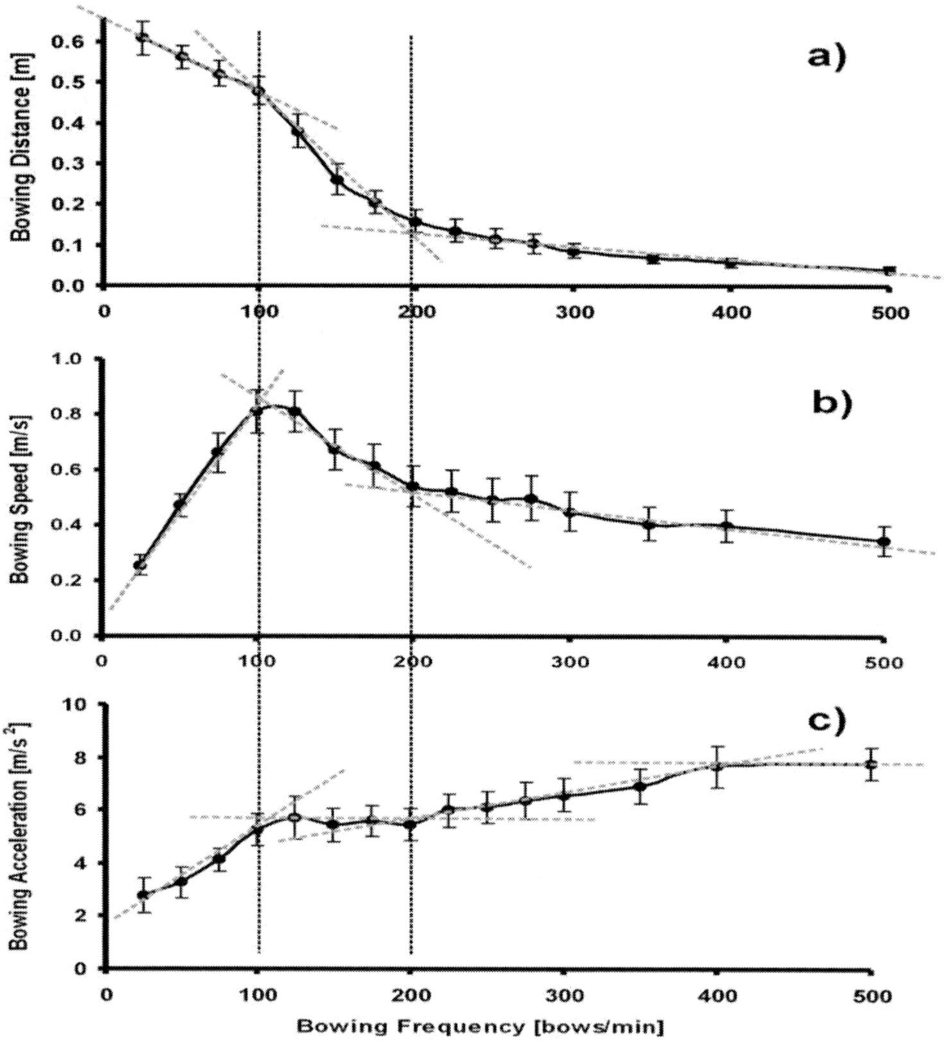

Figure 16. Kinematic factors showing three discrete phases of motor control defined by two critical points (vertical dotted lines). From left to right: increasing physical effort, optimization, and approaching physiological limits (Shan and Visentin, 2003).

Although there are relatively few examples in the pedagogy literature, other sources contain information that also has pedagogic implications. The problem with this is that the venue of the information might preclude reaching those who are most influential in the artistic practice – performers and teachers. One such example comes from the medical literature (G. B. Shan and Visentin, 2003). The study was directed toward quantifying biomechanical factors related to Repetitive Stress Injuries (RSIs). Findings suggested that music education would benefit by applying biomechanics in the teaching and learning process. For example, Figure 17 represents a trial typically found among professional violinists. The elbow heights of the subject were normalized as a percentage of his body height.

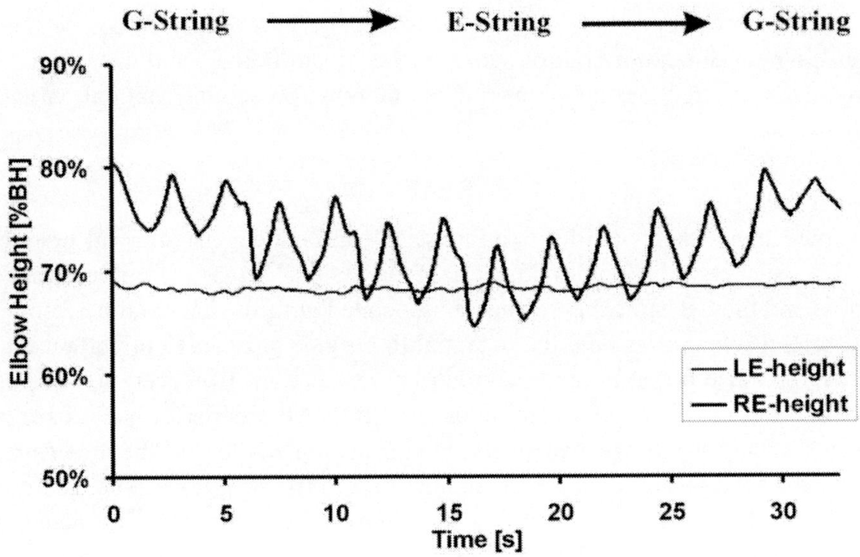

Figure 17. Vertical movement of left and right elbows during a typical 32-second trial. A two-octave G major scale was performed, beginning on the G-string cycling to E-string and back (G. B. Shan and Visentin, 2003).

This kind of kinematic analysis shows asymmetrical control between left and right side of the body. Elbow heights were near static on the left (approximately 68% BH), while the right showed a complex dynamism. Right elbow ROMs were remarkably greater on the E-string (9.3% of body height) than on the G-string (6.9% of body height). The implications for pedagogy are significant. When teaching about bowing, the teacher's efforts need to convey the dynamic nature of the right elbow; its position depending on both string played and part of bow used. Further, teachers need to be aware of how the kinematic chain actually works. The mechanism that raises and lowers the elbow is actually the shoulder. Failure to clearly express such markers can slow the learning process, limit the student's potential, or be a fundamental cause of eventual injuries.

A marker-based teaching approach can provide insight into how it actually feels to play the violin. As these studies suggested, such methodology may have significant implications for development of new pedagogy models for learning performance skills on the violin. Connecting objective data with subjective perception or "feel" is one way to accelerate the motor learning process. Teachers and students alike want to reduce the time required for skill acquisition and an evidence based approach decreases the time students must spend "experimenting" to establish basic coordination control. Clearly, more studies are needed to develop such models, not only for the violin, but for other instruments, for dance and for the related arts. A recent study examining university level trumpet players also alludes to the value of quantitative analyses for the development of new approaches to pedagogy (Devroop and Chesky, 2002).

The selected examples show Biomechanics to be ideally suited for such task. Marker-based approaches to performance pedagogy have the potential to increase the efficiency and effectiveness of skill acquisition. In this way, Arts Biomechanics is a new application of well-established ideas. These ideas can guide the formative stages of learning and influence arts professionals and teachers. But, to be successful, it needs to meet the "Art" on its own terms.

To do so, Arts Biomechanics must be willing to engage in a deeper discourse that explores the relationship between motor-control patterns (the "quantitative") and the subjective aspects of the art – the desired aesthetic outcomes; thereby preserving natural variability and individuality of artistic "style".

Future Directions

Arts Biomechanics is a field in its infancy. In can challenge traditional practices in arts pedagogy by successfully employing quantitative techniques to objectively describe physiological and biomechanical phenomena associated with fundamental performance skills. Systematic scientific inquiries hold the potential to provide answers to important multifaceted question such as those posed at the beginning of this section. However, arts performance is different from most day-to-day human motor activities. All arts performances are influenced by a complex set of factors related to the performers' personal aesthetic sensibilities and perceptions. Quantitative analyses can challenge these self-perceptions. The studies identified in this chapter show ways to establish a science-based approach to pedagogy. Such methodologies can increase the efficiency and effectiveness of skill acquisition, which will be welcomed by the arts industry – but only if they provide practical and convincing scientific evidence, clear definitions, and a means to reevaluate and improve pedagogy in ways that do not compromise artists' perceptions of the end product. It remains to explore the usefulness of Arts Biomechanics in the study of multiple Arts. A more comprehensive body of research could potentially revolutionize how performance skills are taught.

3. Injury Risk Quantification and Prevention Related to Artistic Performance

3.1. Overuse Syndrome

Overuse syndrome is a "catch-all" term for many different types of repetitive stress injuries (RSIs). Among performing artists, a myriad of injuries occur. Because artistic success or failure is often determined by micro-increments of ability, the smallest injury can negatively affect the performer and the artistic outcome. For example, an injury that affects the range of motion of someone's big toe by 10% would go unnoticed by an office clerk. However, for a prima ballerina, it could be the beginning of the end of a career. Pathologies can include: tennis elbow, shoulder injuries, rotator cuff injuries, focal dystonias, tendonitis and lower back pain and more (Fry, 1987; Zaza, 1992, 1998). Although repetitive movement may typically be the mechanism of these injuries, causal factors are deeply rooted in the Arts veneration of their traditions, pedagogic and artistic.

Reported rates of vocational injuries among performing artists are very high. Among musicians, affect between 50% and 76% of professionals (Fry, 1988; Fry et al., 1988; Zaza, 1998). Considering that this applies to musicians with injuries serious enough to affect their vocational activity, often for extended periods and sometime career-ending, these statistics are staggering. There are three general categories of injury commonly diagnosed and RSIs accounts for over half of these. Given the high incidence of RSIs, the industry has recognized the need for developing efficient and effective prevention and risk-management strategies.

Currently most efforts in this regard are still in the early stages of research. Biomechanics has been contributing in the areas of accurate and quantitative description, understanding the nature of the physiological loading, and linking epidemiological studies to the observed activities. The aim has been to minimize or prevent injuries by reducing biological loading. Since Arts performance activities often involve the whole body, and demand high degrees of both fine and gross motor control, an examination of their complex biomechanical processes may lead to generalizations about RSIs that apply to other complex human activities.

Definitions influence research questions, design and structure. Therefore, it is necessary to review traditionally held definitions in order to ensure that they are sufficiently nuanced to direct researchers in the design of their studies. RSI has been traditionally defined as the excessive use of body parts so that the accumulation of micro- trauma exceeds human physiologic limits (D. M. Dawson, Hallett, and Millender, 1990). It has been understood for quite some time that basic factors influencing the etiology of RSIs include: (1) genetic makeup of indificuals; (2) performing technique; and (3) time × intensity of activity (Fry, 1987). Recent studies, identified below, improve upon these basic definitions.

3.2. Biomechanical Modeling as a Platform to Understand OS

The details of this section are drawn from an article published in the journal *Medical Problems of Performing Artists* (G. B. Shan, Visentin, and Schultz, 2004).

Injuries that occur from repetitive use significantly differ from those that typically occur from load-related accidents or during sports. In the case of the latter, injuries are essentially due to load impacts suddenly exceeding human physiological *limits* - limits that are theoretically possible to evaluate, and are relatively easy to diagnose. On the other hand, activities that cause RSIs have characteristic loads well below human physiological limits and injuries result from multiple micro-trauma, the effects of which accumulate over time. A gradual and imperceptible onset of injury shows RSIs to challenge human physiological *tolerance* - a factor that is difficult, if not impossible to quantify. Even though it may be tacitly understood by those who actually deal with such injuries, some current definitions still fail to articulate this distinction. Using the term "physiological limits" in connection with "micro-trauma accumulation" is misleading. RSIs are entirely different than impact injuries because of the time-based and cumulative nature of the pathology. Thus, for meaningful quantitative research, a central analytical tool needs to serve as a nexus for correlating the multiple signals that are generated and for situational comparison.

In the case of sport injuries, biomechanical modeling has successfully served as this nexus for assessment, remediation and prevention purposes (Ballreich and Baumann, 1996; G. B. Shan, 1999). It seems the next logical step to employ biomechanical modeling as a platform in the service of RSI research. However, generalized models must first be proven to be theoretically valid for the human body, and then modified to accommodate the conditions that accompany RSIs. Ultimately, it is hoped that this will lead to research that can clearly explore the concept of physiological tolerance where: 1) a temporal factor (the process of micro-trauma accumulation) is evaluated, 2) rates of recovery during periods of relative inactivity or rest are considered, and 3) increased understanding of the evolutionary nature of OS injuries will result in changed attitudes in the workplace and a focus on the development of prevention strategies.

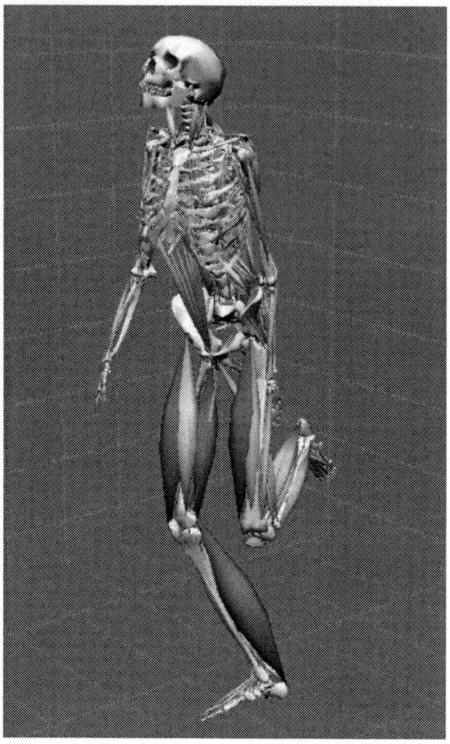

Figure 18. Biomechanical model for quantification of musculo-skeletal loading and muscle lengthening/shortening.

As it is the mechanical work performed by OS patients that obviously causes observed damage, an evaluation of "work" appears to be a logical first step. Biomechanical modeling permits quantification of both the characteristics of the human body during activity and of the workload levels (G. B. Shan and Visentin, 2003; Visentin and Shan, 2003). It is the tool best suited to this task and can be used to make connections between the data generated by multiple assessment techniques (Figure 18). Simply, biomechanical modeling allows us to understand what happens inside the body using information gathered from outside the body.

3.3. The Theory of Biomechanical Modeling as It Applies to OS

The following discussion is based on several biomechanical studies (G. B. Shan et al., 2007; Visentin et al., 2008; Visentin and Shan, 2003, 2004). Using motion capture technology, activities can be analyzed using biomechanical modeling. Anatomical positions can be obtained, the skeletal structure and simple positional data can be translated into skeletal movement using the fundamental precepts of physics. In such modeling, inertial characteristics of the body are estimated using anthropometric "norms" found through statistical studies (G. B. Shan and Bohn, 2003; Winter, 1990). Combined with motion capture information, these allow the calculation of load intensity (which can be described using a variety of parameters measured at the periphery of the body) resulting in factors such as: force, work, moment, (angular) velocity or acceleration and Head Injury Code index (HIC) (C. M. Gadd, 1966; Hennig and Lafortune, 1991; Kosiak, 1959). Biomechanical modeling allows muscles to be added onto the skeletal frame using anatomical knowledge regarding normal attachment points. Thus, muscle lengthening and shortening can be determined in connection with skeletal movement. From these, 1) internal loads (bone-to-bone contact force and muscle force, rotation caused by joint moment) can be quantified and, 2) when correlated with electromyographic signals, muscular work patterns (eccentric, concentric, isometric) can be determined.

All "modeling" involves simplifications of reality. The method's strength is that it provides a means to understand complex systems without becoming lost in detail. In the case of the human body, the key justification for using biomechanical modeling as an analytical tool hinges on the accuracy of *internal load* calculations based on the fundamental precepts of Newtonian physics (classical mechanics). To establish the contextual validity of the calculations, several steps need to be taken:

a) the various forms of mechanical work (W) as they pertain to classical mechanics must be established
b) a generalized form must be devised
c) the generalization must be justifiable and applicable to all cases for biological reactions in the human body
d) the system must be related to RSI specific conditions.

3.3.1. The Physics of Mechanical Work

Depending on the situation, work (W) may be mathematically represented as:

- Work of force F over distance d $W=F\times d$
- Work of force F at velocity v over time period t: $W=F\times v\times t$
- Work of moment M at lever arm length l: $W=M/l\times d$
- Work for accelerating (a) a mass m from 0 to velocity v: $W=\frac{1}{2}m\times v^2$
- Work for accelerating a mass m to momentum p: $W=\frac{1}{2}p^2/m$
- Work for accelerating mass m by force F over time period t: $W=\frac{1}{2}(F\times t)^2/m$
- Work against friction
- (c is a coefficient, n depends on the kind of friction): $W=c\times vn\times d$

Note: for the sake of simplicity, all vectors are considered constant and parallel or orthogonal to each other.

3.3.2. The Generalized Form Related to Work

In order to overcome the diversity of form observed above, it is possible to establish a generalized form covering many (or possibly all) cases through the following steps:

By appropriately substituting $d = v\times t$ and $v = a\times t$ and $p = m\times v$, these expressions can be expressed in alternate forms as follows:

$W=F\times d$	becomes	$W=F\times v\times t$
$W=F\times v\times t$	remains	$W=F\times v\times t$
$W=M/l\times d$	becomes	$W=(M/l)\times v\times t$
$W=\frac{1}{2}m\times v^2$	becomes	$W=\frac{1}{2}m\times a\times v\times t$

$W=\frac{1}{2}p^2/m$	becomes	$W=\frac{1}{2}m\times a\times v\times t$
$W=\frac{1}{2}(F\times t)^2/m$	becomes	$W=\frac{1}{2}F\times v\times t$
$W=c\times v^n\times d$	becomes	$W=c\times v^n\times v\times t$

In summary, unlike variables such as force, which cannot in this case be practicably considered a parameter due to its time-dependent nature, W *may be* used as a parameter, establishing a single value for a given process. In each of the above:

- there is a case specific Variable of Intensity (VI) - which may include force, moment, stress, velocity, acceleration – plus transforming variables
- they are weighted differently by using an activity-specific exponential factor n
- loading duration t is taken into account (which may also have an exponential factor, m).

Thus, the final mathematical representation in all cases is:

$$W=C\times(VI^n\times t)^m$$

where C is a coefficient constant, n and m are integers and VI is the case-specific variable of intensity.

It should be noted that, in addition to Work, three other parameters are also frequently used for predicting biological reactions: momentum (P), load duration (t), and maximum stress (MS). For demonstrating a common structure, they can be written in unusual ways using force as a common factor. Substituting these:

$$P=F^1\times t$$

$$t= F^0\times t.$$

$$MS= c\times\int F(t)^n dt.$$

NOTE: In this last equation, the maximum force determines the value of the integral. With high n, injury occurs as $\Delta t\rightarrow 0$ and mechanical load approaches or exceeds physiological limits (Hennig and Lafortune, 1991). This is the condition of impact injuries.

From the above it can be seen that all have the structure:

$$C\times(\int F(t)^n dt)^m$$

when W, P and t are written as integrals.

3.3.3. Biomechanics and Biological Reactions in the Human Body

As the above are purely mechanical, it may be questionable that they really describe the reaction of a biological system. To demonstrate this, static work is most cited for showing the difference between mechanical and biological systems in terms of loading. When holding a mass, *mechanical work* is zero; *physiological work* is not. But this is an inadequate illustration as it confuses the techniques of "arresting" and "active holding". For example: when standing with bent knees physiologic work is needed, but when standing with straightened (arrested) knees, work is much less.

In consideration of this, empirical in-depth biomechanics studies yield parameters of biological reaction (PBRs), which are remarkably different from those yielded by mechanical ones. Examples from the literature include:

- The Gadd Severity Index $\int a^{2.5} dt = const$, predecessor of the similarly defined Head Injury Criterion HIC is widely used for predicting skull damage, where a certain constant is believed to be related to a certain degree of damage (Administration, 1996; Barth, 2000; C. Gadd, 1966; McLean and Anderson, 1997; USA, ; USGovernment).
- Re-evaluation of data indicating the p/T-limit where ulcers were seen when pressure p was exerted to the skin of dogs over time T revealed that the best PBR describing the limit is $p^{1.2} T = const$. (Kosiak, 1959).
- Load which causes increase of bone density is best described by $\int a^n dt$, where n varies between 4 and 12, depending on the situation. (Nicol, 1998).
- World records in running are described with very high precision by $\int v^n dt = const$. The distance range studied was 12.5m to beyond a marathon and could be divided into four ranges with highly constant n in each range. Between ranges, the n varied from 0.006 to 0.229. As running a world record describes the maximum load that can be bourn in this situation. This formula is a PBR too. (Nicol, 1996, 1998)

In spite of the differences, it can be seen that these studies result in PBRs which have a structure in common with the mechanic ones:

$$(\int VI^n dt)^m = constant$$

The differences are that:

- in mechanical PBRs, m may be an integer larger than zero, whereas biomechanical PBR require m=1
- in the case of biomechanical PBRs, n does not have to be an integer because the empirical data may show a "best fit" to require a non-integer.
- besides PBRs that can be written as integrals over the whole duration of VI, biomechanic PBRs are used which relate to only a part of the activity duration. The criterion of Hennig and Lafortune (1982) is an example. It looks only at the range between the onset and the maximum of the VI.

3.3.4. Characteristics of the PBR related to OS

As it is the mechanical work performed by OS patients that obviously causes observed damage, a work-related model of injury risk has validity. For OS the PBR can be made situation (activity) specific by examining internal and external loads and integrating their influence into risk assessment. Injury risk predictions may thus be a product of this PBR.

Whereas impact injury theory examines loads that exceed physiological *limits*, OS occurs under different circumstances and requires an investigation of physiological *tolerance*.

The following simple model of OS is considered:

- OS is the result of the accumulation of micro-trauma
- activity causes micro-trauma damage over time
- biological repair takes place during periods of relative inactivity
- the cycle of 2 and 3 above is repeated multiple times
- with the result that the micro *damage* accumulates because the repair cycles are insufficient for complete recovery
- finally, a critical limit is exceeded, macroscopic injury may be diagnosed, and the *physiological tolerance* of the individual is surpassed.

In quantitative terms, the following fundamental premises may be used:

1. Since damage (D) = MicroDamage × T, it is obvious to put damage proportional to time T, because microdamage may be considered a constant that depends on the loading situation and on the properties of the biological system (both of which change minimally).
2. As repair (R) = R × T. The repair parameter R may be considered constant too. (As an alternative, R can increase or decrease as total damage changes due to the ability of the biological system to exercise repair under different conditions – i.e. the body may become more or less capable of repair at different levels of injury. Considering R to be constant represents a midpoint between these alternatives.).
3. Cumulative Damage (CD) is thus the difference between damage and repair.

$CD = D \times T - R \times T.$

Figure 9 shows the development of CD over a single cycle where slow repair over a long time leads to complete repair (blue lines), a single cycle where fast repair over a long time leads to complete repair (green) and repeated loading cycles (red) where CD finally exceeds the critical CD value (orange).

From this and other current studies (Visentin et al., 2008; Visentin and Shan, 2004), the following hypotheses may be made:

1. that loads leading to Overuse Syndrome (OS) have *qualitative* and *quantitative* factors, both of which must be considered when undertaking risk assessment and developing prevention strategies
2. that *qualitative* factors are related to the type of load (impact, fundamental, static, quasi-static, or dynamic) and the *quantitative* factor is related to the magnitude and duration aspects of the loading
3. that, at comparable load levels, static or quasi-static loads are more injurious than dynamic ones; making *quality* of load a primary factor to consider when assessing injury risk

4. that the intensity of a series of dynamic load cycles with small intermediate recovery phases may be considerably higher than the intensity of static load of equal duration without causing injury
5. that risk is directly related to the PBR and Intensity^n × Duration of Activity, and
6. that Intensity and n are influenced by *quality* and *quantity* of load.

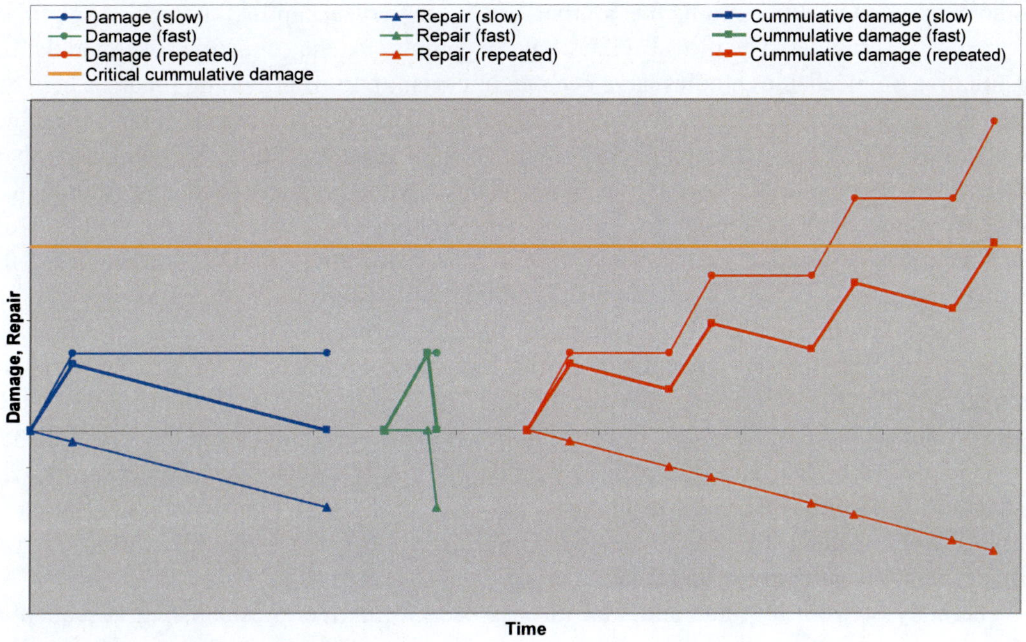

Figure 19. Cumulative damage scenarios from cycles of micro-damage and repair (Visentin et al, 2004).

In quantitative time-dependent terms, the PBR is directly related to activity Intensity (I), which is contingent on the patterns of the load for the activity in question. Therefore:

$$PBR \propto \int (I)^n dt.$$

For RSIs, the *intensity* of activity is very low, work patterns change over time, and duration is the prime factor. Such conditions make RSI research extremely complex since establishing injury risk requires that: 1) intensity (I) be measured for the activity; 2) duration of activity be integrated into the risk calculation, resulting in a context-specific PBR; and 3) that rates of repair during periods of rest be considered. Unlike impact injuries where there is virtually no time between loading and injury ($\Delta t \rightarrow 0$), repair cycles during periods of relative inactivity may not be neglected in the case of OS. In representing this model mathematically, the repair parameter (R) must be considered. Thus, Accumulated Micro-Trauma (AMT) is the net of these two biological processes:

$$AMT = \int (I)^n dt - \int R dt$$

All research must be based on some fundamental theory. This work establishes a more nuanced general form to guide RSI research in the future.

3.4. Current OS Studies Related to Music Performance

Systematic research into artists' vocational injuries is relatively young. The primary venue for knowledge transfer has been the journal *Medical Problems of Performing Artists*. It has a 27-year history examining epidemiology and remedial treatments for artists' injuries. It has long been acknowledged that prevention was an important, but under-addressed area. To promote research in prevention, the journal has been very accepting of cross-disciplinary research that provides vocational views and/or points in the direction of remediation. Quantitative biomechanics studies have been an important part of providing linkages between epidemiology, remediation and prevention since 2003. The earliest of these studies examines biomechanics of violin performance (G. B. Shan and Visentin, 2003; Visentin and Shan, 2003). Prior to this, some qualitative biomechanics work was included and, although it implicitly showed the potential to influence prevention, the results of the studies were essentially just descriptive (Tulchinsky, 1995; Tulchinsky and Riolo, 1994b). One of the missing elements was a theory or framework guiding future steps in the process.

Richard Dawson, one of the founders of *Medical Problems of Performing Artists,* has published two retrospectives documenting the kinds and numbers of studies found in the international literature pertaining specifically to artists' injuries (W. J. Dawson, 2003, 2007). These provide insight into the state of research involving biomechanics and the Arts. Of 2385 studies identified by Dawson (between 1997 and 2006), only 23 involved biomechanics. The vast majority of these used the techniques and technologies of biomechanics to examine causal factors leading to repetitive stress injuries. Below are some examples of Arts Biomechanics research in the literature:

The body of work by Shan and Visentin has been formative in establishing a sequential series of steps and a fundamental theory underpinning the utility of Arts Biomechanics in RSI research (G. B. Shan and Visentin, 2003; G. B. Shan et al., 2004; G. B. Shan et al., 2007; Visentin et al., 2008; Visentin and Shan, 2003, 2004). From phenomenological description to multidimensional signal analysis, Arts Biomechanics was shown to hold significant potential in the areas of loading quantification and injury risk management. Equally important, it was shown to have potential in the areas of changing vocational attitudes and pedagogy. One of these studies identified three factors influencing the relationship between muscle loading and RSIs: load quantity (intensity), quality (static versus dynamic), and duration (time) (Kuipers, 1998; Taylor, 1996; Visentin and Shan, 2004). Regarding load quantity, the study showed muscle activity during violin performance to be within the "safe" range as typically identified by the sports biomechanics literature (Peterson and Renström, 1987). Given the frequency of injury among violinists this is obviously not the case. Shan and Visentin's work also considered load quality and duration. Using multidimensional signal analysis, loading patterns (qualities) in the muscles and joints were correlated with both load quantities and injury rates. Because violin performance is an asymmetrical activity, this provided a means of comparison between static and dynamic loading. Results had implications for the relative risks of static vs dynamic loading and began to quantify relative levels of risk. They also pointed toward pedagogical strategies aimed at prevention. Generally speaking, since static loading is more dangerous than dynamic (for the etiology of RSIs), the study suggested means to minimize static conditions or postures, and encourage dynamic motor control. Further the body of work addressed the issue of disconnects between performers bodily self-awareness during performance and the quantitative reality of the phenomenon. For example, violinists are typically taught that the left elbow should have a degree of lateral movement to accommodate playing on different strings. This improves performance outcomes in the areas of tone production and pitch tuning. Unexpectedly, they found that performers generally locked their shoulders in a highly adducted position; a posture that creates static conditions and increases the risk of injury to the left shoulder (Visentin and Shan, 2003).

In another violin study by Rabuffetti et al. (2007), the effects of using different violin shoulder rests during performance were quantitatively described. The positions of markers placed on the player's body, violin and bow, permitted a kinematic analysis of position changes resulting from the use of different shoulder rests. The study identified that, since the instrument has a fixed geometry, the performer adapts to the instrument throughout the learning process. Shoulder rests are one of the few tools available to adapt the instrument to the performer. Rabuffetti et al concluded that "optimization of the violin setup cannot be based on objective evidence only (anthropometric size, violin features, type of repertoire, etc.) but still requires a trial-and-error approach." Motion analysis was used as means to provide information that might accelerate this process (Rabuffetti, Converti, Boccardi, and Ferrarin, 2007).

In a study of trumpet mouthpieces, Devroop and Chesky (2002) examined intraoralpressures, and compression of the mouthpiece against the embouchure. The basic assumption was that these physical elements contribute to performance related medical problems among trumpeters. Results reinforced concerns regarding risk for performance-related medical problems, including rupture of the orbicularis oris muscle, tooth displacement, head and neck lesions, embouchure dystonia, and temporomandibular joint (TMJ) and jaw pain. In spite of the fact that many pedagogues and performers had attempted to explain some of the mysteries surrounding mouthpiece forces the study contended that their relationship was still not fully understood because these explanations were based upon empirical evidence and not quantitative measurements. The study confirmed that laboratory-based research can be effective in searching for answers to real-world problems of performing artists (Devroop and Chesky, 2002).

In a 3-dimensional biomechanics study of pianists hand and finger movement patterns, Virgilio et al. (2007) analyzed the movements recorded using a multi-camera motion capture system. A complex musical sequence, 16 measures of a Bach Minuet, was used as a test protocol for estimating the actual load imposed on joints during playing, as well as to assess the influence of experience on the actual movements. The central hypotheses blended concepts of injury prevention and pedagogy – that different patterns of movement would be used depending on the player's experience and that pianists could use different techniques to reach the same results. The study identifies that: 1) the movements required for piano playing usually involve low impact loads that do not exceed physiologic limits of human body, but their repetition may provoke microtrauma leading to overuse injuries; 2) Experience may allow a pianist to modify the motor patterns used for a performance, allowing the highest accuracy with minimum effort; and 3) that "didactic techniques should always teach the students how to reduce—or better, to avoid—professional disorders" and "the use of a wrong technique may accelerate the development of overuse stress injuries". Such work underpins the premise that prevention is a matter of risk management during the teaching and learning process (Virgilio, Chiara, Emilia, Paolo, and Sforza, 2007).

Together, these and the other available few examples all directly or indirectly suggest a need for more research in the area of optimizing practice time and pedagogy with regard to both efficiency (skill acquisition time) and effectiveness (learning while managing injury risk).

3.5. Future Directions

Based on the current state of Arts Biomechanics research, there remains a considerable amount of work to do. Bridging the gap between quantitative research and empirical experience – thus providing vocationally meaningful research results – requires a marriage between the quantitative powers of science and the empirical/vocational strengths of the performing Arts. Multidisciplinary collaborations hold the potential to do exactly that. Integrative multidisciplinary research can provide outcomes that have practical application by co-opting the involvement and efforts of performers, teachers, medical practitioners, and biomechanics researchers. Even though term "multidisciplinary" has become catch-phrase of some currency, multidisciplinary research is considerably more difficult to do than might initially be expected. Perhaps this accounts for the dearth of research in the area. In Arts Biomechanics, multidisciplinary research requires the integration of the sciences, fine arts and education. Researchers must engage in dialogue about disciplinary perspectives throughout the process, working directly with practitioners and accounting for perceptual experience at all stages of the work. In short, it is a challenging but rewarding endeavour.

Arts Biomechanics has a particularly important role to play in the area of injury prevention, risk management and remediation. The authors see this as a three stage process. *Context* is established through integrating the multiple perspectives provided by MACRO studies (kinematic, kinetic, and EMG research). Such methods reveal quantitative "external" (motor control pattern and intensity of load) parameters and qualitative internal ones (normalized EMG). They provide the first stage of a holistic approach to establish the true *nature* of biological loading. Additional explorations of OS must include the examination of context-relevant accumulations of micro-trauma on the MICRO level. This second stage of biomechanics research involves applying MACRO conditions in MICRO-level research. We speculate that existing technology can be used to determine the *process* of muscle fibre damage and patterns of activity-specific muscular trauma. The third stage must examine muscular regeneration under different circumstances. Recovery during stages of relative inactivity (rest) is a vital factor for risk management.

Stage 1. Determination of Realistic Loading Conditions

3-D motion capture, biomechanical modeling, and EMG provide multiple viewpoints for researchers and medical practitioners. Multi-dimensional quantitative analysis integrates their several perspectives and allows a systematic evaluation of biological loading. The main difficulties in this stage come from the high number and heterogeneity of signal components in terms of sampling frequencies, units of measure, diversity of generally accepted practice in dealing with the varied raw data, and interpretation of the results. Individually and combinatorially, using several measurement techniques permits quantitative description of factors related to the mechanism of injury. Kinematic modeling confirms group effects among muscles through positional description (an external perspective) and muscle length changes (an internal view). Dynamic and inverse dynamic modeling based on kinematic characteristics establish *quantity* of load, and quantify some factors related to *quality* (impact, fundamental, static, dynamic). EMG determines the presence and general level of loading (e.g. relative muscle activity) with the normalization process pointing toward the physiological limits of the individual. Muscle lengths in combination with EMG and kinetic results create a clearer picture of the working pattern of the muscles. The main task in this phase is to measure diverse Arts performance contexts.

Stage 2 – Micro Examination of Muscle Fibre Under Realistic Conditions from Stage 1

Imaging techniques such as LM, TEM, and MRI are frequently used to examine already injured muscle areas but not the actual process of injury[10] (Brandser, Elkhoury, Kathol, Callaghan, and Tearse, 1995; Komulainen, Takala, Kuipers, and Hesselink, 1998; Rubin et al., 1995; Uchiyama, Tamaki, and Fukuda, 2001). Traditional microscopic techniques have long been used to categorize morphological changes in stressed muscle tissue. Until two decades ago, the modalities available to *image* injuries for muscle and muscular attachments were plain radiography and to a lesser extent ultrasound. Plain radiography is ineffective because of its limited tissue contrast (Sundar and Carty, 1994). Ultrasound is highly operator dependent and has a limited view of soft tissue. More recent developments include computed tomography that can only obtain images in the axial plane and magnetic resonance imaging (MRI). In particular, MRI has recently become the modality of choice for imaging muscles, tendons and ligaments.

Muscle research using animal models shows that injuries can be detected using histological and MRI methods (Biral et al., 2000; Kanbe et al., 1998; Smith, Plyley, Rodgers, and McKee, 1997; Thompson, Balog, Fitts, and Riley, 1999). These studies focus on impact (or near-impact) loading, static loads at high levels of intensity, eccentric (lengthening) conditions, and end-state examinations of the muscles (Koskinen et al., 2002; Stauber and Willems, 2002; Vijayan, Thompson, Norenberg, Fitts, and Riley, 2001). Such methods do not mimic the conditions that lead to OS, where loads are normally far below "impact" levels, load patterns vary over time, and injuries are the result of micro-trauma accumulation. We suggest that, using realistic loading conditions determined in MACRO research, MRI techniques could be used to track the accumulation of micro-trauma using in-vivo methods, examining muscle deterioration over longer loading durations.

Stage 3. Determination of Accumulated Micro-Trauma

Based on the current success of MRI in MICRO-level imaging of existing injuries (Rubin et al., 1995; Steinbach, Fleckenstein, and Mink, 1994), a time course series could be used to determine the effect of repetitive stimulation on a muscle, and the *rate* at which micro-trauma accumulates. Micro-level studies may also be able to explore rates of repair during cycles of rest. In Section 3.3 (The Theory of Biomechanical Modeling as It Applies to OS) OS was mathematically represented time-dependent terms as follows:

$$CD = \int (I)^n dt - \int MR dt$$

Since both of these factors involve internal biological processes, research must focus on determining activity-specific values for I and n in the first of these integrals and on MR for the second.

Prevention first requires meaningful measures and then comparison to the real world. Arts Biomechanics permits the integration of information from multiple inputs with a quantitative, time-based evaluation of load. It provides linking perspectives between the process and the endstate of OS and is the only method currently available that can do so. It remains to fully apply Arts Biomechanics in the service of determining factors associated with OS.

4. Applications of Movement Science Technologies in the Creation of Art

Movement science as a field can sometimes contribute to the arts in unexpected ways. Because innovation is a driving force in both science and art, technology can create conceptual bridges between Art and Science. In the previous sections of this chapter, Arts Biomechanics was seen to provide a means to characterize intersections between body movement and Artistic creation. Technology provided analytic tools and biomechanics provided a means to understand and interpret the data. This relationship can be extended to the creation of art, whether it involves actual movement or provides an analytic perspective of existing artworks. Technology frequently carries potential beyond its intended applications, but artists, composers, performers, and scientists need to work together in order to harness its latent possibilities in the service of the creation of art. Ultimately audiences will be the judge of an artwork's success or failure. Technology's role needs to be carefully considered so that in does not merely serve as a layer of complexity, confounding communication between artist and audience. Rather, the aim of technology should be to enhance the intensity and relevance of the artistic concept. The purpose of this section is to provide discussion of several such links.

In the visual arts, creative motivation typically derives from artists' emphasis of their own individuality and their subjective interpretations of phenomena. As such, the end product can be understood as an effort to represent different aspects of the human experience, as *they* perceive it. This, in fact, is not just an artistic concept, it is a scientific one, too. Both science and art are contingent upon having a means of ordering one's perception of the world and of expressing that perception. Science does so with numbers and formulae while the visual arts does so with images and sculpture. Using the approaches of science, movement analysis technology can provide a link between artistic representations of human movement and our experiential understanding of the bodily condition. The following discussion tracing some of the influence of movement science on the visual arts is based on an in-press manuscript by the authors (Shan, Visentin and Harnett, 2009).

Historically, representing time-based event of human movement in an image or sculpture has been a challenge to artists. Pre-photography, human movement was typically represented by capturing a single critical point during the unfolding of a gesture. Such stop-action representation often exaggerated the phenomenon in order to emphasize an "affect" that would elicit empathy in the viewer; such art typically relies on the viewer's experiential knowledge (of their own bodily condition) or on iconography and/or cultural loading.

Figure 20. Diskobolos of Myron (circa. 460 BC).

For example, Myron's classical representation of a discus thrower (circa. 460 BC) is, from an anthropokinetical standpoint, hardly realistic. Yet, its representation of athleticism and dynamism are understood; the contorted, unnatural body position as a means to show power and effort has become iconographic (Figure 20).

> "What work is there which is as distorted and elaborate as that *Diskobolos* of Myron? But if anyone should criticize this work because it was not sufficiently upright, would he not reveal a lack of understanding of the art, in which the most praiseworthy quality is this very novelty and difficulty?" *Lucian III: The Lover of Lies* (1916) translated by A. M. Harmon (Webpage of the University of Chicago, 2006).

Until the 20th century, affective representations of human movement are the norm. Developments in technology – the advent of photography – inspire some of the most profound changes in 2-D image-making. Photographic techniques were employed by both scientists and artists. The work of Edward James Muybridge may have been scientific investigations, but they had a huge impact on the art world. Muybridge used multi-camera stop action photography (Figure 21 and 22) and multiple exposure techniques (Figure 23) to initiate a scientific approach to human movement studies (Cavanagh, 1990; Solnit, 2003). Further developments using photography to capture and analyze human movement can be found in the work of: 1) Etienne-Jules Marey, who worked with high-speed photography at the same time as Muybridge, capturing multiple images with one camera on one photo plate (Braun, 1992); 2) Wilhelm Braune and Otto Fischer, who created elaborate 3-D models and performed thorough gait analyses using 4 cameras and subjects marked with Geissler tubes which emitted current that could then be seen on film (Cavanagh, 1990); 3) Nikolaj

Bernstein, who developed precise kinematic procedures using cyclography, a technique employing film that was slowly moving through a camera with a mirror placed at a 45 degree angle to the optical axis, thereby creating a picture that gave a frontal and lateral view of the subject (Medved, 2001); and others – using techniques such as strobes, multiple and time exposure (Plagenhoef, 1971). Biomechanical analysis using modern-day, motion capture technology and can be considered a direct outgrowth of this work.

Directly influenced by photography, early 20th-century Futurist painters such as Umberto Boccioni and Gino Severini began to experiment with these ideas, creating works (Figure 24, *Dynamism of a Dancer [1912]*) that compressed the principle of multiple exposure imagery into "kaleidoscope art of force lines, vectored shapes and mobile patterns" (Hunter, Jacobus, and Wheeler, 2004).

Figure 21. Multi-camera stop action recording of a pole-vaulter, by Muybridge.

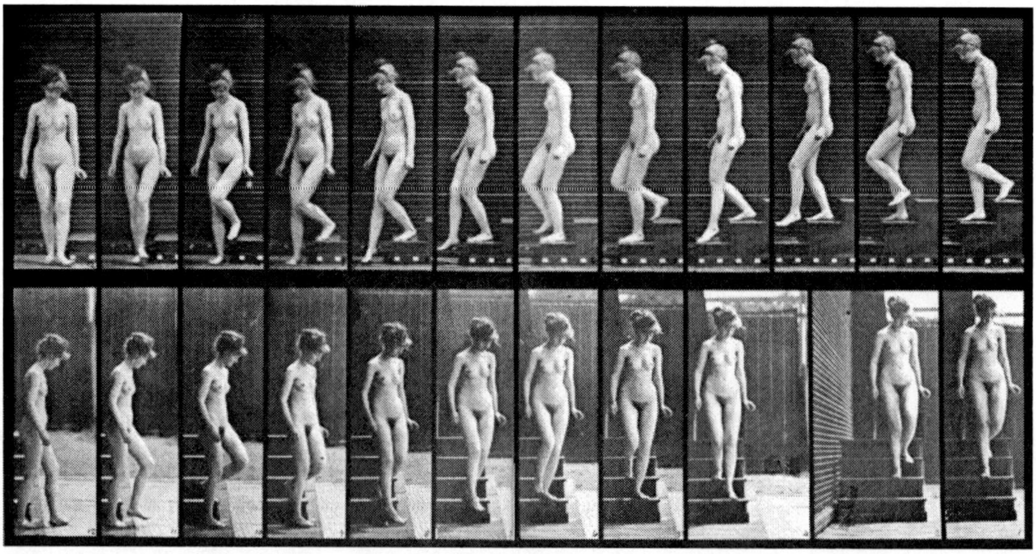

Figure 22. Multi-camera stop action of a nude descending a staircase, by Muybridge.

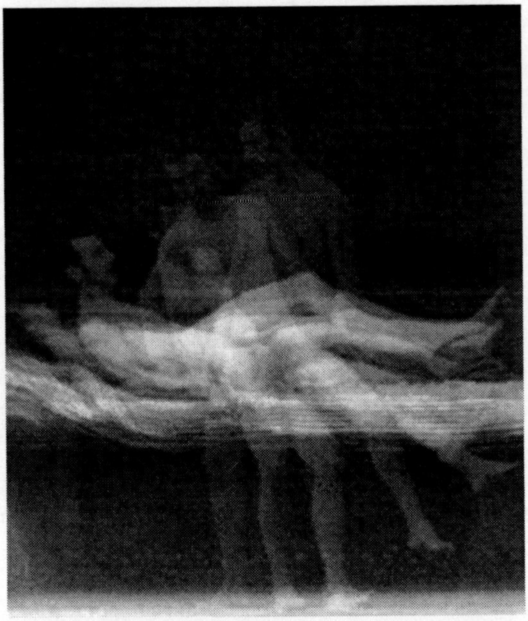

Figure 23. Multiple exposure photography documenting human movement, by Muybridge.

Figure 24. Gino Severini - Painting: Dynamism of a Dancer [1912].

Displaying multiple fragmented essences in a single visual image, these artworks approach an expression of time and position designed to represent a perception of the human experience, by combining the traditional idea of stop-action affective representation with Muybridge's multiple exposure techniques.

In each of these instances, the artwork can be understood to capture a sequential expression of how we experience existence; one where technology has influenced our existential perception of the world. American Modernist art sometimes strips down the imagery to its bare essentials, leaving only gesture. For example, Franz Kline (Kline, 1950) condenses gesture into a single frame of reference (one that may not be anthropokinetically accurate, but is bodily expressive), and contemporary artist Tanya Harnett builds the dynamics of a 3-D space through layered subject matter, gesture, and historical reference, using them as a stage for a photographic image that compresses time, x-y-z dimensionality and cultural reference into a flat-surface gestural work. Harnett's *Nude Descending a Staircase* (Figures 25 and 26) is a direct reference to Marcel Duchamp's work of the same title, who acknowledged the influence of Etienne-Jules Marey's motion capture studies. When such works are viewed, they can be understood as a metaphors and commentaries on the human experience.

Figure 25. *Nude Descending a Staircase* (Tanya Harnett, 2008, reprinted with permission of the artist).

4.1. 3D Motion Capture Technology on Visual Art

If photographic techniques tested the limits of visual culture in the early 20th century, film, video and more recent developments in computers (3-D motion capture, virtual reality, etc.) do so in ours. Cognition in these dynamic environments also relies upon the empathy of the viewers.

Figure 26. *Nude Descending a Staircase* (Tanya Harnett, 2008, reprinted with permission of the artist).

These newer technologies are relatively easy to use and they diminish demands formerly made on content creators; simply, because video media unfolds sequentially, it is easy for viewers to believe in its verisimilitude. With respect to motion capture technology as an extension of video media, four fundamental and consequential questions arise: 1) Can 3-D technologies facilitate analysis and categorization of dynamic, time-based motions? 2) How much representative information is needed in order to evoke an event's unfolding over time (affects of human movement)? 3) What implications does the technology have for artistic image making? And, 4) can we begin to speculate about cultural context as a factor influencing gesture?

4.2. 3D Motion Capture and the Visual Arts: Theoretical Considerations

Gesture communicates information, the meaning of which is driven by empirical experience. Perceptions of gesture as an unfolding of time-based events are intrinsically integrated with the form and function of human movement. Since viewers of a 2-D image have experienced their own bodily gestures as a time-based event, they are inclined to empathize with any such implications in an artwork. The aesthetic experience and understanding of a visual image is both spatial and temporal. Hence, visual communication is linked to the personal experience of the viewer, where culture, iconography and/or visual analogy can play a role.

Motion capture provides the means to accurately record movement in 3-dimensional computer space. The technique differs significantly from actual photography. First, the information gathered is not an actual image, rather it is a collection of trajectories generated by pre-defined, user-determined points. Second, the viewer's choice of perspective is not controlled by camera placement because it can be altered after the motion has been captured. Images synthesized from motion capture trajectories may be used to analyze characteristics that are objectively representational of time and bodiliness. These may be considered signature elements of human movement.

4.3. Exemplary Experiments

Notwithstanding type of activity, trajectory-based images generated using motion capture suggested qualities that may be an aid to establishing signature elements of human movement. The uniqueness of each activity can be understood by relational emphasis on factors such as stability (balance, body position/posture), athleticism (weight, speed, power/vigor), quality of movement (control, smoothness/abruptness), and dimension of the movement (excursion). Selected examples, described below, showed how 3-D motion capture could aid discourse about perceptual characteristics of human movement. Examples are drawn from sports and dance.

Figures 27 and 28 shows trajectories associated with a side kicking technique in Tae Kwon Do. Its main characteristics were:

Stability: trunk trajectories had relatively small, near-linear excursions and they tended to be clumped in one area of the image. The largest trajectories (legs and feet) visibly occurred on a plane rather than in a volume. Clear presence of a centre, circled by rotational trunk trajectories suggested weight, balance and a centripetal centre.

Athleticism: Long trajectory lines of the legs, particularly the feet contrasted the "low-energy", shorter lines of the trunk, lending the impression of a speed differential and a directed power. Orientation of the foot and leg trajectories was in contrast to those of the rest of the image, showing dynamism.

Quality of movement: Continuities of trajectories were smooth, showing controlled gestures, except for acute-angle structures that occurred at the apex of the foot excursion. Abrupt changes of the direction indicated thresholds of the activity or bodily limits where energy of the movement was released. Generally compact excursions of trajectory lines showed the movement to be occurring in a limited space, suggesting bodily control.

Figure 29 shows a soccer kick, viewed from above. Its key characteristics were:

Stability: Most trajectories appeared to be aligned in one general direction and have nearly similar lengths. Relatively parallel to one another, trajectories (particularly trunk and head) showed a state of controlled balance.

Athleticism and dimension: Swirling trajectories of the arms were related to energy and athleticism. The double structure of these swirls indicated building and release phases of the movement. It also gave scale to the overall gesture.

Quality of movement: Abrupt changes in trajectory direction (points of curvilinear discontinuity) showed the release of pent-up power, a quality related to dynamism.

Figure 30 shows a signature generated by a Grand Jeté from classical ballet.

Stability and quality: A dense mass of trajectories, and a planar orientation in the central trajectories of the image (trunk) were evidence of balance and control. These contrasted trajectories (e.g. arm) where excursions had a rotational focal point (i.e. shoulder) and the regularity of the curvilinear path clearly indicated fluidity and speed. These signature characteristics can be related to balancing.

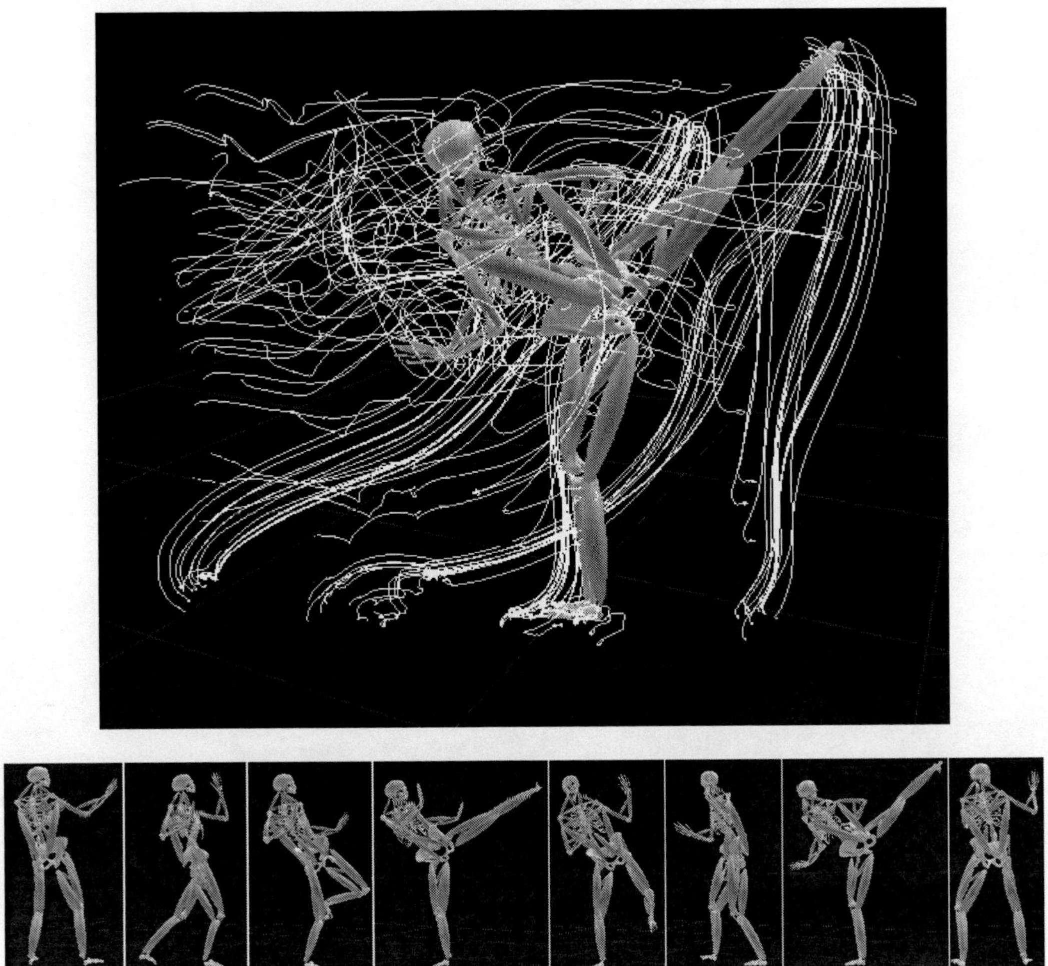

Figure 27. Side view of Tae Kwon Do side kick with body trajectory lines included (above). Sequential stop-frame reconstruction of the movement (below).

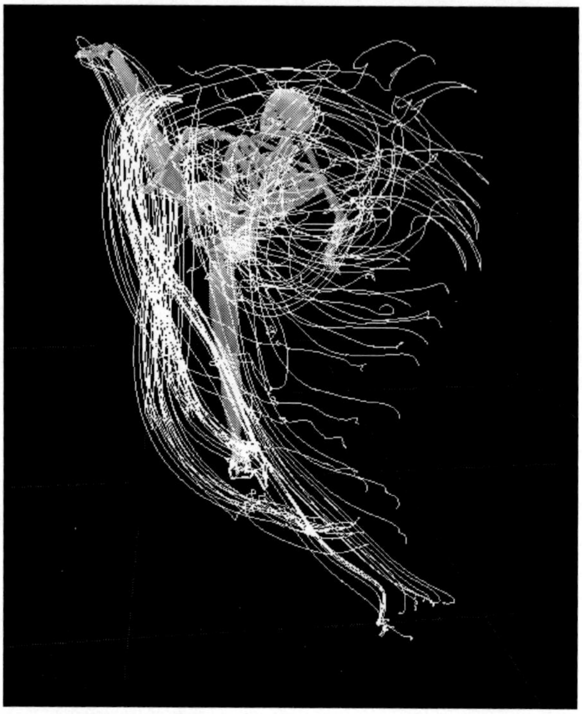

Figure 28. Tae Kwon Do side kick (front-back view) with body trajectory lines.

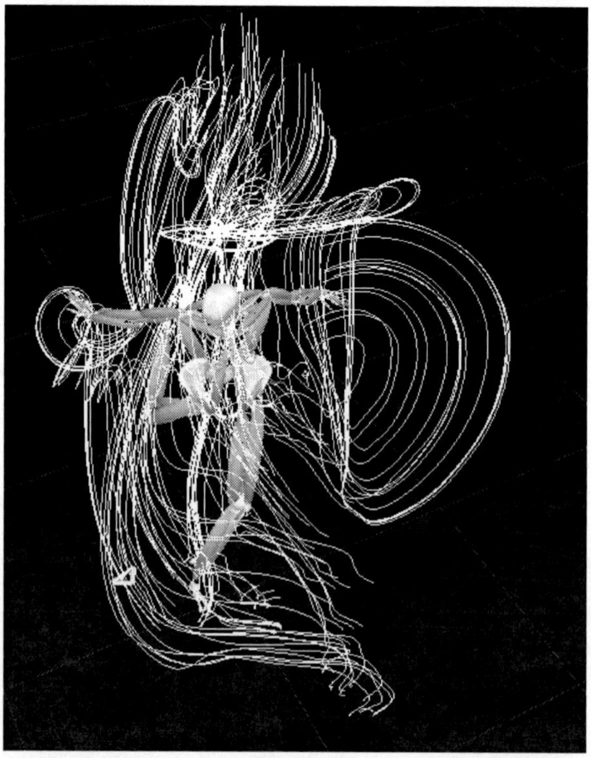

Figure 29. Soccer maximal instep kick (top view) with body trajectory lines.

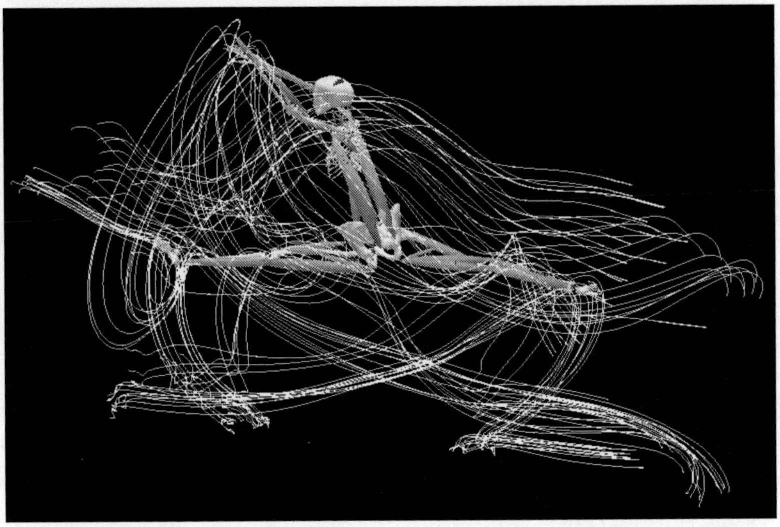

Figure 30. Ballet Grand Jété with body trajectory lines.

Figure 31 shows a fundamental skill of classical Indian dance. In Indian Dance, hand gestures and steps are executed according to highly codified traditions.

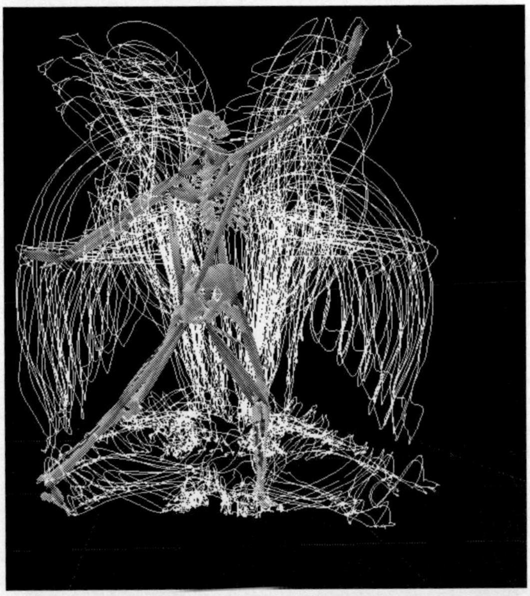

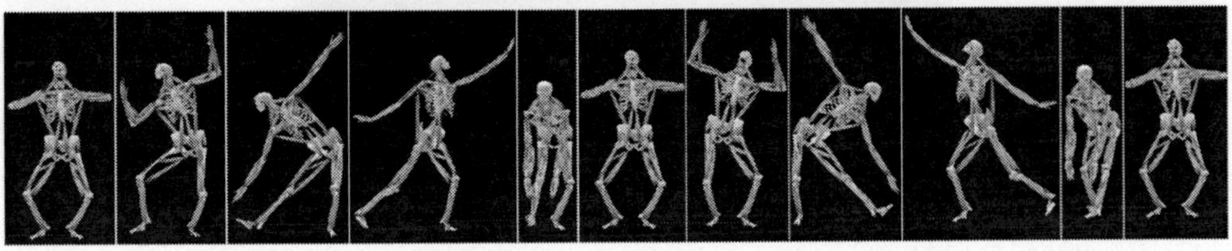

Figure 31. Indian classical dance (front View) with body trajectory lines. Sequential stop-frame reconstruction of the movement (below).

Different combinations of gestures and movements provide a huge vocabulary through which stories can be depicted in surprising detail, perhaps intended to surpass the spoken word (Barboza, 2006).

Stability and control: The clearly formal structure of trajectory lines showed controlled movement. Layered densities of line showed elements of architectural symmetry, balance, and regularity and consistency of movement. Centre of gravity was virtually fixed in a horizontal plane. Such characteristics, particularly the symmetrical, architectural references which were the result of highly codified training, could perhaps be seen to be cultural indicators.

Athleticism and Quality: Continuous gesture trajectories along with no obvious points of curvilinear irregularity indicated dynamism to be less important. Energy expenditure was directed toward control rather than athleticism, which would have more indications of gathering and release. Generally compact excursions of trajectory lines suggested movements in a limited space.

For human movement, the experiential nature of bodiliness is a link connecting the spatial and temporal. Activities such as sports, dance and performance are loaded with cultural mythos, whether it is the embodiment of the warrior class, religious symbolism or the inertia of cultural fashion. Movement signatures necessarily play a role in storytelling. This assertion acknowledges important differences between the concepts of "history" and "aetiology". Whereas history is often viewed as developmental, which should not be conflated with eschatological, aetiology investigates the conditions needed for historical change to take place. Change occurs within the context of culture – whether biologically or philosophically motivated.

Examining sports, the heritages harkens to a surrogate for warrior-class activities, emphasizing power and speed, with additional elements such as control and balance. These can be understood as signature elements of activities derived from a tradition of battle. Culturally, sports teams are an extension of the concept of championing, where selected representatives of the group pit their skills against opponents in proxy for the group as a whole. The signatures of Figures 27 to 29 reveal these very characteristics.

Performing arts activities such as dance represent traditions that can also be understood as a proxy; one for the *values* of the group heritage. In the case of activity 4, Grand Jeté, cultural references are rooted in the Court of Louis XIV, where the foundations of classical ballet originate and dance was used as a political tool. As a matter of note, when he took power after reaching the age of majority, Louis the XIV, a virtuoso dancer, performed in an allegorical ballet, dressed as the Sun God Apollo, symbolizing the rising of the sun to drive away of the darkness of the previous political regime. From this point forward, Louis XIV's noble class involved much of their efforts in learning ballet in order to imitate the noble bearing of the King (Hilton, 1997). Hence, fundamental signature elements - upright bearing, control, naturalness and demonstrated strength through elegant fluidity - become highly codified; controlled symmetrical dance structures and steps analogous with the nobility of the King and his natural, god-given place in the universe. These elements are still present in the fundamentals of ballet today (Figure 30).

Clearly, many activities can be viewed as culturally loaded. The East Indian dance signature (Figure 31) is visually distinctive, with characteristics that display highly codified dance traditions – a stylization that captures something that is meaningful to or representative of the culture. In East Indian dance, iconography is very direct. Much is related to storytelling (religious or historical) through gesture. Formal, architectural structures, stability and control can be related to the rhetorical cohesion needed to tell these stories. Sociological and/or anthropological understanding enhanced by movement analysis technology might be one bridge in the exchange of cross-cultural ideas. Understanding our experience as a temporal bodily phenomenon creates cognitive links between information, still imagery, and cultural mythos as other kinds of metaphors for the human experience.

4.4. Other Contemporary Examples

The potential of movement analysis technology in the creation of art has been explored in several settings. Notably, Paradiso (1999) explored unconventional interfaces for individuals to interact with musical environments. In his "Brain Opera" project responsive environments were built out of "smart objects and sensitive spaces, where any kind of physical activity or motion can result in a sophisticated multimedia reaction" (Paradiso, 1999). The kinds of interfaces explored permitted interactive musical mappings based on the intuitive responses of the participants (who were also the audience). Although the authors claim that several of the Brain Opera's goals were achieved, the experiment has essentially proven to be a "one-off". It may have been an ambitious way of "hinting at the possibilities now opening for large-scale participatory music installations," however the aim seems to have been mostly an exploration of technological capacity and the project fell short of enhancing the intensity and relevance of an artistic concept. In the study, Paradiso explained this to arise from the way in which the "project refuse(d) to fit into any clearly defined musical category". In a separate study, Paradiso et al. (2003) further examined the concept of interactive musical environments using electromagnetic tagging technology, where an ensemble of passively tagged objects was tracked in real time. These were tied to simple notes and musical effects which enabled dynamic control of musical sequences and sonic textures (Paradiso, Pardue, Hsiao, and Benbasat, 2003). Although technologies can advance installation-based interactive music environments, their communicative efficacy still rests in the ears of the audience.

In a multimedia project incorporating music and dance. Bevilaqua et al. (2001) used a 3D motion capture system to produce animations from dance and, more unusually, generate a soundtrack from the dancer's movements. Motion capture data was used to both trigger and modify the timbre of sounds. The project was exploratory, challenging traditional boundaries between the arts (music, dance, studio art and drama) and sciences (motion analysis technology). The authors define their thesis as follows:

> "In electronic music, the relationship between sound generation and touch/gesture is artificially defined, contrary to the way in which sound is produced in acoustic instruments. In fact, the possibility of designing the interface between the gesture (or the touch) and the sound is a fascinating feature of digital music. Most electronic music controllers that have been created are based on existing acoustic instruments (for example the piano keyboard). Such electronic controllers have the obvious advantage to be used relatively easily by 'traditionally' trained musicians" (Bevilacqua, Naugle, and Valverde, 2001).

In summary, such research consistently points to the need and desire for greater collaboration between artists and scientists. In the given examples, technology provides analytic tools and a means to characterize intersections between the body and artistic creation. Biomechanical signals link the artistic conceptualization and its interface with our perceptions. Such explorations are just starting. And, while the benefit may currently seem a bit one-sided – artists taking advantage of technology for the betterment of their art – the benefit really flows both ways. Technology is in a consistent state of rapid change. Artists' demands can be a catalyst for improving it. For example, in the case of 3-D motion capture, some of the first demands for video and sound to be synchronized with the 3-D data came from Arts researchers (VICON motion capture). This development is ubiquitous in the industry today.

In terms of biomechanics research, there continues to be significant unexplored potential. Throughout history, each set of technologies has ushered its own set of changes into the way people interact with their environment and generate understanding. Technology may allow complex mapping functions to be interpreted very sophisticated ways. But, it is sometimes more interesting and useful to consider the artistic gesture, the potential for expression, or the aesthetic design inherent in the digital data generated by it. This kind of thought is a central tenant of the artistic process. And, to reconsider the first question posed in this chapter, it is one of the best reasons to "bother with Arts Biomechanics".

Acknowledgments

This book chapter was financially supported by NSERC (National Sciences and Engineering Research Council of Canada), Westgrid/Canada.

References

Administration, F. A. (1996). Dynamic Evaluation of Seat Restraint Systems andOccupant Protection on Transport Airplanes. Retrieved. from.

Bach, C. P. E. (1759). Essay on the True Art of Playing Keyboard Instruments (W. J. Mitchell, Trans.). New York: W. W. Norton and Co.

Baillot, P. (1835). L'art du violon (L. Goldberg, Trans.). Paris: Northwestern University Press.

Ballreich, R., and Baumann, W. (1996). Grundlagen der Biomechanik des Sports (The basics of biomechanics in sports). Stuttgart: Enke Verlag.

Ballreich, R., and Kuhlow, A. (1980). Beiträge zur Biomechanik des Sports. Schorndorf: Karl Hofmann.

Barboza, K. (2006). Repräsentative und symbolische Zeichen des klassischen Tanzes in Indien und in Europa: Eine interkulturelle, vergleichende Analyse des indischen bharatanāyam und des französischen Balletts. University of Münster, Münster.

Barth, T. (2000). An AIR passenger airbag system for transport aircraft. Retrieved. from.

Bevilacqua, F., Naugle, L., and Valverde, I. (2001). Virtual dance and music environment using motion capture. Paper presented at the *IEEE Multimedia Technology and Applications Conference*, Irvine.

Biral, D., Jakubiec-Puka, A., Ciechomska, I., Sandri, M., Rossini, K., Carraro, U., et al. (2000). Loss of dystrophin and some dystrophin-associated proteins with concomitant signs of apoptosis in rat leg muscle overworked in extension. *Acta Neuropathologica*, 100(6), 618-626.

Brandser, E. A., Elkhoury, G. Y., Kathol, M. H., Callaghan, J. J., and Tearse, D. S. (1995). HAMSTRING INJURIES - RADIOGRAPHIC, CONVENTIONAL TOMOGRAPHIC, CT, AND MR-IMAGING CHARACTERISTICS. *Radiology*, 197(1), 257-262.

Braun, M. (1992). Picturing Time: The Work of Etienne-Jules Marey (1830-1904). Chicago: University of Chicago Press.

Brown, A. N. (1997). Musculoskeletal misuse among youth symphony string players. *Medical Problems of Performing Artists*, 12, 15-18.

Cavanagh, P. (1990). The mechanics of distance running: a historical perspective. In P. Cavanagh (Ed.), Biomechanics of Distance Running (pp. 135-164). Champaign: Human Kinetics.

Chesky, K., Kondraske, G., Henoch, M., Hipple, J., and Rubin, B. (2002). Musicians' health. In R. Colwell and C. Richardson (Eds.), The new handbook of research on music teaching and learning. New York: Oxford University Press.

Chockley, C. (2008). Ground Reaction Force Comparison Between Jumps Landing on the Full Foot and Jumps Landing en Pointe in Ballet Dancers. *Journal of Dance Medicine and Science*, 12(1).

Dawson, D. M., Hallett, M., and Millender, L. H. (1990). *Entrapment Neuropathies*. Boston: Little, Brown and Co.

Dawson, W. J. (2003). The bibliography of performing arts medicine: A five-year retrospective review. *Medical Problems of Performing Artists*, 18(1), 27-32.

Dawson, W. J. (2007). The bibliography of performing arts medicine - A 10 year retrospective review (Part II). *Medical Problems of Performing Artists*, 22(4), 153-159.

Devroop, K., and Chesky, K. (2002). Comparison of biomechanical forces generated during trumpet performance in contrasting settings. *Medical Problems of Performing Artists*, 17(4), 149-154.

Einstein, A. (1931). The World as I See It. Retrieved January 9, 2009, from http://www.aip.org/history/einstein/essay.htm.

Evans, B. (2003). How I Survived My Dance Training. *Medical Problems of Performing* Artists, 18(4), 137-140.

Fotheringham W. (2002). Put Me Back on My Bike: In Search of Tom Simpson. New York: Yellow Jersey Press.

Fry, H. J. H. (1986). Overuse syndrome of the upper limb in musicians. *Medical Journal Austrilia*, 144, 182-185.

Fry, H. J. H. (1987). Prevalence of overuse (injury) syndrome in Australian music schools. *British Journal of Internal Medicine*, 44, 35-40.

Fry, H. J. H. (1988). Patterns of over-use seen in 658 affected instrumental musicians. International Journal of Music Education, 11, 3-16.

Fry, H. J. H., Ross, P., and Rutherford, M. (1988). Music-related overuse in secondary schools. *Medical Problems of Performing Artists*, 3, 133-134.

Gadd, C. (1966). Use of a weighted impulse criterion for estimating injury hazard. Paper presented at the 10th Stapp Car Crash Conference, New York.

Gadd, C. M. (1966). Use of a weighted impulse criterion for estimating injury hazard. Paper presented at the 10th Stapp Car Crash Conference, New York.

Hagglund, K. L. (1996). A comparison of the physical and mental practices of music students from the New England Conservatory and Boston University Music School. *Medical Problems of Performing Artists*, 11, 99-107.

Hartsell, H. D., and Tata, G. E. (1991). A retrospective of music-related musculoskeletal problems occurring in undergraduate music students. *Pysiotherapy Canada* 43(1), 13-18.

Hennig, E. M., and Lafortune, M. A. (1991). Relationships between ground reaction force and tibial bone acceleration parameters. *International Journal of Sport Biomechanics*, 7, 303-309.

Hilton, W. (1997). Dance and Music of Court and Theatre: selected Writings of Wendy Hilton. Stuyvesant NY: Pendragon Press.

Hodgson, P. (1958). Motion Study and Violin Bowing. Urbana: American String Teachers Association.

Hunter, S., Jacobus, J., and Wheeler, D. (2004). Modern Art, Third Edition Revised and Expanded. New York: Prentice-Hall.

Kanbe, K., Hasegawa, A., Takagishi, K., Shirakura, K., Nagase, M., Yanagawa, T., et al. (1998). Analysis of muscle bioenergetic metabolism in rabbit leg lengthening. *Clinical Orthopaedics and Related Research*(351), 214-221.

Kaplinski, V. (2003). Prevention of injuries among keyboard players. Paper presented at the Twenty-First Annual Symposium on Medical Problems of Musicians and Dancers.

Kline, F. (1950). Chief. Retrieved March 22, 2008, 2008, from http://www.moma.org/collection/browse_results.php?object_id=78319

Kolneder, W. (1993). The Amadeus Book of the Violin - Construction, History, and Music. Oregon: Amadeus Press.

Komulainen, J., Takala, T. E. S., Kuipers, H., and Hesselink, M. K. C. (1998). The disruption of myofibre structures in rat skeletal muscle after forced lengthening contractions. *Pflugers Archiv-European Journal of Physiology*, 436(5), 735-741.

Kosiak, M. (1959). Etiology and Pathology of ischemic ulcers. *Archives of Physical Medicine*, 40, 62-69.

Koskinen, S. O. A., Ahtikoski, A. M., Komulainen, J., Hesselink, M. K. C., Drost, M. R., and Takala, T. E. S. (2002). Short-term effects of forced eccentric contractions on collagen synthesis and degradation in rat skeletal muscle. *Pflugers Archiv-European Journal of Physiology*, 444(1-2), 59-72.

Kuipers, H. (1998). Training and overtraining: An introduction. *Medicine and Science in Sports and Exercise*, 30, 1137-1139.

Lehmann, A. C., and Davidson, J. W. (2002). Taking an acquired skills perspective. In R. Colwell and C. Richardson (Eds.), The new handbook of research on music teaching and learning. New York: Oxford University Press.

Levitin, D. (2006). This is Your Brain on Music. New York: Plume.

Lockwood, A. H. (1988). Medical problems in secondary school-aged musicians. *Medical Problems of Performing Artists,* 3, 129-132.

Lundberg, A. (1996). On the use of bone and skin markers in kinematics research. *Human Movement Science*, 15(3), 411-422.

Magill, R. A. (2001). Motor learning concepts and applications (6 ed.). Boston: McGraw-Hill.

McLean, A., and Anderson, R. (1997). *Biomechanics of closed head injuries*. In P. Reilly and R. Bullock (Eds.), Head Injury (pp. 25-37). London: Chapman and Hall.

Medved, V. (2001). Measurement of Human Locomotion. Boca Raton: CRC Press.

Middlestadt, S. E., and Fischbein, M. (1989). The prevelance of severe musculoskeletal problems among male and female symphony orchestra string players. *Medical Problems of Performing Artists*, 4, 41-48.

Moyse, M. (1973). The Flute and its Problems; Tone Development Through Interpretation for the Flute. Tokyo: Muramatsu Gakki Hanbai Co.,Ltd.

Mozart, L. (1756). A treatise on the Fundamental Principles of Violin Playing (E. Knocker, Trans.). Oxford: Oxford University Press.

Nemoto, K., and Arino, H. (2007). Hand and upper extremity problems in wind instrument players in military bands. *Medical Problems of Performing Artists*, 22(2), 67-69.

Nicol, K. (1996). Velocity vs. duration of competition - A tool for characterizing sport events, internal load, and limiting factors. Paper presented at the 9th Biennial Conference, *Canadian Society for Biomechanics*, Vancouver.

Nicol, K. (1998). Gesetzmäßigkeiten von Grenzleistungen. *Leistungssport*, 5 28-35.

Palac, J. A. (1992). Violin bowing technique: an analysis of contemporary pedagogical literature according to principles of human movement. *Medical Problems of Performing Artists*, 7(1), 30-34.

Paradiso, J. A. (1999). The brain opera technology: New instruments and gestural sensors for musical interaction and performance. *Journal of New Music Research*, 28(2), 130-149.

Paradiso, J. A., Pardue, L. S., Hsiao, K. Y., and Benbasat, A. Y. (2003). Electromagnetic tagging for electronic music interfaces. *Journal of New Music Research*, 32(4), 395-409.

Parlitz, D., Peschel, T., and Altenmuller, E. (1998). Assessment of dynamic finger forces in pianists: Effects of training and expertise. *Journal of Biomechanics*, 31(11), 1063-1067.

Peterson, L., and Renström, P. (1987). Verletzungen im sport (Injuries in Sports). Köln: Deutcher Ärte-Verlag.

Plagenhoef, S. (1971). Patterns of Human Movement: a cinematographic analysis. Englewood Cliffs: Prentice-Hall.

Quantz, J. J. (1752). Essay of a Method for Playing the Transverse Flute (R. E. R., Trans.). London: faber and faber.

Rabuffetti, M., Converti, R. M., Boccardi, S., and Ferrarin, M. (2007). Tuning of the violin-performer interface: An experimental study about the effects of shoulder rest variations on playing kinematics. *Medical Problems of Performing Artists*, 22(2), 58-66.

Rubin, S. J., Feldman, F., Staron, R. B., Zwass, A., Totterman, S., and Meyers, S. P. (1995). MAGNETIC-RESONANCE-IMAGING OF MUSCLE INJURY. *Clinical Imaging*, 19(4), 263-269.

Schwartz, B. (1983). Great masters of the violin. New York: Simon and Schuster.

Shan, G. (2005). Comparison of repetitive movements between ballet dancers and martial artists ¾ risk assessment of muscle overuse injuries and prevention strategies. *Research in Sports Medicine*, 13(1), 63-76.

Shan, G. B. (1999). Ein biomechanisches Modell auf Basis von individuellen anthropometrischen Daten fuer das Bewegungslernen der Flugphasen bei sportlichen Bewegungsablaeufen (A biomechanical model for motor learning of air-borne phases based on individual anthropometry). Müster: Lit Verlag.

Shan, G. B., and Bohn, C. (2003). Anthropometrical data and coefficients of regression related to gender and race. *Applied Ergonomics*, 34(4), 327-337.

Shan, G. B., and Visentin, P. (2003). A quantitative three-dimensional analysis of arm kinematics in violin performance. *Medical Problems of Performing Artists*, 18(1), 3-10.

Shan, G. B., Visentin, P., and Schultz, A. (2004). Multidimensional signal analysis as a means of better understanding factors associated with repetitive use in violin performance. *Medical Problems of Performing Artists*, 19(3), 129-139.

Shan, G. B., Visentin, P., Wooldridge, L., Wang, C. D., and Connolly, D. (2007). A frequency-based characterization of spiccato bowing in violin performance. *Perceptual and Motor Skills*, 105(3), 1027-1051.

Shippen, J., and May, B. (2008). Whole body biomechanical analysis and muscle loading of Irish dancers. Paper presented at the International Association for Dance Medicine and Science, 18th Annual Meeting, Cleveland, OH.

Shippen, J., and May, B. (2009). An analysis procedure for the calculation of muscle loading and joint contact forces in dancers. *Journal of Dance Medicine and Science*.

Smith, H. K., Plyley, M. J., Rodgers, C. D., and McKee, N. H. (1997). Skeletal muscle damage in the rat hindlimb following single or repeated daily bouts of downhill exercise. *International Journal of Sports Medicine*, 18(2), 94-100.

Solnit, R. (2003). River of Shadows: Eadward Muybridge and the Technological Wild West. New York: Viking Penguin.

Solomon, R., and Solomon, J. (2004). Publications in dance medicine and science: A bibliographer's perspective. *Medical Problems of Performing Artists*, 19(4), 167-169.

Stauber, W. T., and Willems, M. E. T. (2002). Prevention of histopathologic changes from 30 repeated stretches of active rat skeletal muscles by long inter-stretch rest times. *European Journal of Applied Physiology*, 88(1-2), 94-99.

Steinbach, L. S., Fleckenstein, J. L., and Mink, J. H. (1994). MAGNETIC-RESONANCE-IMAGING OF MUSCLE INJURIES. *Orthopedics*, 17(11), 991-999.

Sundar, M., and Carty, H. (1994). AVULSION FRACTURES OF THE PELVIS IN CHILDREN - A REPORT OF 32 FRACTURES AND THEIR OUTCOME. *Skeletal Radiology*, 23(2), 85-90.

Taylor, J. (1996). Intensity and athletic performance. In J. Van Raalte and B. Brewer (Eds.), A practitioner's guide to sport and exercise psychology. Washington: American Psychological Association.

Thompson, J. L., Balog, E. M., Fitts, R. H., and Riley, D. A. (1999). Five myofibrillar lesion types in eccentrically challenged, unloaded rat adductor longus muscle - A test model. Anatomical Record, 254(1), 39-52.

Tromlitz, J. G. (1791). The Virtuoso Flute Player (A. Powell, Trans.): Cambridge.

Tulchinsky, E. (1995). A BIOMECHANICAL MOTION ANALYSIS OF THE VIOLINISTS BOW ARM (VOL 9, PG 119, YR 1994). *Medical Problems of Performing Artists*, 10(1), 10-10.

Tulchinsky, E., and Riolo, L. (1994a). A biomechanical motion analysis of the violinist's bow arm. *Medical Problems of Performing Artists*, 9(4), 119-124.

Tulchinsky, E., and Riolo, L. (1994b). A BIOMECHANICAL MOTION ANALYSIS OF THE VIOLINISTS BOW ART. Medical Problems of Performing Artists, 9(4), 119-124.

Uchiyama, Y., Tamaki, T., and Fukuda, H. (2001). Relationship between functional deficit and severity of experimental fast-strain injury of rat skeletal muscle. *European Journal of Applied Physiology*, 85(1-2), 1-9.

USA. Airworthiness Standards: Transport Category Airplanes. Retrieved. from.

USGovernment. Airworthiness Standards: Transport Category Airplanes. Retrieved. from.

Vijayan, K., Thompson, J. L., Norenberg, K. M., Fitts, R. H., and Riley, D. A. (2001). Fiber-type susceptibility to eccentric contraction-induced damage of hindlimb-unloaded rat AL muscles. *Journal of Applied Physiology*, 90(3), 770-776.

Virgilio, F. F., Chiara, M., Emilia, B., Paolo, P., and Sforza, C. (2007). Three-Dimensional Analysis of Hand and Finger Movements during Piano Playing. *Medical Problems of Performing Artists*, 22(1), 18-23.

Visentin, P., Shan, G., and Wasiak, E. B. (2008). Informing music teaching and learning using movement analysis technology. *International Journal of Music Education*, 26(1), 73-87.

Visentin, P., and Shan, G. B. (2003). The kinetic characteristics of the bow arm during violin performance: An examination of internal loads as a function of tempo. *Medical Problems of Performing Artists*, 18(3), 91-97.

Visentin, P., and Shan, G. B. (2004). An innovative approach to understand overuse injuries: Biomechanical modeling as a platform to integrate information obtained from various analytic tools. *Medical Problems of Performing Artists*, 19(2), 90-96.

Wilson, F. (1986). Tone deaf and all thumbs? New York: Viking Press.

Winter, D. A. (1990). Biomechanics and Motor Control of Human Movement. New York: Wiley.

Zaza, C. (1992). Playing-related health problems at a Canadian music school. *Medical Problems of Performing Artists*, 7, 48-51.

Zaza, C. (1998). Playing-related musculoskeletal disorders in musicians. *Canadian Medical Association Journal*, 158, 1019-1025.

In: Biomechanics: Principles, Trends and Applications
Editor: Jerrod H. Levy, pp. 57-110
ISBN: 978-1-60741-394-3

Chapter 2

UNDERSTANDING CORNEAL BIOMECHANICS THROUGH EXPERIMENTAL ASSESSMENT AND NUMERICAL SIMULATION

Ahmed Elsheikh
Ocular Biomechanics Group
Division of Civil Engineering, University of Dundee, UK

ABSTRACT

The Ocular Biomechanics Group was established in 2002 with one clear target; to develop a virtual reality model of the human eye that can be used effectively and reliably to predict ocular response to surgery, injury and disease. This ambitious, and seemingly illusive, target helped plan our activities over the last 6 years and will still be focusing our efforts as we strive to create the necessary knowledge using experimental methods, build the predictive tools using programming and analysis means, and validate the findings in both the laboratory and the clinic. This chapter presents an overview of our biomechanical studies from laboratory material characterisation to finite element numerical simulation. The chapter describes what has been achieved and points at the remaining gaps in our knowledge. It explains that while much remains unknown in ocular behaviour, we are now in a good position to use available knowledge to progress predictive modelling and use it in actual applications such as improving the accuracy of tonometry techniques, planning of refractive surgeries and design of contact lenses. The discussion focuses on the cornea, although scleral biomechanics receive some mention. The chapter also refers to microstructural, biomechanical and topographic studies conducted by other research groups. Coverage of these studies has been necessary to provide a more complete image of current understanding of corneal biomechanics.

INTRODUCTION

The transparent cornea is a most important component of the outer ocular tunic. It provides a tough protective envelope for the ocular contents and helps give the eye its general shape, Figure 1. The anterior corneal surface accounts for about two-thirds of the optical power of the eye [1].

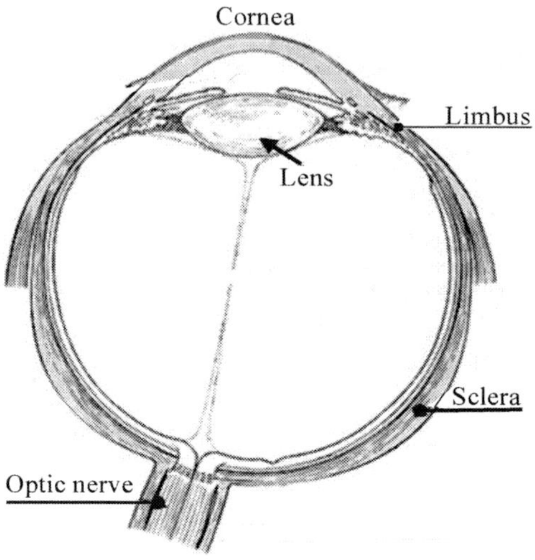

Figure 1. Cross-section of eye globe showing some of the main components [4].

This important role can be defined in terms of corneal shape, regularity and clarity, and is a function of its refractive index. The cornea's contribution to ocular image formation can be degraded by anomalies in shape produced by disease, injury or surgery. The ability to understand and predict corneal response to such conditions is therefore of great clinical importance [2, 3].

The past three decades saw a rapid growth in the use of numerical modelling to improve understanding of corneal biomechanics and to predict ocular response to the conditions met in tonometry and refractive surgery. Numerical models have improved in accuracy and in their representation of the natural conditions of the eye. In the 1980s and 1990s, it was common to assume linear elastic, and homogenous material behaviour and spherical topography in corneal numerical models [5-8]. Since then, and with the better understanding of corneal behaviour and topography achieved experimentally and clinically, these assumptions have become unnecessary [9]. The increase in computing power over these three decades also made it possible to produce and execute highly nonlinear corneal models with fine finite element meshes and with consideration of other ocular components [10, 11]. These highly expensive models, which required specialised mainframe computers in the past, can now be conducted on commonly available personal computers.

However, in spite of this progress, corneal numerical simulation had only a limited role in improving clinical practice. An example of this situation can be seen in the attempts to create numerical models of refractive surgery, where much attention has been given to simulate the procedures [12-23], but as yet no numerical model has been adopted in clinical practice to plan and optimise refractive surgeries. Offering a solution here might not be straightforward since part of the problem is related to the clinicians' reluctance to adopt tools and findings that come through a numerical route. However, much progress can be made by reviewing current modelling approaches, and concentrating on the areas where improvement is needed.

Recent developments in corneal microstructure detailing, material characterisation and topography mapping are offering valuable information for numerical modelling. Microstructural studies have identified the preferential orientation of stromal collagen fibrils [24-27] and the resulting anisotropy in corneal biomechanical behaviour [28, 29]. The corneal nonlinear stress-strain behaviour [14, 30, 31], viscoelasticity [32, 33], hysteresis [34], long-term creep [35] and stress relaxation [36] have been measured experimentally. The epithelium's contribution to corneal biomechanics has been quantified in relation to the stroma [37, 38]. More recently, the inter-lamellar adhesion within the stroma, under both shear [39] and lateral tension [40-42], has been assessed. Much work has also been done to determine the anterior and posterior asphericity [43-45], and the thickness distribution of the cornea [46, 48]. These properties offer new opportunities to make numerical modelling more representative of corneal behaviour, and potentially more reliable. But the question, which this chapter attempts to answer, is how best to use this knowledge, and whether it is essential for a numerical model to consider all these properties regardless of the application at hand.

The chapter starts with an overview of recent microstructural, topographical and experimental advances that provided significant improvements in our understanding of corneal biomechanics. A discussion is then presented on how to embed the experimentally-obtained material behaviour in numerical simulation. Within this discussion, a general introduction to finite element modelling and the simplifications involved in model construction is presented, followed by an assessment of the effects of simplifications on model output and whether these effects are acceptable. The chapter ends with concluding statements on the way forward for corneal numerical simulation.

CORNEAL MICROSTRUCTURE

The human cornea consists of three distinct layers: an anterior epithelium, a posterior endothelium and a central stroma, Figure 2. Between these layers are specialised extracellular structures called Bowman's layer and Descemet's membrane. While the epithelium and endothelium are cellular, composed mainly of keratinocytes, the stroma has a distinctive composite structure with collagen fibrils embedded in an extracellular matrix rich in proteoglycans, glycoproteins and keratocytes.

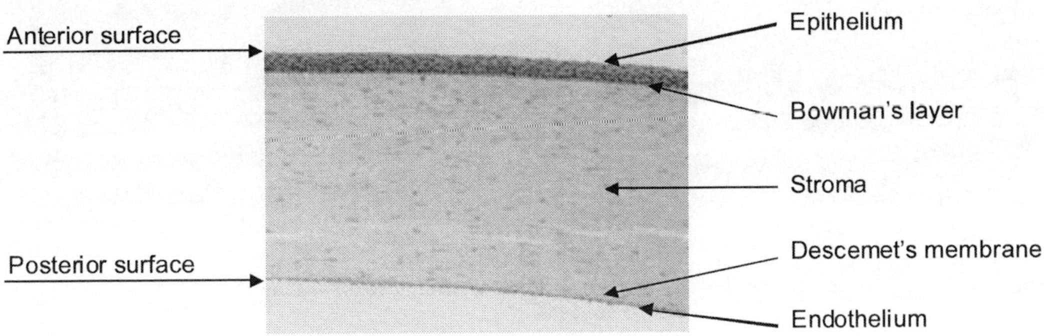

Figure 2. Cross-sectional view of the human cornea's layered construction.

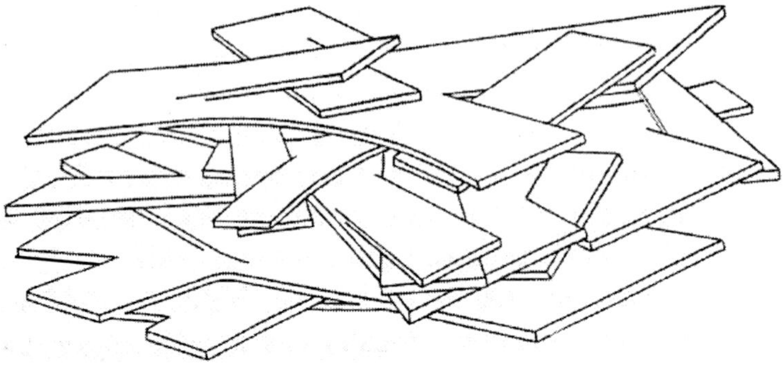

Figure 3. Interweaving of stromal lamellae.

The mechanical behaviour of the human cornea is dominated by the stroma that forms about 90% of its thickness [24, 49]. The stroma has a highly complex structure with 200-400 superimposed lamellae, set mostly parallel to the corneal surface, but increasingly interwoven towards the anterior surface [25, 50-52], Figure 3. The increased lamella interweaving possibly leads to variation in corneal mechanical properties and swelling resistance [53].

Stromal lamellae are composed of fine collagen fibrils regularly spaced and embedded in an extracellular matrix. The fibrils within each lamella run parallel to each other, but make large angles with neighbouring lamellae, Figure 4. Earlier research identified two preferential directions within which most fibrils run in the central human cornea, namely the superior-inferior and the temporal-nasal directions, possibly related to the action of the extraocular muscles [26, 27, 54-57]. Daxer and Fratzl estimated that 49% of fibrils are oriented in these two directions and 66% of the fibrils lie within the 45° sectors surrounding them [58]. Since the fibrils are thought to be the main load carrying elements of the stroma, their preferential orientation is likely to lead to anisotropic mechanical behaviour of the human cornea with the vertical and horizontal directions exhibiting superior stiffness (resistance to deformation) compared with other directions.

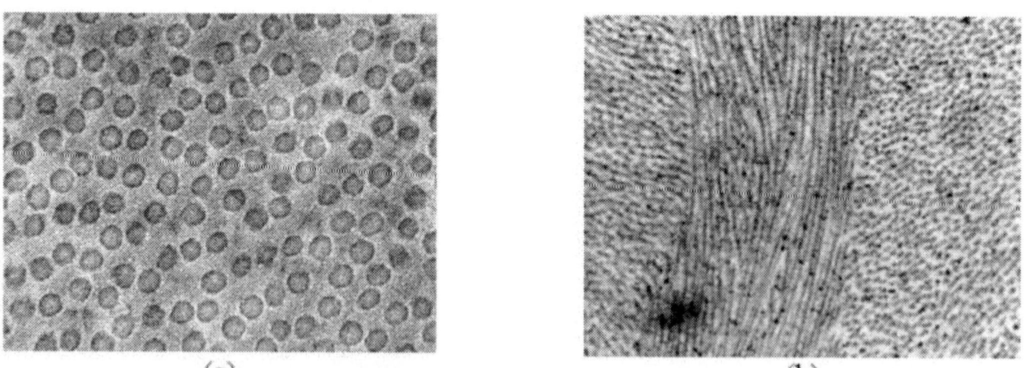

(a) (b)

Figure 4. TEM images of (a) collagen fibril distribution within a stromal lamella, and (b) three lamellae with different fibril orientations.

More recent studies found that corneas from other species exhibited significantly different fibrillar collagen structures [59, 60]. The central corneas of horses, cows and marmosets were found to have dominantly superior-inferior (vertical) orientation, while porcine, rabbit and mouse corneas had mainly circumferential orientation. This variation in structure was thought to be due to differences between species in the length, width and insertion position of the rectus muscles, and possibly also due to the existence of the posterior retractor bulbi muscle in all the above mentioned animals but not in humans [61]. However, despite these differences, a common feature that exists in both human and animal corneas is the gradual change in fibril orientation towards a predominantly circumferential organisation close to the limbus.

Other corneal layers include the endothelium; a monolayer of specialised cells that line the posterior surface of the cornea and face the eye's anterior chamber [62]. The endothelium regulates fluid transport across the corneal posterior surface by allowing leakage of solutes and nutrients from the aqueous humor to the cornea and pumping water from the stroma to the aqueous [63]. The small thickness of the endothelium (≈ 2% of corneal thickness), and its cellular structure suggest it has low mechanical stiffness (resistance to deformation) compared to the stroma, although its stiffness has never been quantified [64].

Lining the other surface of the cornea is the epithelium which comprises about 8-10% of corneal thickness [64]. The epithelium functions primarily to block the passage of foreign material into the eye, absorb oxygen and nutrients from the tear film [1] and control the swelling of the stroma [65]. Similar to the endothelium, the epithelium has a cellular structure, which makes its mechanical stiffness low in comparison to that of the stroma.

Between the stroma and the endothelium lies Descemet's membrane, whose physiological function is unknown although it may be to provide mechanical support to the endothelium, or to act as a fluid barrier. Its low mechanical stiffness and small thickness (ranges between 3 and 10 μm or 0.5 to 2% of corneal thickness) suggest it has low contribution to the cornea's overall mechanical behaviour [66, 67].

Finally, Bowman's layer is located between the epithelium and the stroma, composed of strong collagen fibers and has a thickness of 8-12 μm. Similar to Descemet's membrane, the physiological function of Bowman's layer is unknown.

It appears to provide strong attachment of the epithelium to the stroma in human eyes, but the fact that it does not exist in some mammal eyes has not been explained [68]. Earlier research assessed the mechanical behaviour of Bowman's layer and found it to be insignificant [69].

The discussion in this chapter will assess evidence relating the corneal microstructure, and in particular the stromal lamellae organisation, to overall biomechanical behaviour. The implications of this relationship on our understanding of the effects of disease, injury and surgery will also be covered in the conclusion section.

Corneal Topography

The normal cornea is prolate – steepest at the centre and flattens towards the periphery. A section through the cornea is an ellipse that can be described using Bennett's equation in terms of two parameters; the radius of curvature at corneal apex (apical radius, R_o), and the asphericity or shape factor, p:

$$z = \frac{R_o - \sqrt{R_o^{\,2} - p\,x^2}}{p} \qquad (1)$$

where x and z are coordinates of a general point, a, on the corneal surface, Figure 5. The apical radius, R_o, (which approximately equals the value measured in keratometry) has been measured in a few clinical studies, and examples of the reported average and standard deviation values include 7.87±0.27 and 7.77±0.20 mm [43, 44]. On the other hand, the shape factor, p, provides an indication of how rapidly the surface flattens away from the apex (centre). The average values of p for the human cornea's anterior and posterior surfaces have been found using in-vivo measurements to be 0.82 and 0.62, respectively [43].

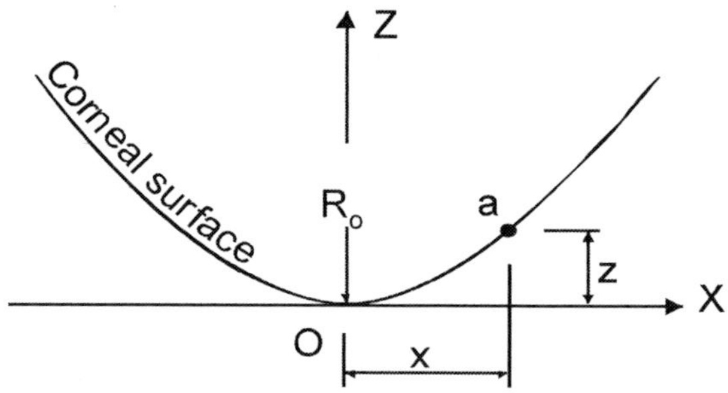

Figure 5. Cross section through a corneal surface with coordinates of a general point a.

Corneal Thickness Distribution

The cornea has a variable thickness that changes from minimum at or close to the centre, to maximum along the limbus, Figure 6. Clinical studies reported different average values, standard deviations and ranges of central corneal thickness, CCT, including 580±54 μm (range = 448-713 μm) [70], 548.2±29.6 (493-617 μm) [71], 535±31 (490-567 μm) [72] and 551.34±40.58 μm for women and 542.34±43.84 μm for men [73].

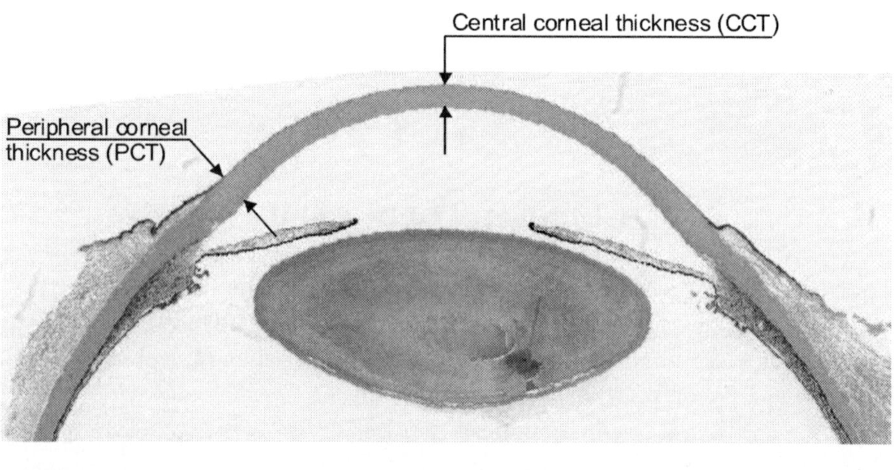

Figure 6. Central and peripheral corneal thickness.

On the other hand, the average values of the peripheral corneal thickness, PCT, reported in the literature varied according to where it had been measured. The values ranged between 600 and 700 μm for measurements taken between 15° to the corneal centre, and at the limbal periphery [74-77]. The relationship between CCT and PCT is yet to be established, although it is known that PCT is usually larger than CCT.

CORNEAL BIOMECHANICAL BEHAVIOUR

The composite microstructure of the cornea makes its biomechanical behaviour quite complex and poses several difficulties including:

1. With the collagen fibrils being the main load carrying components of the stroma, the behaviour becomes hyperelastic with an initially low stiffness (resistance to deformation) increasing gradually under higher loads.
2. The preferential orientation of stromal lamellae leads to anisotropic behaviour with the mechanical stiffness expected to be highest in the preferred directions of lamellae.
3. The change of lamella orientation from the central area to become circumferential close to the limbus affects the directional anisotropy of the tissue at the transition zone.
4. The cellular composition of the epithelium and endothelium is expected to result in significantly lower stiffness levels compared to the fibrous stroma.
5. The increased corneal thickness away from the centre suggests an associated increase in both flexural and membrane stiffness.
6. Stromal swelling, common during sleeping and following prolonged wear of contact lenses, increases the spacing between collagen fibrils and may affect the biomechanical behaviour of the tissue especially under loads creating flexural actions.
7. The increase in lamella interweaving in anterior stroma, and the reported effect of interweaving in increasing swelling resistance, is expected to lead to variation in biomechanical behaviour between anterior and posterior stroma [78].
8. Diseases such as keratoconus are known to affect the cross linking between collagen fibrils in the stroma and lead to deterioration in corneal stiffness in the affected area. The correlation between the reduced cross-linking in keratoconus and change in mechanical properties requires further study [79].
9. The fast remodelling of the epithelium under external loads, such as those experienced while wearing orthokeratology and other types of rigid contact lenses, makes it difficult to study the cornea's long-term behaviour as in this case epithelial remodelling and corneal creep take place simultaneously.

The discussion below presents the results of recent research studies conducted to characterise corneal tissue and to address some of the challenges described above. In spite of the significant advances that have already been made, current knowledge and understanding of corneal biomechanical behaviour remain imperfect and further progress is needed.

Experimental Assessment of Corneal Behaviour

Several experimental studies have been conducted since the 1960s to characterise the biomechanical behaviour of corneal tissue [7, 36, 80-90]. The material constitutive models (stress-strain and Young's modulus-strain relationships) reported in these studies have been compared in Figure 7 and show a wide range of variation [91].

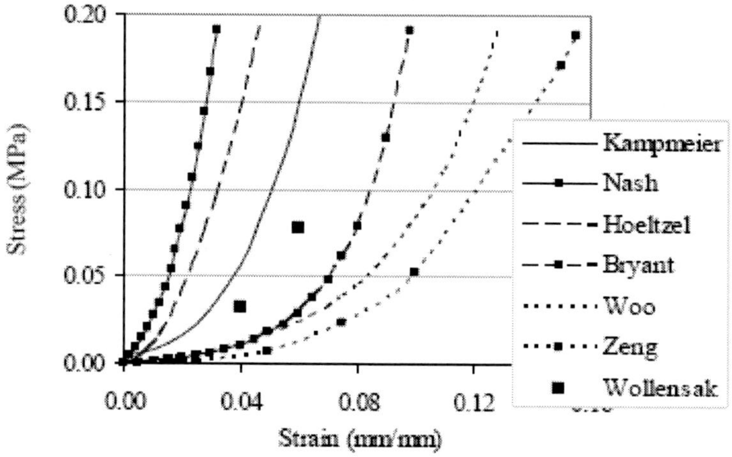

(a)

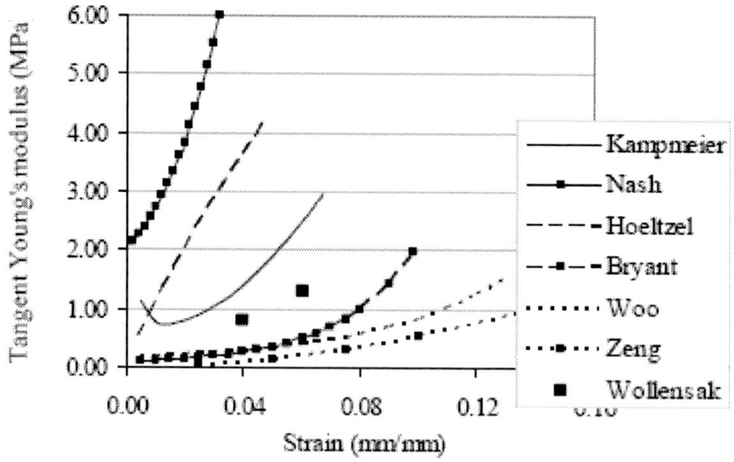

(b)

Figure 7. Biomechanical behaviour of corneal tissue as reported in earlier experimental studies, (a) Stress-strain behaviour, (b) Relationships between tangent Young's modulus and strain – Studies reported include Kampmeier [36], Nash [79], Hoeltzel [92], Bryant [7], Woo [87], Zeng [85] and Woolensak [95].

This wide range, which is far beyond what could be acceptable in any predictive analysis of corneal behaviour, is thought to be related to the viscoelasticity and anisotropy of corneal tissue, and the effects of ageing and hydration. Studies adopting different strain rates or employing different testing techniques reported different material properties [79, 85, 88, 92]. Specimens with varying donor ages were sometimes combined in the same test group despite the known age-related changes in the microstructure of corneal tissue (including an increase in stromal fibril diameter [93] and interfibrillar cross-linking [94]) and the associated effect on the tissue's biomechanical behaviour. Corneal specimens with different swelling extents were also used in earlier studies despite the current incomplete understanding of the effect of hydration and swelling on corneal biomechanical behaviour.

For this reason, it has been essential that the characterisation of corneal biomechanical behaviour be based on experimental techniques that offer tight control of strain rate, hydration and temperature, consider the effect of age on tissue stiffness and test the cornea in a state that closely simulates its normal physiologic conditions. Examples of these techniques are described in the next section.

Experimental Techniques

Two main experimental techniques are currently available to assess corneal behaviour. The first technique, called inflation testing, relies on supporting corneal buttons along the limbus, applying uniform pressure on the cornea's posterior surface, and monitoring the resulting deformation using non-contact methods, Figure 8. The pressure-deformation (p-δ) data obtained experimentally are then converted into stress-strain (σ-ε) behaviour of corneal tissue using shell mathematical analysis through the following steps (see Ref 96 for more details).

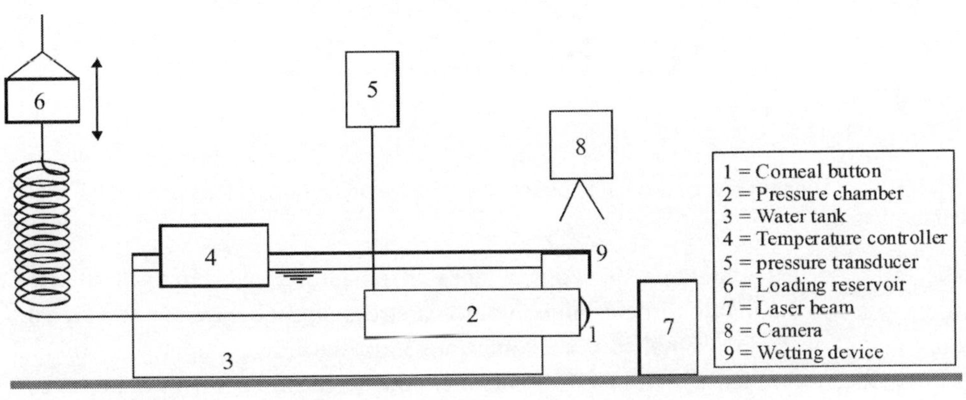

Figure 8. Main components of an inflation test rig.

The secant Young's modulus, E, is first determined for a particular pressure-rise (p-δ) behaviour point as:

$$E = \frac{p\,R_m^{\;2}}{2\,\delta\,t}\,(1-\nu)\left[1-e^{-\beta\eta}\cos(\beta\eta)\right], \qquad (2)$$

where R_m is the radius of the corneal median surface, t the average thickness assumed = ½ (CCT + PCT), η half the central angle of curvature (see Figure 9), $\eta = \sin^{-1}(R_i/R_{ant})$, R_i radius of the corneo-scleral intersection, R_{ant} radius of the cornea's anterior surface, $\beta = \sqrt{R/t} \cdot \sqrt[4]{3(1-\nu^2)}$, and ν Poisson's ratio, taken as 0.49 assuming corneal tissue behaves as an almost incompressible material [5, 7, 36]. The strain at this behaviour point is defined as:

$$\varepsilon = \frac{p\,R_m}{2\,E\,t}(1-\nu)\left(1+\nu\, e^{-\beta\eta}\cos\beta\eta\right), \quad (3)$$

and the stress as:

$$\sigma = \varepsilon\, E. \quad (4)$$

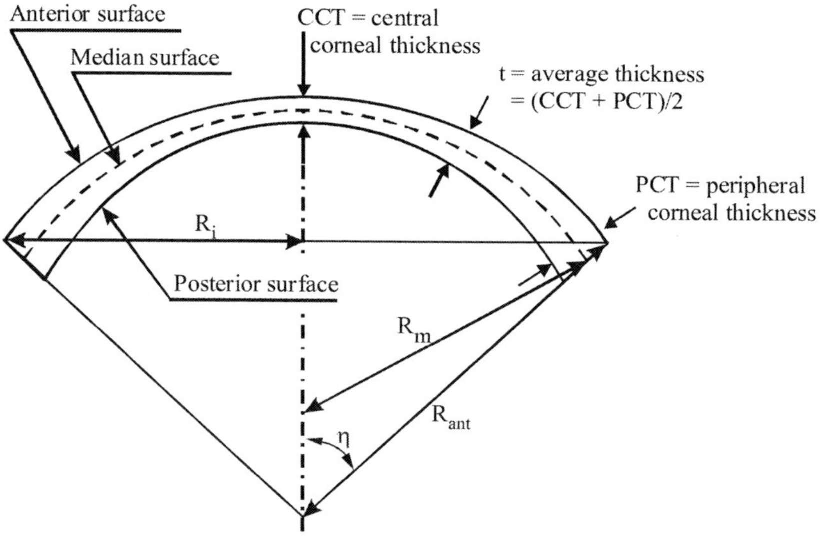

Figure 9. Cross-sectional view of a corneal button showing the main dimensions used in the shell mathematical analysis.

The inflation test maintains the cornea close to its normal working conditions and is therefore considered suitable to determine the global stress-strain behaviour of corneal tissue. The only deviation from normal in vivo conditions is in the development of a pinned edge along the cornea's limbus, which in effect considers the sclera to be stiffer than the cornea to the extent that it can be approximated as a rigid material. This approximation, which is expected to have only a negligible effect on corneal behaviour, has been necessary to make possible the mathematical shell analysis of corneal behaviour, which produced Equations 2-4.

A possible development of inflation testing involves the use of intact eye globes [81], which could be supported along the equator (without preventing radial deformation) and subjected to intraocular pressure loading through a canola inserted into the anterior chamber, Figure 10. However, since no mathematical solution is available to convert the experimentally-obtained pressure-deformation behaviour into stress-strain relationships, the usefulness of this technique is still limited.

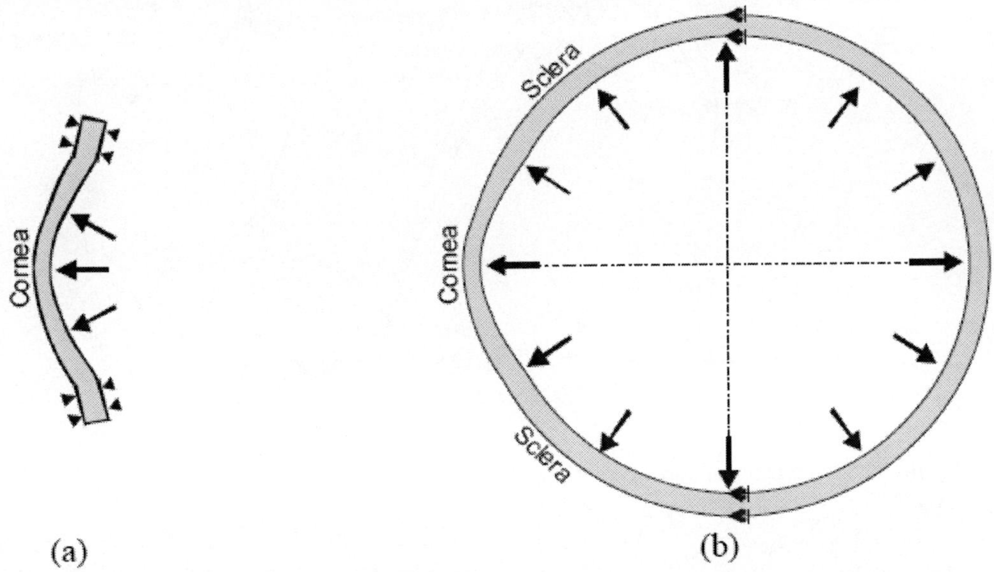

Figure 10. Loading and boundary conditions applied to (a) corneal button specimens, and (b) whole eye globes subjected to inflation tests.

The second experimental technique is strip extensometry testing, in which a strip of corneal tissue with a constant width is extracted and attached to the grips of a tension machine while monitoring its behaviour. The stresses (σ) and strains (ε) are calculated from the load (T) and elongation (δ) data using the following simple equations:

$$\sigma = T / A \text{ and } \varepsilon = \delta / L, \tag{5}$$

where A and L are the specimen's original cross-sectional area and length, respectively. The simplicity of the technique makes it quite common in ocular biomechanics research but a number of inherent deficiencies, which can reduce its reliability, must be recognised and accounted for. The deficiencies originate mainly from the initially curved form of the specimen (Figure 11), the non-uniform corneal thickness and the termination of fibrils which are not aligned with the specimen orientation.

The straightening of the specimen from its curved form results in initial strains that affect the behaviour under subsequent loading [97], Figure 12. The specimen's curved form also means that its edges are slightly shorter than the longitudinal centre line and this length variation leads in turn to non-uniform distribution of stress under loading. The variable corneal thickness, from a minimum at the centre to a maximum at the ends, further complicates the derivation of the stress-strain relationship from the test load-elongation results. Additionally, the relatively large thickness of the specimen poses another difficulty with the potential of unequal clamping of external and internal corneal layers.

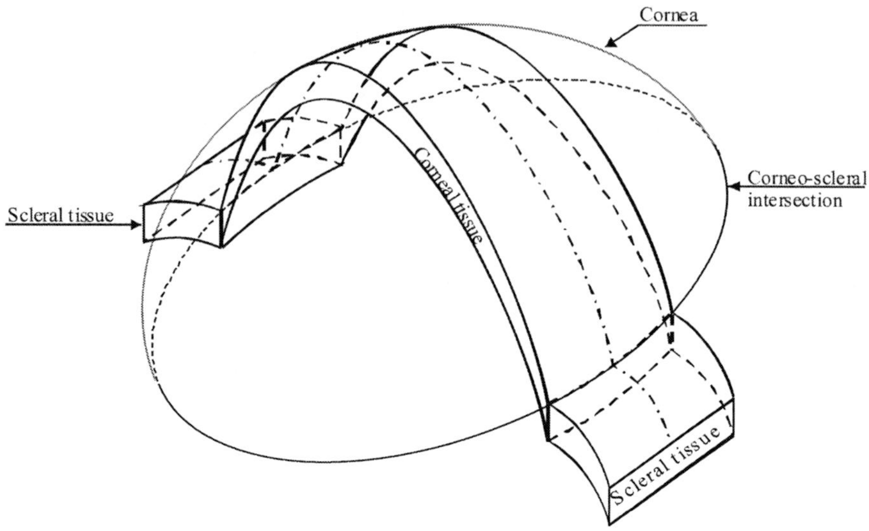

Figure 11. Initial curved form of corneal strip specimens.

Figure 12. Straightening of (a) an initially curved strip specimen produces stresses and strains with the distribution shown in (b).

While these limitations affect the technique's suitability in obtaining the global material properties of the cornea, the technique remains viable for comparative studies such as those on anisotropy (variation in tissue behaviour between different anatomical directions) and viscoelasticity (variation in behaviour with changes in strain rate). In studies on global corneal behaviour, use should be made instead of the above described inflation testing.

However, mathematical techniques have been developed to consider some of the deficiencies of strip testing [98], and these techniques could be useful where strip testing is to be used to produce an approximate estimate of the stress-strain behaviour. The effects of the specimen length variation and the straightening of the initially curved specimen form can be handled first together. Figure 13 shows a 3D view of a strip extracted from a curved cornea (assumed spherical for simplicity) and orthogonal cross-sections showing the variation of specimen length from maximum along the centre line to minimum at the sides.

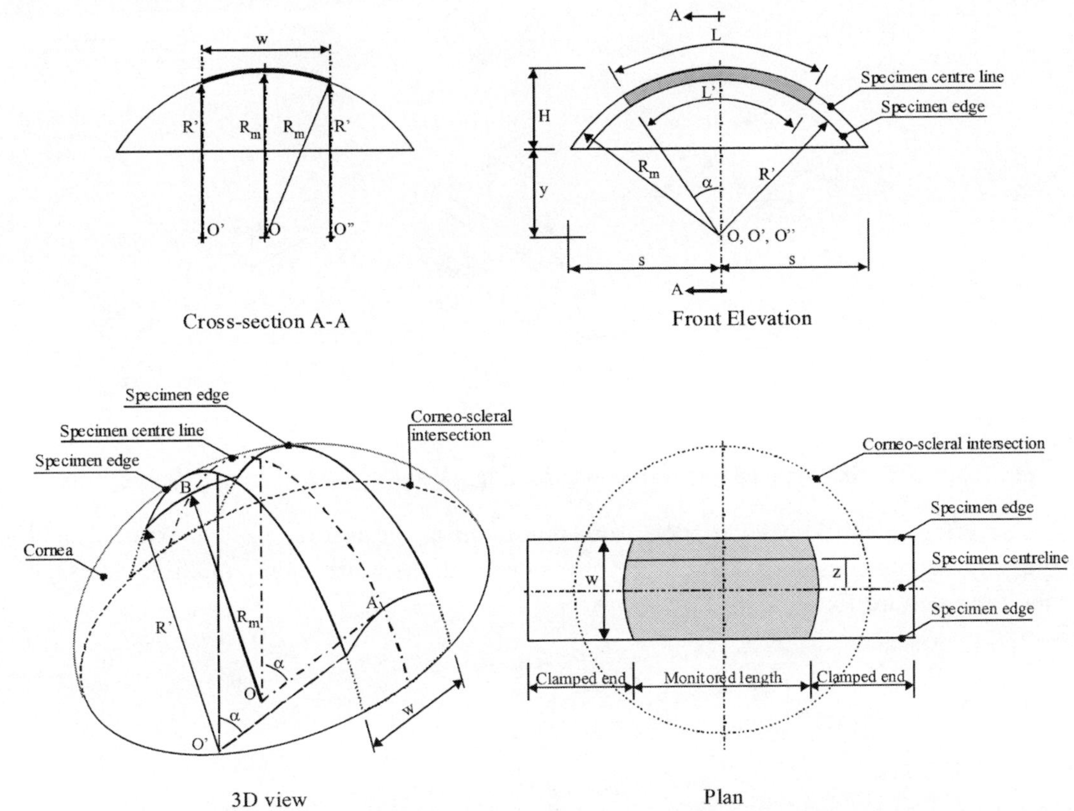

Figure 13. Orthogonal and 3D views of a strip specimen extracted from a corneal button.

Reference is also made to Figure 14 which shows the strain distribution due to (a) the variation in length, (b) initial straightening, and (c) the combined effect. Consideration of these two factors starts with a modified form of the strain, ε (the derivation of Equations 6 to 9 can be found in Ref. 98):

$$\varepsilon = \frac{\delta}{2\,\alpha \times \sqrt{R_m^{\;2} - z^2}} + \frac{h}{R_m} \tag{6}$$

where δ is the measured specimen elongation under tension load T, 2α the angle of curvature of the specimen, Figure 13, and R_m the radius of the corneal median surface. Distances h and z are depicted in Figures 13 and 14b, respectively. This form of strain is then used in an equation relating the applied force, T, and the resulting strain, ε:

$$T = \int_{-t/2}^{+t/2} \int_{-w/2}^{+w/2} \left(a\,\varepsilon^3 + b\,\varepsilon^2 + c\,\varepsilon\right) \cdot dz\,dh \tag{7}$$

where a, b and c are constants to be determined such that the best fit with the experimental data is achieved. With the values of the constants determined, the corresponding stress is then calculated as:

$$\sigma = a\,\varepsilon^3 + b\,\varepsilon^2 + c\,\varepsilon \tag{8}$$

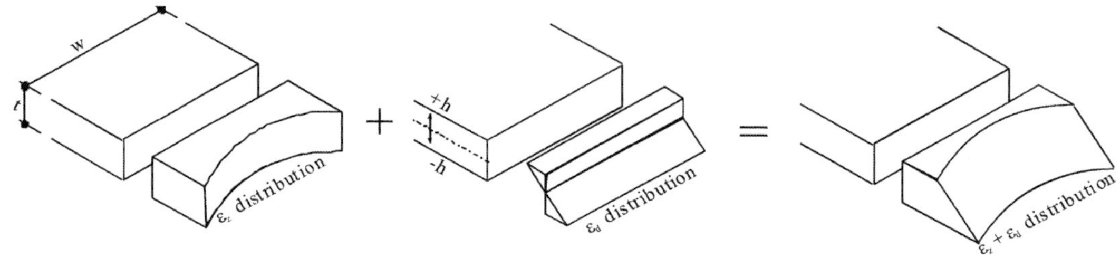

Figure 14. Strain distribution on the cross-section of a strip specimen.

The effect of thickness variation from minimum at the centre, t_1, to maximum at the clamps, t_2, can be considered next in the stress calculations to develop a modified stress, σ', of the form, Figure 15:

$$\sigma' = \sigma \cdot \frac{t_1}{(t_2 - t_1)}\left[\ln\left(\frac{t_2\,L}{2}\right) - \ln\left(\frac{t_1\,L}{2}\right)\right], \tag{9}$$

where L is the specimen's initial length, w the specimen width, and σ the stress derived in Equation 8. An assumption of linear variation between t_1 and t_2 is adopted to simplify the development of Equation 9. The second term of the equation:

$$\frac{t_1}{(t_2 - t_1)}\left[\ln\left(\frac{t_2\,L}{2}\right) - \ln\left(\frac{t_1\,L}{2}\right)\right]$$

represents the correction caused by the variable thickness. The second term is below 1.0 and reaches this value only when $t_2 = t_1$.

With these formulations, some of the deficiencies of strip testing could be addressed in the stress and strain calculations.

Other deficiencies such as the termination of some collagen fibrils along the specimen sides and the difficulty in clamping the layered corneal tissue cannot currently be accounted for using similar mathematical techniques.

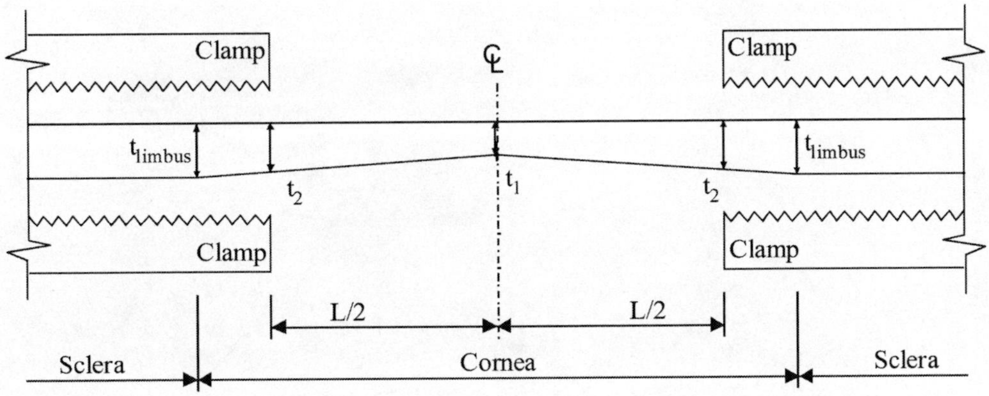

Figure 15. Thickness variation of corneal strip specimens.

INFLATION TEST SETUP

Inflation testing is based on supporting corneal buttons along their limbal tissue, applying uniform posterior pressure that simulates the intraocular pressure and measuring the resulting corneal deformation. The pressure-deformation experimental results are then converted mathematically, using Equations 2-4, to stress-strain tissue behaviour.

An inflation test rig has been developed by the Ocular Biomechanics Group to enable tight control of the conditions thought to have an influence on corneal behaviour including hydration, temperature and load application rate. The rig is totally computer-controlled with all test conditions specified electronically by the user and all behaviour monitoring devices storing their data automatically for later analysis [30].

Fresh and well preserved corneal buttons are mounted onto the custom-built pressure chamber shown in Figure 16 using only mechanical clamps. The pressure chamber is filled either with saline solutions in tests lasting less than 45 minutes, or with the preservation medium Optisol (which can maintain tissue hydration and prevent degradation for up to 14 days post mortem [99-102]) in longer tests. The chamber is connected to a reservoir, which moves vertically up and down using a small motor to control the pressure inside the chamber. The connection between the chamber and the reservoir passes through a large water tank equipped with a temperature controller to set the temperature of the pressure chamber at the required level. The actual pressure in the chamber is measured using an electronic pressure transducer that can be connected to a personal computer. A wetting device is also used to release a slow flow of saline solution on the cornea's anterior surface to prevent its drying.

The motor attached to the reservoir is connected to a variable-voltage power unit that can change the motor speed within a wide range. The ability to change the motor's speed allows the representation of both slow (static) and fast (dynamic) loading conditions to assess the tissue's viscoelasticity. Creep tests can also be conducted by maintaining a constant level of pressure for any reasonable duration of time. Corneal behaviour under loading is monitored using electronic non-contact devices including a laser beam that measures corneal apical deformation with ±1 μm accuracy, and high-resolution digital cameras to record the changes in cornea's profiles caused by pressure changes.

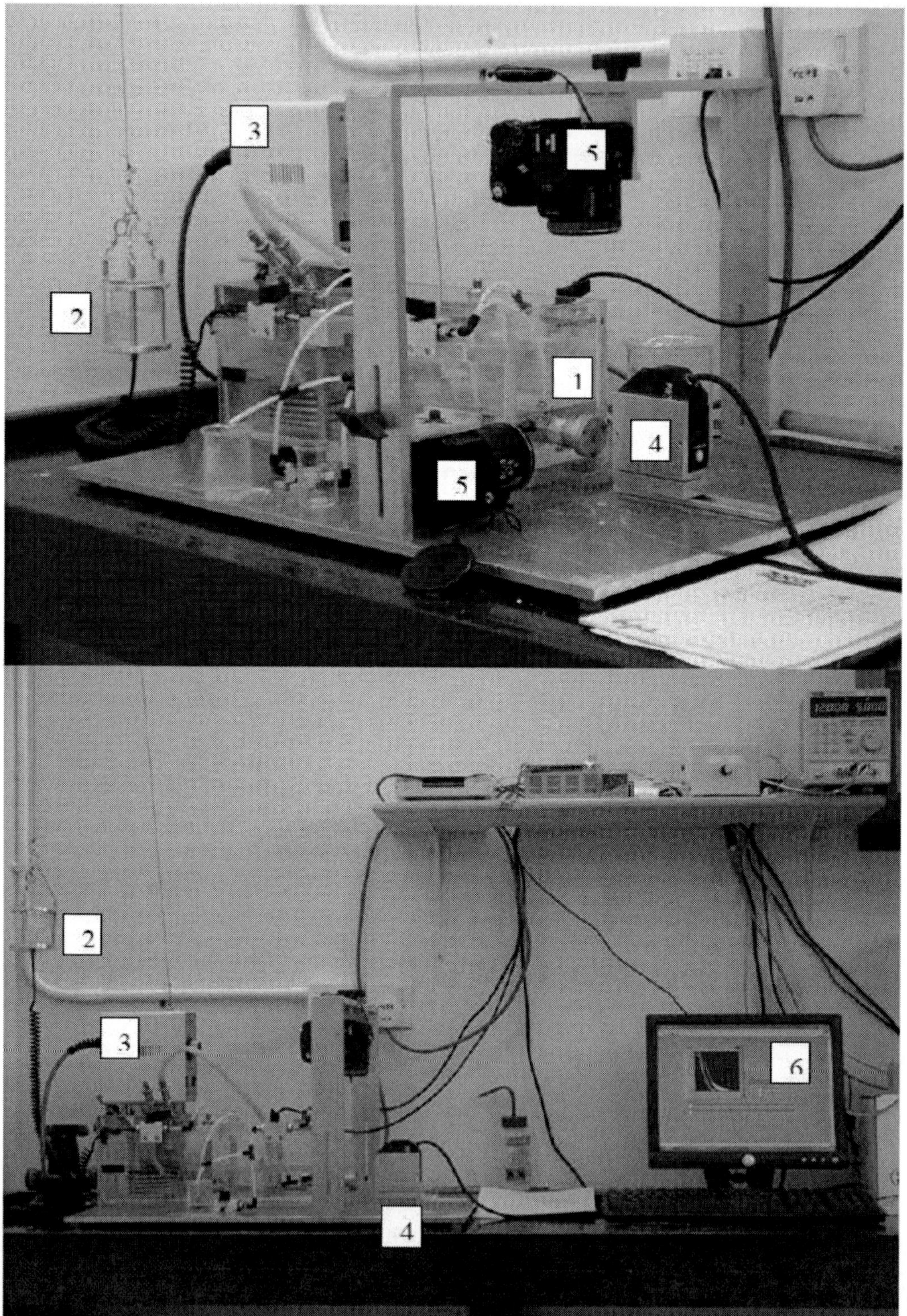

Figure 16. Overall views of inflation test rig showing the main components including (1) the pressure chamber, (2) loading reservoir, (3) temperature controller, (4) laser beam, (5) digital cameras, and (6) control computer.

Controlling the pressure application rate, and the tissue's hydration and temperature, and the use of non-contact monitoring devices have led to experimental results with significantly lower scatter compared to earlier studies which had less control of test conditions. The results obtained using this rig are presented below and used to quantify some of the main biomechanical properties of corneal tissue.

STRIP TEST SETUP

Strip testing is considerably simpler than inflation testing. Tissue strips are extracted from corneal buttons in the desired anatomical direction, connected to mechanical clamps and subjected to uniaxial tension loading in a material testing machine. However, care should be taken to ensure that tissue is fresh and remains well preserved, specimen extraction is in the right location and direction, and connection to the clamps is adequate. Corneal tissue can be preserved in Optisol or Eusol C for up to 14 days post-mortem. Preservation, which also maintains hydration, should continue during the tests as shown in Figure 17. Double blade devices are needed to ensure the uniform width of specimens and the accurate positioning and orientation of the specimen within the corneal button.

Current technology to connect tissue strips to mechanical clamps is inadequate. Too much clamp tightening creates stress concentration in the specimen regions close to the clamps, and can initiate premature failure. On the other hand, insufficient tightening allows slippage and renders the results unreliable.

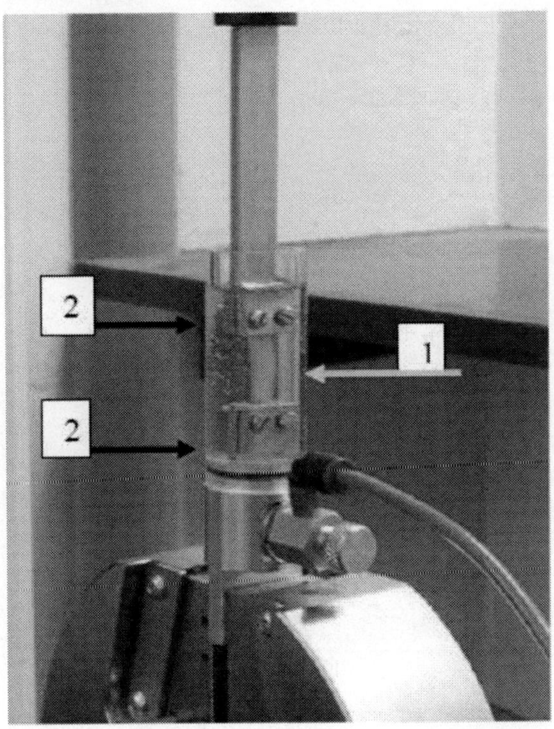

Figure 17. Corneal strip specimen (1) connected to mechanical clamps (2) in a uniaxial tension test.

The fact that the clamps provide non-uniform clamping effect across the thickness of tissue (strong at the outer tissue layers and weaker at mid-thickness) is another cause for concern. However, until a different technique to connect strip specimens to loading machines is developed, mechanical clamps will have to be used with care to minimise the effect of the above described deficiencies.

CORNEAL BEHAVIOUR CHARACTERISTICS

The following sections discuss the corneal behaviour as obtained experimentally using the inflation and strip test rigs described above. In addition to human corneas, the discussion will cover the behaviour of porcine corneas because of their reported biomechanical similarity to, and suitability in some cases as test models for, human corneas. The behaviour characteristics covered include the immediate hyperelasticity, hysteresis and anisotropy of corneal tissue, and the viscoelastic properties represented in the strain-rate effects, and the cornea's creep and stress-relaxation behaviour. Also included are the epithelium's contribution to corneal biomechanics and the effects of the layered stromal construction.

Corneal Hyperelasticity

Under inflation testing, human corneas demonstrate clear hyperelastic (pressure-rise and stress-strain) behaviour with initial low stiffness increasing gradually under higher pressure levels, Figure 18. The end of the initial low stiffness stage coincides with posterior pressure between 15 and 20 mmHg, which is close to the normal intraocular pressure range. The average ratio of final stiffness (observed under IOP higher than 60 mmHg) to initial stiffness (under 5-15 mmHg) has been found close to 11, 9 and 5 for corneas with donors within the age ranges 50-64, 65-79 and 80-95 years, respectively [30].

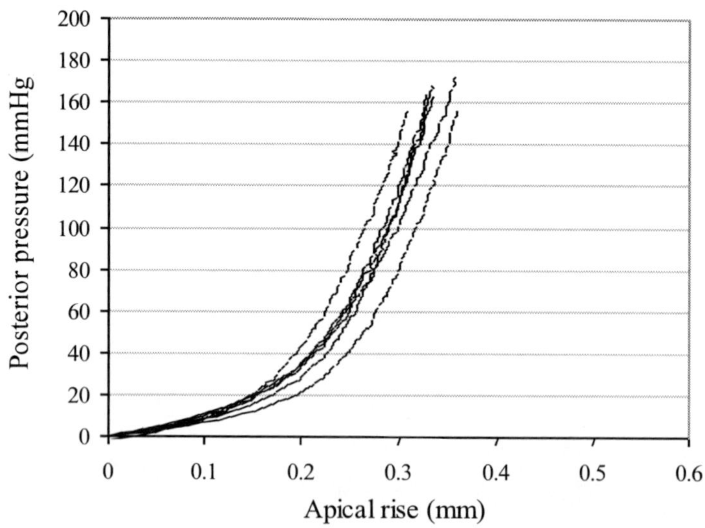

Figure 18. Typical hyperelastic behaviour of human corneas under increasing posterior pressure.

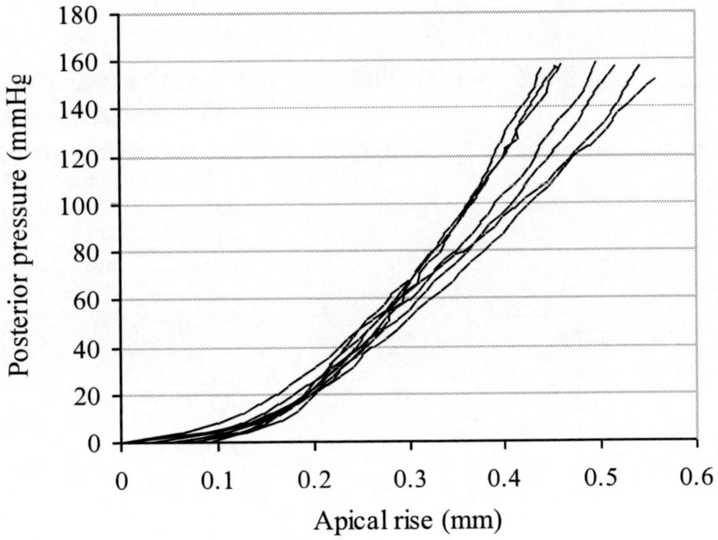

Figure 19. Behaviour of porcine corneas under increasing posterior pressure.

Porcine corneas (aged 4-6 months) exhibit similar hyperelastic behaviour with a number of important differences [35]. The change from the initial low stiffness stage to the final stage with higher stiffness is more sudden and occurs around the lower pressure level of 10 mmHg, Figure 19. The ratio between the final stiffness (under IOP higher than 60 mmHg) and the initial stiffness (under 5-10 mmHg) values is about 5 on average.

Equations with different forms have been tried to fit the stress strain behaviour of human corneas. The best fit was achieved with the exponential equation form $\sigma = A\left(e^{B\varepsilon} - 1\right)$ suggested earlier [7, 87] in which σ and ε are the stress and strain, respectively, and A and B are constants. A clear distinction has been found between the behaviour of specimens within different age groups, and for this reason, three stress-strain equations have been developed of the form:

$$\sigma_{(MPa)} = 0.0070\left(e^{50\varepsilon} - 1\right) \qquad \text{Age 50-64 years} \qquad (10)$$

$$\sigma_{(MPa)} = 0.0077\left(e^{59\varepsilon} - 1\right) \qquad \text{Age 65-79 years} \qquad (11)$$

$$\sigma_{(MPa)} = 0.0083\left(e^{80\varepsilon} - 1\right) \qquad \text{Age 80-95 years} \qquad (12)$$

The ages considered start from 50 years as this was the minimum age of corneal specimens tested. Further work is therefore needed to develop stress-strain relationships for younger human corneas.

The more sudden change in stiffness seen with porcine corneas made it difficult to find an acceptable fit with the above exponential equation form, and an alternative third-order polynomial form was adopted instead:

$$\sigma_{(MPa)} = -350\,\varepsilon^3 + 56\,\varepsilon^2 - 0.21\,\varepsilon \quad (13)$$

The average porcine behaviour described by this equation is compared in Figure 20 against those of human corneas within the three age ranges considered. Overall, porcine corneas appear to have lower initial and final stiffness values compared with all human corneal groups.

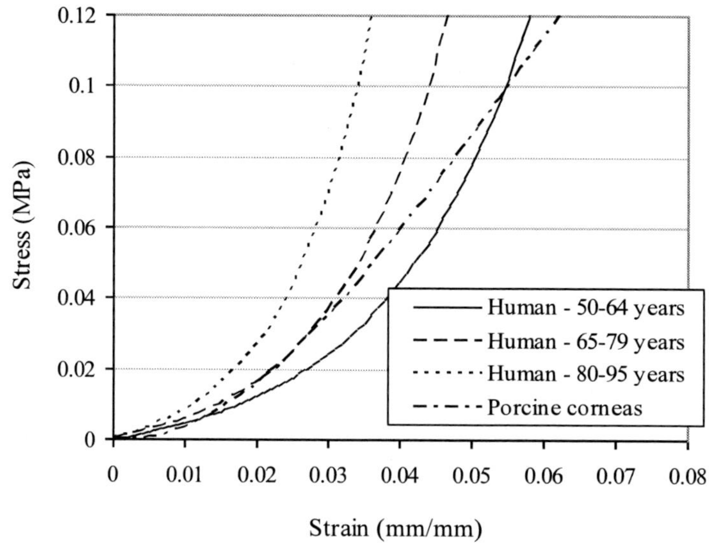

Figure 20. Comparison between the stress-strain behaviour of human and porcine corneas as obtained experimentally using an inflation test rig.

Relationship between Intraocular Pressure and Young's Modulus

The complex hyperelastic form of the cornea's stress-strain behaviour encouraged efforts to find an easier, preferably linear, method to describe the behaviour. One such method is based on presenting the behaviour as relationships between the applied posterior pressure, p, (which represents the intraocular pressure, IOP) and the tangent Young's modulus, E. This method was first attempted by Orssengo and Pye and resulted in the well-known linear relationship [103]:

$$E = 0.0029\,p \quad (14)$$

Similar relationships have been derived from the experimental pressure-rise results discussed above using a modification of Equation 2 of the form:

$$E_{tangent} = \frac{\Delta p \cdot R^2}{2\,\Delta r\,t}(1-\nu)\left[1 - e^{-\beta\eta}\cos\beta\eta\right] \quad (15)$$

where Δp and Δr are the increases in pressure and apical corneal rise within a particular pressure increment, and E the instantaneous (tangent) Young's modulus [31].

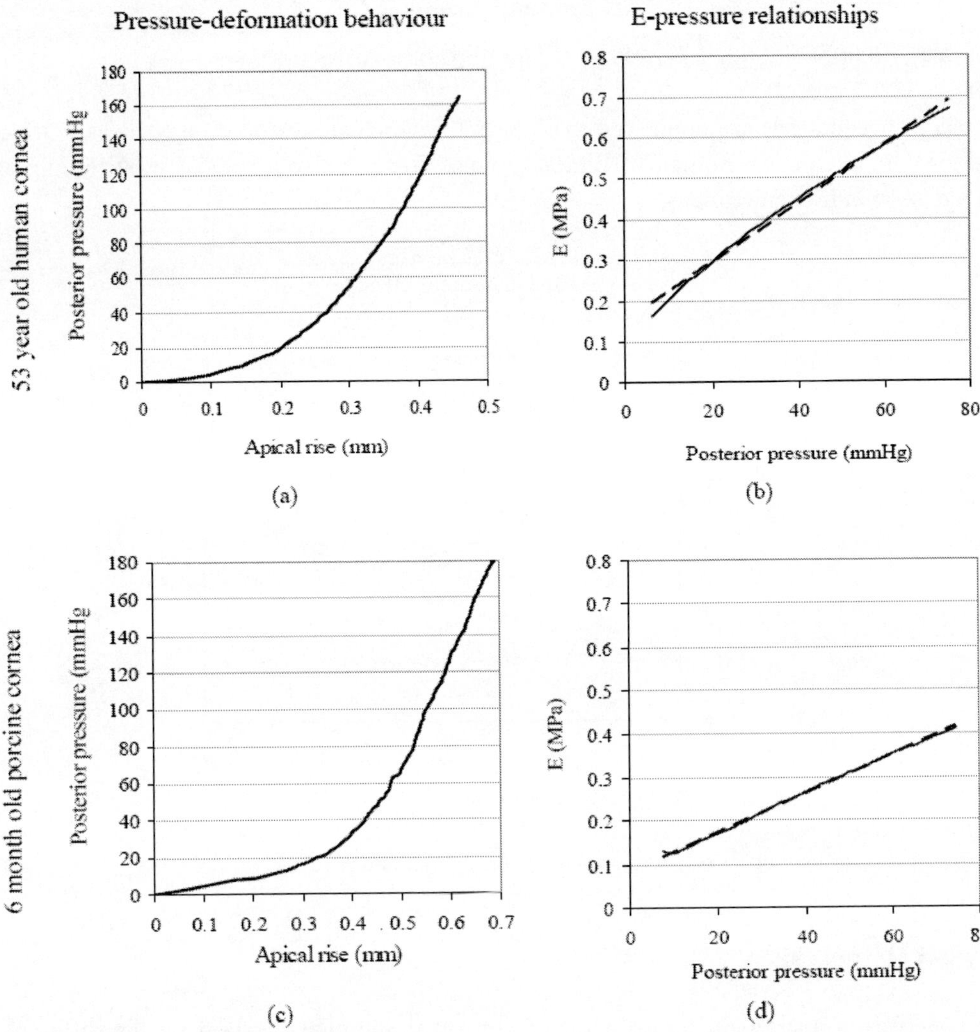

Figure 21. Pressure-deformation and E-p behaviour of example human and porcine corneas, (a,b) 53 year old human cornea, (c,d) porcine cornea.

In spite of the hyperelastic form of both the pressure-rise and stress-strain results, the relationship between Young's modulus, E, and posterior pressure, p, is close to linear. See for example Figure 21 which shows the nonlinear pressure-rise behaviour of a 53 year old human cornea and a porcine cornea, and the corresponding E-p relationships plotted for a pressure range of 0-75 mmHg. In both cases, the E-p relationships are almost linear.

While the E-p results of human specimens are grouped according to the donor age, the porcine results are treated as one group. Within each group, the results are used to find the best fit with an E-p equation of the form: E = C . p + D, where C and D are constants. The four equations obtained are:

$$E_{(MPa)} = 0.0067\, p_{(mmHg)} + 0.1579 \qquad \text{human corneas, age} = 50\text{-}64 \text{ yrs} \qquad (16)$$

$E_{(MPa)} = 0.0088\ p_{(mmHg)} + 0.1914$ human corneas, age = 65-79 yrs (17)

$E_{(MPa)} = 0.0107\ p_{(mmHg)} + 0.3527$ human corneas, age = 80-95 yrs (18)

$E_{(MPa)} = 0.0038\ p_{(mmHg)} + 0.1449$ porcine corneas (19)

The four equations are compared in Figure 22, which shows porcine corneas with lower E values compared to all human specimen groups. There is also a consistent stiffening effect with age seen in human corneas.

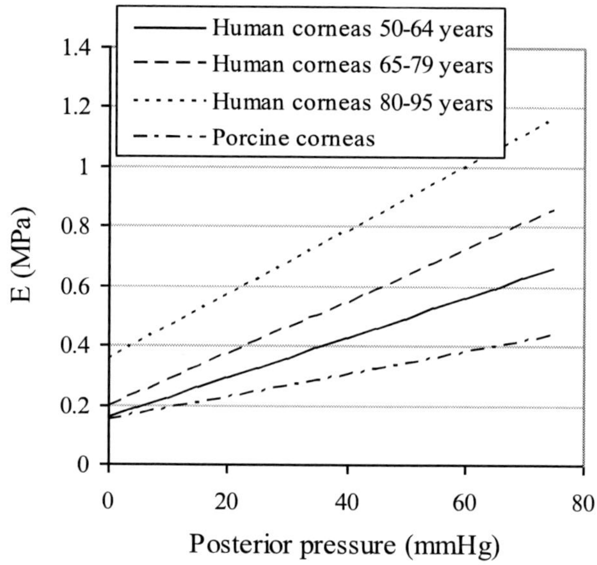

Figure 22. Comparison between E-p relationships for human and porcine corneas.

Corneal Hysteresis

Hysteresis is a viscoelastic property of corneal tissue characterised by the difference in behaviour under loading and unloading conditions, Figure 23. The unloading behaviour has been studied in a similar fashion to the loading behaviour and found to increase in stiffness with age and pressure level. The stress-strain behaviour of the human corneal tissue under unloading conditions can be described using the following equations [34]:

$\sigma_{(MPa)} = 0.00004\left(e^{359\,\varepsilon} - 1\right)$ Age 50-64 years (20)

$\sigma_{(MPa)} = 0.00005\left(e^{427\,\varepsilon} - 1\right)$ Age 65-79 years (21)

$\sigma_{(MPa)} = 0.00025\left(e^{510\,\varepsilon} - 1\right)$ Age 80-95 years (22)

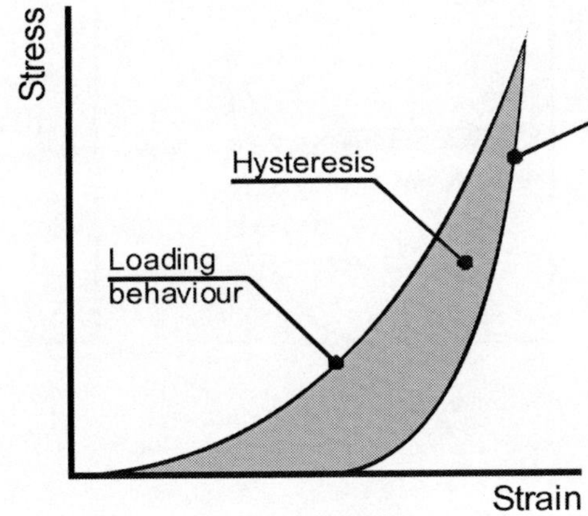

Figure 23. Typical stress-strain behaviour of corneal tissue under both loading and unloading conditions.

An earlier study has considered the unloading behaviour down from different posterior pressure levels, and found the behaviour to be almost independent of the pressure at which unloading starts [34], Figure 24. This particular feature makes the unloading behaviour easy to characterise and implement in the numerical simulation of corneal response.

A direct measure of hysteresis can be obtained experimentally as the difference between the areas under the loading and unloading stress-strain behaviour curves of corneal tissue. Figure 25 depicts experimental results obtained earlier showing a consistent reduction in the magnitude of hysteresis with corneal donor age [34]. The same figure shows a gradual decrease in the areas under both loading and unloading stress-strain curves with age – a finding which can be related to the higher stiffness (and hence lower maximum strain) associated with increased age.

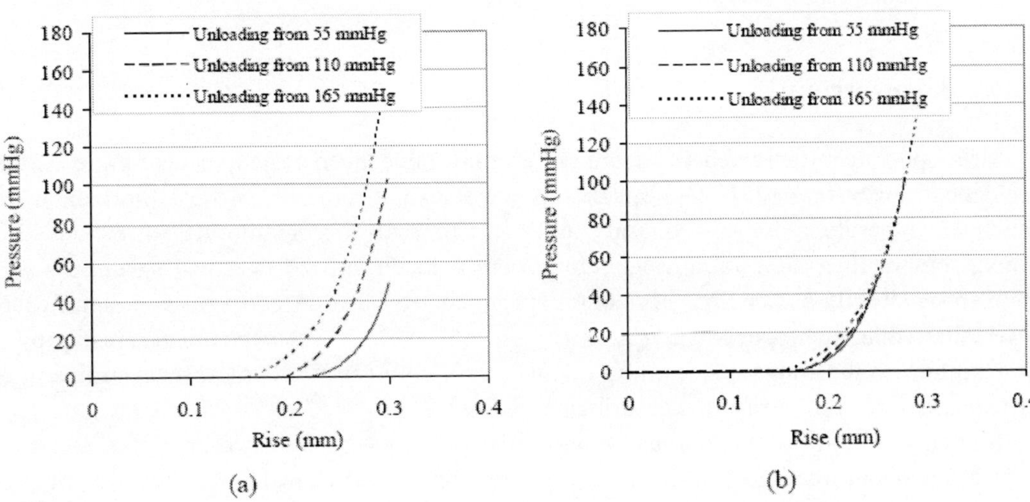

Figure 24. Average unloading behaviour of 80-95 year old corneas, (a) Unloading from three different pressure levels plotted from an arbitrary starting corneal rise of 0.3mm, (b) The three patterns plotted after shifting the behaviour lines with unloading from 55 and 110 mmHg.

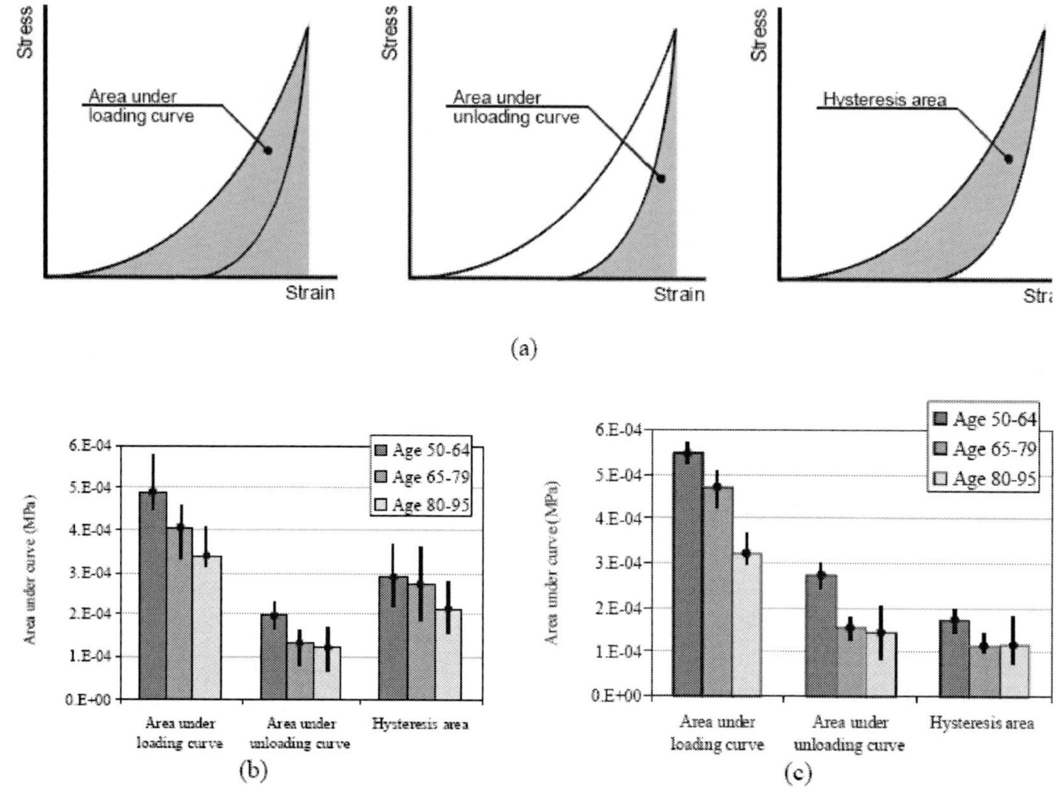

Figure 25. Comparisons between areas under stress-strain loading and unloading curves for corneas with different ages, (a) Sketches showing areas under behaviour curves, (b) Tests under fast pressure application rate of 37.5 mmHg pressure per minute, (c) Tests under slow pressure application rate of 3.75 mmHg pressure per minute – Plots show error bars denoting ranges of variation between specimens within the same group.

Corneal Anisotropy

Strip specimens extracted from human corneas have been tested to determine corneal mechanical anisotropy [28]. The specimens were taken from the vertical, horizontal and diagonal (supcrotemporal-inferonasal and superonasal-inferotemporal) directions and comparisons of their stress-strain behaviour were held, Figure 26. Vertical specimens were found to be slightly stiffer than horizontal specimens (by 8% on average) and significantly stiffer than diagonal specimens (by 49% on average). This mechanical anisotropy is compatible with the known preferential orientation of collagen fibrils of corneal stroma in the vertical and horizontal directions described in Section 2.

Similar work on porcine corneas found little difference between vertical, horizontal and the two diagonal directions [104]. The differences between the mechanical stiffness of strips taken in the four directions were statistically insignificant and below 3% on average. This apparent mechanical isotropy of behaviour was consistent with the microstructure of porcine corneas in which stromal collagen fibrils were dominantly circumferential, and with no preferential orientation in either the vertical, horizontal or diagonal directions.

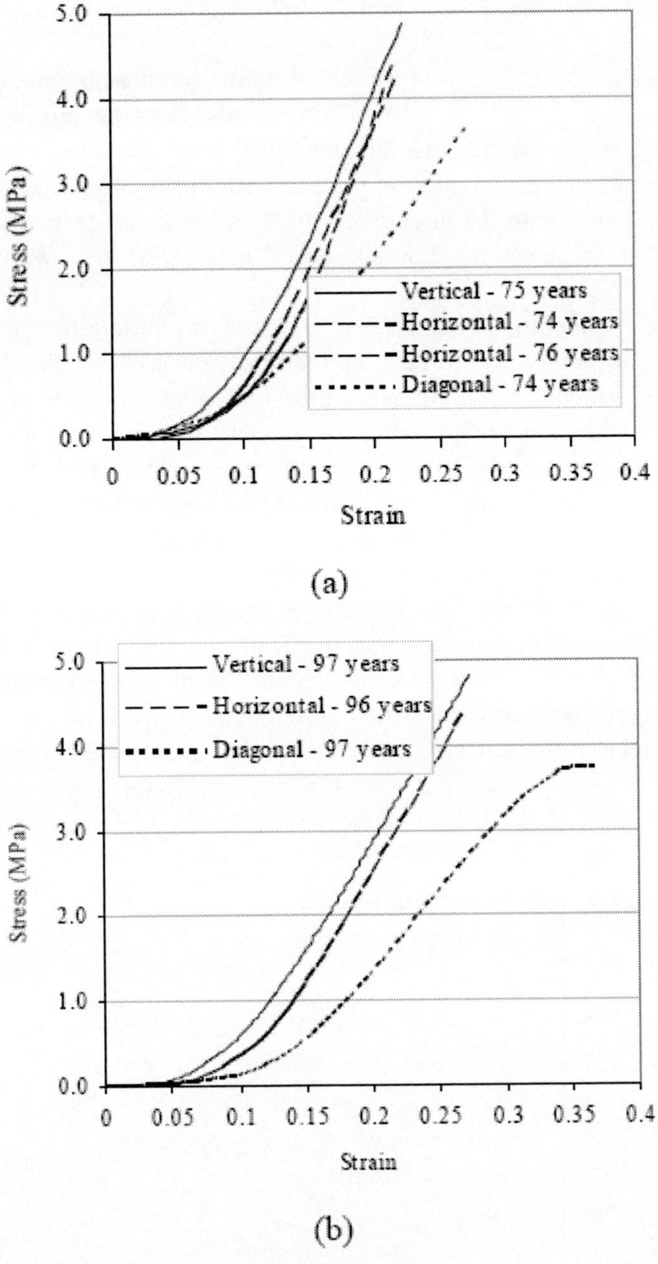

Figure 26. Experimental assessment of the mechanical anisotropy of human donor corneas within different age groups.

Epithelium's Contribution to Corneal Biomechanics

The epithelium is the outermost layer of the cornea, comprising about 10% of its thickness. It functions primarily to block the passage of foreign material into the eye, absorb oxygen and nutrients from the tear film [1] and control the swelling of the stroma [65]. Little work has been done to quantify the epithelium's contribution to corneal biomechanics, although researchers have tended to discard the epithelium's contribution as insignificant [7, 92, 105].

Recent research used experimental testing of human corneas before and after removing the epithelium and found that, on average, the epithelium was approximately 10% the stiffness of stromal lamellae of the same thickness [38].

In order to compare the two sets of pressure-rise relationships (with and without the epithelium), the two lines with the largest and smallest rises were identified within each set and plotted together to mark the boundaries of behaviour seen with and without the epithelium.

These lines were compared in Figure 27a, which shows little difference between the two groups, although specimens with intact epithelium appeared to be slightly stiffer (i.e. experiencing less deformation for the same pressure) than the other group. More precise comparisons were obtained by considering the average stiffness values at specific pressure levels between 10 and 100 mmHg. In this case, stiffness was defined as the posterior pressure in mmHg required to produce an increase in corneal central rise of 1 mm. The stiffness comparisons shown in Figure 27b demonstrate consistently higher average stiffness values associated with the specimens with intact epithelium.

However, the ratios between corresponding values remained within 1-3% in all cases, and these differences were not statistically significant (P>0.05 from t-tests) due to the considerable overlap of the test pressure-rise results. With intact corneas exhibiting higher stiffness by 1-3% than corneas without the epithelium, and with the epithelium being about 10% the corneal thickness, the stiffness of the epithelium is estimated between 9 and 27% that of the stroma.

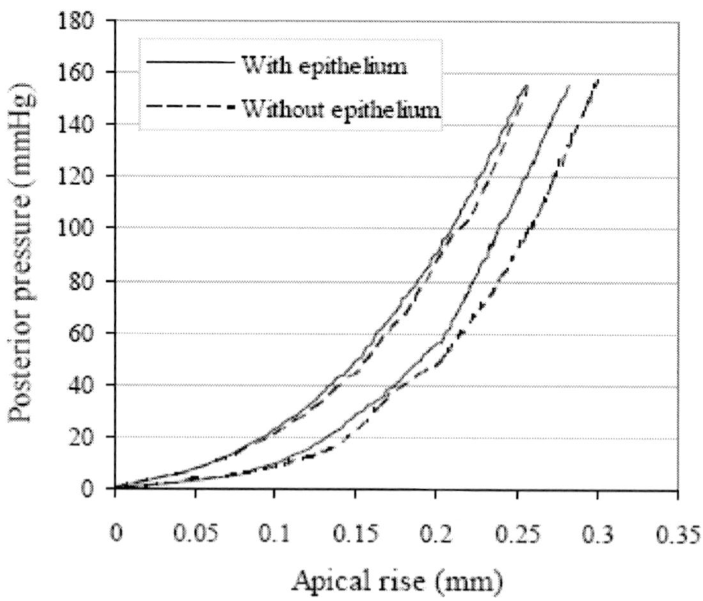

(a)

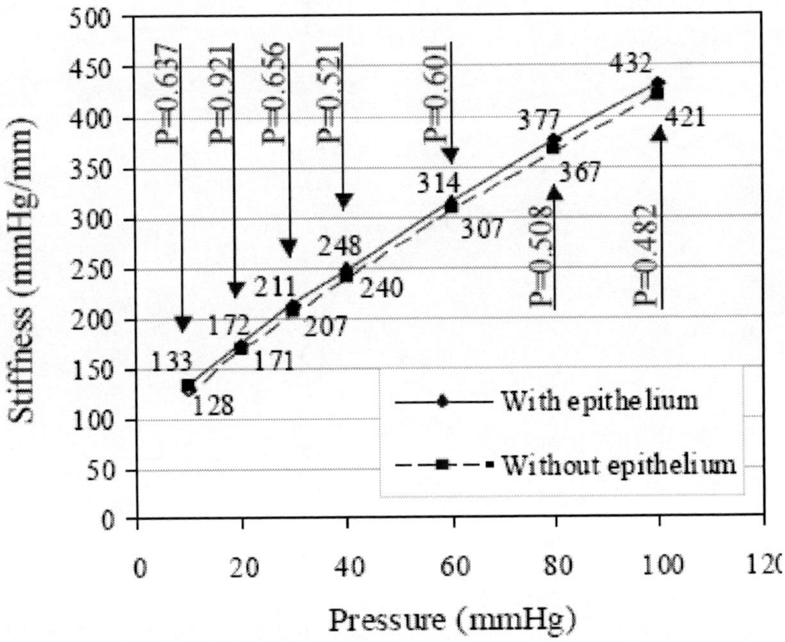

(b)

Figure 27. Comparison of pressure-rise behaviour between corneas with and without the epithelium – Results of statistical analysis of differences between the two specimen groups are included on figure in the form of t-test P values, (a) Range of pressure-rise results for all specimens tested, (b) Average values of stiffness calculated at specific levels of posterior pressure.

Stromal Inter-Lamellar Adhesion

The corneal main layer, the stroma, is analogous to a stack of 300 sheets of copy paper that have some ability to slide relative to each other if subjected to shear [9]. Lack of knowledge of the shear behaviour affecting the relative movement of stromal sheets (or lamellae) led to the exclusion of this feature from numerical simulations [106, 107]. Earlier simulations concentrated on global behaviour, in which it was possible to assume homogenous biomechanical properties (with no consideration of inter-lamellar shear) derived from tests involving intact corneas. These modelling approaches would not be suitable for the simulation of keratoconus for instance, with its abnormal inter-lamellar cross-links [108, 109], and refractive surgery procedures involving the severing of anterior lamellae.

A number of earlier studies determined through experimental means the adhesive tensile strength of the connection between stromal lamellae [40-42]. Strip specimens extracted from human stromal tissue were split at their ends to break the connection between two stromal lamellae, Figure 28. The ends were then clamped and pulled apart, effectively subjecting the interface between stromal lamellae to direct tension. This test, conducted on 68 human specimens resulted in inter-lamellar tensile strength values between 0.069 and 0.077 MPa.

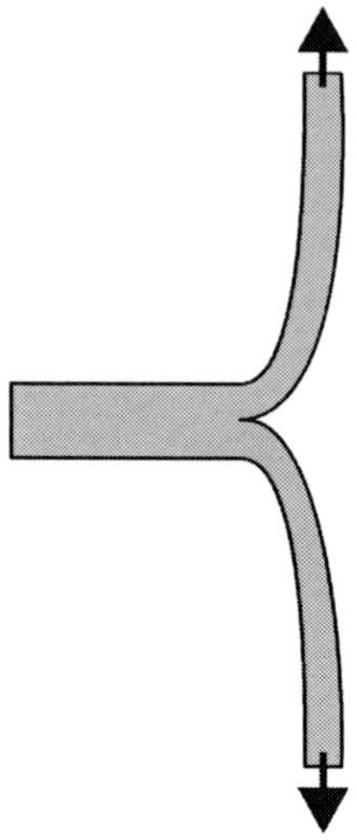

Figure 28. Testing of corneal tissue to determine the adhesive tensile strength of the connection between stromal lamellae.

Further tests subjecting square specimens of stromal tissue to direct shear quantified the shear behaviour of stromal tissue [39]. The behaviour was found to be nonlinear and had the following exponential form:

$$\tau = 0.009\left(e^{1.74\gamma} - 1\right) \tag{23}$$

where τ and γ were the shear stress and shear strain, respectively. This result allowed an indirect assessment of the assumption of stromal isotropy. If isotropy is maintained, the behaviour patterns under shear and surface tension would have been interrelated. This possibility is checked in a simple test, in which an approximate value of shear modulus, G, of 0.03 MPa was obtained from the shear stress-strain Equation 23 (0.02 MPa shear stress divided by 0.65 strain). With an assumed value of Poisson's ratio, ν, of 0.49 (based on considering corneal tissue an incompressible material [7]), the material's Young's modulus, E, would be estimated as $E = G\,[2\,(1 + \nu)] = 0.09$ MPa. This value is considerably lower than those reported in earlier experimental studies [30], indicating that the assumption of stromal isotropy was invalid.

Corneal Long-Term Behaviour

Behaviour under long-term loading can be determined using two main tests, namely; creep and stress-relaxation. In creep tests, specimens are kept for a period of time under a constant level of load or stress while monitoring the gradual increase in deformation. On the other hand, stress-relaxation tests involve subjecting specimens to a certain load, then maintaining the resulting deformation for a period of time while monitoring the gradual decline in stresses, Figure 29.

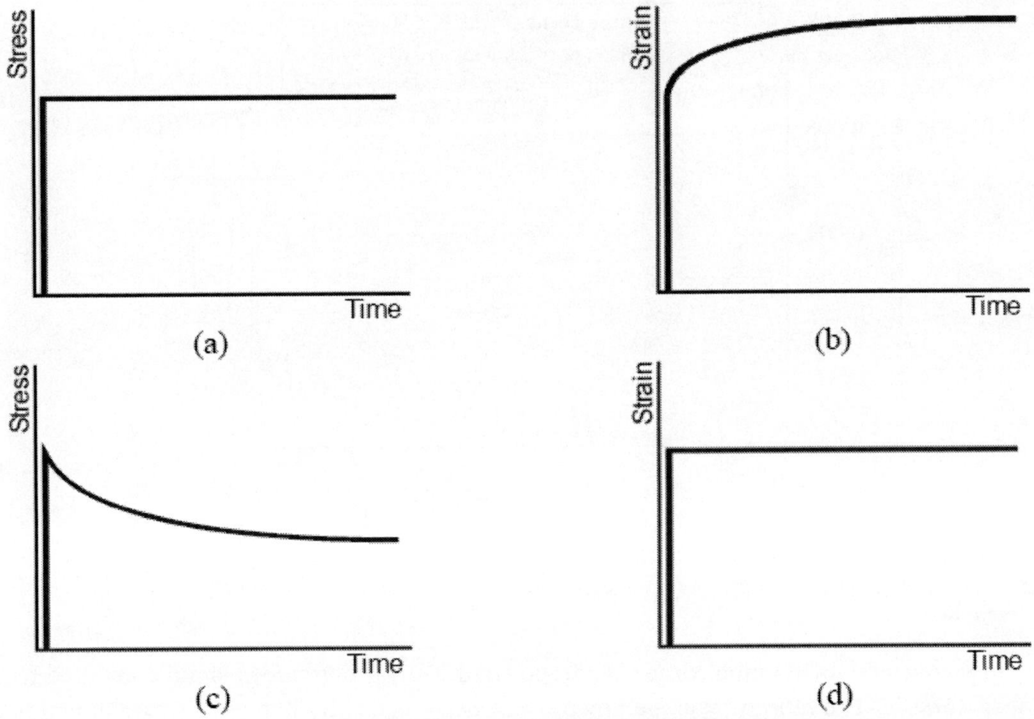

Figure 29. Overview of the stress and strain progression in (a,b) creep and (c,d) stress-relaxation tests.

The creep behaviour of human corneas attached to the inflation test rig and subjected to a constant posterior pressure was found to differ with age as shown in Figure 30 and to follow Equations 24-26 [35]:

$$\varepsilon_{creep} = 16 \times 10^{-6} \left(e^{2.28\,[\log\ t - 1]} - 1 \right) \quad \text{age = 50-64 yrs} \tag{24}$$

$$\varepsilon_{creep} = 16 \times 10^{-6} \left(e^{2.23\,[\log\ t - 1]} - 1 \right) \quad \text{age = 65-79 yrs} \tag{25}$$

$$\varepsilon_{creep} = 18 \times 10^{-6} \left(e^{2.11\,[\log\ t - 1]} - 1 \right) \quad \text{age = 80-95 yrs} \tag{26}$$

where t is the creep time in seconds. A similar equation was developed for porcine corneas of the form:

$$\varepsilon_{creep} = 5 \times 10^{-6} \left(e^{2.02\,[\log\ t - 1]} - 1 \right) \quad \text{porcine corneas} \tag{27}$$

Figure 30 demonstrates a gradual decrease in the average human corneal creep with age, which is more evident between the ages 50-64 and 65-79 years. However, there is a notable overlap of results as demonstrated by the error bars in the figure. A statistical analysis of the results using t-test shows that the association between increased age and reduced creep strain is insignificant.

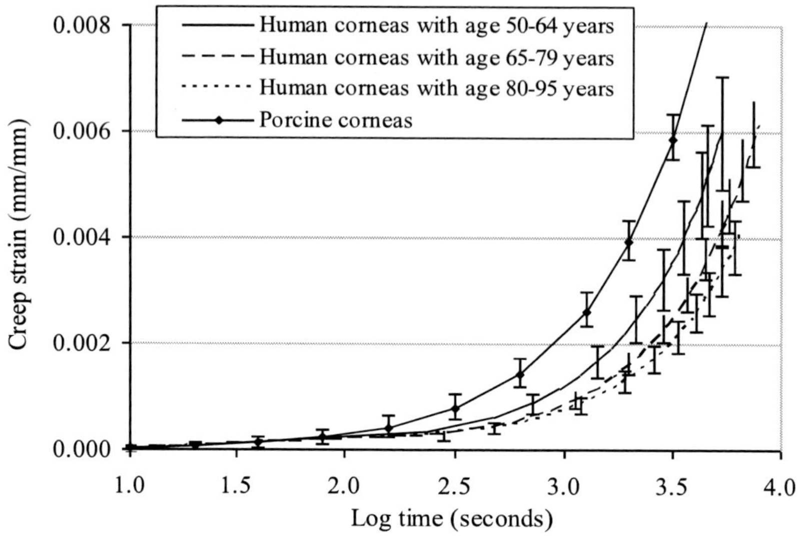

Figure 30. Comparison between average creep behaviour of porcine corneas and human corneas with different ages.

In the mean time, porcine corneas undergo significantly more creep strain compared to all human corneas. The ratio of average creep strain experienced by human and porcine corneas at 300 and 1000 seconds is 49% and 42%, respectively. Figure 30 also shows that the overlap between porcine and human creep-time results is limited to the first stages of the test, up to about 100 seconds, beyond which there is clear separation. A statistical analysis of the two sets of data further returned significant association between creep results and specimen type with $P < 0.001$.

Stress-relaxation tests were also conducted, in which corneal strips were held with the length reached under a uniaxial load of 4N, then 8N and 12N. The sustained load, and hence the stress, were observed to reduce gradually with time. The stress reduction, or relaxation, was highest at the start of the test and reduced gradually with time. The stress-time results obtained in an earlier study were as shown in Figures 31a,c for human and porcine corneas, respectively [35]. The results show a consistent trend in which porcine corneas were observed to lose more of their initial stresses than human corneas. This trend was evident under all three values of the initial load.

Dividing the stresses in each test by the initial stress value under either 4N, 8N or 12N, and obtaining the average behaviour pattern in each case resulted in the plots in Figures 31b,d. Displaying the results in this form reveals two trends including an evidence that porcine corneas relax quicker than human corneas. After 400, 800 and 1200 seconds, porcine corneas respectively lose 49%, 55% and 59% of their initial stresses on average. The corresponding average stress reductions in human corneas are 28%, 30% and 32%. The differences in relaxation rates between human and porcine corneas are statistically significant ($P<0.001$). Further, the stress-relaxation behaviour appears to be largely independent of the initial applied stress. The error bars displayed in Figures 31b,d show significant overlap between the results obtained under different initial applied loads.

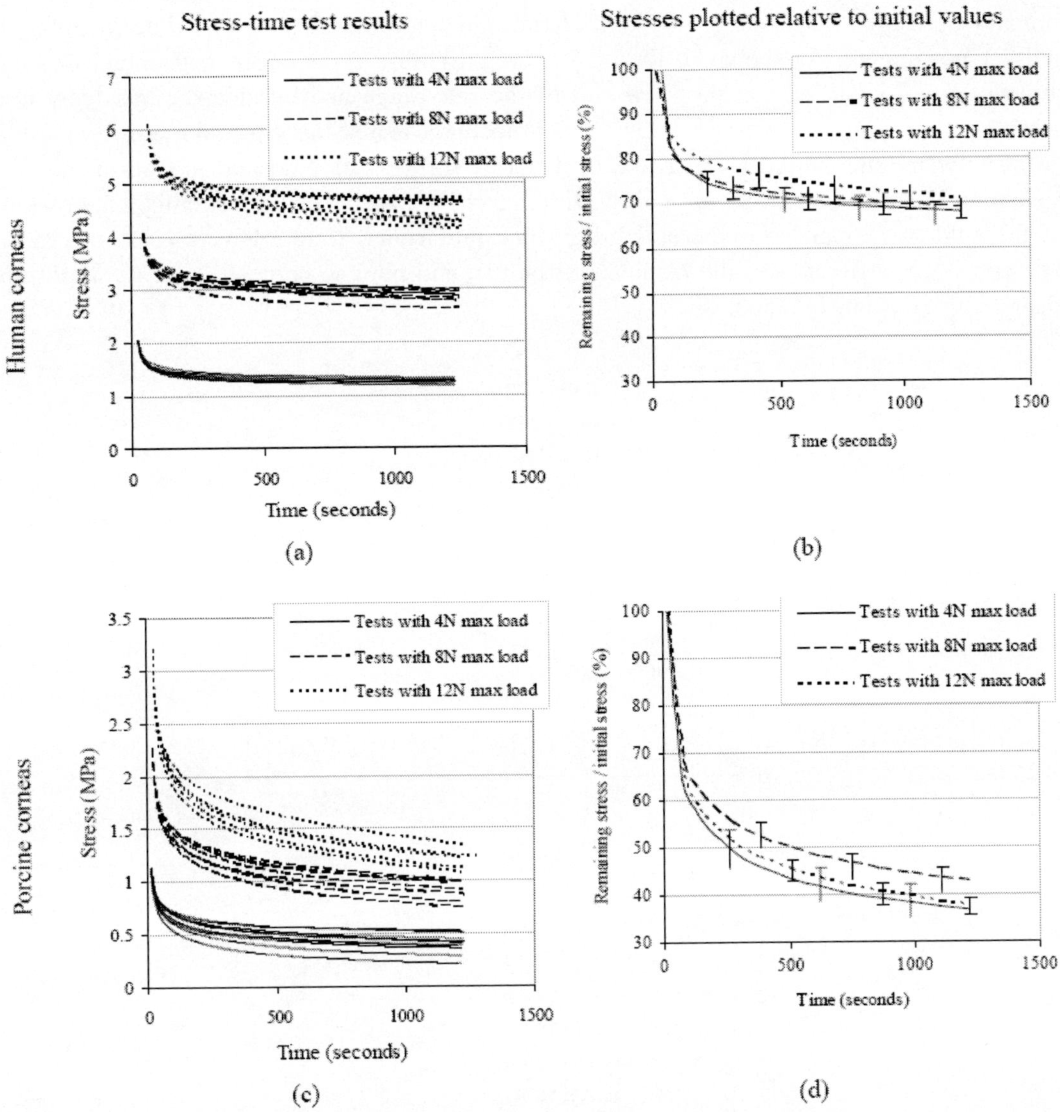

Figure 31. Stress relaxation behaviour of human and porcine corneas, (a,b) Human corneal behaviour, (c,d) Porcine corneal behaviour, (a,c) Stress-time experimental results, (b,d) Results plotted in relation to initial stress values.

The average trend lines are also close and do not show any consistent trends related to the value of initial load. This characteristic simplifies the development of stress-relaxation behaviour patterns for use in numerical simulation applications.

Corneal Dependence on Strain-Rate

As a viscoelastic material, the behaviour of corneal tissue is time-dependent and is expected to vary with the rate at which the material is strained or loaded. Experimental tests involving vertical strips of human and porcine corneal tissue have been conducted under uniaxial tension and with varying strain rates. With specimen length of 12 mm, width of 4 mm and elongation rates between 0.1 and 25 mm/minute, the strain rate varied between 0.008 and 2.1 per minute. As shown in Figure 32, the stiffening experienced with increasing the strain rate was significant at the slow end of the rate range and became progressively less notable as the strain rate increased. Under slow loading, increasing the strain rate from 0.008 to 0.08 per minute caused an increase in material stiffness of 47% and 9% on average in human and porcine corneas, respectively. In contrast, the corresponding average increases in stiffness that corresponded to increasing the strain rate from 0.83 to 2.1 were 1.5% and 1.2%. The consistent difference in the response of human and porcine corneal tissue to strain rate changes remains unexplained.

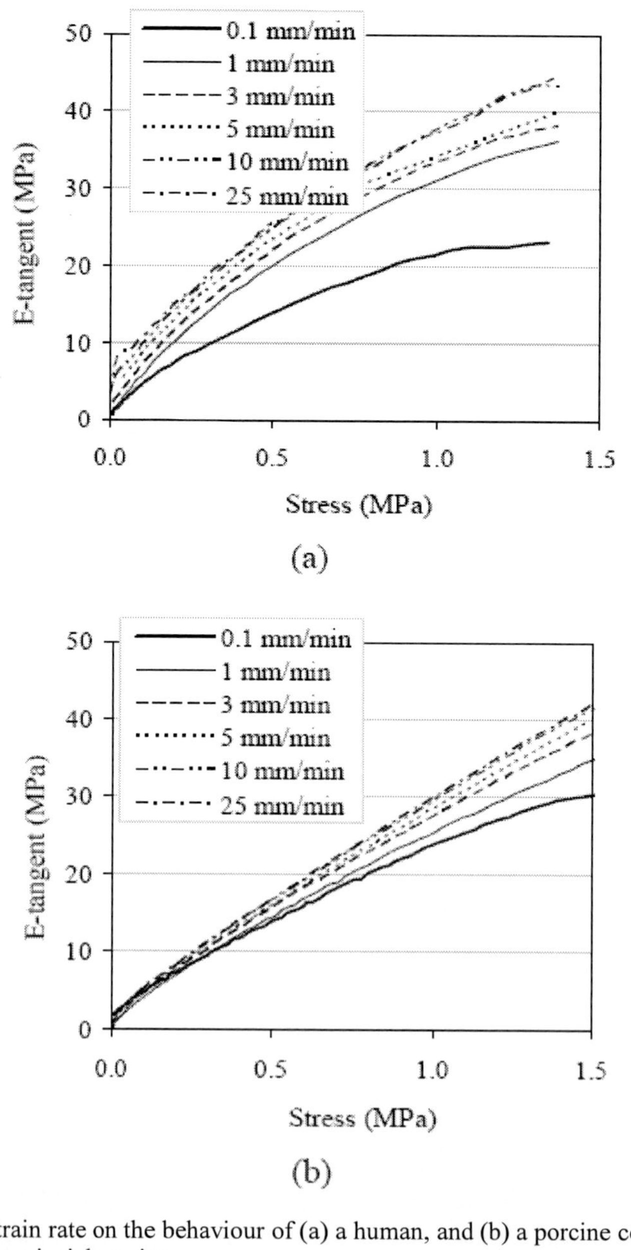

Figure 32. Effect of strain rate on the behaviour of (a) a human, and (b) a porcine corneal strip specimen tested under uniaxial tension.

Concluding Remarks on Corneal Material Behaviour

Characterisation of the biomechanical properties of corneal tissue is key to developing proper understanding of overall ocular behaviour, and response to disease, injury and surgery. It is also essential for the construction of accurate predictive numerical simulations of ocular behaviour, which have several clinical applications including improving the accuracy of tonometry, the planning of refractive surgery procedures, the design of rigid contact lenses, the treatment of myopia and keratoconus, and guiding surgical interventions following eye injuries.

The experimental studies reported above led to fast progress in our understanding of the human cornea's biomechanical behaviour, but several gaps still exist in our knowledge. Overall hyperelastic and hysteretic behaviour, and how it changes with age, has been characterised, but as the corneas tested were aged 50 years or over, similar studies are needed to consider younger corneas. The epithelium's contribution to corneal biomechanics has been quantified and further work is needed to assess the endothelium's and Bowman layer's contributions, and to distinguish between the anterior and posterior parts of the stroma (which differ in lamella interweaving). The correlation between corneal microstructure and biomechanical behaviour has been established in normal healthy eyes, and similar correlations are needed in swollen eyes (e.g. due to diurnal effects) and in eyes with keratoconus. Further work is also needed to build similar knowledge in the biomechanical behaviour and microstructure of the sclera, without which it becomes difficult to understand overall ocular behaviour or to construct predictive numerical models of whole eye globes.

It is also important to have the ability to measure corneal and scleral biomechanical properties in vivo. This is necessary if we are to build accurate and reliable patient-specific ocular numerical simulations, especially with the notable variations in biomechanical behaviour observed experimentally in same-age donor corneas. Recently, Reichert produced the Ocular Response Analyzer (ORA), a tonometer that produces a hysteresis parameter and claims that this parameter provides a measure of corneal biomechanics [110, 111]. However, the true biomechanical origin of the hysteresis measurement, and which biomechanical property it represents, is not yet fully understood.

Overall, and in spite of the current gaps in our knowledge, it remains true that the significant progress made to date in corneal material characterisation, topography mapping and microstructure detailing is offering an unprecedented opportunity to construct accurate predictive numerical simulations of corneal response to disease, injury and surgery. This subject will be discussed in some detail in the remainder of this chapter.

Numerical Modelling of Ocular Biomechanics

In general, a numerical model is a computer-based model that describes the physical behaviour of a structure and is able to predict its response to external loading. The structure is divided into a large number of simple parts called elements, and the behaviour of each element is described in a set of equilibrium equations normally presented in a matrix form. The matrices are then assembled together to form overall equilibrium equations for the structure. The mathematical assembly of individual matrices is done in the same way as the elements can physically be put together to form the actual structure. Solving the overall equilibrium equations under any set of given loads then provides a prediction of the structure's response to the loads including its deformation and stress distribution [112]. This modelling and solution process is commonly described as finite element analysis.

An example corneal model is shown in Figure 33. The model has four element layers including anterior and posterior layers representing the epithelium and endothelium, respectively, and two internal layers representing the stroma. Each layer is divided into 2904 solid prismatic elements, each with 5 faces and 6 corner points, called nodes. The elements are arranged in 6 radial segments and 22 circumferential rings in such a way that ensures all element internal angles remain between 20° and 70°. The model has a thickness that varies from 540 μm at the centre to 690 μm along the edge, central anterior radius of 7.8 mm, anterior shape factor, p = 0.82, and corneo-scleral diameter of 11.5 mm.

For finite element (FE) models to be accurate and predictive of real behaviour, the models need to closely represent the structure's geometry, topography, boundary conditions, loading regimes and material properties. In FE ocular models, these conditions include the non-uniform corneal thickness, the elliptical topography, the connection to the sclera, the uniform effect caused by the IOP, and the hyperelastic, hysteretic, anisotropic, viscoelastic and layered behaviour of corneal tissue. However, since the introduction of these properties can make the construction of FE models quite complex and the analysis costly and time consuming, it becomes important to distinguish between the properties with significant effect on model predictions and those that complicate the analysis without important improvements in accuracy.

It is well known that all FE models involve simplifications in topography, material properties, boundary conditions and/or loads [113-115].

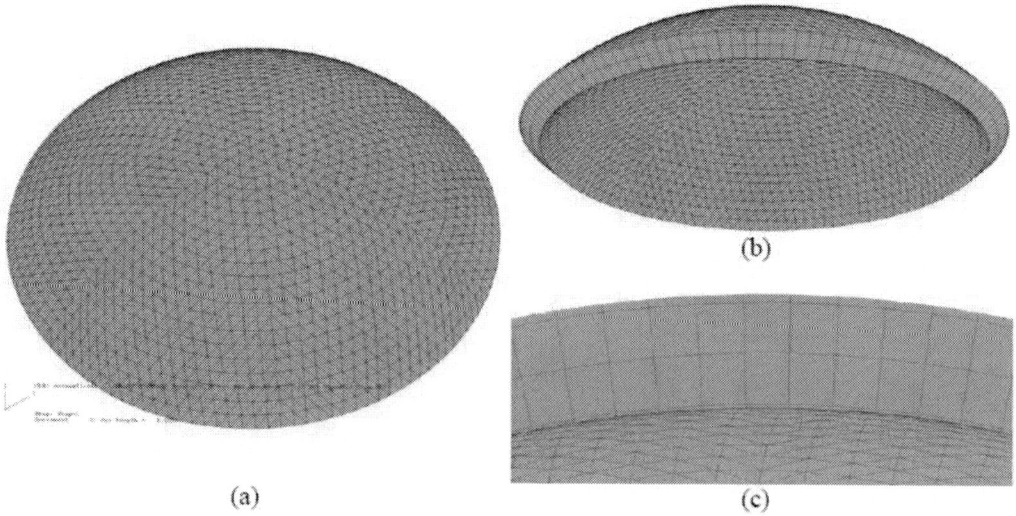

Figure 33. Views of a corneal numerical model with 4 layers, each with 2,904 elements arranged in 6 segments and 22 element rings, (a) Top (anterior) surface of model showing model segments and element rings, (b) View of bottom (posterior) surface and thickness of model, (c) Cross-sectional view showing the 4 layers of the model representing the epithelium, stroma and endothelium.

For instance, the elliptical corneal topography is sometimes approximated as a spherical surface, or the hyperelastic material behaviour idealised using a linear-elastic model. Some simplifications are essential as they reduce the complexity of the model, lower its development and execution costs and make it easier to modify and adapt. However, the more simplifications adopted in FE model construction, the farther away from reality the model becomes, eventually affecting its reliability. It is therefore important in constructing FE models that the user adopts simplifications where they are likely to have a negligible effect and avoid them when they start to erode the accuracy of predictions. This exercise requires experience and validation against known experimental behaviour.

The following discussion aims to assess the effect of simplifications in FE model construction with particular attention to simplifications in corneal topography, material behaviour and boundary conditions.

The discussion refers to numerical results for a corneal model under two simple load cases; a uniform posterior pressure (simulating the intraocular pressure, IOP) and a concentric anterior point load, and the simplifications in a particular parameter are assessed by comparing their effect on the simulation results. This exercise is followed by an overall discussion on how to improve the accuracy of ocular numerical models and to allow them to have a more useful role in clinical practice.

The two loading cases considered have been chosen for their simplicity and because they represent idealised situations where the cornea is subjected to either uniform membrane stresses under posterior pressure, or localised bending stresses under point loads. They have been preferred to subjecting the cornea to more realistic loadings such as tonometric pressure or external impacts since these loads create combinations of membrane and bending stresses, which would make the discussion of results less straightforward. The analysis used the model described above and shown in Figure 33, with some modifications added in the study of each simplification as detailed below.

For simplicity, the following discussion focuses on the deformation recorded at the centre point of the anterior surface, described as the corneal central, or apical, rise. Stiffness as used in the discussion of results means corneal resistance to deformation, and in particular central corneal rise.

SIMULATION OF CORNEAL NON-UNIFORM THICKNESS

The effect of corneal thickness variation is studied by considering two models; one with variable thickness; from 540 μm at the centre to 690 μm along the edge, and onc with constant (average) thickness of 615 μm, Figure 34a. The results shown in Figures 34b, c demonstrate a notable effect of thickness variation in models under point loads. The stiffness increases by 10% on average when adopting a constant thickness of 615 μm, rather than variable thickness. The effect under uniform posterior pressure is considerably lower, and limited to 1.6%.

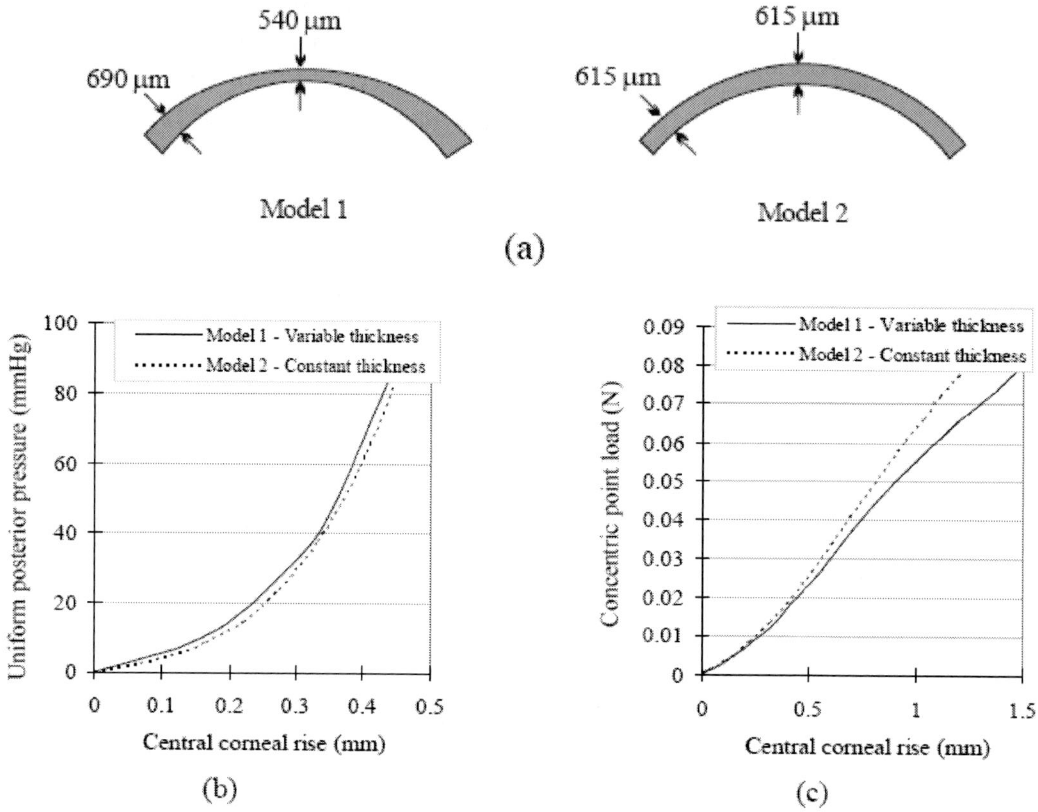

Figure 34. Behaviour predictions obtained using two models (a); one with variable corneal thickness and one with constant thickness, under (b) a uniform posterior pressure, and (c) a concentric point load.

The structural effects of the two loading conditions differ considerably. In the first loading condition, point loads create highly localised bending stresses around the load application point, and this concentration of stress makes the behaviour dependent on the thickness under the load, and almost independent of the thickness variation elsewhere. On the other hand, the behaviour under the tension stresses caused by posterior pressure depends on the thickness in all parts of the corneal surface, and in this case, replacing the real-life variable thickness with an average constant thickness is not expected to lead to notable change in overall corneal behaviour.

SIMULATION OF CORNEAL TOPOGRAPHY

The cornea has a prolate elliptical profile giving it maximum curvature (or smallest radius) at the centre and flatter surface towards the limbus. The effect of elliptical profile is studied under both uniform posterior pressure and anterior point load, and the results are depicted in Figure 35. Models with shape factors, p, between 0.6 and 1.2 are considered to cover the average reported prolate value, 0.8, the spherical case, 1.0, and an oblate case, 1.2 [43-45]. The results show a small, but notable, effect on models subjected to point loads (max effect = 6.3%) and a more evident effect, exceeding 28%, under posterior pressure.

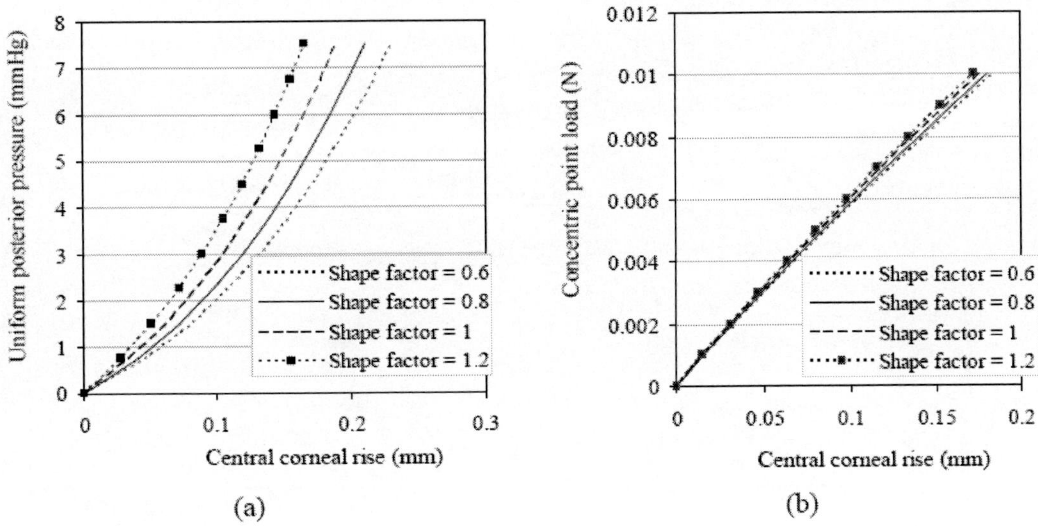

Figure 35. Effect of corneal shape factor, p, on predicted behaviour under (a) a uniform posterior pressure, and (b) a concentric point load.

The highly localised stresses created by point loads are mainly dependent on the thickness under the load, and corneal topography away from the load application point has little effect. On the other hand, the stresses developed in the cornea due to posterior pressure depend on the topography. When the corneal median surface matches the catenary topography, the cornea experiences only tension membrane stresses (i.e. no bending stresses) and hence undergoes small deformations. However, as the cornea deviates from the catenary topography, bending stresses with increasing values arise leading to increased deformation and reduced stiffness.

Simulation of Corneal Boundary Conditions

The connection to the sclera presents a problem in corneal simulation as the sclera is not stiff enough to prevent all deformation along corneal edge and is not flexible enough to allow free edge movement.

This makes the use of pinned supports along corneal edge unrepresentative of real conditions. One possible technique to avoid this problem is to use whole ocular models (involving at least both the cornea and the sclera), but while this solution presents a close representation of the cornea's natural conditions, it enlarges the model significantly and makes it expensive to build and analyse.

Another solution is to provide the cornea with titled supports to consider the expected sclera deformation, Figure 36a.

In this case, the slope of the supports is controlled such that the deformations of the cornea-only model and the whole-globe model almost match. The results of this exercise are demonstrated in Figure 36.

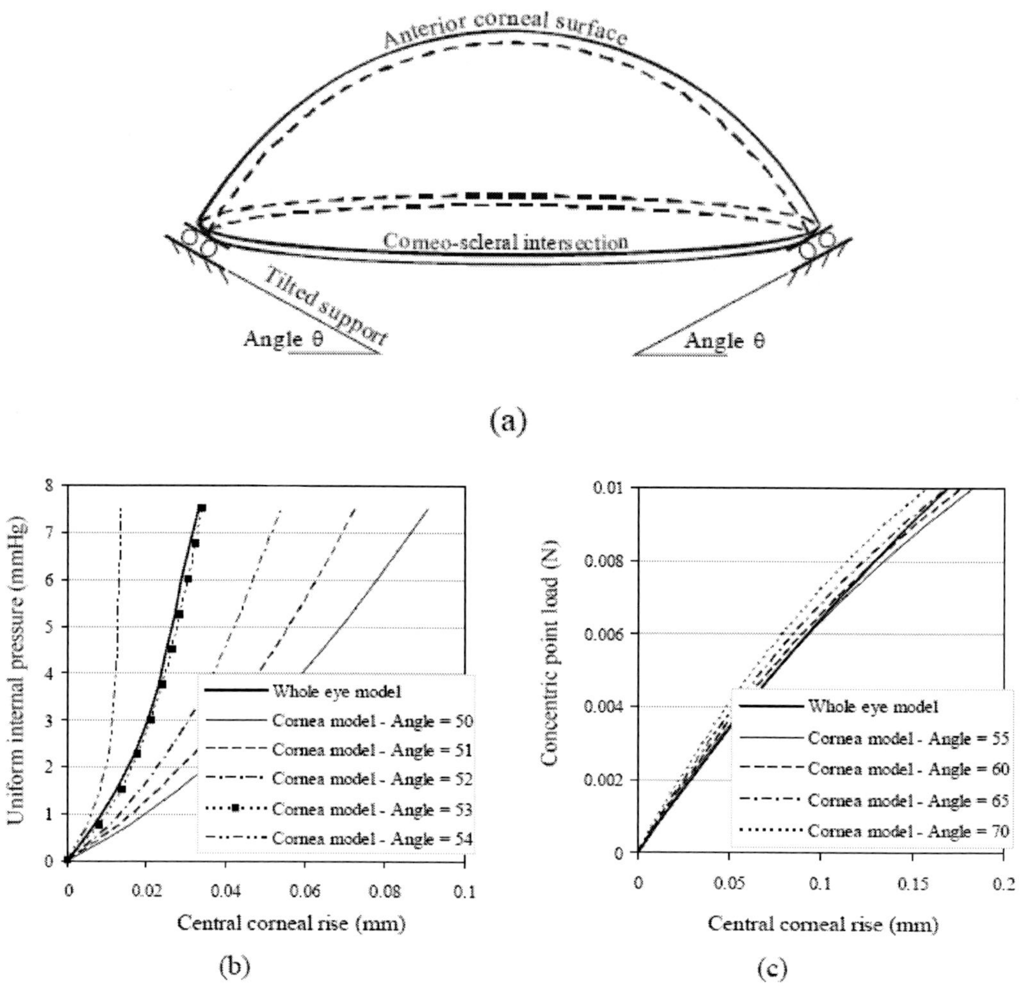

Figure 36. Behaviour of (a) cornea-only models with titled supports under (b) a uniform posterior pressure, and (c) a concentric point load.

The results show first that the support tilts suitable for simulations under posterior pressure and point load were 53° and 60°, respectively. Second, as the behaviour under point loads is dependent mainly on the thickness under the loads and largely insensitive to model changes elsewhere, it is reasonable to expect the model's low sensitivity to variations in support tilt in this case. The results also show the high sensitivity of analysis under posterior pressure to the support tilt. This finding should be considered carefully as it can make cornea-only models in this case less attractive, and much less reliable, than whole eye models.

Simulation of Corneal Hyperelastic Behaviour

Corneal tissue is known to possess hyperelastic behaviour with initial low stiffness, gradually increasing under higher stress levels. Considering this behaviour pattern is a challenge since the engineering materials commonly considered in commercial FE packages have the opposite behaviour, i.e. initial high stiffness, decreasing under higher stresses. For this reason, several earlier modelling studies simplified the corneal behaviour into a linear-elastic pattern, and this was justified by the assumption that common loads on the cornea create low stresses. In Figure 37, this assumption is tested under both posterior pressure and a central point load, and in both cases, two models were considered with hyperelastic and linear-elastic material behaviour patterns, respectively. The hyperelastic model considered the following stress-strain relationship developed earlier for corneas aged 50-65 years [30]:

$$\sigma_{(MPa)} = 0.0070\left(e^{50\,\varepsilon} - 1\right), \tag{28}$$

while the linear-elastic model had a constant Young's modulus, E, equal to the E of the hyperelastic model under low strain, $\varepsilon = 0.01$ ($E = 0.56$ MPa). The results show that the linear elastic model provided reasonable results under low load levels up to about 0.002 MPa (15 mmHg) or 0.01 N, beyond which the deviation between the two models became unacceptably high.

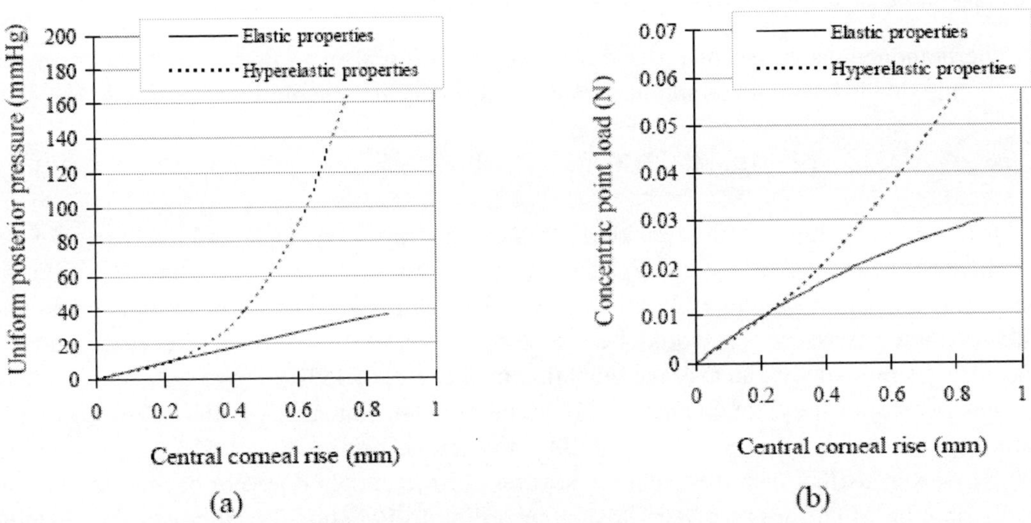

Figure 37. Effect of adopting a linear-elastic material behaviour pattern on the model's response to (a) a uniform posterior pressure, and (b) a concentric point load.

SIMULATION OF CORNEAL LAYERED CONSTRUCTION

The cornea's three main layers, the epithelium, stroma and endothelium, have different material properties. The stroma, with its fibrous structure, has the highest stiffness, while the cellular epithelium and endothelium are much more flexible. An earlier study estimated the epithelial stiffness at 10% that of the stroma [38]. No study has been carried out on endothelial biomechanics, but its cellular structure could justify assuming it had similar properties to the epithelium. The effect of considering the distinctive material properties of each layer on corneal behaviour was assessed and the results are shown in Figure 38.

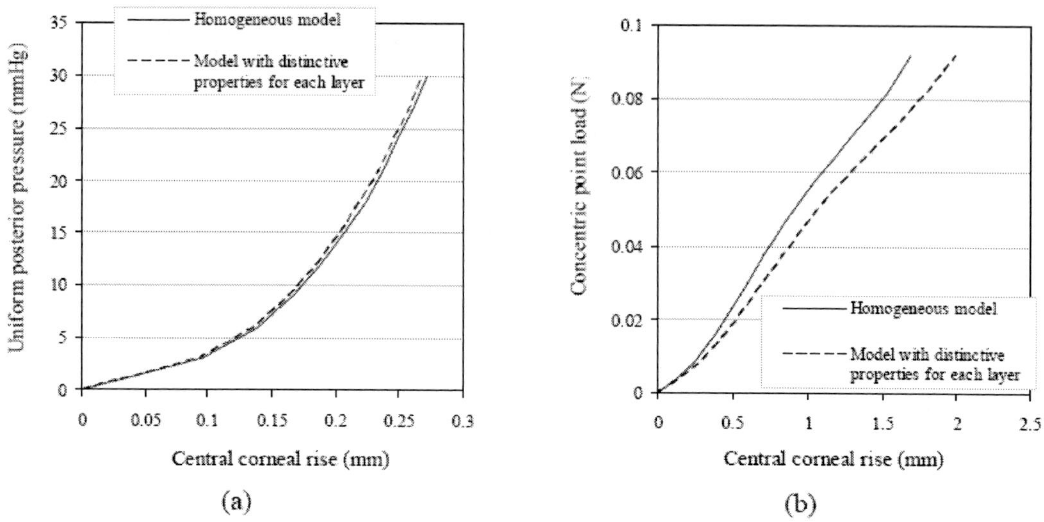

Figure 38. Effect of considering the distinctive material properties of the main corneal layers on predicting the behaviour under (a) a uniform posterior pressure, and (b) a concentric point load.

The homogenous model had a uniform stiffness equivalent to that of the heterogeneous model under loads creating membrane tension. This condition meant that:

$$(E\,.\,t)_{homogenous} = (E\,.\,t)_{epithelium} + (E\,.\,t)_{stroma} + (E\,.\,t)_{endothelium} \quad (29)$$

In this relationship, corneal thickness, t, was split into 8%, 90% and 2% parts for the three layers, respectively [64]. $E_{epithelium}$ and $E_{endothelium}$ were also both taken as 10% of E_{stroma}. The results show almost negligible effect of varying the material properties on the behaviour under posterior pressure, as would be expected. However, the effect was clear under a concentric point load with an average reduction in stiffness of 15%.

Since posterior pressure creates mainly membrane tension stresses, these stresses will be shared by the cornea's three layers according to their respective membrane stiffness values (E.t). However, as the overall membrane stiffness of the cornea does not change by varying the distribution of stiffness between layers (according to Equation 29), the overall behaviour under posterior pressure is not expected to change.

On the other hand, the bending stresses caused by point loads depend on the cornea's bending stiffness calculated as the sum of $E_i\,.\,I_i$ for all corneal layers, where I_i (the second moment of area of layer i) is $t_i^3 / 12 + t_i \times z_i^2$, and z_i the distance between the mid-thickness of layer i and the cornea mid-thickness. Therefore, placing the layers with high Young's modulus, E, nearer the cornea mid-thickness (hence lower z_i), as in the second model, reduces the overall bending stiffness and leads to increased deformation.

Simulation of Stromal Inter-Lamellar Adhesion

The cornea has weak shear stiffness at the surfaces separating stromal lamellae [9]. As was discussed above, the stromal inter-lamellar adhesion was assessed experimentally, and the shear stress-strain, τ–γ, relationship $\tau\,(\mathrm{MPa}) = 0.009\left(e^{1.74\gamma}-1\right)$ was derived [39]. The effect of stromal adhesion on corneal behaviour was considered by reducing the stromal inter-layer contact in the numerical simulations from full to the level observed experimentally. The results depicted in Figure 39 show a consistent trend in which corneal stiffness degraded with reducing stromal inter-lamellar interaction. On average, the stiffness reduced by 6% and 56% for the cases under posterior pressure and point load, respectively.

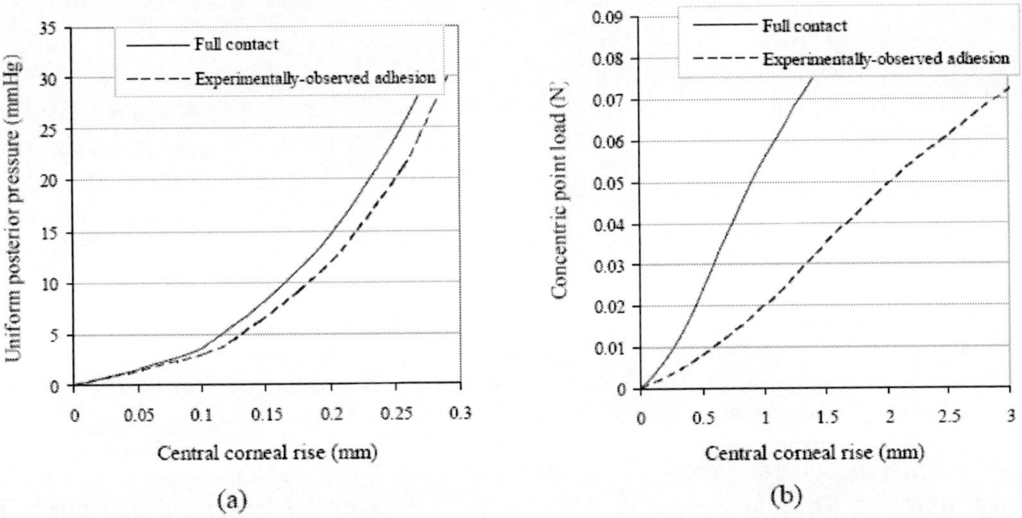

Figure 39. Effect of stromal inter-laminar adhesion on corneal behaviour under (a) a uniform posterior pressure, and (b) a concentric point load.

Since posterior pressure is transferred through membrane tension, which depends on the thickness and stiffness of individual layers, and not on inter-layer contact, reducing the level of contact does not have a considerable effect on overall behaviour under pressure [116]. On the other hand, reducing inter-layer contact reduces the second moment of area (as the stiffness term $t_i \times z_i^2$ is to be multiplied by a reduction factor which depends on the level of interaction) and therefore affects the resistance to deformation under bending actions such as those created by point loads.

SIMULATION OF CORNEAL ANISOTROPY

The human cornea has a preferential orientation of stromal collagen fibrils in the vertical (superior-inferior, SI) and horizontal (temporal-nasal, TN) directions, leading to anisotropy in biomechanical behaviour [24, 26]. The degree by which the SI and TN directions exhibit higher stiffness than diagonal directions has been determined earlier [28] and discussed above. Implementing anisotropy in corneal simulations was studied by considering two multi-layer numerical models; one with anisotropic and another with isotropic behaviour, respectively. The models in this study had 8 element layers representing the stroma, instead of the two layers used in other parts of the study.

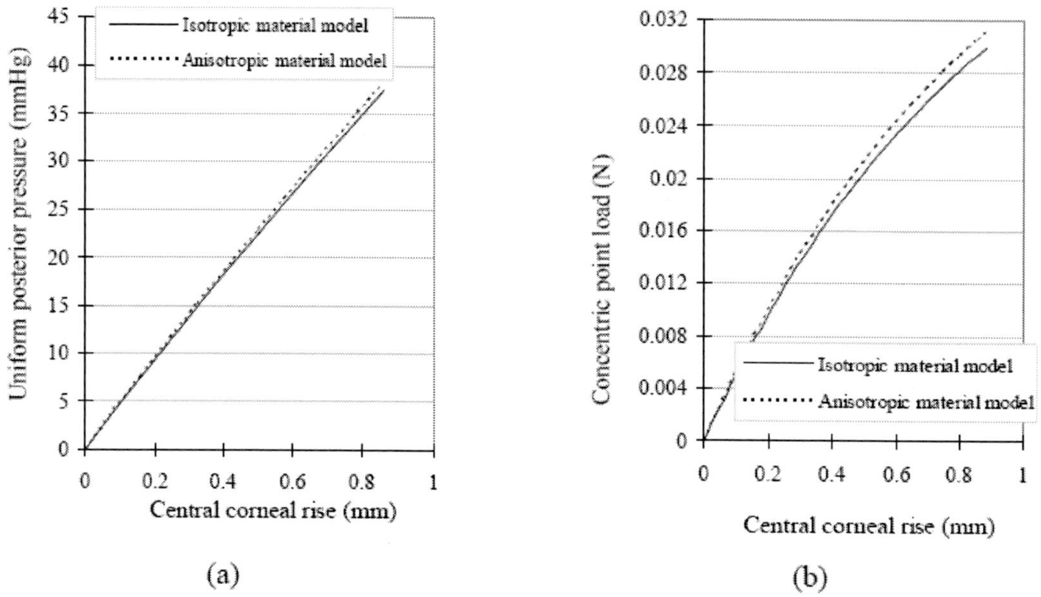

Figure 40. Effect of introducing material anisotropy on predicted behaviour under (a) a uniform posterior pressure, and (b) a concentric point load.

In the first model, odd-numbered stroma layers were assumed to have fibrils in only the vertical direction while even-numbered layers had fibrils in the horizontal direction. The layers were given a high stiffness in the fibril direction and 10% this value in the perpendicular direction.

In the second isotropic model, the stiffness was the same in all directions and taken as the average of the stiffness values in the two directions of the first model. The models were used to predict the behaviour under a posterior pressure and a concentric point load and the results showed little effect, below 2% under posterior pressure, and 4% under point load, caused by introducing anisotropy, see Figure 40.

SIMULATION OF CORNEAL HYSTERESIS

Considering corneal hysteresis in the material definition in FE models is perhaps the most difficult and challenging feature in model construction. This feature requires that the corneal material is defined with both loading and unloading behaviour patterns, and that the simulation can select the pattern that fits the stress history of each element. Following each change in loading, the simulation calculates the stress level within each element and determines whether the element is to undergo further loading or unloading, and based on this decision, the element is allowed to follow the more suitable behaviour pattern from the stress level determined earlier, Figure 41. This process is difficult to program and makes the analysis process quite complex.

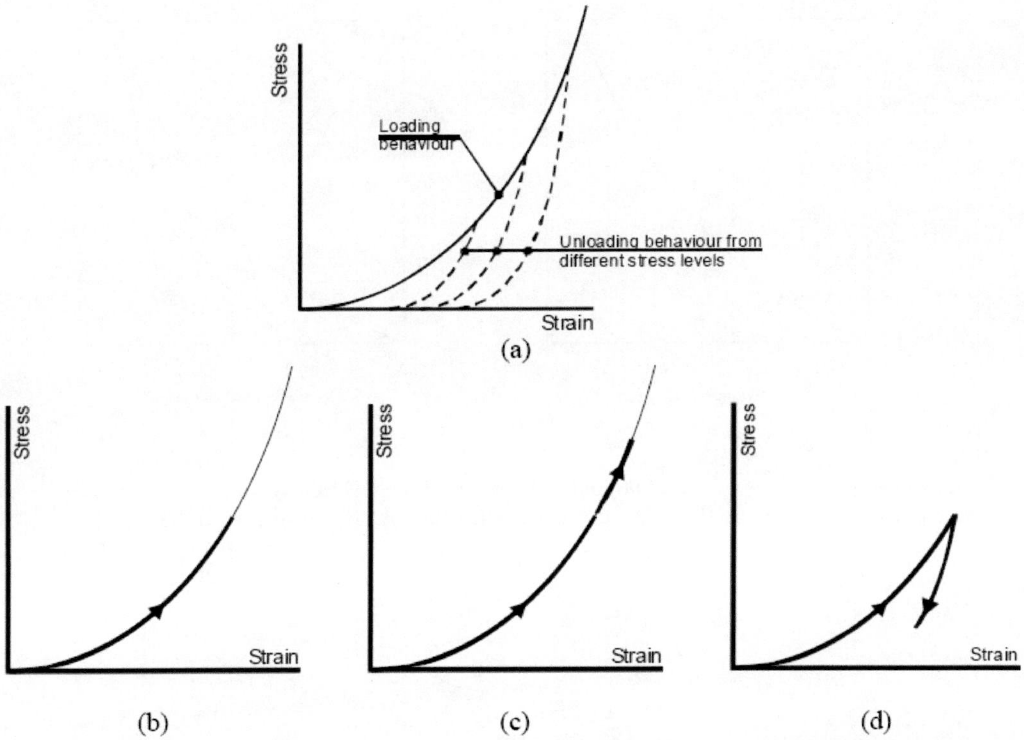

Figure 41. (a) Behaviour of corneal tissue under both loading and unloading conditions, and the stress development in a particular element during an FE analysis including (b) the stress history, and the behaviour adopted if the element is to undergo either (c) further loading or (d) unloading.

Nevertheless, an assessment is needed of whether hysteresis should be incorporated in material definition in the first place. The first trials under monotonically increasing posterior pressure or point loads showed no effect of considering hysteresis. Since all elements underwent further loading in all analysis steps, consideration of the unloading behaviour was not required.

However, further tests showed different outcomes. First, a simple corneal model was subjected to a posterior pressure that increased gradually to 30 mmHg then decreased down to 0. Ignoring hysteresis resulted in pressure-rise predictions that were the same under loading and unloading, Figure 42a. However, when hysteresis was considered, the unloading behaviour differed from the loading behaviour as would be expected in experimental tests, Figure 42b.

Another test involved the simulation of the Ocular Response Analyzer (ORA) – a tonometric technique that applies an air pulse on the cornea in increasing then decreasing intensity. Figures 43b-g show the stress distributions on a corneal cross-section at different stages of loading; first by intraocular pressure and then by external air pulse. The figures illustrate the unloading experienced in the middle region of the model during the second stage (force-out) of the ORA procedure. It is therefore hypothesised that this region should follow the material unloading behaviour in order to trace corneal response during the full ORA procedure.

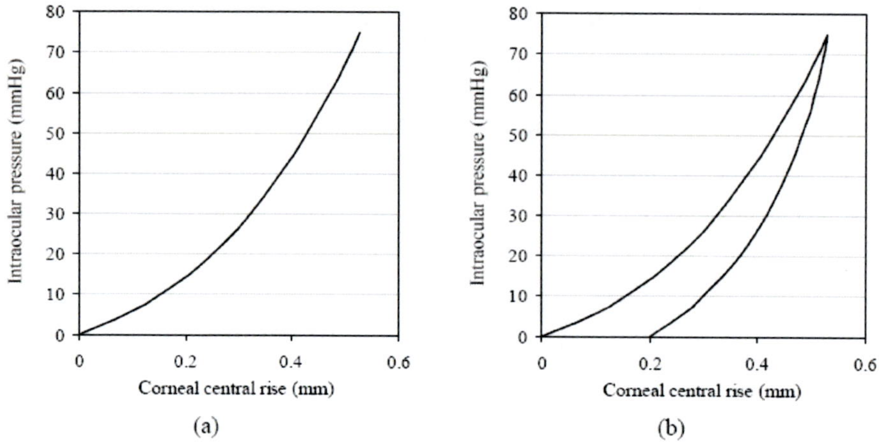

Figure 42. Model predictions of the behaviour of a cornea subjected to a cycle of pressure increase and decrease while (a) ignoring and (b) considering the unloading behaviour of corneal tissue.

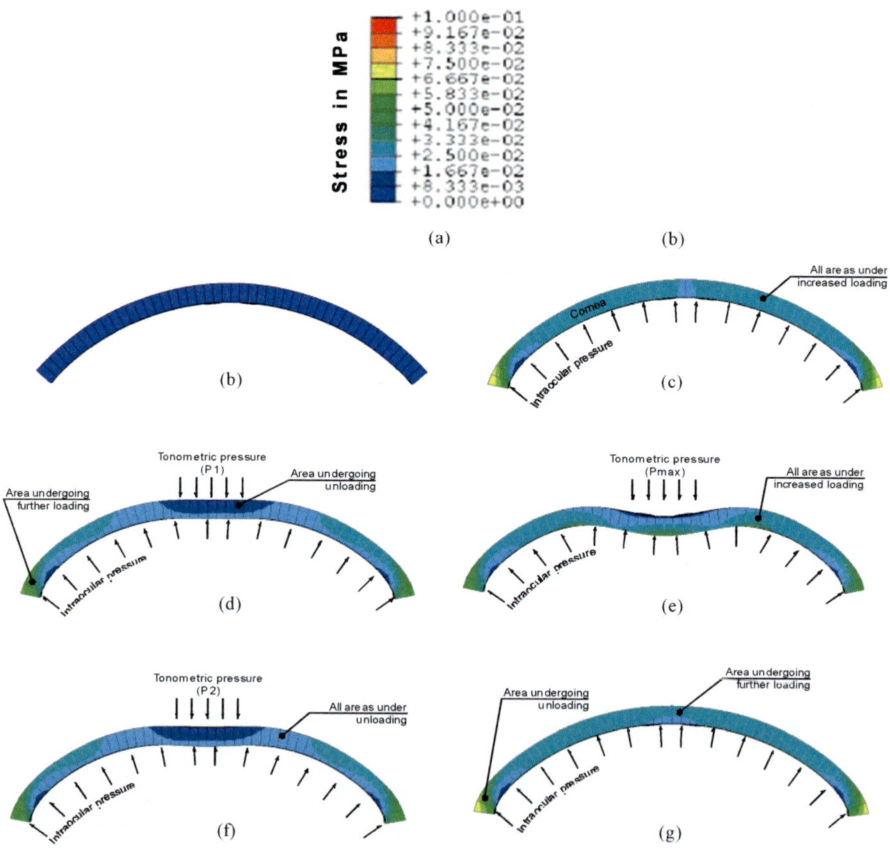

Figure 43. Stress distribution diagrams for a cornea with CCT = 540 μm, PCT = 690 μm, R = 7.8 mm, p = 0.82, age = 57 years and IOP = 15 mmHg, during the progress of the ORA procedure showing areas of loading and unloading, (a) Key showing stress values in MPa, (b) Cross-sectional view at start, (c) Stresses after applying the intraocular pressure **on the model's posterior side, (d) Stresses at first** applanation, (e) Stresses under Pmax, beyond which the air pressure starts to decrease, (f) Stresses at second applanation, (g) Stresses after complete removal of air pressure.

In order to test this hypothesis, the ORA procedure was simulated using two models that considered and ignored corneal hysteresis, respectively. The models were subjected first to an intraocular pressure of 15 mmHg. The ORA air pressure was then gradually applied until applanation was detected at P1 (the first applanation pressure) of 21.3 mmHg. The pressure was further increased to a peak value, Pmax, of 53 mmHg according to the relation: Pmax = 1.1713 P1 + 28.106 [117]. After reaching Pmax, the pressure was decreased gradually while looking for the second applanation pressure, P2.

The deformation was monitored at two points; the corneal centre and a point 0.6 mm away from the centre (the next point on the model's anterior surface away from the centre). Applanation was detected when the two points had the same coordinate in the anterior-posterior direction.

Figure 44 shows that when hysteresis was considered, the deformation under unloading differed from that under loading, as would be expected. As a result, the applanation pressures, P1 and P2, at which the central cornea flattened under air pressure, were different. However, when hysteresis was omitted, the two behaviour patterns were identical, and subsequently P2 was incorrectly equal to P1.

A similar test involving the simulation of the Goldmann Applanation Tonometer (GAT) was conducted. Unlike ORA, GAT is a contact tonometer that pushes a solid cylindrical device against the central cornea until applanation over a circular area with 3.06 mm diameter is achieved.

The behaviour was simulated numerically (including the initial application of intraocular pressure followed by contact with the tonometer), and the stress distribution during the two loading stages is depicted in Figure 45.

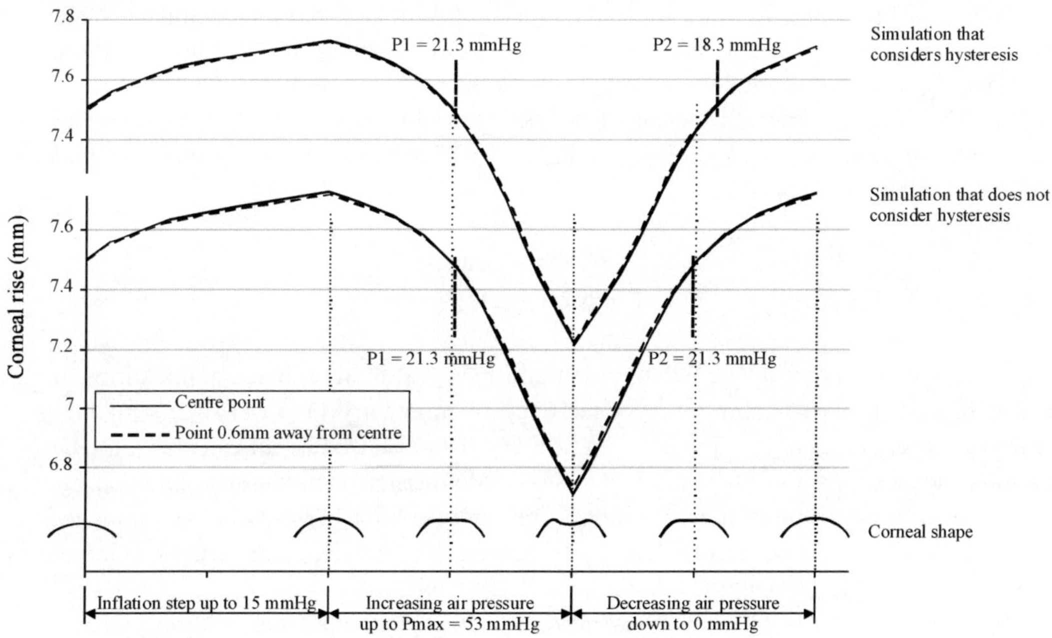

Figure 44. Corneal rise as measured at the centre and a point 0.6mm away from the centre during inflation under 15 mmHg followed by increasing then decreasing air pressure as applied by the Ocular Response Analyzer.

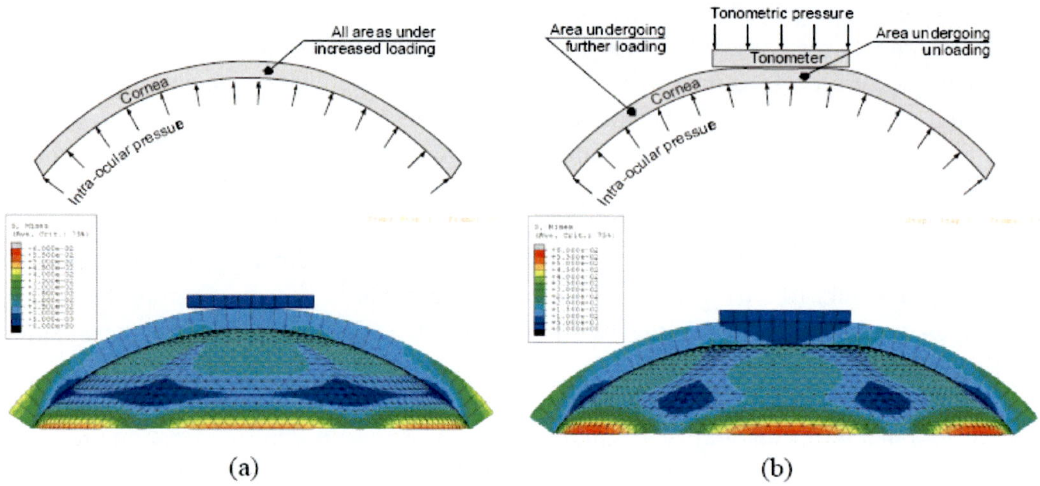

Figure 45. Stress development under IOP and GAT tonometry pressure, (a) Cornea under IOP with all areas following loading behaviour, (b) Cornea under both IOP and tonometry pressure with area under tonometer experiencing unloading.

The figure shows how the central corneal region undergoes unloading during contact with the tonometer – a region which should follow **the material's unloading behaviour in order for** the simulation to provide an accurate representation of the GAT procedure.

An attempt has been made to assess the importance of considering corneal hysteresis in the simulation of GAT. Two cornea models have been developed with hysteresis considered in one and ignored in the other. Similar to ORA simulations, both models have CCT = 540 μm, PCT = 690 μm, R = 7.8 mm, p = 0.82, age = 57 years and IOP = 15 mmHg. While the first model, which considers hysteresis, estimated a GAT reading of IOP (denoted IOPG) of 15.19 mmHg, the second model which ignores hysteresis predicted an IOPG value of 13.96 mmHg. The considerable difference (8%) between the two IOPG values indicate the importance of considering hysteresis although the corneal zone expected to undergo unloading is small in this case.

CONCLUSION

Numerical simulation has a strong potential in corneal biomechanics applications. It can be, and has been, used to improve the accuracy of tonometry [118, 119], the planning of refractive surgery [5,6, 14, 107], the evaluation of damage caused by impacts [11], the design of contact lenses, and the estimation of disease progression. This strong potential can be realised with growing interest from various research groups [120, 121], fast advances in numerical modelling, and significant contributions to knowledge and understanding of corneal microstructure, topography and material characterisation.

However, a number of challenges currently exist and slow the progress of corneal numerical simulation. First, the level of complexity of corneal microstructure with variable interweaving of stromal lamellae, and change in preferential orientation of collagen fibrils from vertical and horizontal at the centre to circumferential near the limbus, make the development of a corneal model that is truly representative of real-life conditions highly complex.

Second, the material properties of corneal tissue pose serious challenges in simulation. On a basic level, the material's hyperelasticity and hysteresis can be considered, albeit with some difficulty, in numerical simulation. However, these properties are not constant, but vary with age, rate of loading and anatomical direction, and therefore a material model which is adequate for a particular case, could be entirely inaccurate in another. One typical example is the difference in speed of loading experienced under contact and non-contact tonometry, which would require widely different material models [118, 119]. The stiffening effect of ageing adds another complication to the behaviour under both loading and unloading [30, 34].

Third and most important is the current lack of understanding of some fundamental corneal properties. These include the effect of swelling (following sleep or contact lens wear) on corneal microstructure and biomechanical behaviour, the changes in microstructure associated with the progression of diseases such as keratoconus, the effects of long term exposure to high levels of IOP in glaucoma, and the wound healing process following surgery and whether stromal tissue is able to regain its mechanical integrity after wound healing. These effects must be quantified experimentally and/or clinically before they can be accurately considered in numerical simulation models.

Fourth, the current level of understanding of other ocular components, most notably the sclera, is not adequate to enable their proper inclusion in ocular simulations. While a few studies conducted over the last 40 years helped improve understanding of scleral thickness variation, regional biomechanical performance and microstructure [122-130], work is still required to quantify the effect of sclera's hyperelasticity, hysteresis, viscoelasticity, anisotropy, layered composition, microstructure, and medical conditions such as myopia on scleral biomechanics. While this work will be important in its own right, it will enable the construction of more reliable whole eye globe models without having to approximate the boundary conditions of the cornea.

The above discussion is intended to point at the areas where corneal biomechanics research is most needed, and to draw attention to the challenges slowing the development and application of corneal numerical simulation. However, it must be emphasised that while work to address these challenges is needed, corneal numerical simulation should not wait and should progress based on current knowledge. What has been achieved over the past few decades in corneal material characterisation, topography mapping and microstructure detailing, is significant and sufficient for several important applications. Normal, healthy and intact corneas can now be reliably modelled under actions such as contact and non-contact tonometry or external impacts. These applications not only address important and relevant clinical problems, they also help demonstrate the potential value of numerical simulation in corneal biomechanics applications. Nevertheless, care should be taken not to expand simulation into areas where knowledge of material behaviour or microstructure has not yet been generated. These attempts would be likely to produce unreliable results, and might affect the general perception of what numerical simulation can do in corneal applications.

Detailed clinical validation is also required to assess the accuracy of numerical simulations and demonstrate their suitability in providing solutions to clinical problems. Clinical validation will also help overcome the traditional reluctance of ophthalmologists and optometrists to embrace predictive tools based on numerical simulations, and prepare the ground for their wide acceptance.

REFERENCES

[1] Fatt I. Physiology of the eye: An introduction to the vegetative functions, Butterworths, London, 1978.

[2] Roberts C. The cornea is not a piece of plastic. *Journal of Refractive Surgery*. 2000;16:407-413.

[3] Ethier CR, Johnson M, Ruberti J. Ocular biomechanics and biotransport. *Annual Review of Biomedical Engineering*. 2004;6:249-273.

[4] Michelacci YM. Collagens and proteoglycans of the corneal extracellular matrix. *Brazilian Journal of Medical and Biological Research*. 2003;36(8):1037-1046.

[5] Vito RP, Shin TJ, McCarey BE. A Mechanical Model of the Cornea: The Effects of Physiological and Surgical Factors on Radial Keratotomy Surgery. *Refractive Corneal Surgery*. 1989;5(2):82-88.

[6] Pinsky P, Datye V. A microstructurally-based finite element model of the incised human cornea. *Journal of Biomechanics*. 1991;10:907-922.

[7] Bryant M, McDonnell P. Constitutive laws for biomechanical modeling of refractive surgery. *Journal of Biomechanical Engineering*. 1996;118:473-481.

[8] Buzard KA. Introduction to bio-mechanics of the cornea. *Refractive and Corneal Surgery*. 1992;8:127-138.

[9] Dupps WJ. Biomechanical modeling of corneal ectasia. J Refractive Surgery. 2005;21(2):186-190.

[10] Velinsky S, Bryant M. On the Computer-Aided and Optimal Design of Keratorefractive Surgery. *Refractive Corneal Surgery*. 1992;8:173-183.

[11] Uchio E, Ohno S, Kudoh J, Aoki K, Kisielewicz LT. Simulation model of an eyeball based on finite element analysis on a supercomputer. *Br. J. Ophthalmology*. 1999;83:1106-1111.

[12] Fernández DC, Niazy AM, Kurtz RM, Djotyan GP, Juhasz T. Finite element analysis applied to cornea reshaping. *Journal of Biomedical Optics*. 2005;10(6):064018-1-11.

[13] Fernández DC, Niazy AM, Kurtz RM, Djotyan GP, Juhasz T. Computational modeling of corneal refractive surgery. *Proceedings of SPIE*. 2004;Vol. 5314. Bellingham, WA.

[14] Alastrué V, Calvo B, Peña E, Doblaré M. Biomechanical modeling of refractive corneal surgery. *Biomechanical Engineering*. 2006;128;150-160.

[15] Hanna, K., Jouve, F., Waring, G., and Ciarlet, P., 1992, "Computer simulation of arcuate keratotomy for astigmatism," *Refract. Corneal Surg*. 8 _1992_ 152-163.

[16] Pinsky, P., and Datye, V., 1991, "A microstructurally-based finite element model of the incised human cornea," *J. Biomech*. 10, pp. 907-922.

[17] Howland, H., Rand, R., and Lubkin, S., 1992, "A Thin-Shell Model of the Cornea and its Application to Corneal Surgery," *Refract. Corneal Surg*. 8, pp.183-186.

[18] Bryant, M. R., Machi, V., and Juhasz, T., 2000, "Mathematical Model of Picosecond Laser Keratomileusis For High Myopia," *ASME J. Biomech. Eng*. 16, pp. 155-162.

[19] Djotyan, G. P., Kurtz, R. M., Cabrera, D., and Juhasz, T., 2001, "An Analytically Solvable Model For Biomechanical Response of the Cornea to Refractive Surgery," *J. Refract. Surg*. 123, pp. 440-445.

[20] Huang, D., Tang, M., and Shekhar, R., 2003, "Mathematical Model of the Corneal Surface Smoothing After Laser Refractive Surgery," *Am. J. Ophthalmol.* 135, pp. 267-278.

[21] Yeh, H.-L., Huang, T., and Schachar, R., 2000, "A Closed Shell Structured Eyeball Model With Application to Radial Keratotomy," *J. Biomech. Eng.* 122, pp. 504–510.

[22] Hanna, K. D., Jouve, F., Bercovier, M. H., and Waring, G. O., 1989, "Computer Simulation of Lamellar Keratectomy and Laser Myopic Keratomileusis," *J. Refract. Surg.* 4, pp. 222–231.

[23] Hanna, K. D., Jouve, F., Ciarlet, P., and Waring, G. O., 1989, "Computer Simulation of Arcuate and Radial Incisions Involving the Corneoscleral Limbus," Eye 3, pp. 227–239.

[24] Boote C, Dennis S, Huang Y, Quantock AJ, Meek KM. Lamellar orientation in human cornea in relation to mechanical properties. *Journal of Structural Biology*. 2005;149:1-6.

[25] Müller LJ, Pelsa E, Schurmans LRHM, Vrensen GFJM. A new three-dimensional model of the organization of proteoglycans and collagen fibrils in the human corneal stroma. *Experimental Eye Research*. 2004;78:493-501.

[26] Meek KM, Newton RH. Organization of collagen fibrils in the corneal stroma in relation to mechanical properties and surgical practice. *Journal of Refractive Surgery*. 1999;15(6):695-699.

[27] Aghamohammadzadeh H, Newton RH, Meek KM. X-ray scattering used to map the preferred collagen orientation in the human cornea and limbus. *Structure*. 2004;12:249-256.

[28] Elsheikh A, Brown M, Alhasso D, Rama P, Campanelli M, Garway-Heath DF. Experimental assessment of corneal anisotropy. *Journal of Refractive Surgery*. 2008;24(2):178-187.

[29] Jayasuriya AC, Ghosh S, Scheinbeim JI, Lubkin V, Bennett G, Kramer P. A study of piezoelectric and mechanical anisotropies of the human Cornea. *Biosens Bioelectron*. 2003;18:381-387.

[30] Elsheikh A, Wang D, Rama P, Campanelli M, Pye D. Assessment of corneal biomechanical properties and their variation with age. *Current Eye Research*. 2007;32:11-19.

[31] Elsheikh A, Wang D, Pye D. Determination of the modulus of elasticity of the human cornea. *Journal of Refractive Surgery*. 2007;23(8):808-818.

[32] Tallec PL, Rahier C, Kaiss A. Three-dimensional incompressible viscoelasticity in large strains: Formulation and numerical approximation. *Comput. Methods Appl. Mech. Eng.* 1993;109:233-258.

[33] Boyce B, Jones R, Nguyen T, Grazier J. Stress-controlled viscoelastic tensile response of bovine cornea. *Journal of Biomechanics*. 2007; 40(11):2367-2376.

[34] Elsheikh A, Wang D, Rama P, Campanelli M, Garway-Heath DF. Experimental assessment of human corneal hysteresis. *Current Eye Research*. 2008;33:205-213.

[35] Elsheikh A, Alhasso D, Rama P. Biomechanical properties of human and porcine corneas. *Experimental Eye Research*. 2008;86:783-790.

[36] Kampmeier J, Radt B, Birngruber R, Brinkmann R. Thermal and biomechanical parameters of porcine cornea. *Cornea*. 2000;19:355-362.

[37] Scherer K, Eggert H, Guth H, Stiller P. Application of simulation techniques in human eye corneal surgery. 13th International Conference of Society for Medical Innovation and Technology. Berlin. 2001.

[38] Elsheikh A, Alhasso D, Rama P. Assessment of the epithelium's contribution to corneal biomechanics. *Experimental Eye Research*. 2008;86:445-451.

[39] Elsheikh A, Ross S, Alhasso D, Rama P. Numerical study of the effect of corneal layered structure on ocular biomechanics. *Current Eye Research*. In Press

[40] Schmack I, Dawson DG, McCarey BE, Waring BO, Grossniklaus HE, Edelhauser HF. Cohesive tensile strength of human LASIK wounds with histologic, ultrastructural and clinical correlations. *Journal of Refractive Surgery*. 2005;21:433-445.

[41] Smolek MK, McCarey BE. Interlamellar adhesive strength in human eyebank corneas. *IOVS*. 1990;31(6):1087-1095.

[42] Smolek MK. Interlamellar cohesive strength in the vertical meridian of human eye bank corneas. *IOVS*. 1993;34:2962-2969.

[43] Dubbelman M, Weeber HA, van der Heijde RGL, Völker-Dieben HJ. Radius and asphericity of the posterior corneal surface determined by corrected Scheimpflug photography. *Acta Ophthalmol. Scand*. 2002;80:379-383.

[44] Read SA, Collins MJ, Carney LG, Franklin RJ. The topography of the central and peripheral cornea. *IOVS*. 2006;47:1404-1415.

[45] Guillon M, Lydon PM, Wilson C. Corneal topography: a clinical model. *Ophthalmic. Physiol. Opt*. 1986;6:47-56.

[46] Aghaian E, Choe JE, Lin S, Stamper RL. Central Corneal Thickness of Caucasians, Chinese, Hispanics, Filipinos, African Americans, and Japanese in a Glaucoma Clinic. *Ophthalmology*. 2004;111:2211-2219.

[47] Cho P, Cheung SW. Central and peripheral corneal thickness measured with the TOPCON specular microscope SP-2000P. Current Eye Research. 2000;21(4):799-807.

[48] Lam AKC, Chan JS. Corneal thickness at different reference points from Orbscan II system. *Clinical and Experimental Optometry*. 2003;86(4):230-234.

[49] Ethier CR, JohnsonM, Ruberti J. Ocular biomechanics and biotransport. *Annu. Rev. Biomed. Eng*. 2004. 6:249-73.

[50] Pouliquen Y. 1985. Fine structure of the corneal stroma. *Cornea* 3:168-77.

[51] Parry DA, Craig AS. Electron microscope evidence for an 80 A° unit in collagen fibrils. *Nature*. 1979;282:213–215.

[52] Komai Y, Ushiki T. The three-dimensional organization of collagen fibrils in the human cornea and sclera. *Invest. Ophthalmol. Vis. Sci*. 1991;32:2244-2258.

[53] Maurice DM, Monroe F. Cohesive strength of corneal lamellae. *Exp. Eye Res*. 1990;50:59.

[54] Kokott W. Über mechanisch-funktionelle Strukturen des Auges. *Albrecht. v Grafes Arch. Ophthalmol* 1938;118:424-485.

[55] Meek KM, Blamires T, Elliot GF, Gyi TJ, Nave C. The organization of collagen fibrils in the human corneal stroma: a synchroton x-ray distraction study. *Current Eye Research,* 1987;6:841-846.

[56] Newton RH, Meek KM. The integration of the corneal and limbal fibrils in the human eye. *Biophysical Journal*, 1998;75:2508-2512.

[57] Newton RH, Meek KM. Circumcorneal annulus of collagen fibrils in the human limbus. *Investigative Ophthalmology and Visual Science*. 1998;39:1125-1134.

[58] Daxer A, Fratzl P. Collagen fibril orientation in the human corneal stroma and its implication in keratoconus. *Investigative Ophthalmology and Visual Science*, 1997;38(1):121-1293.

[59] Hayes, S., Boote, C., Lewis, J., Sheppard, J., Abahussin, M., Quantock, A.J., Purslow, C., Votruba, M., Meek, K.M., 2007. Comparative study of fibrillar collagen arrangement in the corneas of primates and other mammals. *The Anatomical Record* 290:1542-1550.

[60] Boote C, Dennis S, Meek KM. Spatial mapping of collagen fibril organisation in primate cornea—an X-ray diffraction investigation. *Journal of Structural Biology* 146 (2004) 359-367.

[61] Prangen, A., 1928. A study of the comparative anatomy of the extraocular muscles. *Trans. Am. Ophthalmol. Soc.* 26:353-380.

[62] Green K (1991) Corneal endothelial structure and function under normal and toxic conditions. *Cell Biology Review* 25, 169-207.

[63] Johnson M, Shapiro A, Ethier CR, Kamm RD. Modulation of outflow resistance by the pores of the inner wall endothelium. *IOVS*. 1992;33:1670-1675.

[64] Coster DJ. Cornea. BMJ Publishing Group, London. 2001.

[65] Patel, S., Alio, J.L., Perez-Santonja, J.J., 2004. Refractive index change in bovine and human corneal stroma before and after LASIK: A study of untreated and re-treated corneas implicating stromal hydration. *IOVS*. 45(10), 3523-3530.

[66] Beuerman RW, Pedroza L (1996) Ultrastructure of human cornea. *Microscopy Research and Technique* 33, 320-335.

[67] Danielsen CC. Tensile mechanical and creep properties of Descemet's membrane and lens capsule. *Experimental Eye Research* 79 (2004) 343-350.

[68] Merindano MD, Costa J, Canals M, Potau JM, Ruano D. A comparative study of Bowman's layer in some mammals: Relationships with other constituent corneal structures. *Eur. J. Anat.*, 6 (3) : 133-139 (2002)

[69] Seiler T, Matallana M, Sendler S, Bende T. Does Bowman's layer determine the biomechanical properties of the cornea? *Refract Corneal Surg*. 1992;8(2):139-42.

[70] Feltgen, N., Leifert, D., Funk, J., 2001, "Correlation between central corneal thickness, applanation tonometry and direct intracameral IOP readings," *British Journal of Ophthalmology*, 85, pp. 85-87.

[71] Laiquzzaman, M., Bhojwani, R., Cunliffe, I., Shah, S., 2006, "Diurnal variation of ocular hysteresis in normal subjects: relevance in clinical context," *Clinical and Experimental Ophthalmology*, 34, pp. 114-118.

[72] Foster, P. J., Machin, D., Wong, T.-Y., Ng, T.-P., Kirwan, J. F., Johnson, G. J., Khaw, P. T., Seah, S. K. L., 2003, "Determinants of intraocular pressure and its association with glaucomatous optic neuropathy in Chinese Singaporeans: The Tanjong Pagar study," *IOVS*, 44, pp. 3885-3891.

[73] Lleó A, Marcos A, Calatayud M, Alonso L, Rahhal SM, Sanchis-Gimeno JA. The relationship between central corneal thickness and Goldmann applanation tonometry. *Clin. Exp. Optom.* 2003;86(2):104-108.

[74] Shah, S., Spedding, C., Bhojwani, R., Kwartz, J., Henson, D., McLeod, D., 2000, "Assessment of the diurnal variation in central corneal thickness and intraocular pressure for patients with suspected glaucoma," *Ophthalmology*, 107(6), pp. 1191-1193.

[75] Hirji, N. K., Larke, J. R., 1978, "Thickness of human cornea measured by topographic tachometry," *American Journal of Optometry*, 55, pp. 97-100.
[76] Martola, E. L., Baum, J. L., 1968, "Central and peripheral corneal thickness," *Arch. Opththalmol*, 79, pp. 28-30.
[77] Maurice, D. M., Giardini, A. A., 1951, "A simple optical apparatus for measuring the corneal thickness and the average thickness of the human cornea," *British Journal of Ophthalmology,* 35, pp. 169-177.
[78] Hennighausen H, Feldman ST, Bille JF, McCulloch AD. Anterior-posterior strain variation in normally hydrated and swollen rabbit cornea. IOVS. 1998;39:253-262.
[79] Nash IS, Greene PR, Foster CS. Comparison of mechanical properties of keratoconus and normal corneas. *Exp. Eye Res.* 1982; 35:413-24.
[80] Fung, Y.C. Biomechanics: Mechanical properties of Living Tissues. Springer-Verlag, New York, pp. 211, 1981.
[81] Hjortdal, J.O. "Regional elastic performance of the human cornea." *J. Biomechanics*, Vol. 29, No. 7, pp. 931-942, 1996.
[82] Shin, T.J., Vito, R.P. et al. "The distribution of strain in the human cornea." *J. Biomechanics,* Vol. 30, No. 5, pp. 497-503, 1997.
[83] Erkamp R.Q., Wiggins P, Skovoroda, A.R. et al. "Measuring the elastic modulus of small tissue samples." *Ultrasound Imaging*, 20:17–28, 1998.
[84] Chen E.J., Novakofski J., Jenkins W.K., et al. "Young's modulus measurements of soft tissues with application to elasticity imaging." *IEEE Trans. UFFC* 43:191–4, 1996.
[85] Zeng, Y., Yang, J., et al. "A comparison of biomechanical properties between human and porcine cornea." *J. Biomechanics*, Vol. 34, No. 5, pp. 533-537, 2001.
[86] Jue B, Maurice DM. The mechanical properties of rabbit and human cornea. *J. Biomech.* 1986;19:847–853.
[87] Woo SLY, Kobayashi AS, Schlegel WA, Lawrence C. Non-linear Properties of intact cornea and sclera. *Exp. Eye Res.* 1972; 14: 29-39.
[88] Nyquist GW. Rheology of the cornea: Experimental techniques and results. *Experimental Eye Research.* 1968; 7: 183-188.
[89] Hjortdal JO. On the biomechanical properties of the cornea with particular reference to refractive surgery. *The Ophthalmological Journal of the Nordic Countries.* 1998; 76(225): 1-23.
[90] Andreassen TT, Simonsen AH, Oxlund H. Biomechanical properties of keratoconus and normal corneas. *Exp. Eye Research.* 1980; 31: 435-441.
[91] Liu, J., Roberts, C. J., 2005, "Influence of corneal biomechanical properties on intraocular pressure measurement: Quantitative analysis," *J. Cataract Refract Surgery*, 31, pp. 146-155.
[92] Hoeltzel DA, Altman P, Buzard K, Choe K. Strip extensiometry for comparison of the mechanical response of bovine, rabbit, and human corneas. *J. Biomech. Eng.* 1992;114(2):202–215.
[93] Daxer A, Misof K, Grabnerj B, Ettl A, Fratzl P. Collagen fibrils in the human corneal stroma: Structure and aging. *Investigative Ophthalmology and Visual Science*, 1998;39(3):644-648.
[94] Malik NS, Moss SJ, Ahmed N, Furth AJ, Wall RS, Meek KM, Ageing of the human corneal stroma: structural and biochemical changes, *Biochimica et Biophysica Acta.* 1992;1138:222-228.

[95] Wollensak G, Spoerl E, Seiler T. Stress-strain measurements of human and porcine corneas after riboflavin–ultraviolet-A-induced cross-linking. *Journal of Cataract and Refractive Surgery*. 2003;29:1780-1785.

[96] Anderson, K., Elsheikh, A. and Newson, T., Application of Structural Analysis to the Mechanical Behaviour of the Cornea, *Journal of Royal Society – Interface*, 1, 2004, 1-13

[97] Greene, P.R. "Stress-strain behavior for curved exponential strips." *Bulletin of Mathematical Biology*, Vol. 47,No. 6, pp. 757-764, 1985.

[98] Elsheikh A, Anderson K. Comparative study of corneal strip extensometry and inflation tests. *Journal of Royal Society – Interface*. 2005;2:177–185.

[99] Naor J, Slomovic AR, Chipman M, Rootman DS. A randomized, double-masked clinical trials of Optisol-GS vs Chen medium for human corneal storage. *Arch. Ophthalmology*. 2002;120:1280-1285.

[100] Arndt C, Reese S, Kostlin R. Preservation of canine and feline corneoscleral tissue in Optisol GS. *Veterinary Ophthalmology*. 2001;4(3):175-182.

[101] Kaufman HE, Beuerman RW, Steinemann TL, Thompson HW, Varnell ED. Optisol corneal storage medium. *Arch. Ophthalmol*. 1991;109(6):864-868.

[102] Wagoner MD, Gonnah E. Corneal graft survival after prolonged storage in Optisol-GS. *Cornea*. 2005;24(8): 976-979.

[103] Orssengo GJ, Pye DC. Determination of the true intraocular pressure and modulus of elasticity of the human cornea in vivo. *Bull Math. Biol*. 1999; 61:551-572.

[104] Elsheikh, A. and Alhasso, D., Mechanical Anisotropy of Porcine Cornea and Correlation with Stromal Microstructure, Accepted for publication in Experimental Eye Research.

[105] Bryant MR, Szerenyi K, Schmotzer H, McDonnell P. Corneal tensile strength in fully healed radial keratotomy wounds. *IOVS*. 1994;35(7):3022-3031.

[106] Elsheikh A, Wang D. Numerical modelling of corneal biomechanical behaviour. *Computer Methods in Biomechanics and Biomedical Engineering*. 2007;10(2):85-95.

[107] Fernández DC, Niazy AM, Kurtz RM, Djotyan GP, Juhasz T. A finite element model for ultrafast laser-lamellar keratoplasty. *Annals of Biomedical Engineering*. 2006;34(1):169-183.

[108] Radner W, Zehetmayer M, Skorpik C, Mallinger R. Altered organization of collagen in the apex of keratoconus corneas. *Ophthalmic Research*. 1998;30(5):327-332.

[109] Meek KM, Tuft SJ, Huang Y, Gill PS, Hayes S, Newton RH, Bron AJ. Changes in collagen orientation and distribution in keratoconus corneas. *IOVS*. 2005;46:1948-1956.

[110] Luce DA. Determining in vivo biomechanical properties of the cornea with an ocular response analyzer. *Journal of Cataract Refractive Surgery*. 2005;31:156-162.

[111] Shah S, Laiquzzaman M, Bhojwani R, Mantry S, Cunliffe I. Assessment of the biomechanical properties of the cornea with the Ocular Response Analyzer in normal and keratoconic eyes. *IOVS*. 2007;48:3026-3031.

[112] Zienkiewicz OC, Taylor RL. The finite element method. McGraw-Hill Book Company Europe, Berkshire. 2000.

[113] Hamri O, Léon J-C, Giannini F, Falcidieno B. Exploiting features for finite element model generation. *Mathematics in Industry*. 2006;8:585-589.

[114] Armstrong CG. Modelling requirements for finite-element analysis. *Computer-aided design*. 1994;26(7):573-578.

[115] Babuška I, Strouboulis T. The finite element method and its reliability. Oxford Science Publications. 2001.

[116] McPhee TJ, Bourne WM, Brubaker RF. Location of the stress-bearing layers of the cornea. *IOVS*. 1985;26:869-872.

[117] Kotecha A, Elsheikh A, Roberts C, Zhu H, Garway-Heath DF. Corneal thickness- and age-related biomechanical properties of the cornea measured with the Ocular Response Analyzer. *IOVS*. 2006;47(12):5337-5347.

[118] Elsheikh, A., Alhasso, D., Kotecha, A. and Garway-Heath, D.F., Assessment of the Ocular Response Analyzer as a tool for intraocular pressure measurement, *Journal of Biomechanical Engineering*. In Press.

[119] Elsheikh, A., Wang, D., Kotecha, A., Brown, M. and Garway-Heath, D., Evaluation of Goldmann applanation tonometry using a nonlinear finite element ocular model, *Annals of Biomedical Engineering*, 34(10), 2006, 1628-1640.

[120] Pandolfi A, Manganiello F. A model for the human cornea: constitutive formulation and numerical analysis. *Biomechanics and modeling in mechanobiology*. 2006;.5(4):237-246.

[121] Pinsky PM, van der Heide D, Chernyak D. Computational modeling of mechanical anisotropy in the cornea and sclera. *Journal of Cataract and Refractive Surgery*. 2005;31:136-145.

[122] Curtin BJ, Physiopathologic aspects of scleral stress-strain. *Trans. Am. Ophthalmol. Soc*. 1969;67:417-461.

[123] Philips JR, McBrien NA. Form deprivation myopia: elastic properties of sclera. *Ophthal. Physiol. Opt*. 1995;15(5):357-362.

[124] Downs JC, Suh JKF, Thomas KA, Bellezza AJ, Hart RT, Burgoyne CF. Viscoelastic material properties of the peripapillary sclera in normal and early-glaucoma monkey eyes. *IOVS*. 2005;46:540-546.

[125] Downs JC, Suh JKF, Thomas KA, Bellezza AJ, Burgoyne CF, Hart RT. Viscoelastic characterization of peripapillary sclera: material properties by quadrant in rabbit and monkey eyes. *Journal of Biomechanical Engineering*. 2003;125:124-131.

[126] Barraglioli JL, Kamm RD. Measurements of the compressive properties of sclerol tissue. *IOVS*. 1984;25:59-65.

[127] Spoerl E, Boehm AG, Pillunat LE. The influence of various substances on the biomechanical behavior of lamina cribrosa and peripapillary sclera. *IOVS*. 2005;46:1286-1290.

[128] Norton TT, Rada JA. Reduced extracellular matrix in mammalian sclera with induced myopia. *Vision Research*. 1995;35(9):1271-1281.

[129] Guggenheim JA, McBrien NA. Form-deprivation myopia induces activation of scleral matrix metalloproteinase-2 in tree shrew. *IOVS*. 1996;37:1380-1395.

[130] McBrien NA, Gentle A. Role of the sclera in the development and pathological complications of myopia. *Progress in Retinal and Eye Research*. 2003;22(3):307-338.

In: Biomechanics: Principles, Trends and Applications
Editor: Jerrod H. Levy, pp. 111-136
ISBN: 978-1-60741-394-3

Chapter 3

BIOMECHANICS CONCEPTS OF BONE-ORAL IMPLANT INTERFACE

Ahmed Ballo[1,2] and Niko Moritz[3]

[1]Department of Biomaterials, Institute of Clinical Sciences,
The Sahlgrenska Academy, Gothenburg University, Gothenburg, Sweden
[2]Department of Prosthetic Dentistry and Biomaterials Science,
Institute of Dentistry, University of Turku, Turku, Finland
[3]Orthopedic Research Unit, Department of Orthopedic Surgery and Traumatology,
University of Turku, Turku, Finland

ABSTRACT

Osseointegrated implants are actually replacements for natural teeth, and, like natural teeth, they are exposed to various forces. The success of osseointegration is based on the clinical outcome; clinicians must ensure that the stresses that the superstructure, implant, and surrounding bone are subjected to are within the tolerable limits of the various components. Structural compatibility is the optimum adaptation to the mechanical behavior of the host tissues. Therefore, structure compatibility refers to the mechanical properties of the implant material, such as elastic modulus, strength, implant design and optimal load transmission (minimum interfacial strain mismatch) at the implant/tissue interface, which is the key to the successful functioning of the implant device.

This chapter reviews some of the reaction, properties and characteristics of the bone and explains how the bone-implant interface will react under loading condition. The chapter also includes characteristics, properties and other important information about the implant biomaterials and implant coating.

1. INTRODUCTION

Biomechanics is the application of engineering mechanics (statics, dynamics, strength of materials and stress analysis) to the solution of biological problems (Fung, 1981).

Biomechanics pertains to clinical dentistry because the teeth and jaw perform biomechanical activities during mastication.

An osseointegrated oral implant system consists of implants that connect a prosthesis to the jaw bone. In all incidences of functional loading, forces are first introduced to the dental prosthesis and then reach the bone-implant interface via the implant to distribute it over the bone. The primary function of bone-implant interface is to provide a safe and effective load transfer from the dental prosthesis to the bone. This process and the consequences of force transmission into supporting bone depends on the nature of applied force (amplitude, direction, rate and frequency), the design of implants (i.e., shape, surface macro- and micro-architecture), the biological and biomechanical properties of the bone-implant interface and the reaction of bone tissue to the mechanical environment created by loading of the implant.

Understanding the nature of the loading forces and the mechanism of such forces' transfer to the biological tissues have been suggested as factors in the development of implant to tissue interface and, indeed, implant longevity (Lemons and Bidez, 1991).

Bone mechanical properties are basic parameters that reflect the structure and function of bone. The mechanical behavior of bone in normal physiological situations is similar to that of an elastic material with no visible change in external appearance. Also, bone has the ability to repair itself by adaptive mechanisms. Although a dental implant is a dead object, for the transmission of mechanical chewing forces into the bone the implant should interact mechanically with biological tissue.

In order to do this successfully, there is a need to expand the knowledge related to the biomechanics of bone and mechanical parameters of biomaterials that determine load on oral implant units so the implant design will comply with the mechanical principles of the biological tissue.

2. Forces and Moments

Oral forces applied to human dentition are induced due to mastication and biting. Biting is understood as the act of breaking food into large pieces that will be further chewed. In this process, a tooth is subjected to what is called occlusal loads. In mastication, the forces are shorter in duration and less intense.

Force is defined as a "push or pull" that results from physical contact between two objects. Forces acting on both teeth and oral implants are referred to as vector quantities; this means that force quantities have direction as well as magnitude. Mastication mainly induces vertical load in the dentition. The magnitude of masticatory forces applied to teeth varies depending on the location in the jaws, with greater forces in the posterior regions. However, transverse forces are created by horizontal motion of the mandible and the inclination of teeth cusps.

The values of forces acting on teeth have been extensively addressed in the literature (Anderson, 1953; Cimini et al., 2000; Duyck et al., 2000; Gibbs et al., 1981; Howell and Brudevold, 1950; Neill et al., 1989; Waltimo and Kononen, 1993).

The magnitude of occlusal forces varies depends on tooth position in the mouth (Rangert et al., 1995). The mean maximal occlusal forces in an incisal area have been reported to range from 264 N to 370 N (Laurell, 1985; Paphangkorakit and Osbom, 1997), whereas the average maximum biting force in the posterior area can be around 800 N (Van Eijden, 1991).

During clinical loading of an oral implant, the direction of forces almost never coincides along its central long axis, providing two main types of loading: (1) axial force and (2) bending moment.

These two types of loading are completely different in nature. The axial force is more favorable, as it distributes stress more homogenously throughout the implant, while the bending moment exerts stress gradients in the implant as well as in the crestal bone around dental implants which may act as a fulcrum point for lever action when a force is applied (Smedberg et al., 1996; Richter, 1998), indicating that peri-implant tissue could be more susceptible to crestal bone loss by applying forces.

This bending moment is the force times the orthogonal distance between the force direction line and the counter-acting support (Figures 1 and 2). The longer the distance, the greater the bending moment will be (White et al., 1994).

Rangert and his colleagues exemplified the nature of force versus moment by using the "see-saw" model (Rangert et al., 1989). This model illustrates the beam balance of the cantilever prosthesis under the loading; the heavier load should be applied closer to the fulcrum than the lighter load applied on the opposite side.

The factor that determines this equilibrium of different forces is the ratio of the lever arms. The product of force × lever arm must be equal at both sides of the fulcrum.

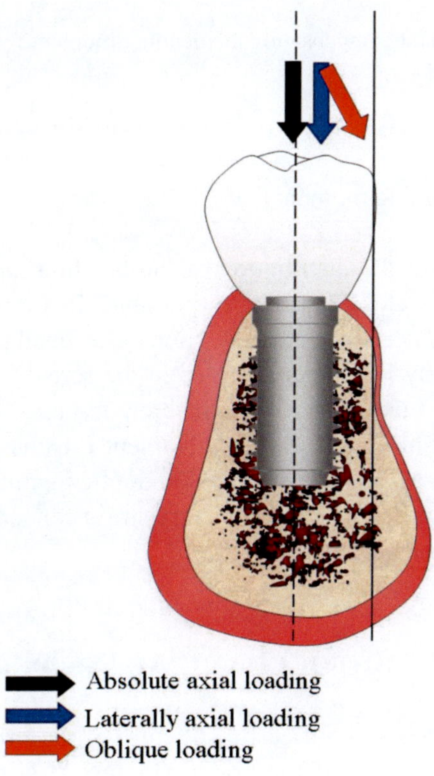

Figure 1. Different loading types from occlusion acting on implant prosthesis.

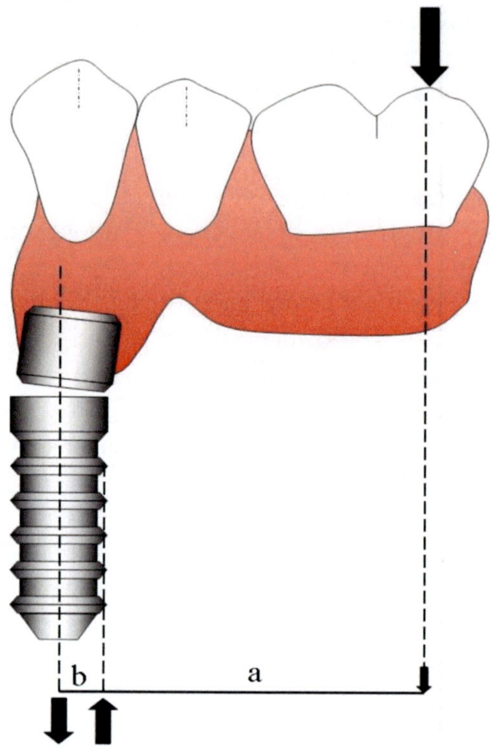

Figure 2. Forces within the implant compensating the bending moment from a vertical occlusal force (bending moment on the implant).

The product of force and the distance to the force axis is called the bending moment, thus

Force × Lever arm = Bending moment

The optimum situation for the distribution of the bending moment along the implant is placement of the implants as close together as possible. In Figure 2, the distal end of the prosthesis thus acts as the fulcrum in this situation. The bending moment on the implant therefore can induce extremely high internal force in the screw of the abutment, which may easily overload the screw and might end up with screw fracture. Although the gold screw is designed to be the weakest link in the implant-abutment-prosthesis connection, its breakage may considerably complicate additional fixed prosthodontic treatment.

Also, the excessive bending moments may cause stress concentration and micro-fractures in alveolar bone and even implant fractures.

3. Bone Biomechanical Properties

The bone structure can be exemplified by structure of the bridges. The visible structure when looking to the distance view of the bridge (Figure 3) is the long beam, arches and pillars. These elements represent the large scale level, while in close view the elements have a hierarchy like the structure that reinforces the bridges.

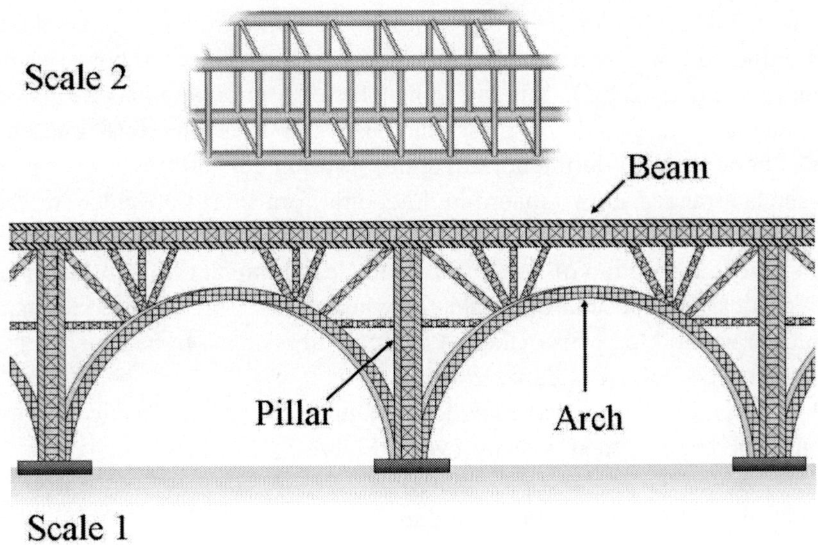

Figure 3. Bridge design applying the principle of structural hierarchy for bridge reinforcement.

Bone tissue is a highly specialized form of connective tissue of mesenchymal origin with a hierarchical structure (Kaplan et al., 1994 and Rubin et al., 2004). The structure is different at different scales (Figure 4). Structure and microarchitecture are determinant aspects of bone strength essential elements for the assessment of bone mechanical properties.

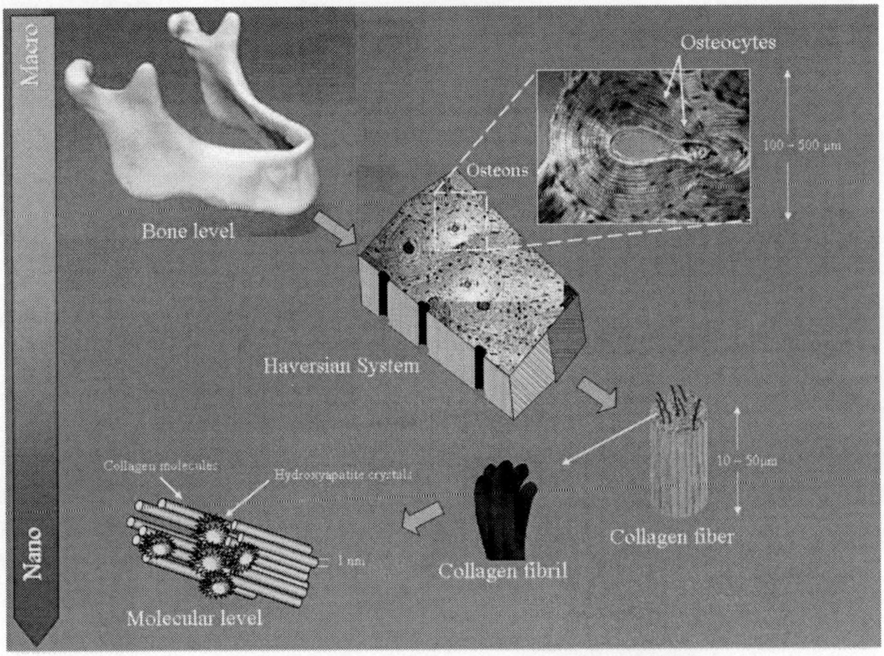

Figure 4. The hierarchical structure of cortical bone.

Bone is not homogenous; its physical properties are highly variable according to species, age, sex, liquid content, type of bone (e.g., femoral, mandibular, cortical, cancellous) and even according to the location within the bone from which the sample is taken (Katz, 1995).

Microarchitecture seems to be a determinant of bone fragility independent of bone density, and important in order to understand the mechanisms of bone fragility (Dalle Carbonare and Giannini, 2004). Among other factors, microarchitecture is affected by loading. Loading also affects bone mass and size and it is therefore important for the maintenance of bone strength during normal aging (Mosekilde, 2000).

Bone tissue is arranged in two macro-architectural forms: (I) trabecular (or cancellous or spongy) and (II) cortical (or compact). These differences in macroarchitecture have been used to derive a clinical classification of bone type in the dental implant field, based on the relative proportion of cortical to trabecular bone (that is, where Class 1 bone is predominantly cortical as in the anterior mandible, while Class 4 bone is almost all trabecular as found in the posterior maxilla) (Leckholm and Zarb, 1985).

Cortical bone is an *anisotropic* material, meaning that its mechanical properties vary according to the direction of load (Currey 1984) (Figure 5). Cortical bone is often considered an *orthotropic* material. Orthotropic materials are a class of anisotropic materials characterized by three different Young's moduli E_1, E_2, E_3 according to the direction of load, three shear moduli G_{12}, G_{13}, G_{23} and six Poisson's ratios ν_{12}, ν_{13}, ν_{23}, ν_{21}, ν_{31}, ν_{32}.

Bone can be either woven (coarse-fiber bone) or lamellar (layered). Woven bone is weak, with a small number of randomly oriented collagen fibers, but forms quickly and without a pre-existing structure during periods of repair or growth. Lamellar bone is stronger, formed of numerous stacked layers and filled with many collagen fibers parallel to other fibers in the same layer.

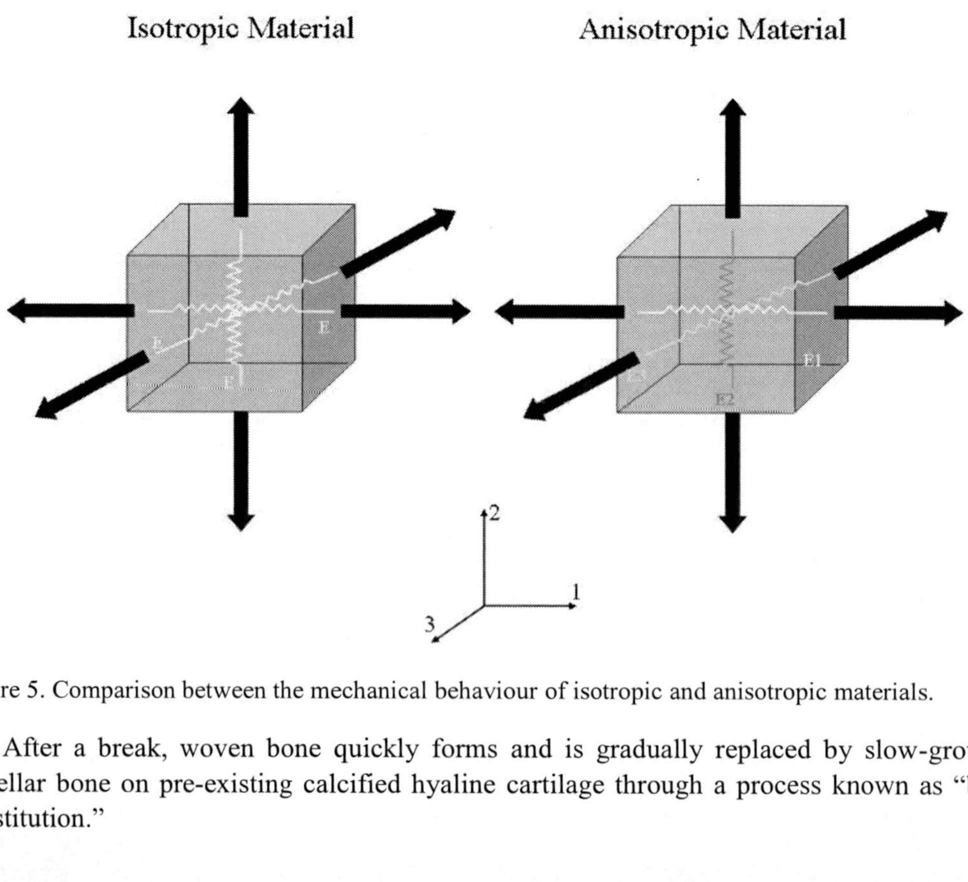

Figure 5. Comparison between the mechanical behaviour of isotropic and anisotropic materials.

After a break, woven bone quickly forms and is gradually replaced by slow-growing lamellar bone on pre-existing calcified hyaline cartilage through a process known as "bony substitution."

It has been observed that there is a relationship between bone structures and applied loads. In 1892, Wolff found that the orientation of trabeculae coincides with the direction of the stress trajectories. He proposed that bone loading is somehow sensed and that the bone adapts its structure accordingly. This principle of functional adaptation is generally known as "Wolff's Law" (Wolff, 1892). It is occurs in conditions of disuse when bone is lost (Bauman et al., 1999; Vico et al., 2000; Zerwekh et al., 1998), and in overloading, which causes a gain in bone mass (Suominen, 1993). It also occurs after fracture healing (Wolff, 1892) and in relation with implant incorporation (Guldberg et al., 1997).

The bone acts as if it has some sensors that can measure the internal load and activate the bone cells to carry out the bone remodeling. They are based on the separate actions of bone resorbing cells called osteoclasts and bone forming cells called osteoblasts. Bone remodeling (Figure 6), appears to be governed by a feedback system in which the bone cells sense the state of strain in the bone matrix around them and either add or remove bone as needed to maintain the strain within normal limits. Osteocytes may play an important role here.

Several studies revealed that these cells respond to mechanical stimulation (Skerry et al., 1989; Klein-Nulend et al., 1995). Together with the lining cells they form a system that seems well equipped for signal transduction (Cowin et al., 1991). It could be that mechanically induced osteocyte signals are transferred through the canaliculi to the bone surface where they control osteoclast and osteoblast activity (Burger and Klein-Nulend, 1999).

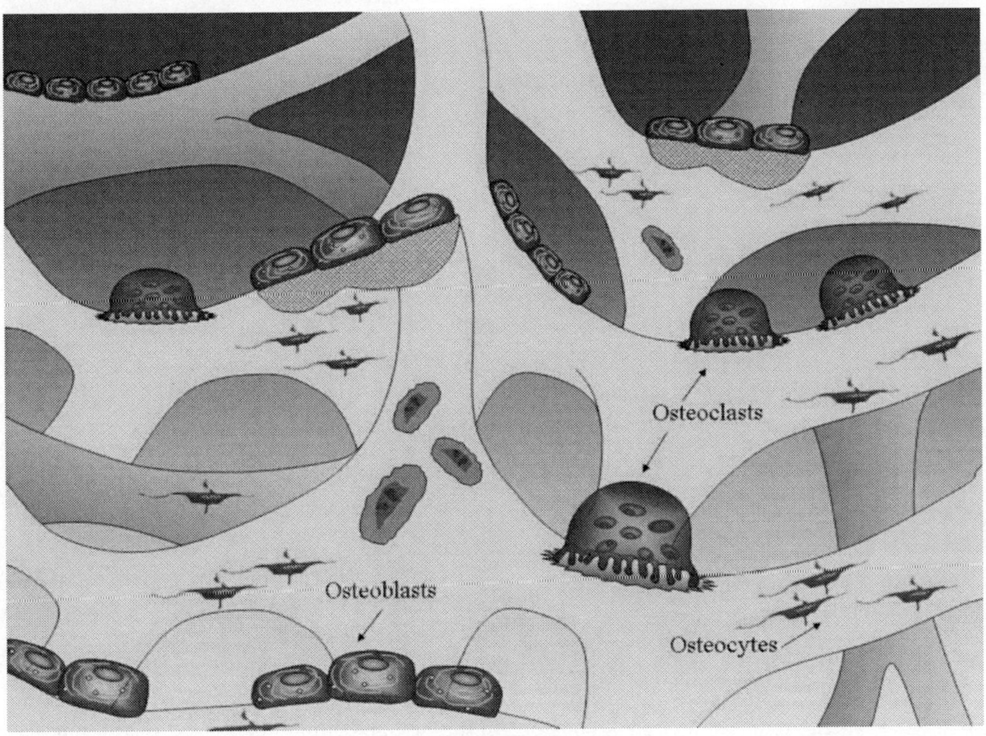

Figure 6. Schematic diagram illustration bone remodeling process.

Bone fractures can be produced by either a single load or the repeated application of load. A fracture will result from the application of a single load if that load produces a stress or strain that exceeds the ultimate strength of the bone. The repeated application of a lower load may also produce a fracture, such a fracture is called a fatigue fracture and is typically produced by either lower repetition of high loads or high repetition of relatively low loads.

4. Principles of Forces Transfer to Bone-Implant Interface

The forces generated during mastication influence the teeth, as well as the periodontium, the jaw bones and the temporomandibular joints. Unlike nature teeth, osseointegrated implants are ankylosed to surrounding bone without the periodontal ligament (Figure 7), which provides mechanoreceptors as well as shock-absorbing function (Schulte, 1995).

Successful implant osseointegration and its clinical longevity depend upon the way mechanical stresses are transferred to the surrounding bone. This force transfer from the implant to the surrounding bone is influenced by the type of loading that occurs (i.e., intermittent, continuous), the bone-implant interface (i.e., direct contact or a gap interface), the length and diameter of the implant, the implant shape, the surface texture of the implant and the quality and quantity of the surrounding bone (Bechtold et al., 2001; Jones et al., 2001; Mouzin et al., 2001; Skripitz and Aspenberg, 2001; Toksvig-Larsen et al., 2001).

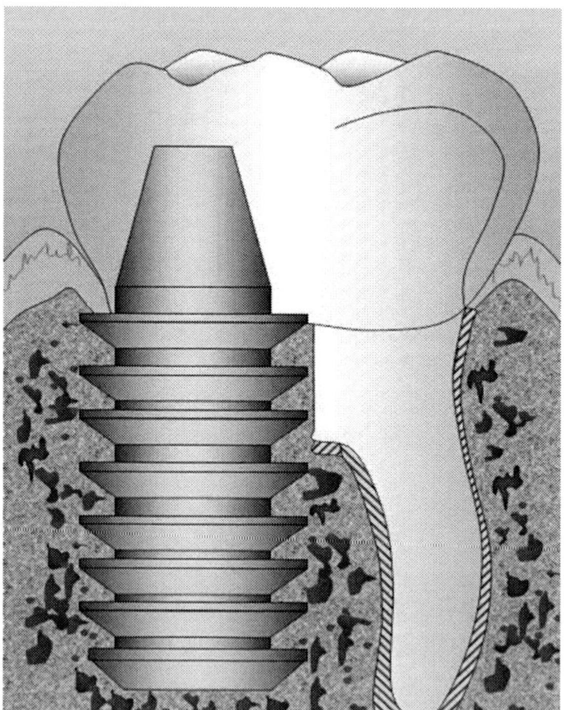

Figure 7. Schematic diagram of an endosseous dental implant and a tooth. Note that the implant lacks the periodontal ligament and is dependent on direct bone support.

The distributions of forces and moments throughout the structure depend on the relative flexural rigidities of the components and on the rigidity of the connections. Some assumptions about rigidity must be made in order to simplify the system but still retain the features necessary to give clinically useful results.

The routine loading protocol established by Branemark and his colleagues was empirically based and used three months for the mandible and six months for the maxilla (Branemark et al., 1985). The primary aim of such an approach is to create a healing environment at the bone–implant interface that is capable of nurturing regeneration and osseointegration rather than repair and fibrous encapsulation. However, clinical and experimental studies demonstrate that osseointegration can be achieved when healing under load is allowed.

The bone tissue response to immediately loaded oral implants has been described by various investigators (Hashimoto et al., 1988; Lum et al. 1991; Akagawa et al., 1993; Piattelli et al., 1993; Sagara et al., 1993; Piattelli et al., 1997; Corso et al., 1999) and Szmukler-Moncler and his colleagues have given a literature review concerning the timing of loading and effect of micromotion on bone-implant interface (Szmukler-Moncler et al., 1998).

Clinical (Quirynen et al., 1992; Naert et al., 1992; Baron et al., 2005) as well as experimental studies (Miyata et al., 2000, 2002) indicate that marginal bone loss at the oral implant may be associated with high occlusal stress on implants. In poor bone conditions, the mismatch of stiffness between bone and metallic implant may lead to implant failure (Lemons, 1998).

This occurs when the tensile or compressive load exceeds the physiological limit of bone tolerance and causes microfracture at the bone-to-implant interface, or initiates bone resorption (Brunski, 1999).

While overloading may be manifested by the application of repeated single loads, which causes micro-fractures within bone tissue, continuous application of low loads may also lead to failure, namely, fatigue fracture. However, stress-shielding (underload) at the interface have also been suggested as a contributing factor for bone resorption. Therefore oral implant should be designed so that it induces a mechanical stimulation in the surrounding bone and high stress peaks do not arise in the bone.

On small scale this micromotion at the implant-bone interface can have two principle effects on cellular and extracellular components of bone. First, the micromotion can lead to a disruption of the bone-cell/implant contacts and therefore has the ability to disturb the cell reaction by a detachment; or second the micromotion can lead to a deformation of osteoblast cells attached to the surface in a strain-related manner.

The bone tissue formation and mineralization by osteoblast cells are dependent on the local mechanical interaction at the cell-implant surface zone is generally accepted (Szmukler-Moncler et al., 1998). Cells generate, transmit and sense mechanical tension, they use these forces to control their shape and behavior. Cells generate tensile forces within contractile microfilament in cytoskeleton by use of an actomyosin filament sliding mechanism (Chirucel et al. 1998).

The cells extend lamellopodia (Figure 8) that adhere to the implant surface formed the sites act as mechanosensory "devices," where internal contractile forces or externally applied force can regulate the assembly of the adhesion site (Geiger and Bershadsky, 2001).

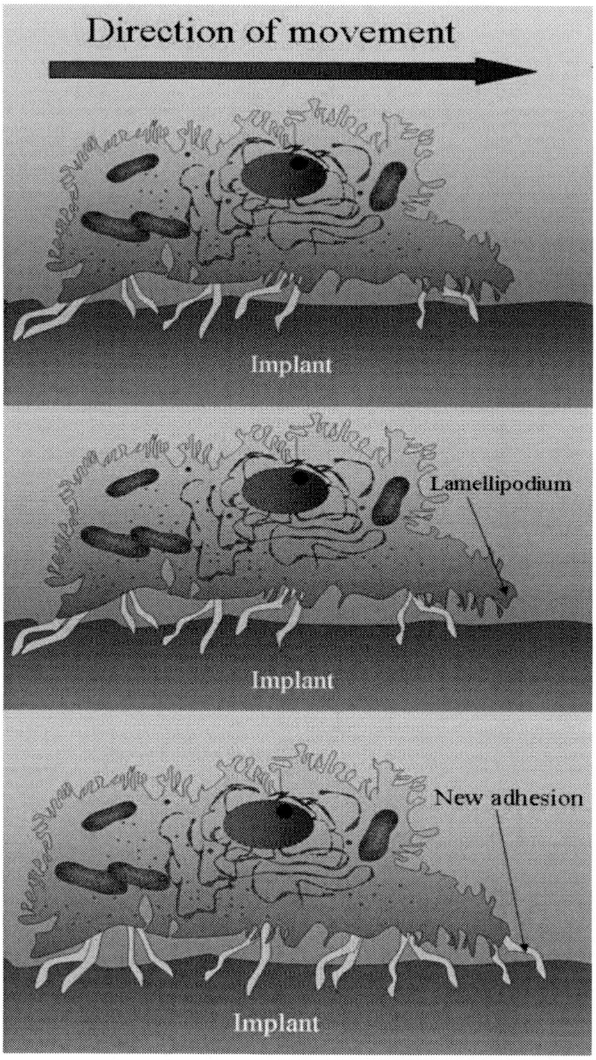

Figure 8. Cell migration. The cells extend lamellopodia that adhere to the implant surface.

The cytoskeleton will subject the adhesion site to a tensile force. The force developed depends on the resilience of the implant material, the stiffer the material, the greater the force. The mechanical strength of both the adhesion site and the part of cytoskeleton under tension are adapted to the force, a consequence of this adaptation of force and strength to the resistance met is that cells tend to migrate in the direction of increasing material stiffness (Lo et al., 2000, Dischler et al., 2005).

Carter and Giori suggested that proliferation and differention of the osteoblast cells responsible for peri-implant tissue formation are regulated by the local mechanical environment according to the tissue differentiation hypothesis proposed by Frost and his colleagues for callus formation (Carter and Giori, 1991).

Recently development in implant surfaces has shown improvement in the mechanical interaction with the bone. Figure 9 illustrates the different scale levels of bone-implant interaction all the way from the overall design down to the molecular level.

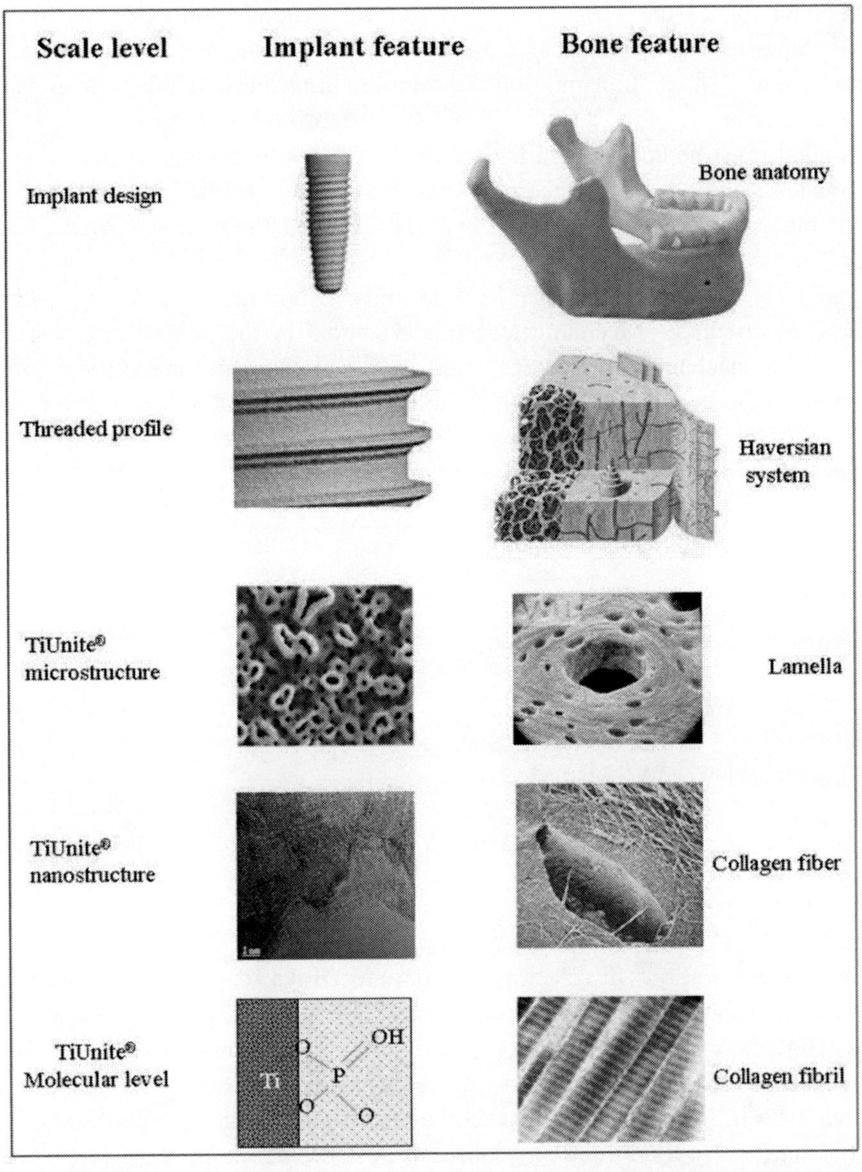

Figure 9. Mechanical interaction between implant and bone on five different scale levels (from the design level to the molecular level).

5. BULK MECHANICAL PROPERTIES OF IMPLANT BIOMATERIALS

The biocompatibility is a descriptive term which indicates the ability of the material to perform with an appropriate host response, in a specific application (Black and Hasting, 1998). This definition has been extended and distinguished between the surface and structural compatibility of an implant (Wintermantel andMayer, 1995). Surface compatibility meaning the chemical, biological, and physical (including surface morphology) suitability of an implant surface to the host tissues. Structural compatibility is the optimum adaptation to the mechanical behavior of the host tissues. Therefore, structure compatibility refers to the mechanical properties of the implant material, such as elastic modulus, strength, implant design and optimal load transmission (minimum interfacial strain mismatch) at the implant/tissue interface (Hedia, 2002).

Any implant must be constructed from a biomaterial. The biological performance of the chosen biomaterial will be of concern. Biomaterials are generally composed of a wide spectrum of materials. These implant biomaterials can be broadly classified into metallic, ceramic, polymeric, and composite materials.

An improperly chosen material can lead not only to failure of the part or a structure but also to unnecessary cost. Mechanical characterization is the classification of materials according to their mechanical properties. Table 1 illustrates some mechanical properties of these materials (O'Brien and William, 2002; Bouillaguet et al., 2006).

Table 1. Comparison of mechanical properties of different implant materials

Property /Materials	Titanium	Alumina	Zirconia	Hydroxyapatite	FRC*
Strength (MPa)	800–1000	400–600	900–1200	115–150	700–1000
Young's modulus (GPa)	110–120	380	210	85	20–40
Hardness (HV)	100–140	2200	1200–1500	300–400	70–80

* Unidirectional fiber-reinforced.

The main considerations in selecting metals and alloys for biomedical applications are biocompatibility, appropriate mechanical properties, and corrosion resistance. The high tensile and fatigue strength of metals, compared with ceramics and polymers, make them the materials of choice for implants that carry mechanical loads. However, implants are made of metals which are stiffer than bone which they contact. This leads to large percent of the load normally transmitted by the bone being borne by the prosthesis instead. This shielding can lead to bone disuse atrophy (Chrisman and Snook, 1968). Using an implant material which is biomechanically more suitable, which it is strong enough but also flexible so that it can reduce the interfacial stresses that causes fatigue fracture or resorption of the bone.

The flexibility or stiffness of a material may be described by the so-called modulus of elasticity (Young's modulus). The elastic modulus is the constant that relates the stress and the strain in the linear elastic region where elastic deformation of a material occurs (Figure 10) (Van Noort, 2002).

Commercial-purity titanium (cp Ti) comes in a number of grades, with the mechanical properties being somewhat different for each grade. Small changes in the oxygen and iron contents greatly influence the mechanical properties, for example, Grade 3 cp Ti has Fe and O contents (maximum) of 0.3 and 0.35 %, respectively, and a yield strength of 380 MPa, while Grade 4 has Fe and O contents 0.5 and 0.4%, respectively, and a yield strength of 483 MPa.

The Ti-6%Al-4%V alloy, however, has a much higher value for yield strength (795 MPa), so this alloy is often chosen for dental implant cases where high strength is necessary (Mishra et al., 1996). Compared with the elastic moduli of either stainless steel or cobalt-chromium, Ti and its alloys have much lower moduli that are still almost an order of magnitude higher than that of bone.

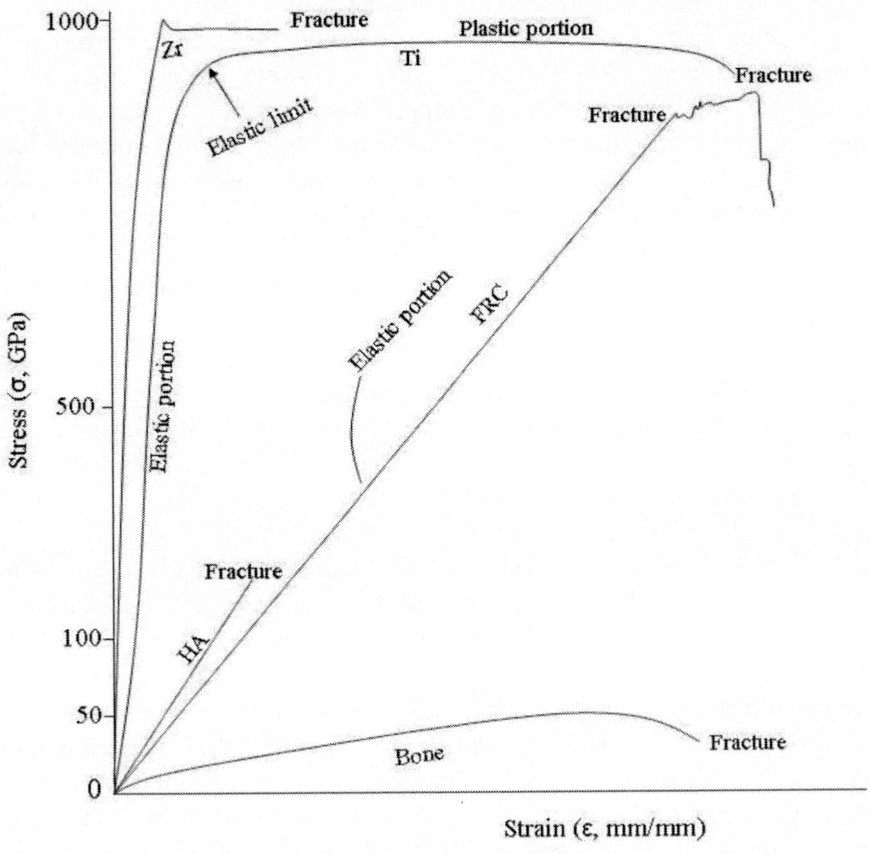

Figure 10. Average stress-strain curve for different materials.

Ceramic are materials composed of metallic and nonmetallic elements held together by ionic and/or covalent bonds. Ceramic material has already been utilized as dental implant material like aluminum oxide (Al_2O_3) (Schulte, 1984). This material osseointegrated well but unfortunately, the biomechanical properties of the implants were not sufficient for long-term load. As a result, this material was withdrawn from the market as dental implant material. Recently, Zirconia has been introduced as a new ceramic dental implant material. As a metal substitute, Zirconia possesses good chemical and physical properties, like low corrosion potential, low thermal conductivity (Drouin and Cales, 1994; Piconi et al., 1998; Richter et al., 1994). Furthermore, its biocompatibility and biomechanical properties as dental implant material has been extensively investigated (Albrektsson et al., 1985; Ichikawa et al., 1992; Kohal et al., 2006). However, these implant materials have very high elastic moduli compared to that of human bone (17–24 GPa).

Hydroxyapatite (HA) Ceramic have been investigated extensively and used for dental implant applications for the past 30 years. HA Properties depend on its porosity. Clinical studies demonstrated that HA ceramics still remain the most biocompatible bone implant material known and possess the added feature of becoming strongly bonded to living bone through natural-appearing bonding mechanisms (Chang et al., 1996), yet the range of problems included fractures during surgery, fractures after loading (Jarcho, 1992). Therefore, HA ceramics are not ideal materials for permanent implant devices. However, bioactive ceramic coatings on metal implants have kept ceramics as a key component in dental implantology (Krauser et al., 1990; Ogiso et al., 1996). However, delamination of the ceramic layer from the metal surface can create serious problems and lead to implant failure.

Polymers are widely used materials in biomedical applications. Polymers are used in drug delivery systems and as a scaffolding material for tissue engineering applications. The polymeric materials have molecular structures completely different from living substances which make them generally more stable in the tissues.

Compared with metal and ceramics, polymers have much lower strength and moduli but they can be deformed to a greater extent before failure. Consequently, polymers are generally not used in biomedical applications that bear loads. Ultra-high-molecular-weight polyethylene is an exception, as it is used as a bearing surface in hip and knee replacements.

In high-load trauma applications such as bone fracture plates, screws and intramedullary nails, the strength of Polyetheretherketone (PEEK) polymer allows it to be a true replacement of metals (Kurtz and Devine, 2007). But polymeric are not commonly seen for dental implants yet.

In 1969, Hodosh placed custom-made polymer implant directly into the extraction socket of tooth for the first time (Brown et al., 1969; Ehrlich and Azaz, 1975; Hodosh et al., 1969; Hodosh et al. 1973). Since then most studies have been done with experimental animals but some work has been published with selected human patients as well (Hodosh et al., 1969). It has been reported that attachment of polymer implant can be achieved with connective tissue capsule resembling the periodontal ligament (Brown et al., 1969). However, due to the high failure rate of 60% after the seven years follow up of 10 patients, this acrylic resin implant was not recommended for clinical use (Hodosh et al., 1970). At the same time, Brånemark introduced good results on the osseointegrated titanium implants (Brånemark et al., 1977).

Recently, fiber-reinforced composite (FRC) has been evaluated as a new material for oral and maxillofacial implants (Ballo et al., 2007; Ballo et al., 2008a and 2008b; Ballo et al. 2009). The pre-clinical studies concerning the mechanical properties of biostable glass FRC implants have shown good performance in the laboratory environment (Ballo et al., 2007 and 2008a). Both in vitro and in vivo studies have shown that FRC material is biocompatible in the cell culture environment and preosteoblasts have been shown to mature and proliferate well on the FRC substrates (Ballo et al., 2008b). The rigid bone-implant interface has been observed in a histological investigation without fibrous connective tissue between the bone and the FRC implant (Ballo et al., 2009). These studies used well-polymerized FRC material in combination with bioactive glass.

Although these are promising results with FRC in pre-clinical studies, many questions remain to be answered before FRC implants can be introduced into clinical use. In the future, further studies are needed to evaluate the bone remodeling process of the FRC implants under loading condition.

8. Implant Surface Coating

Various surface coating s have been extensively evaluated in implant dentistry. Implant coatings have strong impact factor on the strength of bone-implant interface. Chemical bonding is descripting when bioactive ceramics chemically bond to the surrounding bone (Nakamura et al., 1985).

Bioactive ceramic coatings on metal implants have kept ceramics as a key component in dental implantology (Krauser et al., 1990; Ogiso et al., 1996).

Until the introduction of the concept of bioactive materials, surface reactivity was considered an unwanted property for an implant material. However, some surface reactive (biologically active, bioactive) materials have shown the ability to form an interfacial chemical bond with surrounding tissues through a series of biophysical and biochemical reactions, causing "bioactive fixation" of the implant (Cao and Hench 1996). Bioactive materials can be biostable (i.e., synthetic hydroxylapatite) or bioresorbable (i.e., bioactive glasses and glass-ceramics). Some bioactive ceramics like bioactive glasses of certain compositions have been claimed to have a real chemical bonding ability with soft tissues (Wilson, 1981 and 1990). There are indications that titanium oxides are also capable of forming an adherent bond with soft tissues (Areva, 2004; Rossi, 2007; Paldan, 2008).

However, bioactive ceramics in the bulk form are not suitable for load-bearing applications as their flexural strength, strain-to-failure and fracture toughness are less than that of bone and their elastic moduli are greater than that of bone. For these reasons they are usually applied as coatings on metallic implants that possess superior mechanical properties.

Although in the past 30–40 years many studies have been carried out in the field of biomaterials to achieve better tissue attachment, today the majority of implants are still made of biologically inactive materials. This is mainly due to the fact that either these materials have a history of reliable performance or simply because nothing better is available. For example, titanium and its alloys are widely used in clinical practice. Titanium is a highly reactive metal and except for certain artificial conditions is covered by a thin (2–10 nm) surface oxide layer. It is the chemical stability of this oxide layer (mainly TiO_2) that provides the superior corrosion resistance of the implants made of titanium and its alloys. Although Young's modulus of titanium alloys (~100 GPa) differs considerably from that of bone (7–25 GPa for cortical bone, 0.05 GPa–0.5 GPa for cancellous bone) (Hench 1994), these materials are well integrated into bone through morphological or biological fixation and have sufficient strength for load-bearing applications. However absence of a physico-chemical bond between the implant and the host bone especially in the poor bone quality conditions poses a risk of implant failure. Further, in the case of percutaneous or permucosal implants, such as dental implants, lack of bonding to connective tissue and epithelium may help the pathogenic microorganisms to migrate along the surface of the implant leading to infection. Bioactive fixation between an implant and soft tissues could also prevent this (Jansen, 1992).

8.1. Calcium-Phosphate Coatings

To improve attachment between the surface of a medical implant and the surrounding tissues, surface-reactive (bioactive) calcium-phosphate-based coatings can be applied to the surface of an implant by different methods (Leeuwenburgh, 2008; Wolke, 1992; Kangasniemi, 1994; Lacefield, 1988 and 1994,). Calcium-phosphates include a large group of structurally related apatites (tricalcium phosphates, hydroxyapatite, fluorapatite, etc). At present, coating of a medical implant surface with calcium-phospahte by plasma spraying is the only method approved by the FDA (Food and Drug Administration) (Campbell, 2003).

8.2. Bioactive Glass Coatings

Originally introduced by Hench (Hench, 1971), silica-based bioactive glass materials are slowly resorbing synthetic osteoconductive materials which are able to form strong chemical bond with bone. Several methods have been applied to obtain a bioactive glass coating on the load bearing part of the implants (Hench, 1993). Numerous attempts have been made to create bioactive glass coatings on alumina (Greenspan and Hench, 1976; Griss et al., 1976), stainless steel (Piotrowski et al., 1975; Schepers et al., 1989), Co-Cr-Mo alloy (Ducheyne et al., 1984; Lacefield and Hench, 1986; Pajamäki et al., 1995) and titanium alloys (West et al., 1990; Kitsugi et al., 1996). Application of a double glass coating has been suggested to solve the problem of differences in thermal expansion coefficients.

Recently, several attempts were made to prepare bioactive glass coatings on titanium and Ti6Al4V (Bloyer et al., 1999; Saiz et al., 2002). Application of a ground layer prepared from inert glass with a thermal expansion coefficient close to that of Ti6Al4V provided good adhesion to the substrate (Oku, 2001). This ground coating can be used in combination with more surface reactive glass coatings (Gomez-Vega, 2000a), with embedded hydroxyapatite and/or bioactive glass particles (Gomez-Vega, 2000b), or a sol-gel-derived silica coating (Gomez-Vega, 2001). Reactive plasma spraying (Schrooten, 2000) or processing with infra-red laser (Moritz, 2004) have also been attempted to create bioactive glass coatings on titanium and its alloys.

9. Biomechanical Considerations for Coated Implant Design

With respect to orthopaedic hip implants, a lot of knowledge has been accumulated leading to identification of a number of important implant failure scenarios (Huiskes, 1993). Although originally discussed with respect to orthopaedic implants, similar issues are valid for other types of load-bearing medical implants, including dental.

Under repeated dynamic loading there is an accumulation of mechanical damage in materials and materials' interfaces. When the interface stresses exceed the strengths of the materials, the disruption of the implant is initiated by the growth of cracks at the interface. This is followed by interface micromotion, bone resorption, fibrous encapsulation and ultimate failure of the implant (Huiskes, 1993). For uncoated implants, failed bonding is another important issue. If bone ingrowth or osseointegration does not occur, the resulting motion of the interface in excess of 150 μm will always lead to the formation of the fibrous membrane (Huiskes, 1993). The "stress-shielding" phenomenon is often addressed in the literature. After the implantation, it is the bone and the implant together that are engaged in sharing of the load that is normally carried by bone alone. As the implant takes a fraction of the load away from the bone, subnormal loading of the bone according to Wolff's law will lead to its resorption in the form of osteopenia and cortical thinning. The mismatch of the mechanical properties often leads to physiological underloading of the bone, thus increasing the likelihood of adverse bone remodelling. However, it should be noted that, regardless of the mechanical properties of the implant, just the addition of an implant would always cause some degree of stress-shielding. Thus, it is impossible to completely eliminate stress-shielding. Also, stress bypass and destructive wear are examples of other possible failure scenarios (Huiskes, 1993). Moreover, all of these failure scenarios are interlinked, thus there is a need for a systematic approach to implant design so that the entire scope of conflicting requirements are considered. For example, using less-stiff implant materials will obviously decrease the effect of stress-shielding but respectively increase the interfacial stresses, thus promoting accumulated damage and failed ingrowth (Huiskes, 1993).

Regardless of all the possible benefits of bioactive fixation, addition of a bioactive coating will be practical only if the coating is adequately strong to withstand the stresses the implant is subjected to. This applies to implant-coating interface, coating-tissue interface and the coating material itself. Moreover, in real clinical applications, the integrity of the interfaces should be maintained over a long period of time.

Strong bonds between bioactive materials and periimplant bone is a well-known phenomenon reported by many authors (Nakamura et al., 1985; Li et al., 1995; Andersson et al., 1992). In a number of studies by Soballe, performance of hydroxyapatite coatings was tested in different in vivo conditions, including gap-healing, osteopenic bone bed, bone grafting and dynamic loading (Soballe, 1993), demonstrating adequate performance of the coatings. In some studies (Cook, 1992a), selection of torsional loading as a biomechanical testing method is based on the assumption that the torsional loads may influence the longevity of dental implants. Moreover, as it has been demonstrated, in torsional loading the HA-coated cylindrical implants failed mainly at the coating-substrate interface (Cook, 1992a). It has been demonstrated that when load-bearing is not taken into consideration, the implantation of transcortical implants in a femur bone does accurately simulate biological responses to the implants placed in the mandible (Cook, 1992b). Thus, the results were expected to be more clinically relevant than if push-out or pull-out testing was used. The later methods would have also required conically- rather than cylindrically-shaped implants (Andersson, 1992), as misalignment of an implant during testing might easily occur and lead to incorrect results. Further, in the case of macro-textured cylindrical implants such as bioactive glass-coated implants, pull-out or push-out tests were suggested to be unacceptable methods (Dhert and Jansen, 2000).

A large number of testing methods exist to quantify the adhesion strength of the coating to the substrate (Rickerby, 1988; Berndt and Lin, 1993). A number of the testing methods (Manley et al., 1987; Ducheyne and Martens, 1988) are suggested by the FDA alongside standardized testing methods. According to FDA guidelines for testing of the plasma-sprayed coating (FDA 2000), the shear fatigue strength of surface coating should be tested out to at least 10^7 cycles. The static shear strength of the surface/substrate interface should be tested and the adhesion shear strength should exceed 20 Mpa. The static tensile strength of the surface/substrate interface and the coating should exceed 22 Mpa in tensile strength. While these tests are useful for validation of the adhesion quality of a coating, they are not representative of loading conditions of a real implant. Thus, to understand the fracture behavior of a coating under loading conditions, a combination of realistic mechanical testing, acoustic emission measurements and finite element analysis (FEA) were suggested (Schrooten, 1999 and 2000). This allows for qualitative and quantitative evaluation of coatings on a medical implant. The pattern of load transfer depends on several mechanical factors, such implant geometry, properties of the implant material, interfacial strength and loading conditions; thus, finite element analysis has been found to be a valuable tool in development of human joint implants (Huiskes, 1993; Prendergast, 1997). For dental implants, FEA has also been extensively used (van Staden et al., 2006). FEA allows for simulation of real biomechanical testing conditions to study stress distributions at the interfaces of interest. However, FEA of coated implants has received less attention (Mihalko et al., 1992; Evans et al., 1994; Evans and Gregson, 1994; Schrooten 1999 and 2000; Aoki et al., 2006). Introduction of a low modulus interlayer between the implant and the surrounding bone can totally change the pattern of load transfer. When this is done in a controlled way, the performance of the implant can even be improved; however, other aspects of implant design should also be taken into account (Evans and Gregson, 1994). Moreover, a recent FEA study has indicated that when applied to less stiff materials such as fiber-reinforced composites, a bioactive surface component will promote more stress-shielding than an uncoated implant (Zhao et al., 2009).

References

Akagawa, Y., Ichikawa, Y., Nikai, H. and Tsuru, H. (1993). Interface histology of early loaded partially stabilized zirconia endosseous implant in initial bone healing. *Journal of Prosthetic Dentistry*, 69, 599–604.

Albrektsson, T., Hansson, H.A. and Ivarsson, B. (1985). Interface analysis of titanium and zirconium bone implants. *Biomaterials,* 6, 97–101.

Anderson, D.J. (1953). A method of recording masticatory loads. *Journal of Dental Research*, 32, 785-789.

Andersson, Ö.H., Liu, G., Kangasniemi, K. and Juhanoja, J. (1992). Evaluation of the acceptance of glass in bone. *Journal of Materials Science Materials in Medicine*, 3, 145-150.

Aoki, H., Ozeki, K., Ohtani, Y., Fukui, Y. and Asaoka, T., (2006). Effect of a thin HA coating on the stress/strain distribution in bone around dental implants using three-dimensional finite element analysis. *Bio-Medical Materials and Engineering*, 16, 157-169.

Areva, S., Paldan, H., Peltola, T., Närhi, T., Jokinen, M. and Linden, M. (2004). Use of sol-gel derived titania coating for direct soft tissue attachment, *Journal of Biomedical Materials Research Part A*, 70, 169-178.

Areva, S., Peltola, T., Säilynoja, K., Laajalehto, K., Linden, M. and Rosenholm, J.B. (2002). Effect of albumin and fibrinogen on calcium phosphate formation on sol-gel derived titania coatings in vitro. *Chemistry of Materials*, 14, 1614-1621.

Ballo, A.M, Lassila, L.V., Vallittu, P.K. and Närhi, T.O. (2007). Load bearing capacity of bone anchored fiber-reinforced composite device. *Journal of Materials Science Materials* in Medicine, 18, 2025-2031.

Ballo, A.M., Lassila, L.V.J., Närhi, T.O. and Vallittu, P.K. (2008a). In vitro mechanical testing of glass fiber-reinforced composite as dental implants. *Journal of Contemporary Dental Practice,* 9, 41-48.

Ballo, A.M., Kokkari, A.K., Meretoja, V.V., Lassila, L.V., Vallittu, P.K. and Närhi, T.O. (2008b). Osteoblast proliferation and maturation on bioactive fiber-reinforced composite surface. *Journal of Materials Science Materials in Medicine*, 19, 3169-3177.

Ballo, A.M., Akca, E., Ozen, T., Lassila, L., Vallittu, P.K. and Närhi, T.O. (2009). Bone tissue responses to glass fiber-reinforced composite implants—a histomorphometric study. *Clinical Oral Implants Research*, 20, 608-615 .

Baron, M., Haas, R., Baron, W. and Mailath-Pokorny, G. (2005). Peri-implant bone loss as a function of tooth-implant distance. *International Journal of Prosthodontics*, 18, 427-433.

Bauman, W.A., Spungen, A.M., Wang, J., Pierson, R.N. Jr and Schwartz, E. (1999). Continuous loss of bone during chronic immobilization: a monozygotic twin study, *Osteoporosis International*, 10, 123-127.

Bechtold, J.E., Mouzin, O., Kidder, L. and Soballe, K. (2001). A controlled experimental model of revision implants: Part II. Implementation with loaded titanium implants and bone graft. *Acta Orthopaedica Scandinavica*, 72, 650-656.

Berndt, C.C. and Lin, C.K. (1993). Measurement of adhesion for thermally sprayed materials. *Journal of Adhesion Science and Technology*, 7, 1235-1264.

Black, J. and Hasting, G.W. (1998). Handbook of biomaterials properties. London, UK: Chapman and Hall.

Bloyer, D.R., Gomez-Vega, J.M., Saiz, E., Mcnaney, J.M., Cannon, R.M., Tomsia, A.P. (1999). Fabrication and characterization of a bioactive glass coating on titanium implant alloys, *Acta materialia*, 47, 4221-4224.

Bouillaguet, S., Schutt, A., Alander, P., Schwaller, P., Buerki, G., Michler, J., Cattani-Lorente, M., Vallittu, P.K. and Krejci, I. (2006). Hydrothermal and mechanical stresses degrade fiber-matrix interfacial bond strength in dental fiber-reinforced composites. *Journal of Biomedical Materials Research Part B Appl Biomaterials*, 76, 98-105.

Brånemark, P.I., Hansson, B.O., Adell, R., Breine, U., Lindström, J., Hallén, O., Ohman, A. (1977). Osseointegrated implants in the treatment of the edentulous jaw. Experience from a 10 year period. Scandinavian *Journal of Plastic and Reconstructive Surgery and Hand Surgery,* 16, 130-136.

Brånemark, P.I., Zarb, G.A. and Alberktsson, T. (1985). Tissue integrated prostheses: Osseointegration in Clinical Dentistry. Chicago: Quintessence Publishing Company.

Brown, B.H., Neff, P.A. and Tylenda, A. (1969). Acrylic resin tooth implants in rhesus monkeys: a preliminary report. *Journal of Prosthetic Dentistry*, 22, 367-370.

Brunski, J.B. (1999). In vivo bone response to biomechanical loading at the bone/dental-implant interface. *Advances in Dental Research*,13, 99-119.

Burger, E.H. and Klein-Nulend, J. (1999). Mechanosensory transduction in bone – role of the lacuno-canalicular network. *The Journal of the Federation of American Societies for Experimental Biology*, 13, 101-112.

Campbell, A.A. (2003). Bioceramics for implant coatings, *Materials Today*, 11, 26-30.

Cao, W. and Hench, L.L. (1996). *Bioactive materials, Ceramics International*, 22, 493-507.

Carter, D.R., Giori, N.J. (1991). Effect of mechanical stress on tissue differentiation in the bony implant bed. In: J.E. Davies (Ed.), The bone–biomaterial interface (pp. 367-376). Toronto: University of Toronto Press.

Chang, Y.S., Oka, M., Nakamura, T., Gu, H.O. (1996). Bone remodeling around implanted ceramics. *Journal of Biomedical Materials Research*, 30, 117-24.

Chicurel, M.E., Chen, C.S. and Ingber, D.E. (1998). Cellular control lies in the balance of forces. *Current Opinion in Cell Biology*, 10, 232-239.

Chrisman, O.D. and Snook, G.A. (1968). The problem of refracture of the tibia. *Clinical Orthopaedics and Related Research,* 60, 217-219.

Cimini, J.r, Gouvêa, P.H.P, Las Casas, E.B. and Cornacchia, T.P.M. (2000). Loads in teeth: a critical review. In: Proceedings of the METMBS'2000: The 2000 International Conference on Mathematics and Engineering Techniques in Medicine and Biological Sciences (2, pp. 649-655). Las Vegas, NV: CSREA Press.

Cook, S.D., Baffes, G.C., Palafox, A.J., Wolfe, M.W. and Burgess, A. (1992). Torsional stability of HA-coated and grit-blasted titanium dental implants. *Journal of oral implantology,* 18, 354-358.

Cook, S.D., Baffes, G.C., Wolfe, M.W. and Palafox, A.J. (1992). A comparison of femoral and mandibular animal models for the evaluation of HA-coated implants. *Journal of oral implantology*, 18, 359-365.

Corso, M., Sirota, C., Fiorelline, J., Rasool, F., Szmukler-Moncler, S. and Weber H-P (1999). Evaluation of the osseointegration of early loaded free-standing dental implants with various coatings in the dog model: periostest and radiographic results. *Journal of Prosthetic Dentistry,* 82, 428–435.

Cowin, S.C., Moss-Salentijn, L. and Moss, M.L. (1991). Candidates for the mechanosensory system in bone. *Journal of Biomechanical Engineering*, 113, 191-197.

Currey, J. (1984). The Mechanical Adaptations of Bone. New Jersey: Princeton University Press.

Dalle Carbonare, L. and Giannini, S. (2004). Bone microarchitecture as an important determination of bone strength. *Journal of Endocrinoly Investigations*, 27, 99–105.

Dhert, W.J.A., Jansen, J.A. (2000). The validity of a single pushout test. In Y. An, R. Draughn (Eds.), *Mechanical testing of bone at the bone-implant interface* (PP. 477-488). Boca Raton, FL: CRC Press.

Drouin, J.M. and Cales, B. (1994). Yttria-stabilized zirconia for improved hip joint head. In: Ö.H. Andersson, A. Yli-Urpo (Eds.), Bioceramics 7 (pp. 387-394). London, UK: Butterworth-Heinemann Publishers.

Ducheyne, P., Martens, M. and Burssens, A. (1984). Materials, clinical and morphological evaluation of custom-made bioactive-glass-coated canine hip prostheses, *Journal of Biomedical Materials Research*, 18, 1017-1030.

Ducheyne, P. and Martens, M. (1988). Orderly oriented wire meshes as porous coatings on orthopaedic implants II: the pore size, interfacial bonding and microstructure after pressure sintering of titanium (OOWM). *Clinical Materials,* 1, 91-98.

Duyck, J., Van Oosterwyck, H., De Cooman, M., Puers, R., Vander Sloten, J. and Naert, I. (2000). Three-dimensional force measurements on oral implants: a methodological study. *Journal of Oral Rehabilitation*, 27, 744-753.

Ehrlich, J. and Azaz, B. (1975). Immediate implantation of acrylic resin teeth into human tooth sockets. *Journal of Prosthetic Dentistry*, 33, 205-209.

Evans, S.L., Lawes, K.R. and Gregson, P.J. (1994). Layered, adhesively bonded hydroxyapatite coatings for orthopaedic implants. *Journal of Materials Science Materials in Medicine*, 5, 495-499.

Evans, S.L., Gregson, P.J. (1994). Numerical optimization of the design of a coated, cementless hip prosthesis. *Journal of Materials Science Materials in Medicine*, 5, 507-510.

FDA (2000) Guidance for industry on the testing of metallic plasma sprayed coatings on orthopedic implants to support reconsideration of postmarket surveillance requirements. U.S. Department of Health and Human Services, Food and Drug Administration Center for Devices and Radiological Health.

Fung, Y.C. (1981) Biomechanics Mechanical Properties of Living Tissues. New York: Springer-Verlag.

Geiger, B. and Bershadsky, A. (2001). Assembly and mechanosensory function of focal contacts. *Current Opinion in Cell Biology*, 13, 584-592.

Gibbs, C.H., Mahan, P.E., Lundeen, H.C., Brehnan, K., Walsh, E.K. and Holbrook, W.B. (1981). Occlusal forces during chewing and swallowing as measured by sound transmission. *Journal of Prosthetic Dentistry*, 46,443-449.

Gomez-Vega, J.M., Hozumi, A., Saiz, E., Tomsia, A.P., Sugimura, H. and Takai, O. (2001). Bioactive glass-mesoporous silica coatings on Ti6Al4V throgh enameling and triblock-copolymer-templated sol-gel processing, *Journal of Biomedical Materials Research*, 56, 382-389.

Gomez-Vega, J.M, Saiz, E., Tomsia, A.P., Marshall, G.W. and Marshall, S.J. (2000). Bioactive glass coatings with hydroxyapatite and Bioglass particles on Ti-based implants. *Biomaterials*, 21, 105-111.

Gomez-Vega, J.M., Saiz, E., Tomsia, A.P., Oku, T., Suganuma, K., Marshall, G.W., Marshall, S.J. (2000). Novel bioactive glass functionally graded coatings on Ti6Al4V. *Advanced Materials*, 12, 894-898.

Griss, P., Greenspan, D.C., Heimke, G., Krempien, B., Buchinger, R., Hench, L.L. and Jentschura, G. (1976). Evaluation of a bioglass-coated Al2O3 total hip prosthesis in sheep. *Journal of Biomedical Materials Research*, 10, 511-518.

Greenspan, D.C., Hench, L.L. (1976) Chemical and mechanical behavior of bioglass-coated alumina. *J. Biomed. Mater. Res*, 10:503-509.

Guldberg, R.E., Richards, M., Caldwell, N.J., Kuelske, C.L. and Goldstein, S.A. (1997). Mechanical stimulation of tissue repair in the hydraulic bone chamber. *Journal of Bone and Mineral Research*, 12, 1295-1302.

Hashimoto, M., Akagawa, Y., Hashimoto, N., Nikai, H. and Tsuru, H. (1988). Single crystal sapphire endosseous implant loaded with functional stress: clinical and histological evaluation of peri-implant tissues. *Journal of Oral Rehabilitation*, 15, 65-76.

Hedia, H.S. (2002). Stress and strain distribution behavior in the bone due to the effect of cancellous bone, dental implant material and the bone height. *Bio-medical Materials and Engineering*, 12, 111-119.

Hench, L.L. (1994). Bioactive ceramics: theory and clinical applications: Ö.H. Andersson, A. Yli-Urpo (Eds.), Bioceramics 7, (7, pp.3-14). Oxford, UK: Butterworth-Heinmann publisher.

Hench, L.L., Andersson, Ö.H. (1993). Bioactive glass coatings. In: L.L. Hench, J. Wilson (Eds.), An introduction to bioceramics (pp. 239), USA: World Scientific.

Hench, L.L., Splinter, R.J., Allen, W.C. and Greenlee, T.K. (1971). Bonding mechanisms at the interface of ceramic prosthetic materials. *Journal of Biomedical Materials Reseach* Symposium, 2, 117-141.

Hodosh, M., Povar, M. and Shklar, G. (1969). The dental polymer implant concept. *Journal of Prosthetic Dentistry*, 22, 371-380.

Hodosh, M., Povar, M., Shklar, G. (1970). Control of porosity in polymer tooth implants. *Journal of Prosthetic Dentistry*, 24, 453-460.

Hodosh, M., Povar, M., Shklar, G. (1973). Development of the self-supporting polymer tooth implant. *Journal of Biomedical Materials Research*, 7, 205-216.

Howell, A.H. and Brudevold, F. (1950). Vertical forces used during chewing of food. *Journal of Dental Research*, 29,133-136.

Huiskes, R. (1993). Failed innovation in total hip replacement. Diagnosis and proposals for a cure. *Acta Orthopaedica. Scandinavica*, 64, 699-716.

Ichikawa, Y., Akagawa, Y., Nikai, H. and Tsuru, H. (1992). Tissue compatibility and stability of a new zirconia ceramic in vivo. *Journal of Prosthetic Dentistry*, 68, 322-326.

Jansen, J.A., von Recum, A.F., van der Waerden, J.P.C.M. and de Groot, K. (1992). Soft tissue response to different types of sintered metal fiber-web materials. *Biomaterials*, 13, 959-968.

Jarcho, M. (1992). Retrospective analysis of hydroxyapatite development for oral implant applications. *Dent. Clin. North Am.*, 36:19-26.

Jones, L.C., Frondoza, C. and Hungerford D.S. (2001). Effect of PMMA particles and movement on an implant interface in a canine model. *Journal of Bone and Joint Surgery* – British Volume, 83, 448-458.

Kangasniemi, I., Verheyen, C., Velde, E and deGroot, K. (1994). In vivo tensile testing of fluoroapatite and hydroxyapatite plasma-sprayed coatings, *Journal of Biomedical Materials Research*, 42, 295-305.

Kaplan, F.S., Hayes, W.C., Keaveny, T.M., Boskey, A., Einhorn J.P. and Iannotti, J.P. (1994). Form and function of bone. In: S.R. Simon (Ed.), Orthopaedic basic science (pp.127-184). Rosemont, Illinois: American Academy of Orthopaedic Surgeons.

Katz, J.L. (1995) Mechanics of hard tissue. In: Brozino, J.D., ed. The Biomedical Engineering Handbook, 273–289. Boca Raton: CRC Press.

Kitsugi, T., Nakamura, T., Oka, M., Senaha, Y., Goto, T. and Shibuya, T. (1996). Bone-bonding behavior of plasma-sprayed coatings of Bioglass R, AW-glass ceramic, and tricalcium phosphate on titanium alloy. *Journal of Biomedical Materials Research*, 30, 261-269.

Klein-Nulend, J., Van der Plas, A., Semeins, C.M., Ajubi, N.E., Frangos, J.A., Nijweide, P.J. and Burger, E.II. (1995). Sensitivity of osteocytes to biomechanical stress *in vitro*. *The Journal of the Federation of American Societies for Experimental Biology*, 9, 441-445.

Kohal, R.J., Klaus, G. and Strub, J.R. (2006). Zirconia-implant-supported all-ceramic crowns withstand long-term load: a pilot investigation. *Clinical Oral Implants Research*, 17, 565-571.

Krauser, J.T., Boner, C. and Boner, N. (1990). Hydroxyapatite coated dental implants. Biological criteria and prosthetic possibilities. *Les Cahiers de prothèse*, 71, 56-75.

Kurtz, S.M. and Devine, J.N. (2007). *PEEK Biomaterials in Trauma, Orthopedic, and Spinal Implants. Biomaterials,* 28, 4845-4869.

Lacefield, W.R. (1988). Hydroxlapatite coatings. In: P. Ducheyne, J.E. Lemons, (Eds.), Bioceramics. materials characteristics vs in vivo behavior, Vol. 523. New York: Annals of New York Academy of Sciences, 523.

Lacefield, W.R. (1994). Characterization of hydroxylapatite coatings, *Journal of Oral Implantology*, 20, 214-220.

Lacefield, W.R. and Hench, L.L. (1986). The bonding of Bioglass to a cobalt-chromium surgical implant alloy. *Biomaterials,* 7, 104-108.

Laurell, L. (1985). Occlusal forces and chewing ability in dentitions with cross-arch bridges. *Swedish Dental Journal*, 26, 160-7.

Leeuwenburgh, S.C.G, Wolke, J.G.C., Jansen, J.A. and De Groot, K. (2008). Calcium phosphate coatings. In: T. Kokubo (Ed.), From Bioceramics and Their Clinical Applications (pp. 464-484). Cambridge, UK: Woodhead Publishing.

Lemons, J.E. (1998) Unanticipated outcomes: dental implants. *Implant Dentistry*, 7, 351-354.

Li, Z., Kitsugi, T., Yamamuro, T., Chang, Y.S., Senaha, Y., Takagi, H., Nakamura, T. and Oka, M. (1995). Bone-bonding behavior under load-bearing conditions of an alumina ceramic implant incorporating beads coated with glassceramic containing apatite and wollastonite. *Journal of Biomedical Materials Research*, 29, 1081-1088.

Lum, L.B., Beirne, O.R. and Curtis, D.A. (1991). Histological evaluation of Ha coated vs uncoated titanium blade implants in delayed and immediately loaded applications. *International Journal of Oral and Maxillofacial Implants*, 6, 456-62.

Manley, M., Kotzar, G., Stern, L. and Wilde A. (1987). Effects of repetitive loading on the integrity of porous coatings. *Clinical Orthopaedics and Related Research*, 217, 293-302.

Mihalko, W.M., May, T.C., Kajy, J.F. and Krause, W.R. (1992). Finite element analysis of interface geometry effects on crestal bone surrounding a dental implant. *Implant Dentistry*, 1, 212-217.

Mishra, A.K., Davidson, J.A., Poggie, R.A., Kovaac, P. and Fitzgerald, T.J. (1996). Mechanical and tribological properties of diffusion hardened Ti-13Nb-13Zr—A new titanium alloy for surgical implants. In: S. Brown, J. Lemons (Eds.), Medical applications of titanium and its alloys: the material and biological issue (pp. 96-112). West Conshohocken, PA: American Society for Testing and Materials.

Miyata, T., Kobayashi, Y., Araki, H., Ohto, T. and Shin, K. (2000)._The influence of controlled occlusal overload on peri-implant tissue. Part 3: A histologic study in monkeys. *International Journal of Oral and Maxillofacial Implants*, 15, 425-431.

Miyata, T., Kobayashi, Y., Araki, H., Ohto, T. and Shin, K. (2002)._The influence of controlled occlusal overload on peri-implant tissue. Part 4: a histologic study in monkeys. *International Journal of Oral and Maxillofacial Implants*. 17, 384-390.

Moritz, N., Rossi, S., Vedel, E., Tirri, T., Ylänen, H., Aro, H. and Närhi, T. (2004). Implants coated with bioactive glass by CO_2-laser, an in vivo study. *Journal of Materials Science Materials in Medicine*, 15,795-802.

Mouzin, O., Soballe, K. and Bechtold, J.E. (2001). Loading improves anchorage of hydroxyapatite implants more than titanium implants. *Journal of Biomedical Materials Research*, 58, 61-68.

Mosekilde, L. (2000). Age-related changes in bone mass, structure, and strength-effects of loading, *Zeitschrift fur Rheumatologie*, 59, 1–9.

Naert, I., Quirynen, M., van Steenberghe, D. and Darius, P. (1992). A study of 589 consecutive implants supporting complete fixed prostheses. Part II: Prosthetic aspects. *Journal of Prosthetics Dentistry*, 68, 949-956.

Nakamura, T., Yamamuro, T., Higashi, S, Kokubo, T. and Itoo, S. (1985). A new glass-ceramic for bone replacement evaluation of its bonding to bone tissue. *Journal of Biomedical Materials Research,* 19, 685-698.

Neill, D.J., Kydd, W.L., Nairn, R.I. and Wilson, J. (1989). Functional loading of the dentition during mastication. *Journal of Prosthetic Dentistry*, 62, 218-228.

O'Brien, W. J. (2002) Dental materials and their selection (3rd edition). Illinois, USA: Quintessence Publishing.

Ogiso, M., Nakabayashi, N., Matsumoto, T., Yamamura, M. and Lee, R.R. (1996). Adhesive improvement of the mechanical properties of a dense HA-cemented Ti dental implant. *Journal of Biomedical Materials Research*, 30, 109-116.

Oku, T., Suganuma, K., Wallenberg, L.R., Tomsia, A.P., Gomez-Vega, J.M. and Saiz, E. (2001). Structural characterization of the metal/glass interface in bioactive glass coatings on Ti-6Al-4V, *J. Mat. Sci. Mat. Med.*, 12:413-417.

Pajamäki, K.J., Lindholm, T.S., Andersson, Ö.H., Karlsson, K.H., Vedel, E., Yli-Urpo, A. and Happonen, R.P. (1995). Bioactive glass and glass-ceramic-coated hip endoprosthesis: experimental study in rabbit, *Journal of Materials Science Materials in Medicine*, 6; 14-18.

Paldan, H., Areva, S., Tirri, T., Peltola, T., Lindholm, T.C., Lassila, L., Pelliniemi, L.J., Happonen, R.P. and Närhi, T.O. (2008). Soft tissue attachment on sol–gel-treated titanium implants in vivo. *Journal of Materials Science Materials in Medicine*, 19, 1283-1290.

Paphangkorakit, J. and Osbom, J.W. (1987). The effect of pressure on maximum incisal bite force in man. *Archives of Oral Biology*, 2, 11–7.

Piattelli, A., Ruggierie, A., Franchi, M., Romasco, N. and Trisi, P. (1993). A histologic and histomorphometric study of bone reactions to unloaded and loaded non-submerged single implants in monkeys: a pilot study. *Journal of Oral Implantology*, 19, 314-320.

Piattelli, A., Paoloantonio, M., Corigliano, M. and Scarano, A. (1997). Immediate loading of titanium plasma-sprayed screw-shaped implants in man: a clinical and histological report of 2 cases. *Journal of Periodontology*, 68, 591-597.

Piconi, C., Burger, W., Richter, H.G., Cittadini, A., Maccauro, G., Covacci, V., Bruzzese, N., Ricci, G.A. and Marmo, E. (1998). Y-TZP ceramics for artificial joint replacements. *Biomaterials*, 19, 1489-1494.

Piotrowski, G., Hench, L.L., Allen, W.C. and Miller, G.J. (1975). Mechanical studies of the bone Bioglass interfacial bond, Journal of Biomedical Materials Research, 6, 47-61.

Prendergast, P.J. (1997). Finite element models in tissue mechanics and orthopaedic implant design. *Clinical Biomechanics*, 12, 343-366.

Quirynen, M., Naert, I., van Steenberghe, D. and Nys, L. (1992). A study of 589 consecutive implants supporting complete fixed prostheses. Part I: Periodontal aspects. *Journal of Prosthetic Dentistry*. 68, 655-663.

Rangert, B., Jemt, T. and Jorneus, L. (1989). Forces and moments on Brånemark implants. *International Journal of Oral and Maxillofacial Implants*, 4, 241-247.

Rangert, B., Krogh, P.H., Langer, B. and Van Roekel, N. (1995). Bending overload and implant fracture. *International Journal of Oral and maxillofacial Implants*, 10, 326-334.

Richter, H.G., Burger, W. and Osthues, F. (1994). Zirconia for medical implants-the role of strength properties. In: Ö.H. Andersson, A. Yli- Urpo (Eds.), Bioceramics 7 (pp. 401-406). London: Butterworth-Heinemann Publishers.

Richter, E.J. (1998). In vivo horizontal bending moments on implants. *International Journal of Oral and Maxillofacial Implants*, 13, 232-244.

Rickerby, D.S. (1988). Review of the methods for measurement of coating-substrate adhesion. *Surface and Coating Technology*, 36, 541-557.

Rossi, S., Moritz, N., Tirri, T., Peltola, T., Areva, S., Jokinen, M., Happonen, R.P. and Närhi, T. (2007). Comparison between sol-gel-derived anatase- and rutile-structured TiO_2 coatings in soft-tissue environment. *Journal of Biomedical Materials Research Part A*, 82, 965-974.

Rubin, M.A., Rubin, J. and Jasiuk, I. (2004). SEM and TEM study of the hierarchical structure of C57BL/6J and C3H/HeJ mice trabecular bone. *Bone,* 35, 1-20.

Sagara, M., Akagawa, Y., Nikai, H. and Tsuru, H. (1993). The effects of early occlusal loading on one-stage titanium implants in beagle dogs: a pilot study. *Journal of Prosthetic Dentistry,* 69, 281-288.

Saiz, E., Goldman, M., Gomez-Vega, J.M., Tomsia, A.P., Marshall, G.W. and Marshall, S.J. (2002). In vitro behaviour of silicate glass coatings on Ti6Al4V. *Biomaterials*, 23, 3749-3756.

Schepers, E., Ducheyne, P. and De Clerq, M. (1989). Interfacial analysis of fiber-reinforced bioactive glass dental root implants, *Journal of Biomedical Materials Research*, 23, 735-752.

Schrooten, J. and Helsen, J.A. (2000). Adhesion of bioactive glass coating to Ti6A14V oral implant. *Biomaterials*, 21, 1461-1469.

Schrooten, J., Van Oosterwyck, H., Vander Sloten, J., Helsen, J.A. (1999). Adhesion of new bioactive glass coating. *Journal of Biomedical Materials Research*, 44, 243-252.

Schulte, W. (1984). The intra-osseous Al_2O_3 (Frialit) Tuebingen Implant. Developmental status after eight years (I-III). *Quintessence International*, 15, 1-39.

Schulte, W. (1995). Implants and the periodontium. *International Dental Journal*, 45:16-26.

Skerry, T.M., Bitensky, L., Chayen, J. and Lanyon, L.E. (1989). Early strain-related changes in enzyme activity in osteocytes following bone loading in vivo. *Journal of Bone and Mineral Research*, 4, 783-788.

Skripitz, R. and Aspenberg, P. (2001). Early effect of parathyroid hormone (1–34) on implant fixation. *Clinical Orthopaedics and Related Research*, 392, 427-432.

Soballe, K. (1993) Hydroxyapatite ceramic coating for bone implant fixation, *Acta Orthopaedica Scandinavica*, 64, 1-58.

Smedberg, J.I., Nilner, K., Rangert, B., Svensson, S.A. and Glantz, S.A. (1996). On the influence of superstructure connection on implant preload: a methodological and clinical study. *Clinical Oral Implants Research*, 7, 55-63.

Suominen, H. (1993). Bone mineral density and long term exercise. An overview of cross-sectional athlete studies. *Sports Medicine*, 16, 316-330.

Szmukler-Moncler, S., Salama, S., Reingewirtz, Y., Dubruille, J.H. (1998). Timing of loading and effect of micro-motion on bone–implant interface: a review of experimental literature. *Journal of Biomedical Materials Research*, 43, 193-203.

Toksvig-Larsen, S., Jorn, L.P., Ryd, L. and Lindstrand A. (2000). Hydroxyapatite-enhanced tibial prosthetic fixation. *Clinical Orthopaedics and Related Research*, 370, 192–200.

Van Eijden, T.M. (1991). Three-dimensional analyses of human bite-force magnitude and moment. *Archives of Oral Biology*, 36, 535-539.

Van Noort R. (2002). Introduction to dental materials (2nd edition). London, UK: Mosby.

Van Staden, R.C., Guan, H. and Loo, Y.C. (2006). Application of the finite element method in dental implant research. *Computer Methods in Biomechanics and Biomedical Engineering,* 9, 257-270.

Vico, L., Collet, P., Guignandon, A., Lafage-Proust, M.H., Thomas, T., Rehaillia, M. and Alexandre, C. (2000). Effects of long-term microgravity exposure on cancellous and cortical weight-bearing bones of cosmonauts. *Lancet* 355, 1607-1611.

Waltimo, A. and Kononen, M. (1993). A novel bite force recorder and maximal isometric bite force values for healthy-young adults. *Scandinavian Journal of Dental Research*, 101, 171-175.

West, J.K., Clark, A.E., Hall, M.B. and Turner, G.E. (1990). In vivo bone-bonding study of Bioglass-coated titanium alloy. In: T. Yamamuro, L.L. Hench, J. Wilson (Eds.), Handbook of bioactive ceramics (pp. 161-166). Boca Raton FL: CRC Press.

White, S.N., Caputo, A.A. and Anderkvist, T.A. (1994). Effect of cantilever length on stress transfer by implant-supported prostheses. Journal of Prosthetic Dentistry, 71, 493-499.

Wilson, J., Schoen, F.J., Pigott, G.H. and Hench, L.L. (1981). Toxicology and bioactivity of Bioglasses. *Journal of Biomedical Materials Research*, 15, 805-817.

Wilson, J. and Noletti, D. (1990). Bonding of soft tissues to Bioglass. In: T. Yamamuro, L.L. Hench, J. Wilson (Eds.), Handbook of bioactive ceramics (pp. 235-244). Boca Raton: CRC Press.

Wintermantel, E. and Mayer, J. (1995). Anisotropic biomaterials strategies and developments for bone implants. In: D.L. Wise, D.J. Trantolo, D.E. Altobelli, J.D. Yaszemiski, J.D. Gresser, and E.R. Schwartz (Eds.), Encyclopedic Handbook of Biomaterials and Bioengineering (1st, pp. 3-42). NewYork: Mearcel Dekker.

Wolff, J. (1892). *Das Gesetz der Transformation der Knochen*. Berlin A. Hirchwild. Translated as: *The Law of Bone Remodeling*. Maquet P. and Furlong R. (1986). Berlin: Springer-Verlag.

Wolke, J.G.C., deBlieck-Hogervorst, J.M.A., Dhert, W.J.A., Klein, C.P.A.T. and deGroot, K. (1992) Studies on the thermal spraying of apatite bioceramics, *Journal of Thermal Spray Technology*, 1, 75-85.

Yaszemski, M.J., Payne, R.G., Hayes, W.C., Langer, R., and Mikos, A.G. (1996). Evolution of bone transplantation: molecular, cellular, and tissue strategies to engineer human bone. *Biomaterials*. 17, 175-185.

Zerwekh, J.E., Ruml, L.A., Gottschalk, F. and Pak, C.Y.C. (1998). The effects of twelve weeks of bed rest on bone histology, biochemical markers of bone turnover, and calcium homeostasis in eleven normal subjects. *Journal of Bone and Mineral Research*, 13, 1594-1601.

Zhao, D.S., Moritz, N., Laurila, P., Mattila, R., Lassila, L.V.J., Strandberg, N., Mäntylä, T., Vallittu, P.K. and Aro, H.T. (2009). Development of a Multi-Component Fiber-Reinforced Composite Implant for Load-Sharing Conditions. *Medical Engineering and Physics*, 31, 461-469.

In: Biomechanics: Principles, Trends and Applications
Editor: Jerrod H. Levy, pp. 137-162
ISBN: 978-1-60741-394-3

Chapter 4

BIOMECHANICAL REMODELING OF THE DIABETIC GASTROINTESTINAL TRACT

Jingbo Zhao*, Donghua Liao, Jian Yang and Hans Gregersen
Mech-sense, Aalborg Hospital Science and Innovation Centre (AHSIC),
9000 Aalborg, Denmark

ABSTRACT

Gastrointestinal tract sensory-motor abnormalities are common in patients with diabetes mellitus with symptoms arising from the whole GI tract. Common complaints include dysphasia, early satiety, reflux, constipation, abdominal pain, nausea, vomiting, and diarrhea. The pathogenesis of GI symptoms in diabetes mellitus is complex in nature, multi-factorial (motor dysfunction, autonomic neuropathy, glycemic control, psychological factors, etc.) and is not well understood. Histologically, many studies have demonstrated prominent proliferation of different GI wall layers during diabetes. During the past several years, several studies demonstrated that experimental diabetes induces GI morphological and biomechanical remodeling. Following the development of diabetes, the GI wall becomes thicker and the stiffness of the GI wall increases in a time-dependent manner. It is well known that mechanosensitive nerve endings exist in the GI tract where they serve a critical role for tissue homeostasis and symptom generation. Mechanoreceptor-like structures such as intraganglionic laminar nerve endings and intramuscular arrays have been identified. The changes of stress and strain in the GI wall will alter the biomechanical environment of the mechanosensitive nerve endings, therefore, the structure as well as the tension, stress and strain distribution in the GI wall is important for the sensory and motor function. Biomechanical remodeling of diabetic GI tract including alterations of residual strain and increase in wall stiffness will alter the tension and stress distribution in the vicinity of the mechanosensitive afferents with consequences for perception and motility of the GI tract.

Keywords: Diabetes, gastrointestinal tract, mechanosensory, biomechanics, remodeling.

* Telephone: +45 99326907, E-mail: jz@rn.dk, jzhao@hst.aau.dk

INTRODUCTION

Diabetes mellitus (DM) is a chronic disease requiring lifelong medical attention in order to limit the development of potentially devastating late complications and to manage them if they occur. DM affects almost 6% of the world's population, means that approximately 246 million people worldwide were affected (International Diabetes Federation [http://www.idf.org/]) and the number of diabetic patients will reach 300 million in 2025 [Adeghate et al., 2006]. Therefore, it is becoming one of the major public health problems because a great proportion of the healthcare expenditure has been spent on the treatment of its associated morbidity and mortality. In addition, persons with DM have a lower quality of life compared with persons without DM.

Chronic complications of DM include nerve disease, heart disease, vascular disease, kidney disease, eye disease as well as digestive disease [Nair, 2007]. Gastrointestinal (GI) disorders are common in diabetic patients [Verne and Sninsky, 1998; Folwaczny et al., 1999]. The entire GI tract from the esophagus to the anorectal region can be affected. The pathogenesis of the GI abnormalities is complex of nature, multi-factorial (motor dysfunction, autonomic neuropathy, glycemic control, psychological factors, etc.) and is not well understood [Horowitz and Samsom, 2004]. A number of motor dysfunctions have been described in different segments of the GI tract in diabetic patients: esophagus (dysmotility), stomach (dysmotility, delayed emptying) and small and large bowel (dysmotility, delayed transit, bacterial overgrowth, and diarrhea) [Horowitz and Samsom, 2004]. A few studies have addressed the visceral sensory function in DM [Kamath et al., 1998, 2000; Rathmann et al., 1991]. Although the hyperglycemia and neuropathy seem to be the main mechanisms to the motor-sensory dysfunction in the GI tract in DM, many studies have however demonstrated morphological changes [Mayhew et al., 1989; Tahara and Yamamoto, 1988; Zoubi et al., 1995; Charlton et al., 2000] and biomechanical remodeling in the GI tract in DM [Jorgensen et al., 2001; Liao et al., 2006; Sha et al., 2006; Yang et al., 2004, 2006; Zeng et al., 2004; Zhao et al., 2002, 2003a, 2003b, 2006, 2007, 2008; Frokjaer et al., 2007a, 2007b]. Therefore, the morphological and biomechanical remodeling of the GI wall may also be an important factor in the pathogenesis to the GI motor-sensory dysfunction in the diabetic patients. Data on the biomechanical properties are crucial for the understanding of the motor function of the GI tract because, 1) peristaltic motion that propels the food through the GI tract is a result of interaction of the passive and active tissue forces and the hydrodynamic forces in the food bolus and 2) remodeling of the mechanical properties reflects the changes in the tissue structure that determine a specific motor dysfunction.

SYMPTOMS FROM GI TRACT IN DM

Motor and sensory abnormalities in DM can affect the entire GI tract or part hereof, and the perceived symptoms may originate from one or several parts of the GI tract. The prevalence of GI symptoms is high in both insulin dependent DM (IDDM) and non-insulin dependent DM (NIDDM) [Ko et al., 1999; Maleki et al., 2001; Ricci et al., 2000; Schvarcz et al., 1996; Spangeus et al., 1999; Talley et al., 2001, 2002]. The symptoms relating to the esophagus, stomach, small intestine and colon are summarized in Table 1. Most symptoms are non-specific of nature, therefore when dealing with GI symptoms; some specific issues need to be addressed. Chest pain may relate to reflux or esophageal motor disorders, but ischemic heart disease and other causes of non-cardiac chest pain are also possible causes and must be excluded. Dysphagia, the most characteristic symptom of impaired esophageal transit, may be caused by motility disorders of the esophagus. A number of other incidental conditions must be excluded when diabetic gastroparesis is suspected: gastric outlet obstruction caused by tumors and ulcer disease; metabolic abnormalities such as diabetic ketoacidosis or uremia and side effects of pharmacotherapy. Diabetic patients frequently report abdominal pain and this may be the only symptom of diabetic gastroparesis; however, abdominal pain can also be seen in diabetic ketoacidosis and severe metabolic acidosis [Umpierrez and Freire, 2002]. Diabetic patients with thoracic polyradiculopathy, a rare condition, may also suffer from abdominal pain [Longstreth, 1997]. Diarrhea in DM patients may be induced by a number of factors. These may include food composition, abnormal intestinal motility, small intestinal bacterial overgrowth, excessive loss of bile acids, pancreatic insufficiency and more [Samsom and Verhagen, 2004]. Abnormal small intestinal motility (rapid or delayed transit) is a frequent condition in the diabetic patients as described below. Rapid transit may induce an increase in intra-luminal contents that reach the caecum, whereas delayed transit may cause bacterial overgrowth, both potentially resulting in diabetic diarrhea. Celiac disease is overrepresented in IDDM and a cause of severe diarrhea to be excluded when dealing with diabetic diarrhea [Anderson, 2005]. Disordered defecation, characterized by fecal incontinence and constipation occurs frequently in patients with DM [Sun and Read, 2004]; the patients frequently fail to report disturbances in defecation unless the symptoms are severe. Constipation has been cited as the most common gastrointestinal complaint of diabetic patients [Batte et al., 1983]. Profound constipation with massive fecal impaction may occur [Batte et al., 1983]. These patients may develop marked abdominal distention and severe nausea and vomiting associated with electrolyte disturbances. Prolonged fecal impaction may lead to stercoral ulcerations and perforation. Steatorrhea has generally been attributed to pancreatic insufficiency, celiac disease, or bacterial overgrowth of the small intestine [Battc et al., 1983]. In these patients the colon appears to play a secondary or permissive role in the production of diarrhea. In the absence of steatorrhea, colonic dysfunction may be a primary contributor to diabetic diarrhea.

Table 1. Symptoms of GI tract in DM

Organs	Symptoms
Esophagus	Heartburn, dysphagia and chest pain
Stomach	Postprandial nausea, vomiting, early satiety, bloating, weight loss and abdominal pain
Small intestine	Diarrhea, discomfort, pain and pseudo-obstruction
Large intestine	Faecal Incontinence, constipation, diarrhea

Mechano-Sensory Dysfunction of GI Tract in DM

Many studies have demonstrated that DM patients have disorders of transit and motility. However, the data related to the sensory dysfunction of the GI tract are sparse compared with those relating to the motor dysfunction of the GI tract. The most frequent motility disorders of GI tract are shown in Table 2.

Esophagus: Impaired esophageal transit has been reported both in IDDM and NIDDM patients [Horowitz et al., 1991; Jorgensen et al., 1991; Tsai et al, 1995]. The esophageal transit appears to be delayed in about 50% of patients with long-standing DM [Annese et al., 1999]. The retarded esophageal transit in the DM usually reflects either peristaltic failure or focal low-amplitude pressure waves [Holloway et al., 1999]. Esophageal manometric abnormalities occur in over 50% of patients with DM [Smout, 2004] (see Table 2). The reduced amplitude of lower esophageal sphincter pressure is in accordance with the increased prevalence of gastroesophageal reflux disease in DM. More recently the evoked esophageal contractile activity to standardized bag distension was assessed using a specialized ultrasound-based probe by Frokjaer et al [2007]. A balloon-like bag was positioned 10 cm above the lower esophageal sphincter and inflated. It was demonstrated, both at the bag and 6 cm proximally, that the distension induced hyperreactivity and impaired coordination of the contractions in the diabetic patients.

Elevation of perception thresholds to esophageal electrical stimulation has been observed in patients with DAN and different severity of GI symptoms [Rathmann et al., 1991]. Increased vagal tonus and abnormal evoked brain potentials to mechanical and electrical stimulation of the esophagus has also been shown [Kamath et al., 1998].

Table 2. Motility disorders of GI tract in DM

Organs	Motility disorders
Esophagus	1) amplitude and number of peristaltic contractions↑
	2) number of spontaneous and non-propagated contractions↑
	3) amplitude of lower esophageal sphincter pressure↓
	4) multi-peaked contractions
Stomach	1) antral IMMC↓
	2) post-prandial antral activity and the number of antral contractions↓
	3) pyloric dysmotility
Small intestine	1) frequency and amplitude of the antropyloroduodenal contractions↑↓
	2) duration of MMC cycle↑
	3) early recurrence of the MMC and clusters of contractile activity
Colon	1) enhanced spontaneous activity
	2) Increase in stretch-induced rhythmic motor activity

Stomach: Several animal studies reported a slowing gastric emptying in IDDM and NIDDM rats [Mehta et al., 2002; Ogata et al., 1997; Liu et al., 2004], whereas other studies on IDDM and NIDDM animals demonstrated that gastric emptying increased [Nowak et al., 1994; Ogata et al., 1996; Young et al., 1995]. Using radionuclide measuring techniques it has been demonstrated that gastric empting of solid, or a liquid meal was abnormally slow in 30-50% of patients with long-standing IDDM and NIDDM [Horowitz et al., 1996]. Gastroparesis in DM has been known clinically for more than 50 years [Smith and Ferris, 2003]. The gastric emptying delay in DM is related to both slow transit with increased retention of food in the proximal and distal stomach [Jones et al., 1995], and abnormal motility of the gastric wall [Samsom et al., 1998]. Motility disorders of the fundus and pylorus have been demonstrated in the diabetic animals [James et al., 2004; Korolkiewicz et al., 1998]. In human studies, it is recognized that disordered gastric contractile activity as assessed by manometry and gastric emptying occurs frequently in DM [Bjornsson et al., 1994; Samsom et al., 1996]. The motility disorders may include three aspects: Inter-digestive migrating motor complex (IMMC), amplitude and frequency of contractions, and pyloric dysfunction (Table 2).

Rayner et al [2000] did isovolumetric and isobaric distensions of the proximal stomach in ten randomly selected patients with IDDM. They demonstrated that the perception of gastric distension during euglycemia was increased compared with healthy controls. To study mechanisms behind postprandial symptoms in patients with diabetes, the gastric accommodation of the meal was assessed by abdominal ultrasound [Undeland., 1998]. In DM patients, a large proximal stomach was associated with perception of fullness and a large antrum was associated with perception of pain after a meal.

Small intestine: Delayed and rapid transit in the small intestine was observed in animal diabetic models. The GI transit was rapid in non-obese diabetic mice [el Salhy, 2001] and was slower in obese diabetic mice [el Salhy, 2002]. Anjaneyulu and Ramarao [2002] reported an increase in intestinal transit and a decrease in intestinal tone due to increased cholinergic and decreased beta-adrenergic receptor activities in DM rats. Slow small intestinal transit in DM patients have been documented using breath hydrogen appearance time after the ingestion of lactulose [de Boer et al., 1993], by using radiopaque markers [Iida et al., 2000] and by use of metal-detector test [Folwaczny et al., 1995]. On the other hand, Keshavarzian and Iber [1986] investigated intestinal transit in IDDM patients after the ingestion of both liquid and solid meals and showed abnormal fast intestinal transit in their sample of diabetic patients. Nguyen et al [1997] used intraluminal multiple impedance measurements to identify the postprandial duodenal chyme transport in patients with long standing IDDM. They demonstrated that the patients had disturbed propulsive chyme transport through the duodenum and the duodenal chyme clearance activity was decreased. Camilleri and Malagelada [1984] reported that small intestinal motility was abnormal in about 80% patients of long-standing DM with delayed gastric emptying. Both postprandial and fasting small intestinal dysmotility in the DM was reported (Table 2). Dooley et al [1988] studied fasting GI motility by manometry for a mean of 210 min in a group of 12 NIDDM patients with diarrhea and DAN. The patients showed grossly disordered motility. The migrating motor complex (MMC) disorders reflect the prolongation of phase II without change in the duration of phase I and III. The results of studies on postprandial motility at the level of the small intestine are inconsistent. However, abnormal motility patterns were observed in the diabetic subjects [Samsom et al., 1996].

More recently, Frøkjær et al [2007] used a multimodal stimulation device to investigate the visceral sensitivity to mechanical, thermal and electrical stimulation in the esophagus and duodenum in IDDM patients with DAN and GI symptoms. This study demonstrated that the patients had decreased sensitivity to the stimulations of the esophagus and duodenum. This indicates that the affection of the sensory nerves is widespread in the GI tract. As the multimodal approach is thought to stimulate the mucosa, submucosa and muscle layers differentially, the disease seems to be generalized to nerves in all layers of the GI tract.

Colon: Delayed transit was frequent in males with IDDM of greater than 10 years duration and the symptoms related closely to the region impaired [Iber et al., 1993]. The mean total colonic transit time of the diabetic patients was significantly longer than that in healthy subjects. Furthermore, the diabetic patients with constipation showed longer total, left and recto-sigmoid colonic transit times than those without constipation [Jung et al., 2003]. Diabetic patients with mild constipation had a postprandial increase in colonic motility; however, the response was delayed to 60-90 min after eating. Diabetic patients with severe constipation had no postprandial increase in colonic motility. These studies suggest that patients with diabetes mellitus and severe constipation may have a DAN which leads to an absent postprandial gastrocolonic response [Battle et al., 1980]. A marked alteration of motility in the colon was also observed in diabetic animals; however, the change in motility is unlikely to be due to a change in inhibitory control mechanisms and may be due to an increased excitability [Forrest and Parsons, 2003]. In diabetic rats, the stretch-induced rhythmic contractile activity remained robust and of similar frequency but was significantly higher in amplitude compared with that in control rats. This robust presence of low-frequency contractions is consistent with the unaffected pacemaker, the ICC associated with Auerbach's plexus, and the increased amplitude correlates with loss of and injury to ICC of the submuscular plexus and intramuscular ICC [Forrest et al., 2008]. Hyperglycemia blunts mechanoreceptor-mediated gastrocolonic responses and ascending contractions but not descending components of the peristaltic reflex in humans, effects not caused by hyperinsulinemia or direct muscle actions. These inhibitory effects on long and short neural reflexes that modulate colonic motility may contribute to constipation in diabetes [Sims et al., 1995].

Morphological and Biomechanical Remodeling of GI Tract in Diabetes Mellitus

Although the hyperglycemia and neuropathy seem to be the main mechanisms to the motor-sensory dysfunction in the GI tract in DM, the question remains whether the disordered motor and sensory functions of the GI tract are only due to the neuronal changes and dysfunction or if primary diabetes-induced structural and biomechanical changes in the GI tract also play a role? Histologically, many studies have demonstrated prominent proliferation of different GI wall layers during diabetes.

During the past several years, several studies demonstrated that experimental diabetes induces GI morphological and biomechanical remodeling. Following the development of diabetes, the GI wall becomes thicker and the stiffness of the GI wall increases in a time-dependent manner. Therefore, the morphological and biomechanical remodeling of the GI wall may also be an important factor in the pathogenesis to the GI motor-sensory dysfunction in the diabetic patients.

Morphological and Biomechanical Remodeling of Different Parts in GI

Esophageal remodeling: Many studies have shown that DM causes morphological changes and biomechanical remodeling in the esophagus. Yang et al [2004] in the *in vitro* study on the STZ-induced diabetic rat esophagus found that the wall thickness and cross-sectional wall area increased after the induction of diabetes. During development of the diabetes, the esophagus became stiffer in the circumferential (Figure 1A) and the longitudinal as well as in the shear directions [Yang et al., 2004].

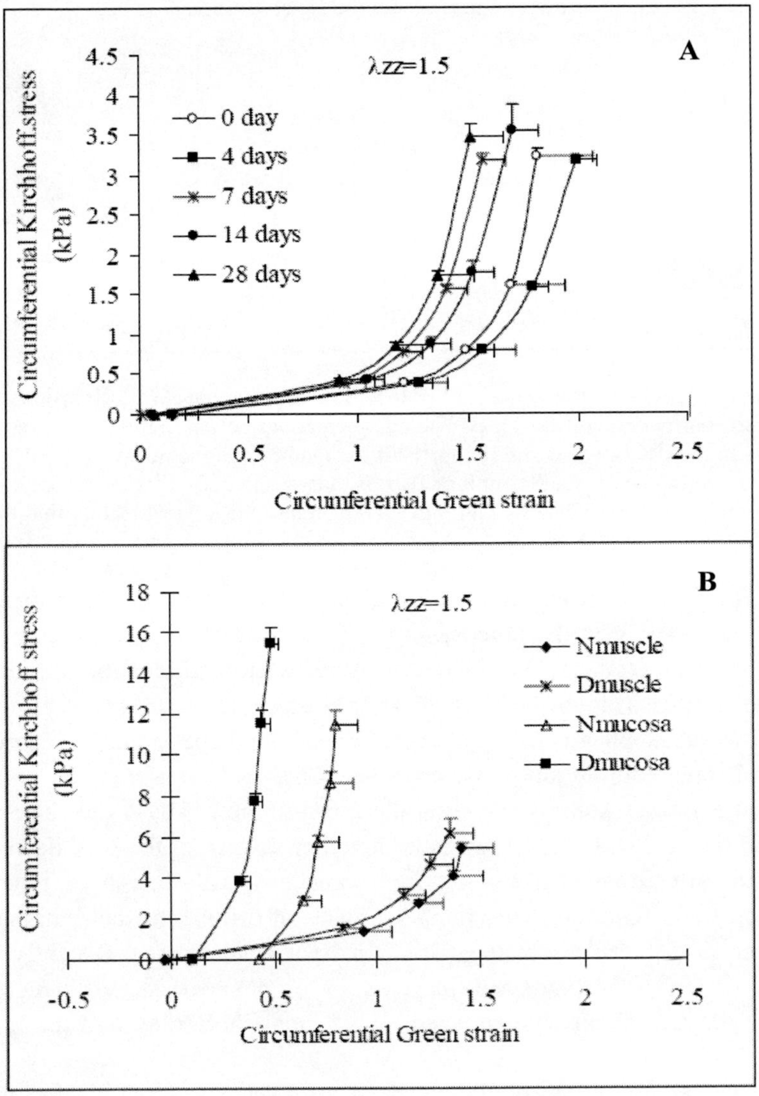

Figure 1. A: Effect of diabetes on the circumferential stress and strain of intact oesophageal wall during the 4 weeks study period. Diabetes shifted the curve to the left indicating increased circumferential stiffness. B: Circumferential stress-strain relations in the muscle layer and mucosa-submucosa layer at a longitudinal stretch ratio of 1.5. The curve for the mucosa-submucosa layer during diabetes shifted to the left in the circumferential direction compared to the control group indicating that the mucosa-submucosa layer became stiffer in the diabetic rats. N: Normal control, D: 28 days diabetes.

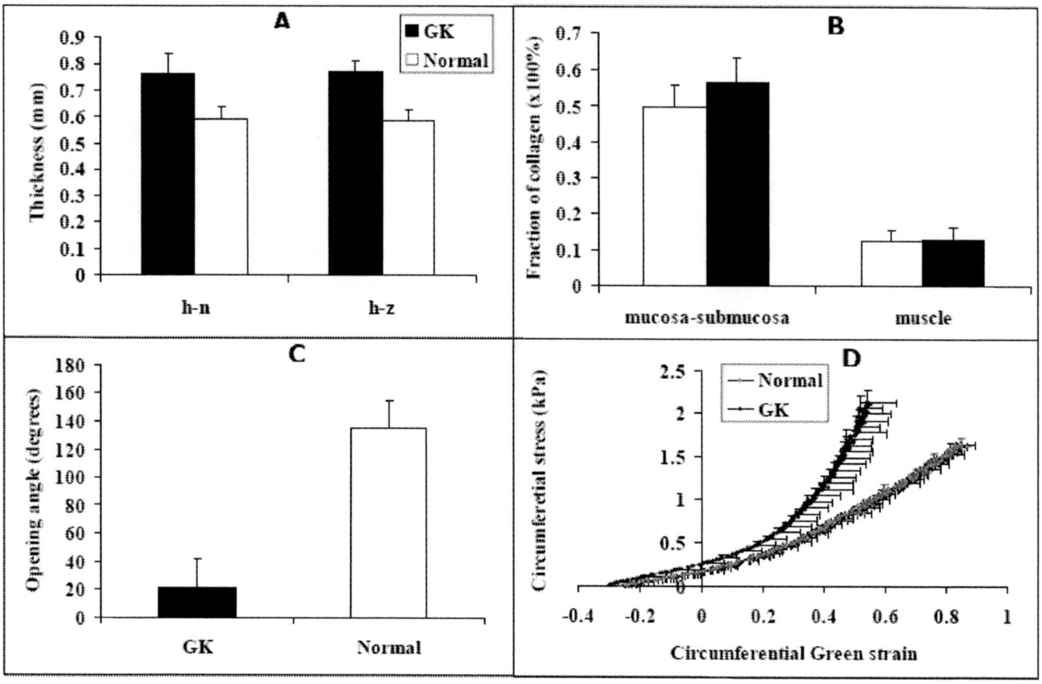

Figure 2. A: The wall thickness in the no-load state and zero-stress state were significantly bigger in the GK rats than in the normal rats ($P<0.01$). B: The collagen area fraction in the mucosa-submucosa layer of the esophagus in the GK rats was significantly higher than that in the normal rats ($P<0.01$). C: The opening angle was significantly smaller in the GK rats than in the normal rats ($P<0.01$). D: The circumferential stress-strain curve in the GK rats was shifted to the left, indicating that the esophageal wall stiffness increased in the GK rats.

Detail layered wall biomechanical analysis showed that the morphometric and biomechanical remodeling in the mucosa-submucosa layer of the twenty-eight days diabetic esophagus was characterized by increased circumferential length, luminal area and circumferential stiffness (Figure 1B) [Yang et al., 2006]. The esophageal remodeling was also found in NIDDM rat esophagus [Zhao et al., 2007]. In the diabetic rats the esophageal weight per length, wall thickness (figure 2A), cross-sectional wall area and the collagen fraction (figure 2B) in the mucosa-submucosa layer increased in the NIDDM rats. The opening angle (figure 2C) and residual strain distribution in the outer surface of the wall decreased, whereas the passive circumferential stiffness of the esophageal wall increased (figure 2D). One previous human study reported that among six cases of diffuse muscular hypertrophy of the esophagus, four cases were associated with DM [Iyer et al., 1986]. Recently, Frøkjær et al [2007] in a human study found an increase in esophageal wall thickness and altered deformation to the distension with reduced longitudinal shortening and the radial stretch in IDDM patients.

Stomach remodeling: A morphological study in DM rats demonstrated that the gastric mucosa thickness increased in DM rats compared with controls [Watanabe et al., 1995]. Remodeling of the interstitial cells network of Cajal in the stomach were found both in animals and humans with DM [Forster et al., 2005; Horvath et al., 2005; Long et al., 2004]. A histopathological study of the human stomach in DM patients with severe gastroparesis showed prominent collagenization and smooth muscle atrophy of the muscle layer [Ejskjaer et al., 1999]. Regarding the biomechanical remodeling of the stomach only one report by Liao et al is available [Liao et al., 2006]. The rat stomach was distended *in vitro*. Gastric compliance, the surface tension, and circumferential and longitudinal deformation-pressure curves were calculated based on three-dimensional ultrasound reconstructions of non-diabetic and diabetic stomach models. In experimental DM, gastric compliance was lowered both in the non-glandular stomach (proximal part) and the glandular stomach (distal part) (figure 3). Furthermore, the circumferential stiffness in the non-glandular part increased (figure 3). The structural changes of the stomach due to DM may together with the sensory and motor nerve dysfunction contribute to the delayed gastric emptying and the symptoms in diabetic patients.

Small intestinal remodeling: Many human [Folwaczny et al., 1999; Charlton et al., 2000; He et al., 2001; Secondulfo et al., 2004] and animal [Folwaczny et al., 1999; Verne and Sninsky, 1998; Zoubi et al., 1995; Bhor et al., 2004; Noda et al., 2001; Tormo et al., 2002; Adachi et al., 2003] studies have shown that DM causes morphological alterations in the small intestine. However, only few data exist on biomechanical remodeling of the small intestine in DM [Jorgensen et al., 2001].

Recently we did a series of studies on the morphological and biomechanical remodeling of small intestine in the STZ-induced diabetic rats [Sha et al., 2006; Zhao et al., 2002, 2003a, 2003b]. The major findings were that the opening angle and residual strain were lower in the duodenum and larger in the jejunum and ileum in diabetic rats compared to normal rats (Figure 4).

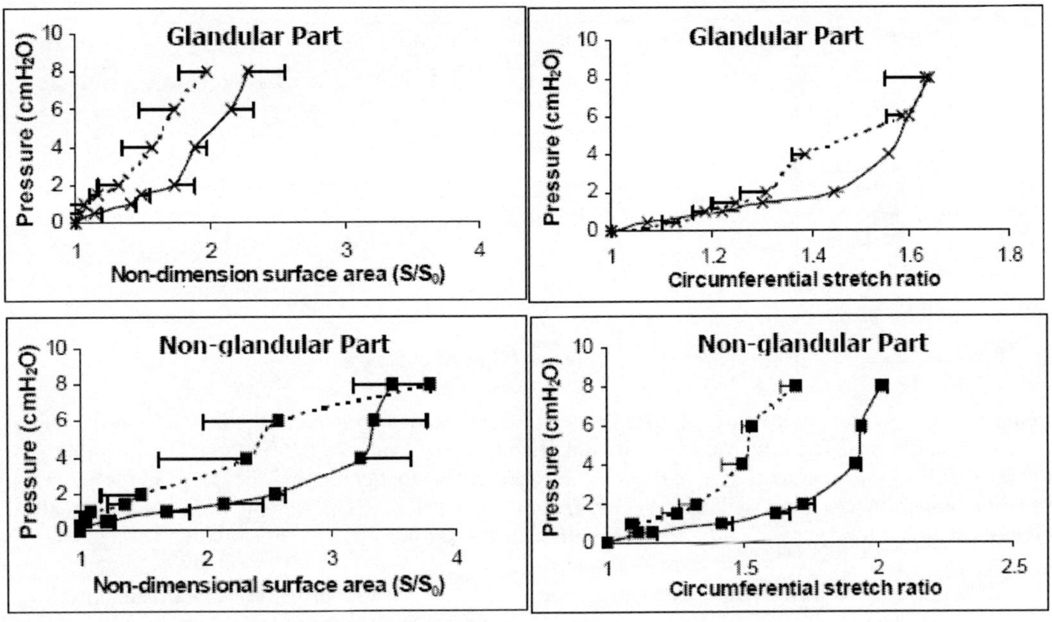

Figure 3. Left: Non-dimensional surface area-pressure curve for the glandular (top) and the non-glandular stomach part (bottom) in the non-diabetic (solid lines) and the diabetic (dashed lines) rats. Surface area for the non-glandular stomach in the non-diabetic group increased more than the diabetic group ($P<0.02$). Right: The circumferential stretch ratio-pressure for the glandular (top) and the non-glandular part (bottom) in the non-diabetic (solid lines) and the diabetic (dashed lines) rats. The circumferential deformation of the non-glandular part in the diabetic group was smaller than that in the non-diabetic group ($P<0.01$).

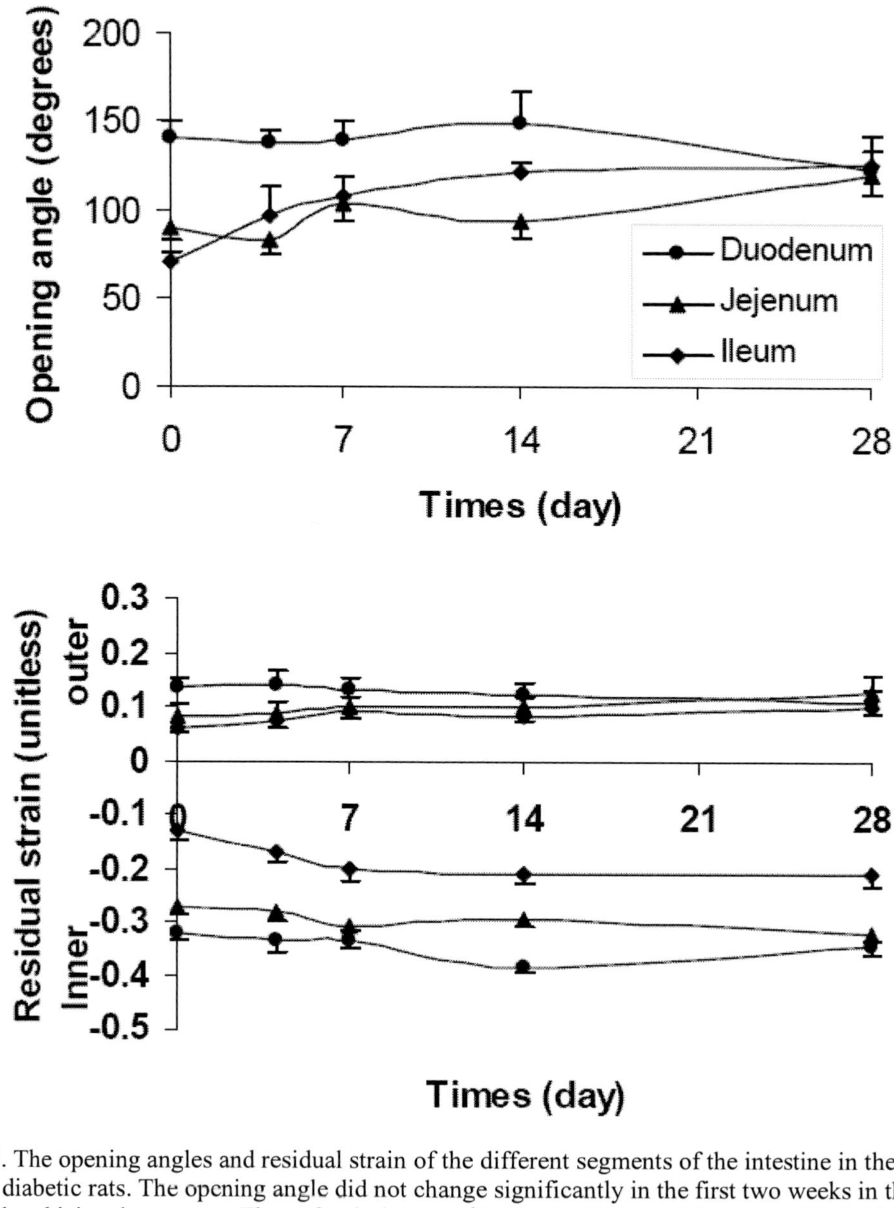

Figure 4. The opening angles and residual strain of the different segments of the intestine in the STZ-induced diabetic rats. The opening angle did not change significantly in the first two weeks in the duodenal and jejunal segments. Thereafter it decreased in the duodenum and increased in the jejunum. The opening angle was increased after 4 days diabetes in the ileum. The same patterns as the opening angle were found for the residual strain at the mucosa and serosa.

Stress-strain relationship analysis showed that both in the circumferential and longitudinal directions the stiffness of the intestinal wall increased with the duration of diabetes (Figure 5). The viscoelastic behavior of intestinal wall also changed during the development of diabetes (Figure 6). Furthermore, we confirmed previous data that the intestinal weight, weight per-unit length and wall thickness and area increased in the diabetic rats (Figure 7).The stress-strain distribution and viscoelastic behavior mainly reflects the elastic properties of intestine. The changes in elastic properties reflect the structural remodeling of the intestinal wall during the diabetic development.

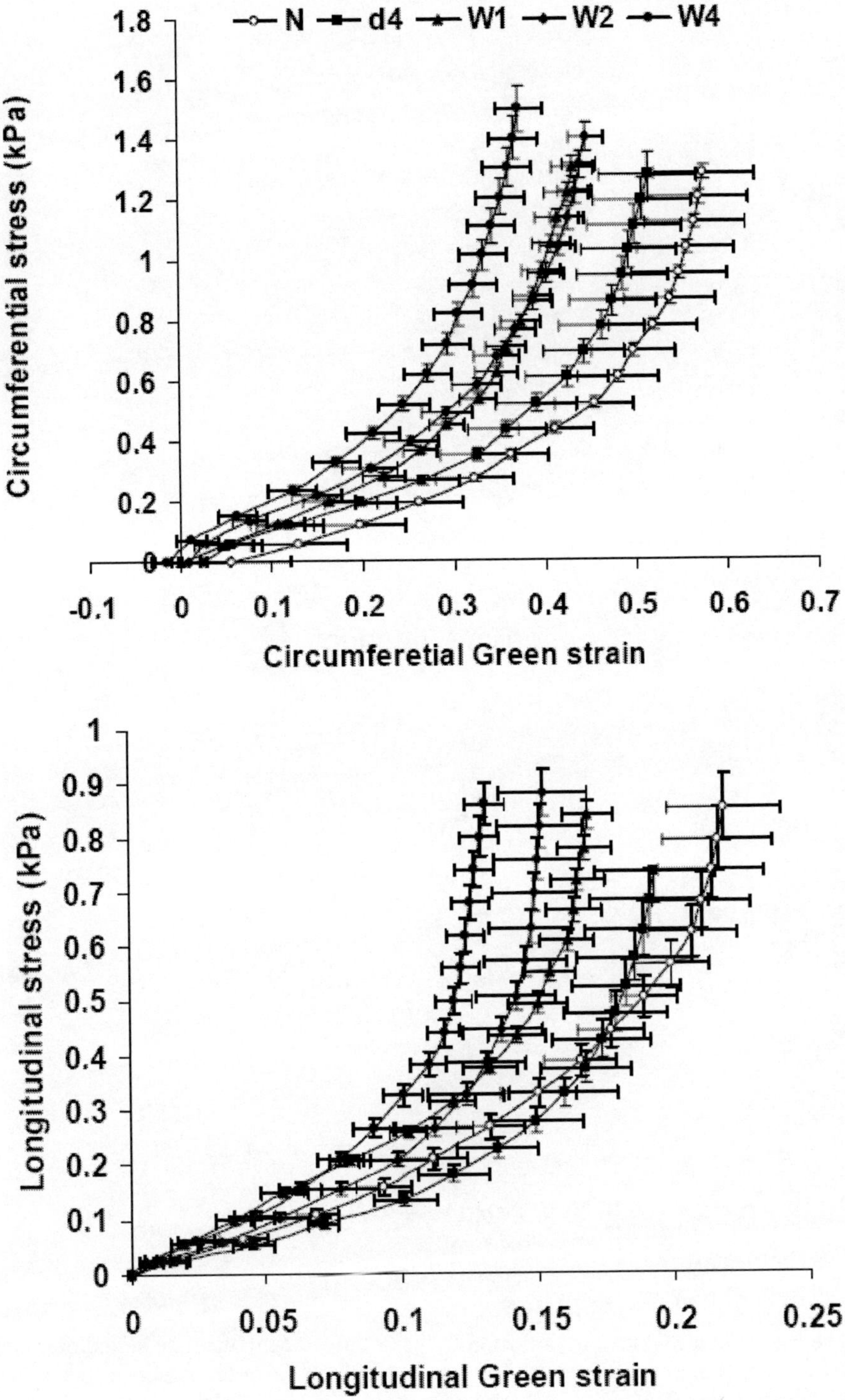

Figure 5. The circumferential (top) and longitudinal (bottom) stress-strain relations of the duodenum a in the STZ-induced diabetic rats. The stress-strain curves in both directions shifted to the left during experimental diabetes indicating the wall became stiffer.

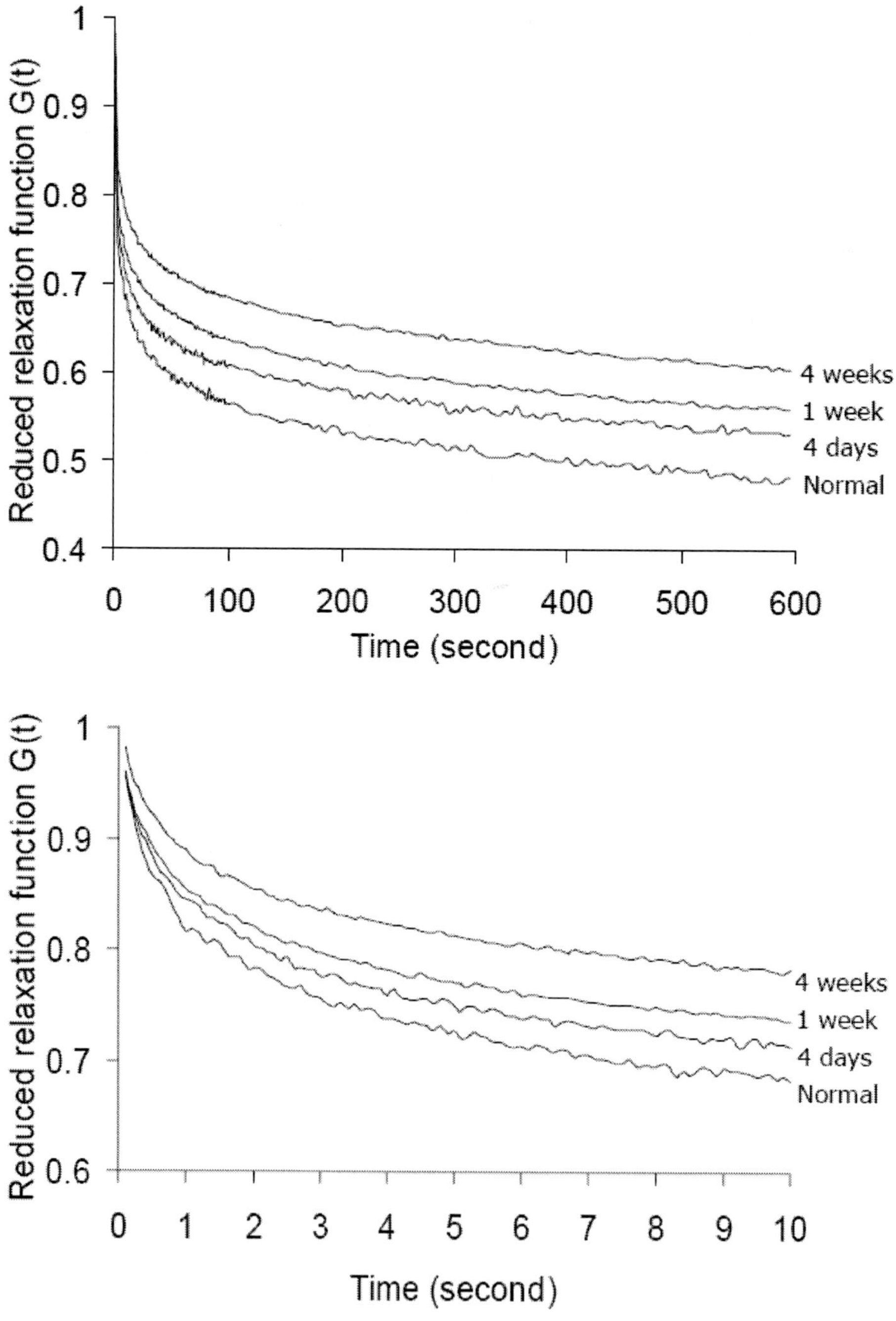

Figure 6. The mean reduced relaxation function vs linear time scale in the time period of 600 seconds (top) and 10 seconds (bottom) was shown. The curves appear in the order of largest-to-smallest *G* (t) as W4, W1, 4d and Normals. The stress relaxation decreased with the development of experimental diabetes (all correlation coefficients > 0.99).

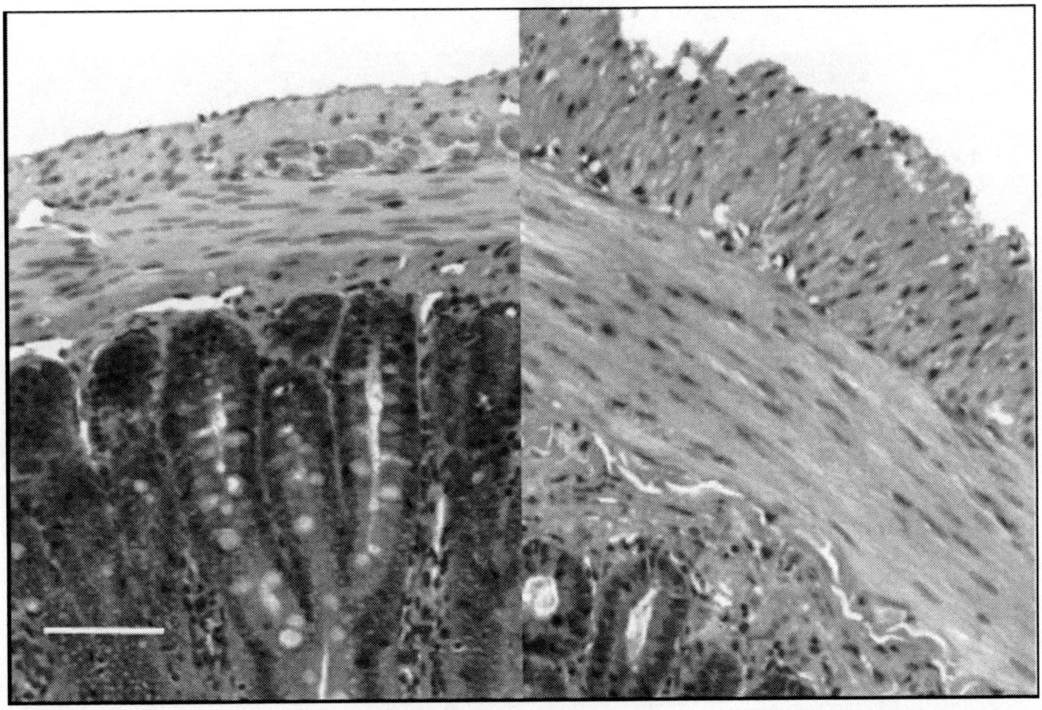

Figure 7. The micro-photographs show the normal (left) and 4 weeks diabetic (right) duodenal histological sections. It clearly demonstrated that the muscle and submucosa layers in the diabetic duodenum became much thicker than in the normal duodenum. The bar is 100 um.

Colon remodeling: Some studies have demonstrated the morphological and histological remodeling in the colon due to DM, for example, 1) changes in the innervation of the colon of the streptozotocin-diabetic model shortly after the induction of diabetes [Lincoln et al., 1984]; 2) abnormalities of endocrine cells in the colon of type 2 diabetic animal model [Spangeus et al., 1998, 1999]; 3) impairment of nitrergic enteric neurons in spontaneously diabetic rats [Yoneda et al., 2001]; 4) deficiency of KIT-positive cells in the colon of patients with DM [Nakahara et al., 2002]; 5) early activated the apoptosis cascade in DAN patients [Guo et al., 2004]; 6) reduced density of ghrelin-immunoreactive cells in animal models of human diabetes type 1 and 2 [Rauma et al., 2006]; and 7) increased the thickness of the subepithelial collagen layer in the patients with DM [Unal et al., 2008]. However, it lacks the data of biomechanical remodeling in the colon due to DM. Recently we investigated the time-dependent biomechanical and morphometric properties of the colon in diabetic rats [Zhao et al., 2008]. The colon weight per-unit length, wall thickness and area, and the thickness of different layers increased in the diabetic rats. For most parameters the biggest changes was observed during the first four weeks where after it leveled off. The opening angle and residual strain were bigger in the colon in diabetic rats compared to normal rats. Stress-strain analysis showed that both in the circumferential and longitudinal directions the stiffness of the colon wall increased with the duration of diabetes (Figure 8).

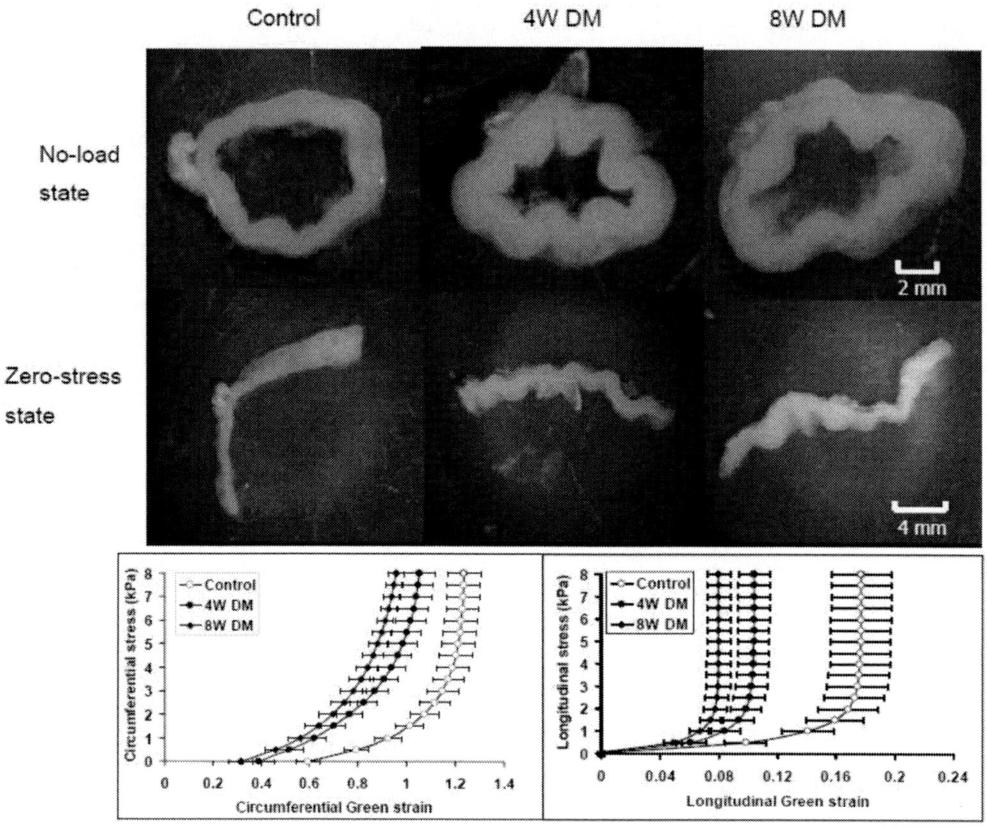

Figure 8. The top figure shows the no-load state and zero-stress state in control, 4W and 8W streptozotocin-induced diabetic rats, the wall thickness and opening angle increased during the development of diabetes. The bottom figure shows the relation between circumferential (left) and longitudinal (right) stress and strain. Both in the circumferential direction and the longitudinal directions, the stress-strain curves shifted to the left in the 4W and 8W diabetic groups compared to those in the control group. Thus, the colon wall stiffness increased in both directions during the development of diabetes.

Mechanisms of Morphological and Biomechanical Remodeling of GI Tract in Diabetes

Untreated diabetic rats eat 2-3 times as much as controls, and it has been suggested that this hyperphagia account for the observed GI growth seen in these animals [Jervis and Levin RJ, 1966]. However, diabetic rats' pair-feed diets isocaloric with those of controls still exhibit significantly greater small bowel mass and enhanced crypt cell DNA synthesis [Miller et al., 1977; Granneman and Stricker, 1984]. This indicates that at least part of the growth response of GI adaptation in diabetes is independent of increased nutrient consumption. The mechanism by which hyperphagia stimulates GI growth is unclear. One study demonstrated increased concentrations of both plasma and tissue glucagon-like peptide 2 (GLP-2) that correlate with GI growth and adaptation to STZ diabetes. Furthermore, both the elevation in GLP-2 levels and the bowel growth response were reverted by insulin therapy. Interestingly, the increments in plasma levels of GLP-2 preceded any changes in intestinal weight, thereby providing further evidence for a relationship between GLP-2 and the induction of GI growth in diabetes [Fischer et al., 1997]. Their study supports the idea that an increased nutrient load may stimulate release of hormones trophic to the GI tract. The combination of hyperphagia and increased gastric emptying [Granneman and Stricker, 1984] observed in diabetic rats results in a greater nutrient load being delivered to the small intestine. Because luminal nutrients, including fat and carbohydrate, are physiological L cell secretagogues *in vivo* [Brubaker et al., 1991; Vahl and D'Alessio, 2004] , this provides a possible mechanism whereby increased luminal nutrients stimulate release of GLP-2 into the circulation, thereby exerting a trophic action on the intestinal epithelium. It is also well known that the epithelial homeostasis is balanced by regulation of cell proliferation and cell death. The active form of caspase-3 was attenuated 1 week after streptozotocin treatment. Suppression of apoptosis in the early days of STZ-induced diabetes was responsible for the increased mucosal height in the small intestine in STZ-induced diabetic animals [Noda et al., 2001]. GLP-2 also inhibits apoptosis in the small intestine [Drucker, 2001]. Although no direct evidence of intestinal mucosa proliferation was reported on human diabetics, small bowel hypertrophy was described in a patient carrying a proglucagon-expressing tumor that secreted the intestinal PGDPs [Gleeson et al., 1971].

Advanced glycation end products (AGEs) produced during the development of DM are likely associated to the morphological changes and biomechanical remodeling of the GI tract in DM. AGEs are generated by the sequential non-enzymatic glycation of protein amino groups and by oxidation reaction [Singh et a., 2001]. The accumulation of AGEs in tissues alters the structure and function of matrix proteins [Monnier et al, 1992]. Studies on DM and ageing show that AGEs are causing cross-linking of collagen molecules responsible for basement membrane thickening and loss of matrix elasticity [Ulrich et al., 2001; Sanchez et al., 2000; Reddy, 2004]. The GI wall is rich in collagen [Gregersen, 2002]. Therefore, alterations in the orientation, configuration, and content of collagen are likely to affect the biomechanical properties of the GI tract based on remodeling studies in other disease models. Some studies on the arteries in the diabetes have demonstrated that the non-enzymatic glycation of collagen have a close relation with the increased arterial wall stiffness [Sims et al., 1996; Wolffenbuttel et al., 1998]. We believe that the same way apply to the collagen in the GI wall. AGEs may contribute to the diabetic GI morphological and biomechanical remodeling by at least two major mechanisms. The first is receptor-independent alteration of the extracellular matrix architecture by non-enzymatic glycation and the formation of protein cross-links. The second mechanism is receptor-dependent and consists of modulation of cellular functions through ligation of specific cell surface receptors [Bierhaus et al., 2005; Stern et al., 2002; Wautier et al., 1996]. Sanchez SS and coworkers demonstrated that the expression of small intestinal extracellular matrix proteins changed in STZ-induced diabetic rats [Sanchez et al., 2000]. Furthermore, the different components in GI wall have different relaxation functions [Fung, 1993]. For example, collagen has a large Young's modulus and it relaxes less. Smooth muscle has a low modulus of elasticity and relaxes more. Purslow and co-workers have studied the effect of collagen reorientation on creep and stress relaxation in soft connective tissues [Purslow et al., 1998]. They demonstrated that the collagen reorientation is not a primary source of their viscoelastic properties. The relaxation processes within the collagen fibers or at the fiber-matrix interface may be responsible for their viscoelastic nature. Following the changes of structural components in the GI wall during diabetes, the viscoelastic properties of GI wall change as well, as we demonstrated that the viscoelastic behavior of intestinal wall changed during the development of diabetes. Therefore, the nonenzymatic glycation of the GI tissue induced by long-term hyperglycemia seems to be an important mechanism behind the GI wall remodeling in DM.

Implication of GI Remodeling on Mechanosensory Assessments

Mechanosensation is of fundamental importance for the GI function. The mechanosensitive nerve endings exist extensively in the GI tract where they serve a critical role in homeostasis. Several mechanoreceptor-like structures have been identified such as IGLE and IMA in the GI tract [Patterson et al., 2003., Phillips and Powley, 2000; Raab et al., 2004; Swithers et al., 2002; Zagorodnyuk et al., 2001, 2003, 2005]. The mechanosensitive afferents in the intrinsic and extrinsic pathways were described as low-, wide-dynamic- or high-threshold tension-receptors. The GI wall structure or deformation changes in the DM will alter the relative positions of the mechanosensitive afferents (zero setting of the mechanosensitive afferents) [Pederse et al., 2003]. The biomechanical remodeling in the DM such as alterations of residual strain and stress distribution and increase the wall stiffness will alter the tension and stress distribution of the mechanosensitive afferents. As results, the perception and motility of the GI tract will change as well. Hence, the morphological changes and biomechanical remodeling of GI tract in the DM is likely to affect the function of mechanosensitive afferents in the GI wall and further affect the motor and sensory function. However, so far data are sparse on the association between the morphological and biomechanical remodeling of GI tract and the motor-sensory dysfunction in the DM [Frokjaer et al., 2007a, 2007a]. The multimodal stimulations have been proven accurate and reliable in the assessment of visceral sensation in several studies [Drewes et al., 2002, 2003; Pedersen et al., 2004]. The geometry data of GI tract can be obtained by impedance and cross-sectional imaging such as ultrasound [Drewes et al., 2003; Frokjaer et al., 2005; Pedersen et al., 2005]. Combined with pressure recordings, biomechanical parameters such as tension, stress and strain can be obtained and correlated to the symptoms and mechanosensory data in the DM subjects [Frokjaer et al., 2007b]. Therefore, combined studies the GI motor-sensory dysfunction and morphological and biomechanical remodeling in the diabetic GI tract will improve our understanding about the pathophysiology of GI disorders in the DM patients.

Future Perspectives

The pathogenesis for GI motor disorders in diabetes is not well understood. However, as mentioned above biomechanical remodeling of the GI tract during the development of diabetes may contribute to the GI motor disorders. Therefore, the study of the relation between the GI motor dysfunction and morphological and biomechanical remodeling in the diabetic GI tract may shed more light to understand the mechanism of GI dysfunction in the diabetic patients.

Combine Study of Active and Passive Biomechanical Properties of GI Tract

According to the well known Hill's model in muscle mechanics and physiology, we consider the GI tissue to consist of passive and active elements (Figure 9). The first step is to analyze the passive elements belonging to the connective tissue because the passive elastic properties reflect the state of the structural components determining a specific function or malfunction. When the passive properties are known, future work can be directed to complete the Hill's three-element model so that we may have an understanding of the interaction between the parallel and contractile elements in diseases and in remodeling of the GI tissues. In the future experiments will be designed to mimic physiological conditions both *in vivo* and *in vitro* in normal and diabetic rats. The pressure in the lumen of the GI segment and the rate of flow through the lumen can be controlled. At the same time the motility pattern of GI tract can be measured. Through these combined measurements, we can get more useful physiological data and further understand the mechanism of diabetes and the GI remodeling and function.

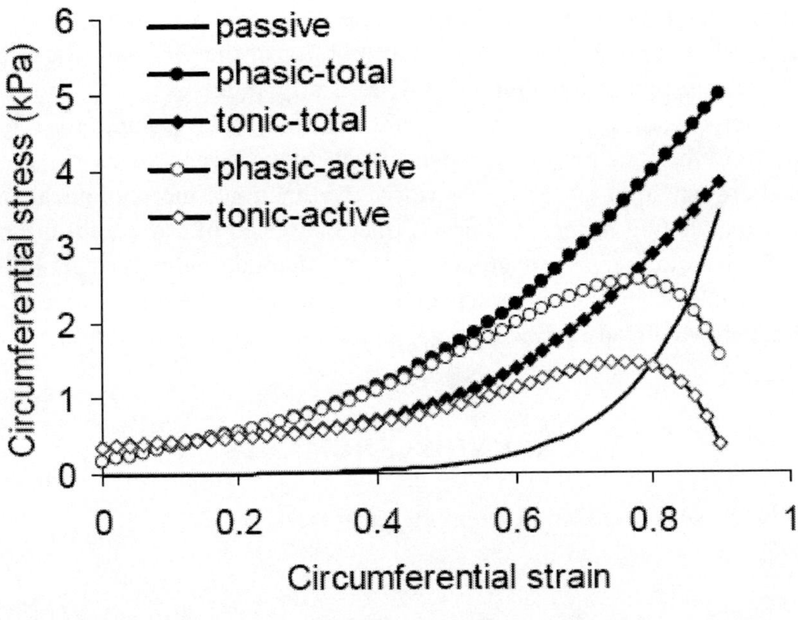

Figure 9. The typical active phasic and tonic stress-strain relation curves from rat jejunal segment are shown. The isometric length-tension diagram known from *in vitro* studies of muscle strips can be obtained in intact intestinal segments. The total and passive stress increased in an exponential-like way as function of strain. The active phasic and tonic stress reached a maximum of 2.54 and 1.46 kPa at a strain of 0.78 and 0.76.

Investigation of the Mechanism for the GI Remodeling in Diabetes

The studies showed that diabetes alters the stress-strain relation of the GI tract in a way consistent with a stiffer wall. However, no evidence of which structure in the wall causes the increased stiffness is given. The GI submucosa is mainly composed of collagen. As we postulated above alterations in the orientation, configuration, and content of collagen are likely candidates based on remodeling studies in other disease models. In order to address this issue, histochemical, immunohistochemical and electron microscopic work must be done to study the contents, ultrastructure and orientation of collagen in the intestine of diabetic rats. The analysis of advanced non-enzymatic glycation ending products (AGEs) and receptors of AGEs expression at diabetic Gl tissues will also be subject of study.

The morphological remodeling of GI tract in the diabetes is likely to affect the zero-setting of mechanoreceptors in the GI wall. Therefore, the identification of mechanoreceptors will obviously be beneficial to understand the mechanisms of GI remodeling during the diabetes. Furthermore, as mentioned in the literature, the motility disorder of GI tract is likely closely associated with the nervous dysfunction in the GI wall. Therefore, the next step is to investigate the enteric nerve function combined with the biomechanical properties of GI tract in the diabetic animal model.

Conclusion

GI symptoms are frequent in the diabetic patients and are associated with sensory-motor abnormalities, such as impaired perception and motility of the GI tract. The pathogenesis of abnormal GI sensory-motor function in DM is clearly multi-factorial. The morphological changes and biomechanical remodeling of the GI wall may compromise the GI motor function and affect the function of the mechanosensitive afferents in the GI wall. Studies of the relation between the GI motor-sensory dysfunction, morphological changes and biomechanical remodeling in the diabetic GI tract may shed of more light to understand the mechanism of GI motor-sensory dysfunction in the diabetic patients. This knowledge may prove to be valuable in the development of new treatment strategies, such as using agents which can break down already formed glycation end product protein-protein crosslinks.

Acknowledgments

Karen Elise Jensens Foundations supported this work.

References

Adachi T, Mori C, Sakurai K, Shihara N, Tsuda K, Yasuda K. Morphological changes and increased sucrase and isomaltase activity in small intestines of insulin-deficient and type 2 diabetic rats. *Endocr J.* 2003; 50: 271-279.

Adeghate, E., P. Schattner, and E. Dunn. "An update on the etiology and epidemiology of diabetes mellitus." *Ann. N Y Acad. Sci.* 2006; 1084: 1-29.

Anderson RP. Coeliac disease. *Aust Fam Physician.* 2005; 34: 239-242.

Anjaneyulu M, Ramarao P. Studies on gastrointestinal tract functional changes in diabetic animals. Methods Find. *Exp. Clin. Pharmacol.* 2002; 24: 71-75.

Annese V, Bassotti G, Caruso N, De Cosmo S, Gabbrielli A, Modoni S, Frusciante V, Andriulli A. Gastrointestinal motor dysfunction, symptoms, and neuropathy in noninsulin-dependent (type 2) diabetes mellitus. *J. Clin. Gastroenterol.* 1999; 29: 171-177.

Battle WM, Snape WJ Jr, Alavi A, Cohen S, Braunstein S. Colonic dysfunction in diabetes mellitus. *Gastroenterology*. 1980; 79: 1217-1221.

Battle WM, Cohen JD, Snape WJ Jr. Disorders of colonic motility in patients with diabetes mellitus. *Yale J. Biol. Med.* 1983; 56: 277-283.

Bhor VM, Raghuram N, Sivakami S. Oxidative damage and altered antioxidant enzyme activities in the small intestine of streptozotocin-induced diabetic rats. *Int. J. Biochem. Cell Biol*. 2004; 36: 89-97.

Bierhaus A, Humpert PM, Morcos M et al. Understanding RAGE, the receptor for advanced glycation end products. *J. Mol. Med.* 2005; 83: 876-886.

Bjornsson ES, Urbanavicius V, Eliasson B, Attvall S, Smith U, Abrahamsson H. Effects of hyperglycemia on interdigestive gastrointestinal motility in humans. *Scand J. Gastroenterol*. 1994; 29: 1096-1104.

Brubaker PL, Stobie KM, Roberge JN, Lui EY, Drucker DJ. Proglucagon-derived peptides in the neuroendocrine system. *Adv. Exp. Med. Biol*. 1991: 291: 143-159.

Camilleri M, Malagelada JR. Abnormal intestinal motility in diabetics with the gastroparesis syndrome. *Eur. J. Clin. Invest*. 1984; 14: 420-427.

Charlton M, Ahlman B, Nair KS. The effect of insulin on human small intestinal mucosal protein synthesis. *Gastroenterology*. 2000:118: 299-306.

de Boer SY, Masclee AA, Lam WF, Schipper J, Jansen JB, Lamers CB. Hyperglycemia modulates gallbladder motility and small intestinal transit time in man. *Dig. Dis. Sci.* 1993; 38: 2228-2235.

Dooley CP, el Newihi HM, Zeidler A, Valenzuela JE. Abnormalities of the migrating motor complex in diabetics with autonomic neuropathy and diarrhea. *Scand J. Gastroenterol.* 1988; 23: 217-223.

Drewes AM, Schipper KP, Dimcevski G, Petersen P, Andersen OK, Gregersen H, Arendt-Nielsen L. et al. Multimodal assessment of pain in the esophagus: a new experimental model. *Am. J. Physiol. Gastrointest Liver Physiol.* 2002; 283: G95-103.

Drewes AM, Gregersen H, Arendt-Nielsen L. Experimental pain in gastroenterology: a reappraisal of human studies. *Scand J Gastroenterol*. 2003; 38: 1115-1130.

Drewes AM, Pedersen J, Liu W, Arendt-Nielsen L, Gregersen H. Controlled mechanical distension of the human oesophagus: sensory and biomechanical findings. *Scand. J. Gastroenterol*. 2003; 38: 27-35.

Drucker DJ. Glucagon-like peptide 2. *J. Clin. Endocrinol Metab*. 2001; 86: 1759-1764.

Ejskjaer NT, Bradley JL, Buxton-Thomas MS, Edmonds ME, Howard ER, Purewal T, Thomas PK, Watkins PJ. Novel surgical treatment and gastric pathology in diabetic gastroparesis. *Diabet Med.*1999; 16: 488-495.

el Salhy M. Gastrointestinal transit in nonobese diabetic mouse: an animal model of human diabetes type 1. *J. Diabetes Complications*. 2001; 15: 277-284.

el Salhy M. Gastrointestinal transit in an animal model of human diabetes type 2: relationship to gut neuroendocrine peptide contents. *Ups J. Med. Sci.* 2002; 107: 101-110.

Fischer KD, Dhanvantari S, Drucker DJ, Brubaker PL. Intestinal growth is associated with elevated levels of glucagon-like peptide 2 in diabetic rats. *Am. J. Physiol.* 1997; 273: E815-E820.

Folwaczny C, Hundegger K, Volger C, Sorodoc J, Kuhn M, Tatsch K, Landgraf R, Karbach U. Measurement of transit disorders in different gastrointestinal segments of patients with diabetes mellitus in relation to duration and severity of the disease by use of the metal-detector test. *Z. Gastroenterol.* 1995; 33: 517-526.

Folwaczny C, Riepl R, Tschop M, Landgraf R. Gastrointestinal involvement in patients with diabetes mellitus: Part I (first of two parts). Epidemiology, pathophysiology, clinical findings. *Z. Gastroenterol.* 1999: 37: 803-815.

Forrest A, Parsons M. The enhanced spontaneous activity of the diabetic colon is not the consequence of impaired inhibitory control mechanisms. *Auton Autacoid. Pharmacol.* 2003; 23: 149-158.

Forrest A, Huizinga JD, Wang XY, Liu LW, Parsons M. Increase in stretch-induced rhythmic motor activity in the diabetic rat colon is associated with loss of ICC of the submuscular plexus. *Am. J. Physiol. Gastrointest Liver Physiol.* 2008; 294: G315-G326.

Forster J, Damjanov I, Lin Z, Sarosiek I, Wetzel P, McCallum RW. Absence of the interstitial cells of Cajal in patients with gastroparesis and correlation with clinical findings. *J. Gastrointest Surg* 2005; 9: 102-108.

Frokjaer JB, Andersen SD, Gale J, Arendt-Nielsen L, Gregersen H, Drewes AM. An experimental study of viscero-visceral hyperalgesia using an ultrasound-based multimodal sensory testing approach. *Pain.* 2005; 119: 191-200.

Frokjaer JB, Andersen SD, Ejskaer N, Funch-Jensen P, Arendt-Nielsen L, Gregersen H, Drewes AM. Gut sensations in diabetic autonomic neuropathy. *Pain.* 2007a; 131(3): 320-329.

Frokjaer JB, Andersen SD, Ejskjaer N, Funch-Jensen P, Drewes AM, Gregersen H. Impaired contractility and remodeling of the upper gastrointestinal tract in diabetes mellitus type-1. *World J. Gastroenterol.* 2007b; 13: 4881-4890.

Fung YC. *Biomechanics. Properties of living tissues*. Berlin: Springer-Verlag, 1993.

Gleeson MH, Bloom SR, Polak JM, Henry K, Dowling RH. Endocrine tumour in kidney affecting small bowel structure, motility, and absorptive function. *Gut.* 1971; 12: 773-782.

Granneman JG, Stricker EM. Food intake and gastric emptying in rats with streptozotocin-induced diabetes. *Am. J. Physiol.* 1984: 247: R1054-R1061.

Gregersen H. *Biomechanics of the Gastrointestinal Tract.* London: Springer-Verlag, 2002.

Guo C, Quobatari A, Shangguan Y, Hong S, Wiley JW. Diabetic autonomic neuropathy: evidence for apoptosis in situ in the rat. *Neurogastroenterol Motil.* 2004; 16: 335-345.

He CL, Soffer EE, Ferris CD, Walsh RM, Szurszewski JH, Farrugia G. Loss of interstitial cells of cajal and inhibitory innervation in insulin-dependent diabetes. *Gastroenterology.* 2001; 121: 427-434.

Holloway RH, Tippett MD, Horowitz M, Maddox AF, Moten J, Russo A. Relationship between esophageal motility and transit in patients with type I diabetes mellitus. *Am. J. Gastroenterol.* 1999; 94: 3150-3157.

Horowitz M, Maddox AF, Wishart JM, Harding PE, Chatterton BE, Shearman DJ. Relationships between oesophageal transit and solid and liquid gastric emptying in diabetes mellitus. *Eur. J. Nucl. Med.* 1991; 18: 229-234.

Horowitz M, Wishart JM, Jones KL, Hebbard GS. Gastric emptying in diabetes: an overview. *Diabet Med.* 1996; 13: S16-S22.

Horowitz M, Samsom M. *Gastrointestinal function in diabetes mellitus*. Chichester: John Wiley and Sons, Ltd, 2004.

Horvath VJ, Vittal H, Ordog T. Reduced insulin and IGF-I signaling, not hyperglycemia, underlies the diabetes-associated depletion of interstitial cells of Cajal in the murine stomach. *Diabetes*. 2005; 54: 1528-1533.

Iber FL, Parveen S, Vandrunen M, Sood KB, Reza F, Serlovsky R, Reddy S. Relation of symptoms to impaired stomach, small bowel, and colon motility in long-standing diabetes. *Dig. Dis. Sci.* 1993; 38: 45-50.

Iida M, Ikeda M, Kishimoto M, Tsujino T, Kaneto H, Matsuhisa M, Kajimoto Y, Watarai T, Yamasaki Y, Hori M. Evaluation of gut motility in type II diabetes by the radiopaque marker method. *J. Gastroenterol Hepatol*. 2000; 15: 381-385.

Iyer SK, Chandrasekhara KL, Sutton A. Diffuse muscular hypertrophy of esophagus. *Am J Med*. 1986; 80: 849-852.

James AN, Ryan JP, Crowell MD, Parkman HP. Regional gastric contractility alterations in a diabetic gastroparesis mouse model: effects of cholinergic and serotoninergic stimulation. *Am. J. Physiol. Gastrointest Liver Physiol*. 2004; 287: G612-G619.

Jervis EL, Levin RJ. Anatomic adaptation of the alimentary tract of the rat to the hyperphagia of chronic alloxan-diabetes. *Nature*. 1966; 210: 391-393.

Jones KL, Horowitz M, Wishart MJ, Maddox AF, Harding PE, Chatterton BE . Relationships between gastric emptying, intragastric meal distribution and blood glucose concentrations in diabetes mellitus. *J. Nucl. Med* 1995; 36: 2220-2228.

Jorgensen F, Boesen F, Andersen EB, Hesse B. Oesophageal transit in patients with autonomic dysfunction. The effect of treatment with fludrocortisone. *Clin. Physiol*. 1991; 11: 83-92.

Jorgensen CS, Ahrensberg JM, Gregersen H, Flyvberg A. Tension-strain relations and morphometry of rat small intestine in experimental diabetes. *Dig. Dis. Sci.* 2001: 46: 960-967.

Jung HK, Kim DY, Moon IH, Hong YS. Colonic transit time in diabetic patients--comparison with healthy subjects and the effect of autonomic neuropathy. *Yonsei Med. J.* 2003; 44(2): 265-272.

Kamath MV, Tougas G, Fitzpatrick D et al. Assessment of the visceral afferent and autonomic pathways in response to esophageal stimulation in control subjects and in patients with diabetes. *Clin. Invest. Med*. 1998: 21: 100-113.

Kamath MV, May A, Hollerbach S et al. Effects of esophageal stimulation in patients with functional disorders of the gastrointestinal tract. *Crit Rev Biomed Eng*. 2000: 28: 87-93.

Keshavarzian A, Iber FL, Dangleis MD, Cornish R. Intestinal-transit and lactose intolerance in chronic alcoholics. *Am. J. Clin. Nutr*. 1986; 44: 70-76.

Ko GT, Chan WB, Chan JC, Tsang LW, Cockram CS. Gastrointestinal symptoms in Chinese patients with Type 2 diabetes mellitus. *Diabet Med*. 1999; 16: 670-674.

Korolkiewicz R, Rekowski P, Szyk A, Kato S, Yasuhiro T, Kubomi M, Tashima K, Takeuchi K. Effects of diabetes mellitus on the contractile activity of carbachol and galanin in isolated gastric fundus strips of rats. *Pharmacology*. 1998; 57: 65-78.

Liao D, Zhao J, Gregersen H. Three-dimensional geometry analysis of the stomach in type II diabetic GK rats. *Diabetes Res. Clin. Pract*. 2006; 71: 1-13.

Lincoln J, Bokor JT, Crowe R, Griffith SG, Haven AJ, Burnstock G. Myenteric plexus in streptozotocin-treated rats. Neurochemical and histochemical evidence for diabetic neuropathy in the gut. *Gastroenterology*. 1984; 86 654-661.

Liu J, Qiao X, Micci MA, Pasricha PJ, Chen JD. Improvement of gastric motility with gastric electrical stimulation in STZ-induced diabetic rats. *Digestion*. 2004; 70: 159-166.

Long QL, Fang DC, Shi HT, Luo YH. Gastro-electric dysrhythm and lack of gastric interstitial cells of cajal. *World J. Gastroenterol*. 2004; 10: 1227-1230.

Longstreth GF. Diabetic thoracic polyradiculopathy: ten patients with abdominal pain. *Am. J. Gastroenterol*. 1997; 92: 502-505.

Maleki D, Locke GR, Camilleri M 3rd, Zinsmeister AR, Yawn BP, Leibson C, Melton LJ 3rd. Gastrointestinal tract symptoms among persons with diabetes mellitus in the community. *Arch. Intern. Med*. 2000; 160: 2808-2816.

Mayhew TM, Carson FL, Sharma AK. Small intestinal morphology in experimental diabetic rats: a stereological study on the effects of an aldose reductase inhibitor (ponalrestat) given with or without conventional insulin therapy. *Diabetologia*. 1989: 32: 649-654.

Mehta N, Veliath S, Thombre DP. Effect of experimental diabetes and vagotomy on gastric emptying in rats. *Indian J. Physiol. Pharmacol*. 2002; 46: 441-448.

Miller DL, Hanson W, Schedl HP, Osborne JW. Proliferation rate and transit time of mucosal cells in small intestine of the diabetic rat. *Gastroenterology*. 1977; 73: 1326-1332.

Monnier VM, Sell DR, Nagaraj RH, Miyata S, Grandhee S, Odetti P, Ibrahim SA. Maillard reaction-mediated molecular damage to extracellular matrix and other tissue proteins in diabetes, aging, and uremia. *Diabetes*. 1992; 41 Suppl 2: 36-41.

Nair, M. "Diabetes mellitus, part 1: physiology and complications." *Br. J. Nurs*. 2007; 16: 184-188.

Nakahara M, Isozaki K, Hirota S, Vanderwinden JM, Takakura R, Kinoshita K, Miyagawa J, Chen H, Miyazaki Y, Kiyohara T, Shinomura Y, Matsuzawa Y. Deficiency of KIT-positive cells in the colon of patients with diabetes mellitus. *J. Gastroenterol. Hepatol*. 2002; 17: 666-670.

Nguyen HN, Silny J, Wuller S, Marschall HU, Rau G, Matern S. Abnormal postprandial duodenal chyme transport in patients with long standing insulin dependent diabetes mellitus. *Gut*. 1997; 41: 624-631.

Noda T, Iwakiri R, Fujimoto K, Yoshida T, Utsumi H, Sakata H, Hisatomi A, Aw TY. Suppression of apoptosis is responsible for increased thickness of intestinal mucosa in streptozotocin-induced diabetic rats. *Metabolism*. 2001; 50: 259-264.

Nowak TV, Roza AM, Weisbruch JP, Brosnan MR. Accelerated gastric emptying in diabetic rodents: effect of insulin treatment and pancreas transplantation. *J. Lab. Clin. Med*. 1994; 123: 110-116.

Ogata M, Iizuka Y, Murata R, Hikichi N. Effect of streptozotocin-induced diabetes on cyclosporin A disposition in rats. *Biol. Pharm. Bull*. 1996; 19: 1586-1590.

Ogata M, Uchimura T, Iizuka Y Murata R, Suzuki S, Toyota T, Hikichi N. Effect of non-insulin dependent diabetes on cyclosporin A disposition in Goto-Kakizaki (GK) rats. *Biol. Pharm. Bull*. 1997; 20: 1026-1029.

Patterson LM, Zheng H, Ward SM, Berthoud HR. Vanilloid receptor (VR1) expression in vagal afferent neurons innervating the gastrointestinal tract. *Cell Tissue Res*. 2003; 311: 277-287.

Pedersen J, Gao C, Egekvist H, Bjerring P, Arendt-Nielsen L, Gregersen H, Drewes AM. Pain and biomechanical responses to distention of the duodenum in patients with systemic sclerosis. *Gastroenterology*. 2003; 124: 1230-1239.

Pedersen J, Reddy H, Funch-Jensen P, Arendt-Nielsen L, Gregersen H, Drewes AM. Cold and heat pain assessment of the human oesophagus after experimental sensitisation with acid. *Pain*. 2004; 110: 393-399.

Pedersen J, Drewes AM, Gregersen H. New analysis for the study of the muscle function in the human oesophagus. *Neurogastroenterol. Motil*. 2005; 17: 767-772.

Phillips RJ, Powley TL. Tension and stretch receptors in gastrointestinal smooth muscle: re-evaluating vagal mechanoreceptor electrophysiology. *Brain Res. Brain Res. Rev*. 2000; 34: 1-26.

Purslow PP, Wess TJ, Hukins DW. Collagen orientation and molecular spacing during creep and stress-relaxation in soft connective tissues. *J. Exp. Biol*. 1998: 201: 135-142.

Raab M, Neuhuber WL. Intraganglionic laminar endings and their relationships with neuronal and glial structures of myenteric ganglia in the esophagus of rat and mouse. *Histochem. Cell Biol*. 2004; 122: 445-459.

Rathmann W, Enck P, Frieling T, Gries FA. Visceral afferent neuropathy in diabetic gastroparesis. *Diabetes Care*. 1991; 14: 1086-1089.

Rauma J, Spångeus A, El-Salhy M. Ghrelin cell density in the gastrointestinal tracts of animal models of human diabetes. *Histol. Histopathol*. 2006; 21: 1-5.

Rayner CK, Verhagen MA, Hebbard GS, DiMatteo AC, Doran SM, Horowitz M. Proximal gastric compliance and perception of distension in type 1 diabetes mellitus: effects of hyperglycemia. *Am. J. Gastroenterol*. 2000; 95: 1175-1183.

Reddy GK. AGE-related cross-linking of collagen is associated with aortic wall matrix stiffness in the pathogenesis of drug-induced diabetes in rats. *Microvasc. Res*. 2004; 68: 132-142.

Ricci JA, Siddique R, Stewart WF, Sandler RS, Sloan S, Farup CE. Upper gastrointestinal symptoms in a U.S. national sample of adults with diabetes. *Scand. J. Gastroenterol*. 2000; 35: 152-159.

Samsom M, Jebbink RJ, Akkermans LM, Berge-Henegouwen GP, Smout AJ. Abnormalities of antroduodenal motility in type I diabetes. *Diabetes Care*. 1996; 19: 21-27.

Samsom M, Roelofs JM, Akkermans LM, Berge Henegouwen GP, Smout AJ. Proximal gastric motor activity in response to a liquid meal in type I diabetes mellitus with autonomic neuropathy. *Dig. Dis. Sci*. 1998; 43: 491-496.

Samsom M, Verhagen MAMT. Intestinal function. In: Horowitz M, Samsom M, editor. *Gastrointestinal function in diabetes mellitus*. Chichester: John Wiley and Sons, Ltd, 2004; 177-218.

Sanchez SS, Genta SB, Aybar MJ, Honore SM, Villecco EI, Sanchez Riera AN. Changes in the expression of small intestine extracellular matrix proteins in streptozotocin-induced diabetic rats. *Cell Biol. Int*. 2000; 24: 881-888.

Schvarcz E, Palmer M, Ingberg CM, Aman J , Berne C. Increased prevalence of upper gastrointestinal symptoms in long-term type 1 diabetes mellitus. *Diabet. Med*. 1996; 13: 478-481.

Secondulfo M, Iafusco D, Carratu R, deMagistris L, Sapone A, Generoso M, Mezzogiomo A, Sasso FC, Carteni M, De Rosa R, Prisco F, Esposito V. Ultrastructural mucosal alterations and increased intestinal permeability in non-celiac, type I diabetic patients. *Dig. Liver Dis.* 2004; 36: 35-45.

Sha H, Zhao JB, Zhang ZY, Zhou SP, Tong XL, Zhuang FY, Gregersen H. Effect of Kaiyu Qingwei Jianji on the morphometry and residual strain distribution of small intestine in experimental diabetic rats. *World J. Gastroenterol.* 2006; 12: 7149-7154.

Sims MA, Hasler WL, Chey WD, Kim MS, Owyang C. Hyperglycemia inhibits mechanoreceptor-mediated gastrocolonic responses and colonic peristaltic reflexes in healthy humans. *Gastroenterology*. 1995; 108: 350-359.

Sims TJ, Rasmussen LM, Oxlund H, Bailey AJ. The role of glycation cross-links in diabetic vascular stiffening. *Diabetologia*. 1996: 39: 946-951.

Singh R, Barden A, Mori T, Beilin L. Advanced glycation end-products: a review. *Diabetologia*. 2001; 44: 129-146.

Smith DS, Ferris CD. Current concepts in diabetic gastroparesis. *Drugs*. 2003; 63: 1339-1358.

Smout AJPM. Oesophageal function. In: Horowitz M, Samsom M, editor. *Gastrointestinal function in diabetes mellitus*. Chichester: John Wiley and Sons, Ltd, 2004; 97-116.

Spangeus A, El-Salhy M. Large intestinal endocrine cells in non-obese diabetic mice. *J. Diabetes Complications*. 1998; 12: 321-327.

Spangeus A, el Salhy M, Suhr O, Eriksson J, Lithner F. Prevalence of gastrointestinal symptoms in young and middle-aged diabetic patients. *Scand. J. Gastroenterol.* 1999; 34: 1196-1202.

Spangeus A, Kand M, El-Salhy M. Gastrointestinal endocrine cells in an animal model for human type 2 diabetes. *Dig. Dis. Sci.* 1999; 44: 979-985.

Stern DM, Yan SD, Yan SF, Schmidt AM. Receptor for advanced glycation endproducts (RAGE) and the complications of diabetes. *Ageing Res. Rev.* 2002; 1: 1-15.

Sun WM, Read NW. Anorectal function. In: Horowitz M, Samsom M. editor. *Gastrointestinal function in diabetes mellitus*. Chichester: John Wiley and Sons, Ltd, 2004; 219-246.

Swithers SE, Baronowsky E, Powley TL. Vagal intraganglionic laminar endings and intramuscular arrays mature at different rates in pre-weanling rat stomach. *Auton Neurosci.* 2002; 102: 13-19.

Tahara T, Yamamoto T. Morphological changes of the villous microvascular architecture and intestinal growth in rats with streptozotocin-induced diabetes. *Virchows Arch. A. Pathol. Anat. Histopathol.* 1988: 413: 151-158.

Talley NJ, Young L, Bytzer P, Hammer J, Leemon M, Jones M, Horowitz M. Impact of chronic gastrointestinal symptoms in diabetes mellitus on health-related quality of life. *Am. J. Gastroenterol.* 2001; 96: 71-76.

Talley NJ, Howell S, Jones MP, Horowitz M. Predictors of turnover of lower gastrointestinal symptoms in diabetes mellitus. *Am. J. Gastroenterol.* 2002; 97: 3087-3094.

Tormo MA, Martinez IM, Romero dT, Gil-Exojo I, Campillo JE. Morphological and enzymatic changes of the small intestine in an n0-STZ diabetes rat model. *Exp. Clin. Endocrinol. Diabetes*. 2002; 110: 119-123.

Tsai SC, Kao CH, Pan DY, ChangLai SP, Wang SJ. Effects of oral erythromycin on esophageal motility in patients with noninsulin-dependent diabetes mellitus. *Gaoxiong Yi Xue Ke Xue Za Zhi*. 1995; 11: 430-435.

Ulrich P, Cerami A. Protein glycation, diabetes, and aging. *Recent Prog. Horm. Res*. 2001; 56: 1-21.

Umpierrez G, Freire AX. Abdominal pain in patients with hyperglycemic crises. *J. Crit. Care*. 2002; 17: 63-67.

Unal A, Guven K, Yurci A, Torun E, Gursoy S, Baskol M, Ozturk F, Arsav V. Is increased colon subepithelial collagen layer thickness in diabetic patients related to collagenous colitis? An immunohistochemical study. *Pathol. Res. Pract*. 2008; 204: 537-544.

Undeland KA, Hausken T, Gilja OH, Aanderud S, Berstad A. Gastric meal accommodation and symptoms in diabetes. A placebo-controlled study of glyceryl trinitrate. *Eur J Gastroenterol Hepatol*. 1998; 10: 677-681.

Vahl TP, D'Alessio DA. Gut peptides in the treatment of diabetes mellitus. *Expert. Opin. Investig Drugs*. 2004; 13:177-188.

Verne GN, Sninsky CA. Diabetes and the gastrointestinal tract. *Gastroenterol Clin. North Am*. 1998: 27: 861-874.

Watanabe T, Asanuma A, Tanaka M, Akiba T, Koga T. [Morphological study on the gastric mucosa in diabetes mellitus rats induced by streptozotocin]. *Exp. Anim*. 1995; 43: 693-696.

Wautier JL, Zoukourian C, Chappey O, Wautier MP, Guillausseau PJ, Cao R, Hori O, Stern D, Schmidt AM. Receptor-mediated endothelial cell dysfunction in diabetic vasculopathy. Soluble receptor for advanced glycation end products blocks hyperpermeability in diabetic rats. *J. Clin. Invest*. 1996; 97: 238-243.

Wolffenbuttel BH, Boulanger CM, Crijns FR et al. Breakers of advanced glycation end products restore large artery properties in experimental diabetes. *Proc. Natl. Acad. Sci. U S A*. 1998: 95: 4630-4634.

Yang J, Zhao J, Zeng Y, Gregersen H. Biomechanical properties of the rat oesophagus in experimental type-1 diabetes. *Neurogastroenterol Motil*. 2004; 16: 195-203.

Yang J, Zhao J, Liao D, Gregersen H. Biomechanical properties of the layered oesophagus and its remodelling in experimental type-1 diabetes. *J Biomech*. 2006; 39: 894-904.

Yoneda S, Kadowaki M, Kuramoto H, Fukui H, Takaki M. nhanced colonic peristalsis by impairment of nitrergic enteric neurons in spontaneously diabetic rats. *Auton Neurosci*. 2001; 92: 65-71.

Young AA, Gedulin B, Vine W, Percy A, Rink TJ. Gastric emptying is accelerated in diabetic BB rats and is slowed by subcutaneous injections of amylin. *Diabetologia*. 1995; 38: 642-648.

Zagorodnyuk VP, Chen BN, Brookes SJ. Intraganglionic laminar endings are mechano-transduction sites of vagal tension receptors in the guinea-pig stomach. *J. Physiol*. 2001; 534: 255-268.

Zagorodnyuk VP, Chen BN, Costa M, Brookes SJ. Mechanotransduction by intraganglionic laminar endings of vagal tension receptors in the guinea-pig oesophagus. *J. Physiol*. 2003; 553: 575-587.

Zagorodnyuk VP, Lynn P, Costa M, Brookes SJ. Mechanisms of mechanotransduction by specialized low-threshold mechanoreceptors in the guinea pig rectum. *Am. J. Physiol. Gastrointest Liver Physiol*. 2005; 289: G397-G406.

Zeng YJ, Yang J, Zhao JB, Liao DH, Zhang EP, Gregersen H, Xu XH, Xu H, Xu CQ. Morphologic and biomechanical changes of rat oesophagus in experimental diabetes. *World J. Gastroenterol*. 20041; 10: 2519-2523.

Zhao J, Sha H, Zhou S, Tong X, Zhuang FY, Gregersen H. Remodelling of zero-stress state of small intestine in streptozotocin-induced diabetic rats. Effect of gliclazide. *Dig. Liver Dis*. 2002; 34: 707-716.

Zhao J, Liao D, Yang J, Gregersen H. Viscoelastic behavior of small intestine in streptozotocin-induced diabetic rats. *Dig. Dis. Sci*. 2003a; 48: 2271-2277.

Zhao J, Yang J, Gregersen H. Biomechanical and morphometric intestinal remodelling during experimental diabetes in rats. *Diabetologia*. 2003b; 46: 1688-1697.

Zhao J, Frøkjaer JB, Drewes AM, Ejskjaer N. Upper gastrointestinal sensory-motor dysfunction in diabetes mellitus. *World J. Gastroenterol*. 2006; 12: 2846-2857.

Zhao J, Liao D, Gregersen H. Biomechanical and histomorphometric esophageal remodeling in type 2 diabetic GK rats. *J. Diabetes Complications*. 2007; 21: 34-40.

Zhao J, Nakaguchi T, Gregersen H. Biomechanical and Histomorphometric Colon Remodelling in STZ-Induced Diabetic Rats. *Dig. Dis. Sci.* 2009; 54: 1636-1642

Zoubi SA, Mayhew TM, Sparrow RA. The small intestine in experimental diabetes: cellular adaptation in crypts and villi at different longitudinal sites. *Virchows Arch*. 1995: 426: 501-507.

In: Biomechanics: Principles, Trends and Applications ISBN: 978-1-60741-394-3
Editor: Jerrod H. Levy, pp. 163-206

Chapter 5

BIOMECHANICS OF THE GASTROINTESTINAL TRACT IN HEALTH AND DISEASE

Jingbo Zhao*[*1], Donghua Liao[1] and Hans Gregersen[1,2]
[1]Mech-Sense, Aalborg Hospital Science and Innovation Centre (AHSIC), 9000 Aalborg, Denmark
[2]La Jolla Bioengineering Institute, La Jolla, California, USA

ABSTRACT

The gastrointestinal (GI) tract is functionally subjected to dimensional changes. Hence, biomechanical properties such as the stress-strain relationships are of particularly importance. These properties vary along the normal GI tract and remodel in response to growth, aging and disease. The biomechanical properties are crucial for GI motor function because peristaltic motion that propels the food through the GI tract is a result of interaction of the passive and active tissue forces and the hydrodynamic forces in the food bolus and remodeling of the mechanical properties reflects the changes in the tissue structure that determine a specific motor dysfunction. Therefore, biomechanical data on the GI wall are important to understand the pathogenesis to the GI motor-sensory function and dysfunction. Moreover, biomechanical studies of the GI tract pave the way for further mathematical and computational modelling. Biomechanical studies of the GI tract will advance our understanding of GI physiological function, diseases such as dyspepsia and visceral pain, and GI dysfunction due to systemic diseases. Furthermore, integrated GI simulation models will be beneficial for medical education and for evaluation of the efficacy and safety of new drugs on GI function.

Keywords: Gastrointestinal tract, biomechanics, remodeling, diseases, modeling.

* Telephone: +45 99326907, E-mail: jz@rn.dk, jzhao@hst.aau.dk

INTRODUCTION

The major function of the gastrointestinal (GI) tract is digestion facilitated by motility, secretion and absorption. The movement of luminal contents is facilitated by changes in the geometry of the GI wall and lumen due to the motility. Therefore, the biomechanical properties of the GI tract are important for its function [Gregersen, 2002a]. The mechanical properties of the GI tract are important determinants of flow, the stress distribution in the wall, and the remodeling process of the wall under physical environment change [Gregersen and Kassab 1996; Gregersen 2002a]. In order to arrive at useful approximations, it is important to understand the basic relation between stress and strain. In addition to determining the distensibility of the GI tract, a more general relationship between stress and strain is desirable for a more thorough understanding of the mechanical properties of the GI tract. The stress-strain relationship must ideally be determined in reference to the zero-stress state. Furthermore, the biomechanical properties will remodel due to normal growth [Gregersen et al., 2004; Christensen et al., 1992a], aging [Christensen et al., 1992b], surgical invention [Storckholm et al., 2007], systemic diseases [Zhao et al., 2006], local disorders [Yang et al., 2009] and etc. Understanding the biomechanical properties of GI tract and its remodeling aids the understanding of the physiology and pathophysiology of GI tract.

Medical imaging methods such as ultrasonography, magnetic resonance imaging (MRI) and endoscopic ultrasound (EUS) are well known stand-alone clinical methods that can disclose structural and functional abnormalities of the GI tract [Berstad et al., 1996, 1998; Frokjaer et al., 2005, 2007a; Gilja et al., 2006, 2007]. However, a thorough study of the GI tract requires understanding of the interactions between cells, tissues and GI organs in health and disease. This depends on knowledge, not only of cellular ionic current mechanisms and signal transduction pathways but also of large scale GI tissue structures and the distribution of the nervous network. A unique way of coping with this complexity is mathematical and computational modelling, providing a computational framework for the multilevel modelling and simulation of the human GI anatomy and physiology [Clapworthy et al., 2007; Hunter et al., 2002, 2005]. In GI tract studies, this approach is not common, mainly because we still lack models that could emulate the behaviour of the human body. Nevertheless, exploration of the GI tract has dramatically improved by the introduction of cross sectional imaging modalities such as Computed Tomography (CT) and MRI that have revolutionised the way in which many conditions are diagnosed and treated. The ability to examine in detail structures inside the GI tract, without resorting to surgery, has allowed clinicians to diagnose problems and plan corrective procedures with a minimum of risk to the patient [Gregersen, 2006; Liao et al., 2008]. In order to continue this exploration, it will be necessary to complement the traditional approach with an integrative approach that combines observations, theories and predictions across the temporal and dimensional scales, across scientific disciplines, and across the anatomical subsystems, all of which reflect the rather artificial divisions described.

The aim of this review is to describe the normal biomechanical properties, the biomechanical remodelling due to various causes, and the currently developed biomechanical modelling works in the GI tract on humans and animals, that can be further used to integrate the physiological, anatomical and medical knowledge of the GI system.

BIOMECHANICAL PROPERTIES OF NORMAL GI TISSUES

The GI tract consists of a series of organs which resemble one another in constitution, being variously arranged as cylinders, spheroids, or intermediate forms. The main functions of the GI tract are transport and digestion of food. The different segments show a big variation in morphology and muscle mechanical properties. The esophagus mainly serves to quickly transport the food bolus from the mouth to the stomach where the food in the stomach is stored for some time and simultaneously being broken down into smaller components. The small intestine is the longest section of the digestive tract and the chyme is mixed, absorbed and transported to large intestine. Finally the large intestine absorbs the remaining water and electrolytes from indigestible food matter, stores food remains that were not digested and eliminate solid waste (feces) from the body. The movement of luminal contents in the GI tract is facilitated by changes in the geometry of the GI wall and lumen due to the motility. It is well known that a bolus propelled in front of a peristaltic contraction distends the gut wall [Siegle et al., 1987, 1989, 1990]. Therefore, in order to understand the morphological and biomechanical GI remodeling due to normal growth, aging and different diseases, it is important to know the biomechanical properties of the normal GI tract.

Biomechanical Properties of Normal Esophagus

The mechanical properties of the esophagus are important for its function because the esophagus is subjected to changes in wall stress and strains caused by the passage of boli and the action of peristalsis.

Using impedance planimetry (IP) the cross-sectional area and pressure changes of the esophagus can be measured at the same time and the esophageal wall tension and stiffness can be obtained *in vivo* or *in vitro*. Using this novel technique it was demonstrated that the circumferential wall tension and wall stiffness of the esophageal wall increased exponentially as function of pressure [Orvar et al., 1993; Rao et al., 1995; Patel et al., 1998M; Vanags et al., 2003]. The tension distribution were non-uniform along the esophagus [Patel et al.,1998; Vanags et al., 2003] and the LES and the proximal esophagus had greater wall tension and were less deformable than the mid- or distal esophagus [Patel et al., 1998]. Combining with ultrasound the wall thickness of the esophagus also can be obtained and the stress-strain relationship of the esophageal wall can be computed [Assentoft et al., 2000; Frokjaer et al., 2006a., Zhao et al., 2007a]. Using IP combining the ultrasound, it was demonstrated that the esophageal wall stress-strain curves are also exponential [Zhao et al., 2007a] meaning that the tissue is soft at physiologic pressures and stiffer in the supraphysiologic pressure range. This mechanism seems to prevent overstretch of the esophageal wall when luminal loading becomes supraphysiologic. Furthermore it was also shown that the stress-strain distribution is non-uniform across the esophageal wall, i.e., the stiffness increased throughout the wall and was highest at the outer surface [Frokjaer et al., 2006a]. However, Takeda et al [2002, 2003] showed that the stress-strain relationship of the human esophagus in vivo was linear during both isovolumic and isobaric distension, before as well as after injection of the antimuscarinic drug atropine. The stress-strain relationship of the active component (muscle contraction) was different during isovolumic and isobaric distensions but the passive components were similar.

Furthermore, the method of esophageal distension significant influences on the active but not on the passive biomechanical properties due to a strain-rate effect. Investigating the degree of stretch and tension to initiate motor and sensory responses of the esophagus showed that the stretch rather than tension appears to be the major factor influencing sensory responses to esophageal distension [Barlow et al, 2002; Drewes et al, 2003a].

Esophageal function is usually evaluated by means of manometry which is a proxy of the force in the radial direction. However, force measurements in the axial direction will provide a better measure of esophageal transport function. Gravesen et al [2008] recently developed a probe based on electrical impedance measurements to quantify the axial force generated by esophageal contractions, i.e. probe elongation was associated with the axial force. The probe showed good reproducibility and the dispersion was <0.04. The swallowed bolus size correlated with the axial force measurements (P = 0.038) but not with manometric measurements. Therefore, the new technique measuring axial force in the esophagus may in the future provide valuable information about esophageal function.

Experimental and diagnostic procedures like distension of a balloon catheter, bougie, and esophagogastroduodenoscopy can induce shear deformation in the esophageal wall. Hence, knowing the shear modulus of the esophagus is important. Yang et al [2004a] has studied the shear modulus of elasticity of the esophagus in the rat. Results were obtained using a triaxial instrument to perform simultaneous torsion, inflation, and longitudinal stretching tests (Figure 1). The shear modulus varied with the inflation pressure and the longitudinal stretch ratio. The mechanical constant of the esophagus showed that the esophageal wall was anisotropic with different stiffness in the circumferential, longitudinal, and the shear directions. The stiffness in the longitudinal direction was higher than in the circumferential direction.

To understand esophageal function, it is necessary to know the stress-strain relationship for the different layers, because the esophageal wall is composed of four layers. Studies in layered wall of the esophagus showed that the stress distribution in the different layers were nonlinear and anisotropic [Lu and Gregersen, 2001; Gregersen et al, 2008; Fan et al, 2004; Liao et al, 2003a, 2007; Zhao et al, 2007b; Yang et al, 2007a]. The submucosa-mucosa layer has the highest stiffness [Fan et al., 2004; Liao et al.,2003a]. The residual strain differs

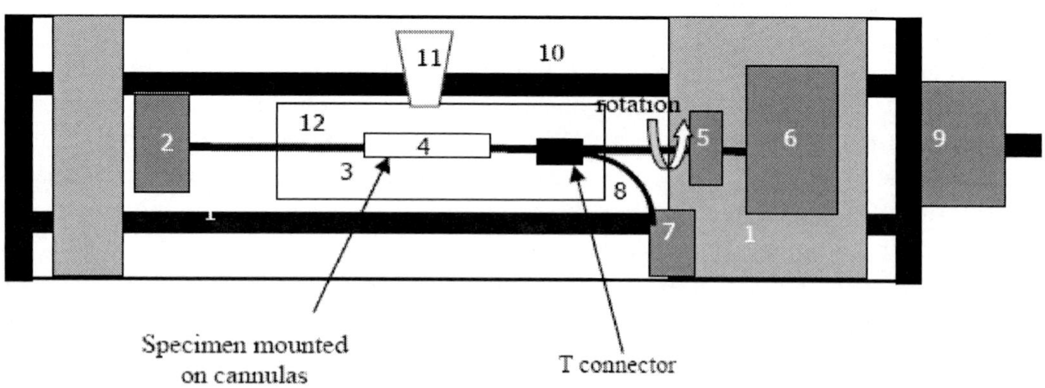

Figure1. Torsion machine setup. 1: Linear stage, 2: Torque transducer, 3: Organ bath, 4: Specimen, 5: Force transducer, 6: Motor for rotation, 7: Pressure transducer, 8: Infusion channel, 9: Motor for linear stage, 10: Rails for linear stage, 11: CCD camera, 12: Plastic rod.

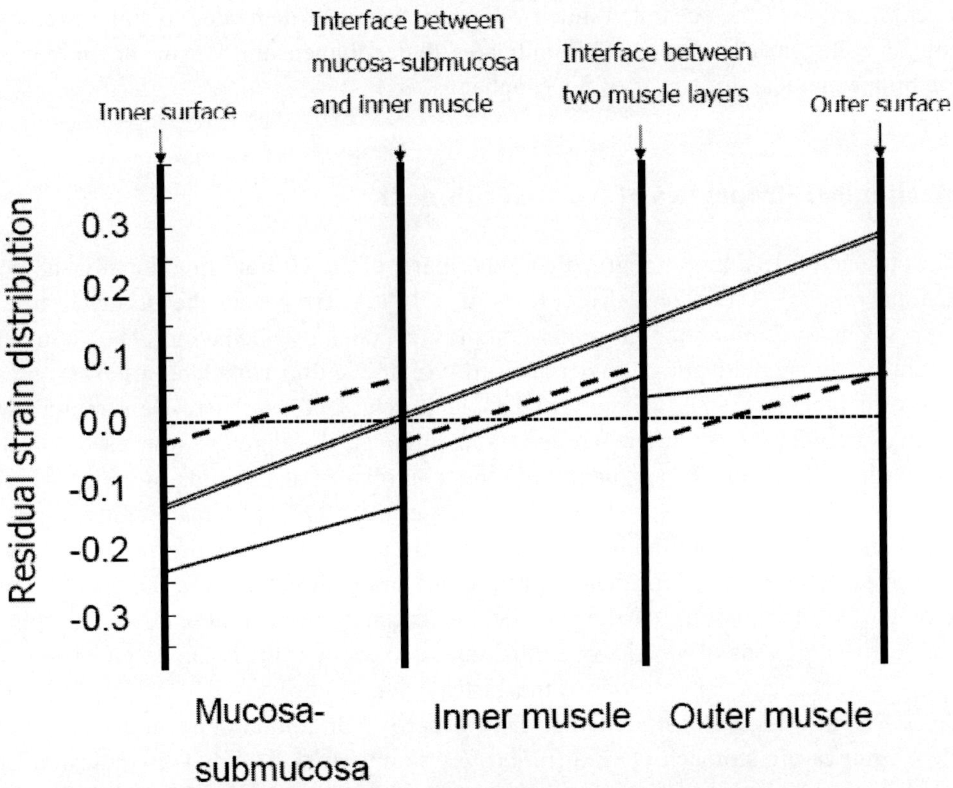

Figure 2. Residual strain distribution in the intact wall and layers of pig oesophagus. Residual strain distribution of the intact wall referenced to the zero-stress state of the intact wall Residual strain distribution of layered wall in intact wall referenced to the zero-stress state of layered sectors Residual strain distribution of separated layers referenced to the zero-stress state of layered sectors. Referenced to the zero-stress state of the intact sectors, the inner residual strain of intact rings was -0.13 ± 0.04 and outer residual strain was 0.31 ± 0.03. Referenced to the "true" zero-stress state of separated three layered sectors, the inner residual strain of intact rings was -0.22 ± 0.02 and outer residual strain was 0.07 ± 0.02. The residual strain distribution of the layers in intact rings referenced to the "true" zero-stress state is also shown. The inner surface residual strain was negative at mucosa-submucosa and inner muscle layers and was positive at outer muscle layer, whereas the outer surface residual strain was negative at the mucosa-submucosa layer and positive at the inner and outer muscle layers. Referenced to the "true" zero-stress state, the inner residual strain was negative and outer residual strain was positive for the separated layered rings.

between the layers, and the residual strain distribution was more uniform after the layers were separated (Figure 2) [Zhao et al, 2007b]. Furthermore, experimental studies have demonstrated esophageal mucosal folds in the no-load state. This indicates that mucosal buckling must be considered in the analysis of the mechanical reference state since the material stiffness drops dramatically after tissue collapse [Liao et al, 2007; Yang et al, 2007a].

The esophageal wall is composed of different material components that contribute to the esophageal biomechanical properties. Fan et al [2005] studied the effect of collagenase and elastase on the morphological and biomechanical properties in the no-load and zero-stress states in the rat esophagus. It was demonstrated that the collagenase and elastase caused the opening angle and the residual strain in the separated mucosa-submucosa layer to decrease.

The opening angle of the separated mucosa-submucosa layer depended to some extent on the fraction of collagen and elastin. This indicates that collagen and elastin are important for passive biomechanical properties of the esophagus.

Biomechanical Properties of Normal Stomach

The stomach is to a larger degree than other parts of the GI tract functionally subjected to dimensional change [Schulze-Delrieu K et al., 1998]. However, the stomach has more complex geometry, indicating a more complex mechanical behavior. The stomach has traditionally been studied using *in vivo* pressure-volume and *in vitro* length-tension measures [Haffner and Stadaas, 1972; Milenov and Golenhofen, 1982; Schulze-Delrieu and Shirazi, 1987]. In a number of studies in animals and humans, the barostat was used to measure pressure-volume relations for evaluation of tone, compliance and tension in the gastric fundus [Azpiroz and Malagelada, 1985, 1987; Whitehead et al., 1997]. The main assumptions made in such studies were that the fundus is geometrically a sphere with uniform tension. In other words it is assumed that the mechanical properties in all directions are the same. However, the geometry of the fundus of the stomach is much more complex than the spheroidal geometry. Investigations of gastric tone must take into account the complex geometry and the anisotropic biomechanical properties of the gastric wall.

In 1975, it was demonstrated that the properties of longitudinal and circumferential muscle of guinea-pig stomach were different [Kuriyama et al, 1975]. This indicated that the topical differences of the motility in the stomach may be due not only to the activity of nervous elements, but also to differences in the properties of the muscle fibers themselves. Egorov et al [2002] showed that the values of maximal stress and destructive strain of stomach strips were direction-dependent. In order to investigate the anisotropic properties of the stomach wall, Zhao et al [2005] studied uniaxial stress-strain properties of gastric strips obtained from rats and rabbits. It was demonstrated that the biomechanical properties of the gastric strips are location-dependent, direction-dependent and species-dependent. The assumption in physiological pressure-volume studies that the stomach is a sphere with uniform tension is not valid. More recently, Zhao et al [2008a] confirmed the anisotropic biomechanical properties in pig stomach wall. Furthermore, they made an in-depth biomechanical test on the layered wall of the stomach in different regions. It was demonstrated that the uni-axial stress-strain curves of pig stomach were location-, direction- and layer-dependent (Figure 3). The stiffer wall in the corpus is likely due to its thicker mucosa, i.e., the stiffness of the mucosa-submucosa layer seems can explain the intact wall stiffness. Since the structure and function of the pig stomach are somewhat similar to the human stomach, the data obtained from this study can be extended to humans.

Detailed biomechanical mapping of the stomach will likely help us to understand physiological functions of the different parts of the human stomach, such as gastric accommodation and mechanosensation. A better understanding of gastric accommodation and gastric perception requires knowledge of regional gastric geometry and local gastric tension throughout the stomach. Based on medical imaging data, Liao et al [2005] described the three-dimensional (3D) rat stomach geometry and tension distribution and showed that the surface principal radii of curvatures and tension distribution in the stomach were non-homogeneous.

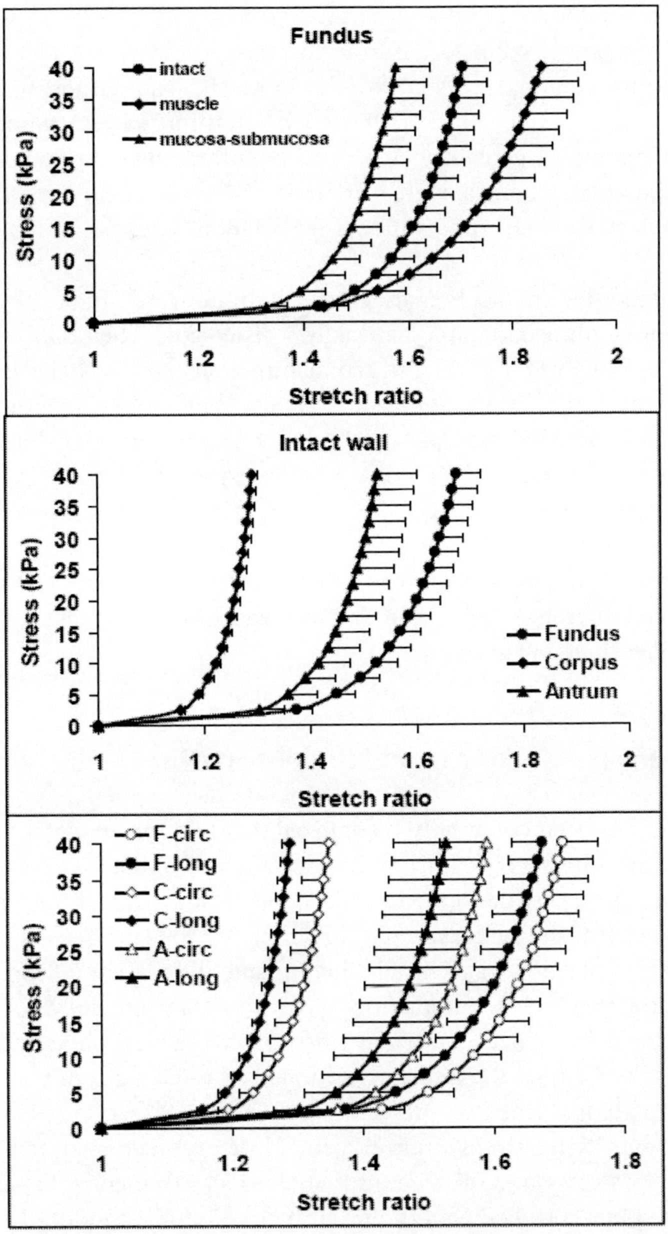

Figure 3. Gastric strips presented as layer-dependent properties (top), Location- dependent properties (middle) and Direction-dependent properties (bottom). The Top figure shows circumferential uniaxial stress-strain relationships of gastric strips in fundus. The stress-strain curve for the mucosa-submucosa strips was located to the left whereas the curve of the muscle strips was located to the right. It indicated that the mucosa strips were the stiffest and the muscle strips were the softest. The Middle figure shows longitudinal uniaxial stress-strain relationships of the intact gastric strips. The curves of the corpus strips were located to the left and the curves of the fundus strips were located to the right. It indicated that the corpus strips were the stiffest and the fundus strips were the softest. The Bottom figure shows uniaxial stress-strain relationships of the intact gastric strips in different location. The longitudinal stress-strain curves were located to the left, indicating that those strips were the stiffest, whereas the curves for the circumferential strips were located to the right, indicating they were softest. Notes: F, fundus; C, corpus; A, antrum; circ, circumferential direction; long, longitudinal direction.

Later, Liao et al [2004a] analyzed the surface geometry of the human stomach using real-time 3-D ultrasonography in vivo to describe the three-dimensional (3-D) geometry of the gastric antrum, gastric fundus and the whole stomach. The principal curvatures spatial distributions were non-homogeneous in the gastric antrum, gastric fundus and the whole stomach due to their complex geometry. The maximum longitudinal principal curvature in the antrum, fundus and total stomach were different. As consequence providing an average tension for the stomach does not reflect the large variation in tension throughout the stomach wall.

Mechanical properties in the human gastric antrum were studied by Gregersen et al [2002b] using B-mode ultrasonography and antral distension. The strain was positive in the circumferential direction, negative in the radial direction, and with no deformation in the longitudinal direction. Later, Gregersen et al [2006b] studied mechanosensory properties in the human gastric antrum using the same method. The bag distensions elicited contractions in the antrum and sensory responses below the pain threshold. Butylscopolamine abolished the contractions and significantly reduced the sensory response. The muscle length-tension diagram and typical preload-afterload curves as modem the Frank-Starling cardiac law can be obtained in the human gastric antrum. The sensory responses were most closely associated with the luminal circumference, indicating that the sensation during antral distension depends on deformation rather than on tension.

Biomechanical Properties of Normal Small Intestine

For a century it has been commonly recognized that two fundamental patterns of motility are conducted by the intestine [Szurszewsky, 1998]. 1. Propulsion- peristalsis: A ring of muscle contraction appears on the oral side of a bolus of ingesta and moves toward the anus, propelling the contents of the lumen in that direction. As the ring moves, the muscle on the other side of the distended area relaxes, facilitating smooth passage of the bolus. 2. Mixing-Segmentation contraction: This contraction is a common type of mixing motility seen especially in the small intestine. Segmental rings of contraction chop and mix the ingesta. Like other parts of the GI tract the intestine is subjected to changes in wall stress and strains caused by these two kinds of contraction.

Using IP with and without combing ultrasound, it was demonstrated that the intestinal wall tension-strain or stress-strain distributions showed an exponential behavior [Jorgensen et al., 1991; Gregersen et al, 1992a; Storkholm et al, 1995; Jorgensen et al, 1995; Duch et al, 1996; Gao et al, 2003; Frokjaer et al, 2006b]. The stiffness was different between the duodenal, jejunal and ileal segments [Storkholm et al., 1995; Duch et al., 1996]. It is now well known that the biomechanical analysis should be referenced to the zero-stress state [Gregersen et al., 2000]. Gregersen et al [1999] first time studied morphometry and strain distribution in guinea pig duodenum with reference to the zero-stress state. Afterwards, several studies were carried out to analyze the biomechanical properties of the normal small intestine with reference to the zero-stress state [Gao et al., 2000a; Dou et al., 2003; Zhao et al., 2002a]. It was found that a large axial variation of opening angle and residual strain existed along the small intestine (Figure 4) [Zhao et al., 2002a]. The axial variation of residual strain from duodenum to ileum correlated to the morphometric variation along the small intestine [Gabella and Blundell, 1981].

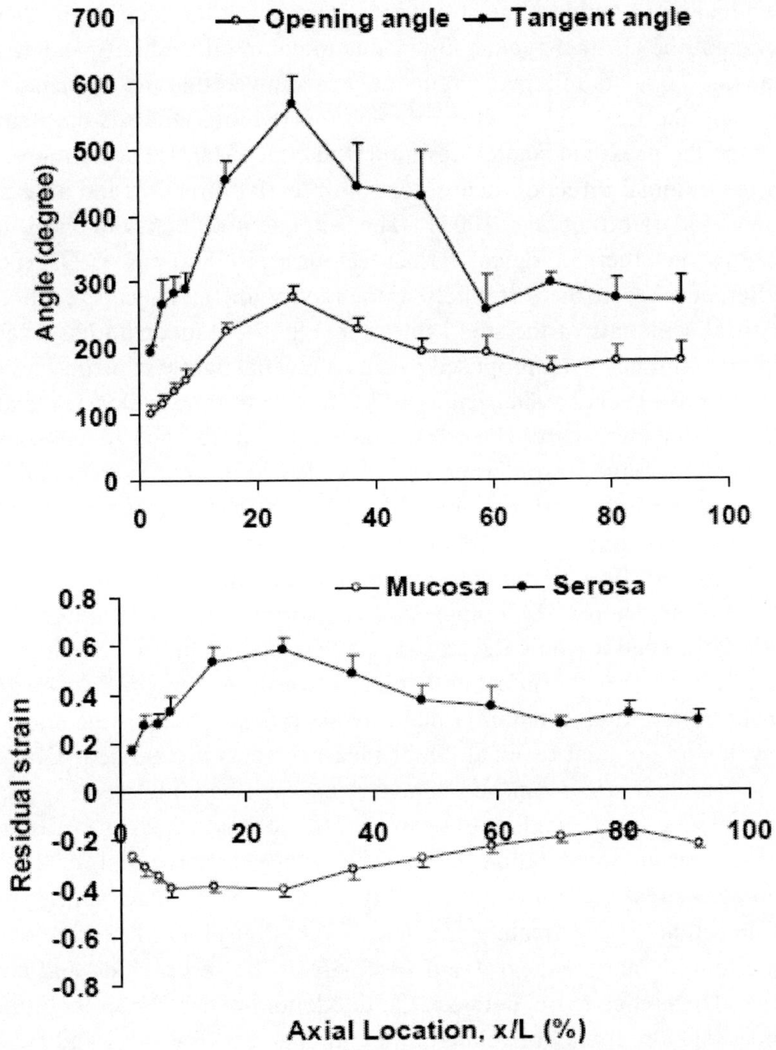

Figure 4. Top: The distribution of the opening angle and tangent rotation angle in the rat small intestine. Both angles increased along the length of the duodenum and had its highest value 30% down the intestine. They decreased again further down the intestine. Bottom: Residual strains distributed at the mucosal and serosa surface. The serosa residual strain was tensile and mucosa residual strain was compressive with highest absolute values also close to the 30% down the intestine.

The residual strain is non-uniformly distributed across the intestinal wall, it was negative at the mucosa surface and positive at the serosa surface [Gao et al., 2000a; Dou et al., 2003; Zhao et al., 2002a]. Hence the serosal residual strains are tensile and the mucosal residual strains are compressive. Distension experiments showed that the residual strain makes the stress distribution through the wall more uniform in the pressurized state [Gao et al., 2000a]. From the duodenum data it appears that the neutral axis (where residual strain is 0) is located closest to the mucosa. This is even more pronounced in the distal parts of the small intestine where it is located about 20% away from the mucosal surface [Dou et al., 2003]. The implications of these data are clear. The zero-stress state is better to be accounted for in biomechanical studies of the small intestine. Furthermore, as it has been demonstrated in arteries, the residual stress reduces the stress concentration at the inner wall, thereby offering a better protection at the GI tract against injury due to contractile activity and against the flow of luminal contents. The data of stress-strain analysis showed that the duodenal wall is stiffer than the jejunal and the ileal wall in circumferential direction, whereas the ileum was softer longitudinally than the proximal small intestine [Dou et al., 2003]. Furthermore, all segments were stiffer in longitudinal direction than in circumferential direction and all curves show an exponential behavior [Dou et al., 2003]. The exponential behavior is consistent with mechanical studies on other biological tissues [Dobrin, 1978; Fung, 1991; Fung and Liu, 1992]. The differences in stiffness is likely associated with the specialized functions of the proximal and distal segments of the small intestine, e.g. the duodenum has a large influence on gastric emptying and has been proposed to act as a capacitative resistor whereas the distal ileum acts as a reservoir [Schulze-Delrieu, 1991]. The transit time is slower distally [Bueno et al., 1975, 1994] which may partly be explained by the difference in the viscosity of the chyme. However, the passive elastic properties may also contribute to differences in intestinal flow patterns: a passing bolus will be slowed to a lesser degree in the duodenum where the wall stiffness is high whereas the compliant ileal wall bulges leading to pooling of luminal contents and decreased flow. The deformation of the intestine during contraction and relaxation is three-dimensional. To obtain a more thorough understanding of the intestinal transport function, we need to know the stresses and strains in multiple directions. In previous studies it was observed that the intestine bends outwards in longitudinal direction when a radial cut is made. Therefore, to obtain a more detailed description of the stress-free state of the intestinal wall, longitudinal residual strain measurements are needed [Dou et al., 2006]. The longitudinal residual stress can be characterized by a bending angle (unit: degrees per unit length and positive when bending outwards). The no-load state can be used as reference for the longitudinal strain computation because the mid-wall length did not change from the no-load state to zero-stress state in longitudinal direction. The average bending angle is in the whole small intestine approximately 55°/mm. As the circumferential residual strain distribution, the longitudinal residual strain is tensile at the serosal side and compressive at the mucosal side. Difference exists between the duodenum and ileum in longitudinal residual strain at the mucosal side. It may be related to the thicker mucosal layer and faster passage of food in the duodenum than in the ileum. By comparison, the residual strains in longitudinal direction measured in this study are smaller than those in circumferential direction [Dou et al., 2003], especially at the mucosal side. This could be due to that the largest tensile stress and strain are distributed in the circumferential direction when the intestine is distended during the passage of food. Hence, it is reasonable that larger residual strain exists in circumferential direction. Smith et al [2003] measured the time-dependent (viscoelastic) behavior in the change of the small intestinal opening angle and tested how well the behavior could be described by the Kelvin model for a standard linear solid. The change of opening angle over time for all the small intestine segments fits well to the standard linear spring-dashpot model. This viscoelastic constant of the rat small intestine is fairly homogenous along its length. The data obtained from this study add to a base set of biomechanical data on the small intestine and provide a reference state for comparison to other tissues, diseased intestinal tissue or intestinal tissue exposed to drugs or chemicals.

In the small intestine, submucosa is rich in collagen and is believed important for the passive biomechanical properties. Fackler et al [1981] observed the effect of stress on rat and bovine intestine using polarizing optical microscope, especially focus on the orientation of collagen in the submucosa. Collagen fibres were found to be biaxially oriented at approximately +30 degrees and -30 degrees to the longitudinal direction. The fibres are arranged in layers with the fibres in each layer densely packed in parallel undulating arrays. The initial response to stress is straightening of the fibres. Gradual straightening of the fibres is related to the increasing stiffness of the tissue observed in the stress--strain relationship. Once the fibres are straightened, the biaxial orientation of the fibres produces higher strength in the longitudinal direction than in the transverse direction. Storkholm et al [1998] compared the stress-strain distributions obtained from isolated segments of the guinea pig duodenum, jejunum, and distal ileum, and the relation between the elastic properties and the collagen content. An association was found between the collagen content and the incremental elastic modulus (stiffness). Recent Chen et al [2008] first provided direct experimental evidence that the villi are important for the biomechanical properties of guinea pig small intestine in circumferential direction, because the villi not only affect the zero-stress state configuration but also partially affect the stress-strain distribution in the intestinal wall (Figure 5).

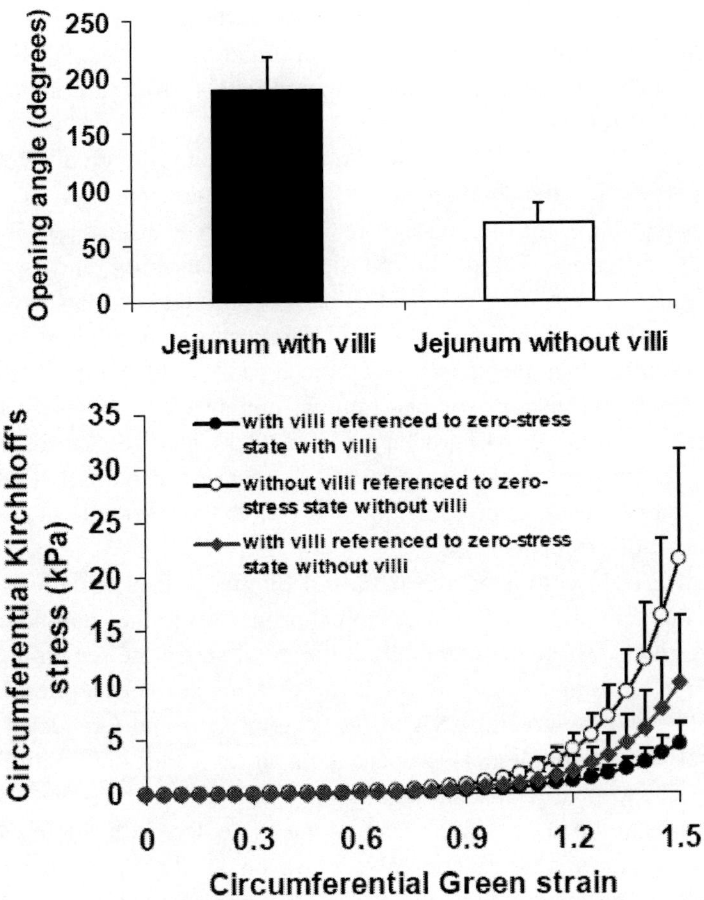

Figure 5. The opening angle (top) was smaller in the segments without villi than in the segments with villi. Stress-strain analysis (bottom) showed that the circumferential stress-strain curve of the segment with villi located at the right compared to the segment without villi, indicating the villi do not support tensile stretch. The curve was located in the middle if the villi were not taken into account in the computation of the stress-strain in the intact segment.

Therefore, the villi should be taken into account in the analysis of biomechanical properties of the intestinal wall. Li et al [2008] recently investigated the effect of smooth muscle tone on morphometry and residual strain in rat duodenum, jejunum and ileum. It was demonstrated that the intestinal mechanical properties are affected by intestinal SM tone. The tone can be evaluated by measuring the opening angle and residual strains of sectors in intestinal segments with and without SM relaxation.

Biomechanical Properties of Normal Colon and Rectum

Same as other parts of GI tract, the biomechanical properties are also important for colon physiological function. Tensile stretch test showed that rat colon has a tensile strength of around 50 g/mm^2 and increased in strength from proximal to distal [Watters et al.,1985a]. Later, Watters et al [1985b] compared the features of the African and European colon in vitro. The tensile strength in a Kampala group was greater than in an Edinburgh one. The width at burst was greater in the Kampala group and declined with age. Stress-relaxation was similar in both groups. In view of the similar properties in childhood of colons from Edinburgh and Kampala, the strength of the adult African compared with European colons may derive later from environmental factors such as diet.

The quasi-linear viscoelastic law could be applied to the in vivo determination of the mechanical properties of the human rectal wall [Bouchoucha et al., 1994]. It was demonstrated that the shapes of the pressure-relaxation curves at different level of distension were significantly different. Tonic and phasic colonic motility and compliance of the transverse and sigmoid colon were assessed using a combined barostat-manometry assembly in healthy subjects [Ford et al., 1995]. They demonstrated that there are quantitatively different but qualitatively similar phasic and tonic responses to the meal in the two colonic regions. Differences in the viscoelastic and luminal dimensions may partly account for these differences in tonic responses. Bharucha et al [2001] studied the colonic viscoelastic properties of the human descending colon by assessing pressure-volume (P-V) relationships. Quasi-static P-V curves were approximated to a power exponential function and revealed hysteresis, indicative of viscoelasticity.

Petersen et al [2001] compared the pain and biomechanical responses to standardized distension of the human colon. The biomechanical properties in the sigmoid colon and rectum were alike. For a given wall tension and circumferential strain the sensibility seems equal in the rectum and the sigmoid colon. The observed difference in perception between the two segments was related to the greater CSA in the rectum. Gao and Gregersen [2000b] studied the biomechanical and morphological properties in rat large intestine with referenced to the zero-stress state. The wall thickness and wall thickness-to-circumference ratio increased in the distal direction. The opening angle varied along the colon with the highest values in the beginning of the proximal colon. The residual strain at the inner surface was negative indicating that the mucosa-submucosal layers of the large intestine in no-load state are in compression. The stress-strain distributions were exponential. All segments were stiffer in longitudinal direction than in the circumferential direction. The transverse colon was stiffest both in the circumferential and longitudinal directions. Frokjaer et al [2007a] used magnetic resonance imaging (MRI) to evaluate the three-dimensional geometry and mechanosensory properties of the sigmoid colon. The geometry of the distended sigmoid colon was complex and the spatial distributions of the biomechanical parameters were non-homogeneous. The circumferential length, strain, pressure and wall stress increased as a function of bag volume. In response to butylscopolamine, the pressure and wall stress were reduced and the stress-strain curves were shifted to the right. The sensory response was a linear function of the biomechanical parameters and decreased in response to butylscopolamine as a function of volume.

Rao et al [1997] compared incremental elastic modulus (IEM) with rectal compliance during ramp inflation. Because slope of stress-strain relation can reflect rectal wall stiffness, the study emphasized the importance of rectal wall viscoelastic evaluation based on stress and strain rather than compliance. IEM had a linear relation with strain in contrast to compliance. The series elastic component (SEC), a measure of passive viscoelasticity, was rate independent elasticity and a rate dependent viscosity. Qiao et al [2005] showed that the mechanical properties of rectal wall exhibited pronounced nonlinear and anisotropic characteristics. However, Dall et al [1993] demonstrated that the circumferential wall tension of human rectum increased in a linear way. Rectal compliance decreased in a non-linear way with no further decline between 30 and 40 cmH_2O. The pressure elastic modulus increased steeply until a distension pressure of 35 cmH_2O with no further increase to 40 cmH_2O. This suggests that rectal tone is reduced as the muscle fails to resist further distension at 35 cmH_2O and higher pressures.

Biomechanical Properties of Normal Sphincter Regions in GI Tract

Several sphincters, such as upper esophageal sphincter (UES), esophageal-gastro junction (EGJ) including lower esophageal sphincter (LES), pyloric sphincter, sphincter of Oddi (SO), ilealcolic sphincter, and internal and external anal sphincters (IAS and EAS) distributed along the GI tract. UES and LES control entry into the esophagus and passage into the stomach, the latter normally prevents reflux of gastric contents. The pyloric sphincter controls passage into the duodenum and the ileocolic sphincter controls movement between the ileum and caecum. The SO regulates bile and pancreatic flow into the duodenum, makes diversion of hepatic bile into the gallbladder and prevents reflux of duodenal contents into the pancreaticobiliary tract. The IAS and EAS control stool out of the body. Therefore, the mechanical properties to maintain the opening and closing of the sphincters are important for their physiological functions.

Studies with biomechanically analyzed the mechanism of UES opening showed that the swallow-related UES opening are subject to volitional augmentation [Kahrilas et al., 1991]. Jacob et al [1989] showed that trans-UES transport of increasing swallow bolus volumes is accomplished by modulating sphincter diameter, opening interval, and flow rate. High-resolution manometry could accurately, comprehensively, and highly reproducibly depict the pharyngo-esophageal (PE) segment space-time-pressure structure and specific physiological events related to UES opening and transsphincteric flow during normal swallowing [Williams et al., 2001]. Intrabolus pressure variations are highly dependent on position within the PE segment and time. Biomechanical and bolus flow aspects of swallowing changed when people performed effortful swallows with boluses [Hind et al., 2001]. Using simultaneous high-resolution endoluminal sonography and manometry, Miller et al [2004] evaluate the motion, morphology, and pressure of the UES. The UES has a greater muscular CSA and resting pressure than the upper esophageal body. In the cadaver studies, the UES was imaged in conjunction with a significant increase in pressure, indicating that the pressure is due to passive mechanical conformational changes.

Force-velocity analysis [Cohen and Green, 1973]demonstrated that both the work and power generated during esophageal and LES muscle contraction were determined by: (a) the initial muscle length produced by the preload, (b) the afterload against which the muscle was contracting, and (c) the contractility of inotropism of the muscle. Biancani et al [1973] showed that pressure-diameter curve of LES was sigmoid in shape; at first it showed a decline and then an increase followed by decline in pressure again as function of increasing inside diameter. The ratio of wall thickness to inside radius varied with inside diameters and this ratio increased steeply at small inside diameters. The tension diameter curves of the sphincter muscle showed that optimal tension development occurred not near sphincter closure but at a much wider diameter and that this muscle developed tension even at small luminal diameters. Massey et al [2006] demonstrated that transient LES relaxations must occur before the gastroesophageal junction can open. Swallow and esophageal distension can induce the relaxations of the LES due to axial stretch [Dogan et al, 2007]. Accurate monitoring of LES and GEJ pressures is important for the diagnostic assessment of many patients and for advancing research into several aspects of the mechanical function of the LES and its surrounding structures [Dent, 2007]. Recently, McMahon et al [2007a] evaluated the function of the EGJ using a functional lumen imaging probe (FLIP). The FLIP measured eight cross-sectional areas (CSAs) 4 mm apart together with the pressure inside a saline-filled cylindrical bag. The geometric profile of the EGJ was reconstructed in a video animation of its dynamic activity. A plot of curve-fitted data for the smallest CSA vs. pressure after balloon distension indicated that the pressure increased from 18 cmH_2O at a CSA of 38 mm^2 to a pressure of 37 cmH_2O at a CSA of 230 mm^2 for the healthy controls. Therefore, FLIP represents the first dynamic technique to profile the function and anatomy of the EGJ.

In vitro force-length curves of rings from the gastroduodenal junction were showed that basal, total, passive, and active forces were higher for pyloric rings than for duodenal and antral rings [Biancani et al.,1980]. Furthermore, in the pylorus the force-length curves were shifted to the left along the length axis due to a narrowing at the gastroduodenal junction. These data indicated that the higher forces and pressures observed in the pylorus are due to a combination of higher active and passive components. Keinke et al [1984] showed in a dog study that gastric emptying is controlled by (1) the depth of the antral waves, (2) the pyloric sphincter opening, (3) the receptive relaxation of the duodenum and (4) the type of the duodenal contractions. Shirazi et al [1984] showed in a cat study that duodenum provided a resistance to gastric emptying over and above that generated by the pylorus, and that the duodenal resistance is caused by limitations in duodenal capacity. The mechanical activity of the isolated stomach and duodenum is affected by the chemical composition of their luminal contents. Mandrek and Kreis [1992] showed that mechanical responses of gastric and pyloric smooth muscles to acetylcholine, histamine, substance P, noradrenaline and adrenaline were different. The spatial patterning of antropyloric pressure waves sequences is diverse [Sun et al., 1997], and may explain the differing mechanical outcomes among individual gastric contractions. Shafik [1998] showed that pyloric sphincter contraction and antral dilatation depended on duodenal distension, suggested duodenopyloric reflex is existed. This reflex appears to prevent duodenopyloric reflux. Indireshkumar et al [2000] demonstrated that gastric emptying of nutrient liquids is primarily through the "pressure pump" mechanism controlled by pyloric opening during periods of relative quiescence in antral contractile wave activity.

The Intramural neural control and motility of the SO in opossum were studied by Helm et al [1985, 1989]. The SO is innervated by intramural cholinergic excitatory nerves and nonadrenergic noncholinergic inhibitory nerves. The frequency of SO phasic contractions is determined by control wave frequency. Mechanical properties of isolated smooth muscle from rabbit SO and duodenum were studied by Elbrønd et al [1991]. The difference was found between circular smooth muscle strip of duodenum and the SO. Measurements of SO electrical and mechanical activity are feasible and can be done as part of ERCP [Abell et al, 1998]. The frequency (cycles/min) of electrical signal was about 4.9 and tonic pressure was about 15 mmHg in SO. It is important to study the dynamic changes of geometry and pressure in the SO associated with other parts if GI tract, such as the stomach, duodenum and bile ducts.

The associations existed between volume of rectal distension and the parameters of IAS relaxation [Penninckx et al., 1992]. Because the IAS ring cannot be completely closed, the anal mucosa and the haemorrhoidal plexuses fill the gap. By compressing these tissues, the IAS perfectly closes the anal canal to retain not only solids but also fluid stool and gas. Acute rectal distension and rectal activity, mainly through intramural pathways, induce reflex IAS relaxation, permitting the rectal contents to be sampled by receptors in the upper anal canal while continence is temporarily maintained by EAS activity and by expansion of the haemorrhoidal cushions. Glavind et al [1993] studied the mechanical properties of isolated smooth muscle from human rectum and IAS. The muscle resting active tension was different between rectum and IAS. The mechanical responses of smooth muscle in the IAS were modulated by alpha-adrenergic excitatory and non-adrenergic, non-cholinergic inhibitory nerves. During defecation, the relaxation of IAS smooth muscle was associated with a characteristic motility pattern of the colon and anorectum [Matsufuji and Yokoyama, 2003]. A three-dimensional (3-D) physiological model of the IAS bioengineered in vitro from isolated smooth muscle cells was developed by Hecker et al [2005]. The bioengineered sphincter rings showed striking functional differences when compared with bioengineered rings made from isolated colonic smooth muscle cells. The EAS is a triple-loop system; each loop can function as a separate sphincter through voluntary inhibition action and mechanical compression [Shafik, 1987]. Tension-stress can effect the growth of the EAS [Li et al., 1995]

Summary of the Biomechanical Properties of GI Tract

As mentioned above the GI wall has unique and complex mechanical properties [Gregersen and Kassab, 1996]: (1) anisotropic properties. The GI wall has complex three-dimensional structures with different material properties in different directions. This feature is called anisotropy and it implies that a large set of material constants have to be specified in order to completely describe the mechanical behaviour. (2) non-linear stress-strain relationship. The non-linear mechanical behaviour between stress and strain facilitates stretch in the physiological pressure range and prevents overstretch and damage to the tissue at higher stress levels. Owing to the non-linearity of the material properties it is necessary to use more complex constitutive equations than Hooke's law. (3) large strain (finite deformation). This means that large deformation elastic theory based on non-linear principles should be applied to the study of GI mechanics. (4) high degree of viscoelasticity, including the time-dependent nature of the stress-strain properties of materials. Furthermore, knowing the zero-stress configuration is essential in any mechanical analysis since it serves as the reference state for computing stress and strain under physiological and pathophysiological conditions.

BIOMECHANICAL REMODELLING OCCUR IN THE GI TRACT

Data on the biomechanical properties are crucial for the understanding of the dysfunction due to disease because, 1) peristaltic motion that propels the food through the GI tract is a result of interaction of the passive and active tissue forces and the hydrodynamic forces in the food bolus and 2) remodelling of the mechanical properties reflects the changes in the tissue structure that determine a specific sensory-motor dysfunction.

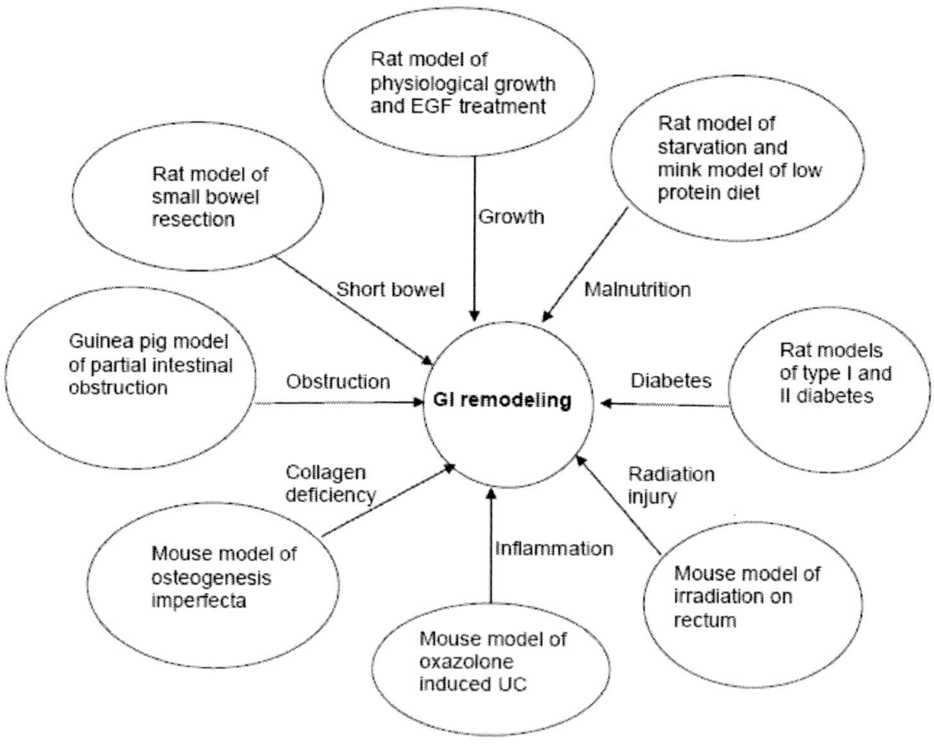

Figure 6. Disease induced GI remodeling in animal models.

Human studies documented that diabetes mellitus (DM) [Frokjaer et al., 2007b] and systemic sclerosis [Pedersen et al., 2003] induced biomechanical GI remodeling. Using different animal models (Figure 6) we have demonstrated that biomechanical and histomophological remodeling occurred in the GI tract duo to normal physiological growth [Gregersen et al., 2004; Lu et al., 2005], malnutrition [Dou et al., 2001, 2002a], inflammation [Yang et al., 2009], obstruction [Gregersen et al., 1992b., Stockholm et al., 2007, 2008], bowel resection [Dou et al., 2002b], DM [Jørgensen et al., 2001; Liao et al., 2006a; Yang et al., 2004b, 2006; Zhao et al., 2002b, 2003a, 2003b, 2006, 2007b, 2008b], radiation injury

[Gregersen et al., 2002c], collagen changes [Gregersen et al., 2001; Fan et al., 2005] and EGF treatment [Liao et al., 2003b; Yang et al., 2003a, 2003b; Zhao et al., 2002c, 2003c].

As mention above, the morphometric properties are best described at the zero-stress state where no internal or external forces deform the tissue. With referenced to the zero-stress state, combined the morphometry data and pressure data, we can computer the stress-strain relationship of the GI wall. The stress-strain distribution mainly reflects the elastic properties of the GI tract. Therefore, we consider the opening angle of the zero-stress state, residual strain and stress-strain relationship as the most relevant biomechanical parameters to describe diseases causing GI remodelling. Generally diseases and factors inducing tissue overgrowth, such as DM, obstruction and EGF treatment, increase GI wall stiffness whereas factors reducing tissue growth, such as fasting and low protein diet decrease GI wall stiffness (Table 1).

Table 1. Disease caused biomechanical remodeling of GI tract

Diseases	Species	Test organs	Biomechanical remodeling		
			OA	RES	Stiffness
Type I DM	Human	Esophagus	ND	ND	Circ NC Long↑
		Duodenum	ND	ND	Circ NC Long↑
	Rat	Esophagus	↓	↓	Circ↑ Long ND
		Duodenum	↓	↓	Circ↑ Long↑
		Jejunum	↑	↑	Circ↑ Long↑
		Ileum	↑	↑	Circ↑ Long↑
		Colon			Circ↑ Long↑
Type II DM	Rat	Esophagus	↓	↓	Circ↑ Long ND
		Stomach	ND	ND	Circ↑
Systemic sclerosis	Human	Duodenum	ND	ND	Circ↑
Ulcerative colitis	Mice	Colon	↑	↑	Circ↑ Long↑↑
Fasting	Rat	Duodenum	↑	↑	Circ↓ Long↓
		Jejunum	↑	↑	Circ↓ Long↓
		Ileum	↑	↑	Circ↓ Long↓
Low protein diet	Mink	Duodenum	NC	NC	Circ↓ Long NC
		Jejunum	↓	↓	Circ↓ Long NC
		Ileum		↓	Circ↓ Long NC
Partial Obstruction	Opossum	esophagus	ND	ND	Circ↑ Long ND
	Guinea pig	Jejunum	↓	↓	Circ↑ Long ND
Osteogenesis imperfecta	Mice	Esophagus	↑	↑	ND
Irradiation	Mice	Rectum	↑	↑	ND
Small bowel resection	Rat	Jejunum	↑	↑	NC
		Ileum	↑	↑	NC
EGF treatment	Rat	Esophagus	↑	↑	Circ↑ Long ND
		Duodenum	↑	↑	Circ↑ Long NC
		Jejunum	↑	↑	Circ↓ Long↑
		Ileum	↑	↑	Circ↓ Long↑
		Colon	↑	↑	ND

Notes: OA: opening angle, RES: residual strain, Circ: circumferential direction, Long: longitudinal direction, ND: not done, NC: no change.

The more collagen in the GI wall, the stiffer the wall is [Pedersen et al., 2003], and visa versa. However, the effect of different factors on the opening angle and residual strain of GI tract depend on the changes of the layered structure. Fung's hypothesis of non-uniform remodelling states that if the inner wall grows more than the outer wall, the opening angle will increase [Fung, 1993] whereas if the outer wall grows more than the inner wall, the opening angle will decrease. The biomechanical remodelling in the GI tract reflects the histological structural remodelling.

Esophageal Biomechanical Remodeling

During normal growth, the esophageal changed its biomechanical properties [Gregersen et al., 2004]. The weight per unit length, the wall cross-sectional area, the inner and outer circumferential length of the mucosa and muscle, and the thickness and area of the layers increased as function of age. The opening angle was approximately 140 degrees at age 1 and 2 weeks and gradually decreased to approximately 80 degrees after 16 weeks. The stress-strain analysis showed that the esophagus become more compliant during the first 4 weeks of life in circumferential direction and the first 16 weeks of life in longitudinal direction (Figure 7). Rao et al [2003] showed in human study that aging was associated with a larger lumen and a stiffer but less sensitive esophageal wall. EGF Treatment induced pronounced growth of rat esophagus morphometrically and histologically [Zhao et al., 2003c]. At the same time, the biomechanical properties were remodeled. The opening angle and residual strains increased with the highest value after 14 days EGF treatment. The change in opening angle depended largely on the change in mucosa thickness. Furthermore, the circumferential stiffness of the esophagus reached a maximum after 7 days EGF treatment [Zhao et al., 2003c].

Biomechanical remodeling of the rat esophagus occurs in experimental type-1 DM [Yang et al., 2004b]. Following the development of DM, for 1 week the circumferential stiffness increased and after 4 weeks the longitudinal and shear stiffness increased. Furthermore, the biomechanical properties of the layered esophagus were also remodeled in experimental type-1 DM [Yang et al., 2006].

DM induced pronounced increase in the outer perimeter, inner perimeter and lumen area in both the muscle and mucosa-submucosa layer. The growth of the mucosa-submucosa layer was more pronounced than the muscle layer. Furthermore, the circumferential stiffness of the mucosa-submucosa layer increased 28 days after STZ treatment. Zhao et al [2007b] recently demonstrated that the histomorphometric and biomechanical remodeling is a feature of the esophageal wall in type 2 diabetic GK rats. DM induced pronounced increasing of the wall thickness and wall cross-sectional area. Histologically, the circumference, thickness, and area of the muscle layer and the collagen fraction of mucosa-submucosa layer were significantly increased in the GK rats. The opening angle and outer residual strain decreased significantly in the GK rats. Furthermore, the circumferential stiffness of the esophagus is significantly higher in the GK rats than in the normal rats. Frokjaer et al [2007b] showed in human study that the esophageal wall and especially the mucosa-submucosa layer had increased thickness in the diabetic patients and the longitudinal and radial compressive stretch was less in diabetics.

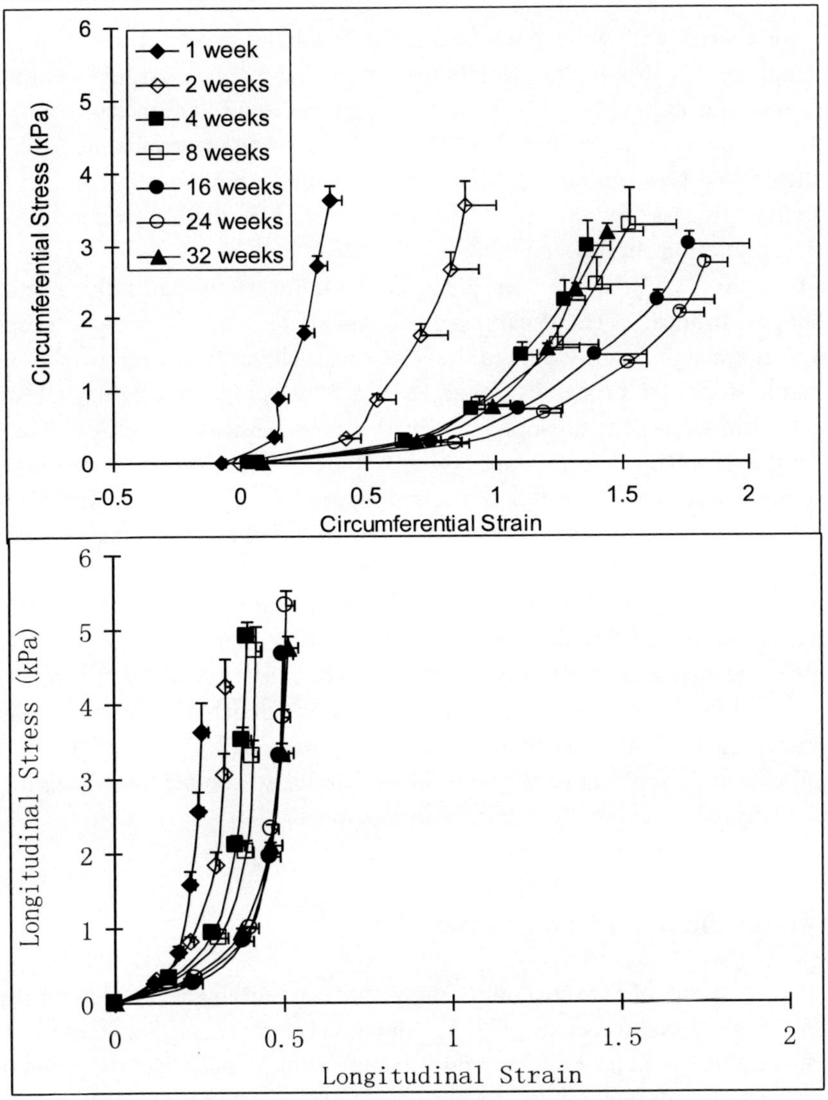

Figure 7. The graphs show the stress-strain relations for circumferential (top) and longitudinal direction (bottom). The circumferential and longitudinal stress-strain curves were exponential. Circumferential stress-strain curves shifted from left to the right up to four weeks of age where after no further change was observed. Longitudinal stress-strain curves shifted from left to the right up to 16 weeks of age where after no further change was observed. The esophageal wall was stiffer in longitudinal direction than in circumferential direction.

The rabbit esophageal wall with varices showed a significant loss of distensibility and recovery of the wall distensibility when the varices were treated [Gregersen et al., 1988]. The partially obstructed opossum esophagus expressed dilatation and decreased ability of the esophageal wall to resist intraluminal pressure [Gregersen et al., 1992b]. By means of IP, Juhl et al [1994] showed that endoscopic sclerotherapy entailed an inversion of the normal esophageal configuration, with narrowing 5 cm above the GEJ and increased CSAs 10 cm above the GEJ. Gregersen et al [2001] investigated esophageal morphometry and residual strain in a mouse model of osteogenesis imperfecta murine (oim) mice with collagen deficiency. Compared with wild-type mice, the esophagi in oim mice had smaller layer thicknesses and areas with the largest reduction in the submucosa, larger opening angles, and larger residual strains at the serosal and mucosal surfaces and the mucosa-submucosal-muscle layer interface. The data suggested that the residual stress in esophagus is caused by the tension in the muscle layer rather than the stiffness of the submucosa in compression and that the remodeling process in the oim esophagus is due mainly to morphometric and biomechanical alterations in the submucosa.

Holland et al [1993] studied esophageal compliance in naturally occurring canine megaoesophagus. In most cases of canine megaoesophagus the viscoelastic properties of the esophageal wall are significantly altered, however in such cases the disorder is unlikely to be purely dynamic and that processes other than the duration of esophageal dilatation are responsible for the alteration in esophageal wall biomechanical properties. Rao et al [1996] showed that the esophageal compliance was reduced in unexplained chest pain patients. Sensory and biomechanical properties of the esophagus in non-erosive reflux disease (NERD) were studied by Reddy et al [2007]. The NERD patients were less sensitive to mechanical stimulation as assessed by the cross-sectional area and volume and hypersensitive to heat stimuli of the esophagus, with an increase in referred pain to the evoked visceral pain. The data indicate that peripheral sensitization of heat-sensitive pathways together with facilitation of central pain mechanisms is important in the pathogenesis of NERD. Remes-Troche et al [2007] demonstrated that the esophagus was more reactive and less distensible in gastroesophageal reflux disease patients than controls. After Nissen fundoplication, in patients with persistent symptoms, the sensory thresholds were unchanged but the esophageal wall reactivity decreased and distensibility improved.

Stomach Biomechanical Remodeling

Gastroparesis is one of the most common motility disturbances in DM and it affects up to 58% of diabetics [Horowitz et al., 1991; Malagelada et al., 1980; Verne et al., 1998]. However, the etiology is poorly understood. Patients with diabetic gastroparesis have delayed gastric emptying and abdominal pain [Rothstein, 1990], suggesting that the sensitivity of gastric mechanoreceptors, gastric accommodation, and the mechanical properties may be important in the etiology of symptoms [Moragas et al., 1993; Rayner et al., 2000; Schwizer et al., 2002]. Barostat studies have shown that the compliance in the proximal stomach may be altered by DM [Karakida and Homma, 1989; Noriyuki et al., 1999; Rayner et al., 2000]. More recently, Kumar et al [2008] performed intermittent, phasic balloon distensions of the stomach using a gastric barostat device in diabetic patients and normal subjects. The accommodation response to a meal was significantly impaired in patients with DM when compared to controls. However, several studies have demonstrated that perception of gastric distension depends on wall tension rather than on intraluminal pressure or volume [Gregersen et al., 1992c, 1992d; Distrutti et al., 2000]. Hence evaluation of tension and its surface variation seem to be the best approach in designing gastric sensitivity and function studies. Liao et al [2006a] made three-dimensional geometry analysis of the stomach in type II diabetic GK rats. The diabetic stomach exhibited lower material compliance not only in the proximal stomach but also in the entire stomach. The biomechanical properties differed between regions and in different directions in both the non-diabetic and the diabetic stomach.

Therefore, the changes due to DM may contribute to the delayed gastric emptying and the symptoms in diabetic patients.

Salet et al [2006] studied the responses to gastric distension in functional dyspepsia. Patients with functional dyspepsia show slightly higher compliance to mechanical distension. Their visceral perception of mechanical stress is enhanced. In contrast with the balloon distension, relaxation after a meal was less.

Intestinal Biomechanical Remodelling

Intestinal physiological growth induced morphometric and biomechanical remodeling [Lu et al., 2005]. Small intestinal mass and dimensions increased many-fold with age. The opening angle of duodenum was approximately 220 degrees and 290 degrees during the first and second week after birth and decreased to 170 degrees at other ages. The opening angle of ileum ranged between 120 degrees and 150 degrees. The residual strain of duodenum at the mucosal surface did not vary with age whereas the residual strain of ileum at the mucosal surface decreased with age. The stress-strain analysis showed that the small intestine became more compliant during early life.

Vinter-Jensen et al [1996] demonstrated that EGF treatment for 1 to 4 weeks caused a time-dependent increase in intestinal weight. The growth was characterized by increased wall thickness, increased cross-sectional area and reduced wall stiffness. Zhao et al [2002c] demonstrated that EGF treatment was associated with pronounced morphometric and histological changes in the rat small intestine. The opening angle and residual strain increased in all segments of the small intestine during EGF treatment with the highest value in the 14 days EGF treatment group. Linear regression analysis demonstrated that the opening angle mainly depended on the mucosa thickness and area. Furthermore, the circumferential stiffness increased in the duodenum and decreased in the jejunum and ileum during EGF treatment. A plateau was reached after 7 days where after it started to normalize. In the longitudinal direction, all intestinal segments became stiffer after EGF treatment for 7 days. After 14 days the curve started to normalize in duodenum and jejunum but not in the ileum. Yang et al [2003a] also showed that EGF caused significant changes both in the morphology and in the passive mechanical properties of the rat ileum. Liao et al [2003b] has shown that the longitudinal modulus of rat intestine after 4 days EGF treatment was significantly larger than in a control group. The cross modulus decreased during the first 4 days of EGF treatment thereafter it increased to a maximum at 7 days.

Nutrition plays an important role in the maintenance of normal gut structure and function. The small intestine undergoes functional changes when food is withheld. To explore the morphological and biomechanical remodeling during starvation, intestinal segments from the fed and fasted rat duodenum, jejunum, and ileum were investigated by Dou et al [2002a]. Fasting decreased the outer circumferential length, wall thickness, wall area, inner circumferential length, and luminal area in the three intestinal segments. Histological examination showed that the mucosal and the submucosal thickness decreased during fasting, whereas the muscle layers were unchanged. Fasting shifted the stress-strain curves to the right in both circumferential and longitudinal directions at the three segments, indicating that the intestinal wall became soft due to fasting. However, the mechanism behind the fasting remodeling is not yet clear. In order to increase our understanding about the mechanism of the effect of fasting on GI remodeling, Chen et al [2008] studied the intestinal remodeling in mink fed with reduced protein content. Feeding the low protein diet induced decreasing the wet weight per unit length, wall thickness and area. Histologically, the wall thickness, mainly the mucosa, was decreased in the jejunum and ileum. The stress-strain curves were shifted to the right in the circumferential direction in all three intestinal segments, indicating decreased stiffness of the wall in the circumferential direction. This study demonstrated that feeding low protein diet in mink induces histomorphometric and biomechanical remodelling in intestine comparable to the previously observed changes during fasting (Figure 8). Therefore, the study provided evidence that protein malnutrition is an important reason for fasting-induced intestinal remodelling. In order to study the recovery of starvation-induced structural and biomechanical remodeling in the rat small intestine, Dou et al [2001] studied the effect of re-feeding on the biomechanics of fasting intestine. Fasting-induced biomechanical and structural remodeling could be normalized by re-feeding in a time- and location-dependent way.

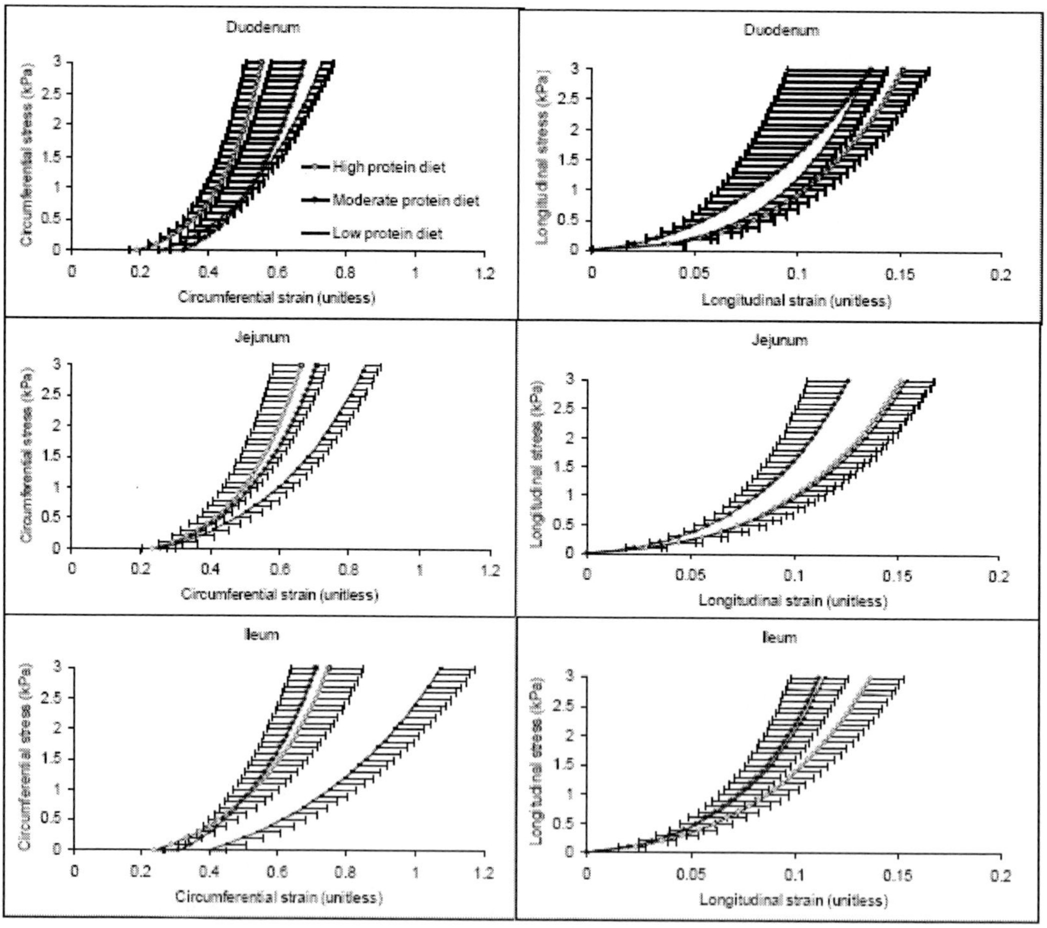

Figure 8. The relationship between circumferential and longitudinal stress and strain. In the circumferential direction, the stress–strain curves in all segments of the L group were shifted to the right compared with those of H and M groups. Thus, the stiffness of intestinal wall decreased (all $P<0.01$ compared with normal group). No significant difference was observed in the longitudinal direction in all three intestinal segments.

Many human [Folwaczny et al., 1999; Charlton et al., 2000; He et al., 2001; Secondulfo et al., 2004] and animal [Folwaczny et al., 1999; Verne and Sninsky, 1998; Zoubi et al., 1995; Bhor et al., 2004; Noda et al., 2001; Tormo et al., 2002; Adachi et al., 2003] studies have shown that DM causes morphological alterations in the small intestine. However, only few data exist relating to biomechanical remodeling of the small intestine in DM. Jorgensen et al [2001] demonstrated that the intestinal wall stiffness increased during the development of STZ-induced rat DM. Zhao et al [2003a] studied the biomechanical and morphometric intestinal remodeling during experimental DM in rats. The blood glucose concentration increased four- to fivefold in the diabetic rats. STZ-induced DM generated pronounced increase in the weight per centimeter length, wall thickness and wall cross-sectional area in all intestinal segments during DM. Histological analysis showed that the thickness of the intestinal layers was increased in all segments during DM. In the duodenum the opening angle and residual strain did not change in the first 2 weeks and decreased after 4 weeks. In the jejunum and ileum the opening angle increased after 1 week in the diabetic group. Furthermore, it was found that the circumferential and longitudinal stiffness of the intestinal wall increased with the duration of DM. Moreover, Zhao et al [2003b] showed the viscoelastic behavior of rat intestinal wall was also changed, i.e., stress relaxation decreased during the development of DM. Gliclazide [Zhao et al., 2002b] and KYQWJJ [Sha et al., 2006] treatment could partly restore the changes of blood glucose level and the remodeling of morphometry and residual strain of small intestine in diabetic rats. The linear regression analysis demonstrated that the effect of Gliclazide and KYQWJJ on intestinal opening angle and residual strain is partially through its effect on the blood glucose level.

Small intestinal obstruction is a frequently encountered clinical problem. To understand the mechanisms behind obstruction and the clinical consequences, data are needed on the relation between the morphologic and biomechanical remodeling that takes place in the intestinal wall during chronic obstruction. Stockholm et al [2007] demonstrated that the luminal cross-sectional area, wall thickness, and elastic modulus in circumferential direction increased in a time-dependent manner proximal to the obstruction site. The circumferential stress-strain curves of the proximal segments in 2- and 4-week groups shifted to the left, indicating the intestinal wall became stiffer. Histological examination revealed a massive increase in the thickness of the muscle layer especially the circular smooth muscle layer. The collagen content proximal to the obstruction site was significantly larger in the partially obstructed animals compared to controls. Strong correlation was found between the collagen content and the elastic modulus at stress levels of 70 kPa and 10 kPa proximal to the obstruction site suggesting that alteration of collagen has great impact on the mechanical remodeling.

Dou et al [2002b] studied morphometric and biomechanical remodeling in the intestine after small bowel resection in the rat. It was demonstrated that resection of the majority of the small bowel results in significant remodeling in structural and residual strain properties in the rat small intestine. The remodeling seems to be guided by the need for a greater absorptive surface area rather than for a change in the stress-strain properties. Abnormalities of the small intestine have also been indicated in systemic sclerosis. Pedersen et al [2003] showed in a human study that systemic sclerosis resulted in increased stiffness and impaired muscle function of the duodenum.

Colon and Rectal Biomechanical Remodelling

The elasticity of the rectal wall after accommodation to distension was significantly greater in patients with Hirschsprung's disease [Arhan et al., 1978], and the time taken by the rectum to accommodate was longer. The increased elasticity correlated well with severity of the illness. The quasilinear viscoelastic law was applied to the in vivo determination of the mechanical properties of the rectal wall in controls and in children with constipation by Bouchoucha et al [1997]. The elastic response was similar in the 2 groups. The reduced relaxation function was significantly different between the 2 groups, with the absence of an asymptotic value in the group with constipation. The effects of a diverting colostomy on weight, collagen content, and biomechanical strength of the left colon was investigated in rats by Kissmeyer-Nielsen et al [2003]. It demonstrated decrease in collagenous proteins and mechanical strength of distal colon after diverting colostomy in rats.

Radiotherapy of malignancies in the pelvis is associated with both early and late gut reactions with subsequent alterations in rectal function because this part of the gut often is included in the radiation field. Therefore, the biomechanical properties of rectum may be affected by irradiation. Early and late effects of irradiation on morphometry and residual strain of mouse rectum were studied by Gregersen et al [2002c]. The irradiated groups had 25-75% smaller opening angles than the normal groups. The early irradiated group had higher values of residual strain than the control group. Furthermore, proximal to the irradiated zone in the late irradiated group had significantly higher values of residual strain compared to the normal group, indicating a higher stress level proximal to the irradiated zone. Therefore, radiation therapy remodelled the morphometry and zero-stress state of rectum.

Age-related alterations in the strength and collagen content of left colon in rats were studied by Christensen et al [1992a]. An accumulation of collagenous proteins was found in old rats compared with middle-aged rats and this was accompanied by a decrease in the strength, which may deteriorate the functional integrity of the left colonic wall with age. The effect of growth hormone treatment on the left colon was investigated in 4-month-old Wistar rats by Christensen et al [1990]. Growth hormone increased the mass, the collagenous proteins, and the strength of rat colon. Studied in 27-month-old male rats showed that the maximum load and stiffness of the left colonic specimens in the b-hGH-treated group were enhanced by 27 and 60%, respectively, and the strain at maximum load was 24% less than in the control group [Christensen et al., 1992b]. The systemic EGF treatment caused remodelling of the morphology of the zero-stress state in the large intestine in a time-dependent manner [Yang et al., 2003b]. The growth was most pronounced in the ascending and transverse colon and involved mainly the mucosal layer.

The pain, urgency, and incontinence in ulcerative colitis may be related to changes in viscoelastic properties of the gut wall or to alterations of the sensory pathways. The profound organic changes in UC indicate that the mechanical and mechanosensory functions are affected. Reduced compliance and increased sensation have been reported in several clinical studies [Rao et al., 1987; Loening-Baucke et al., 1989]. Used an advanced rectal probe, Drewes et al [2006] studied the mechanosensory and smooth muscle properties in patients with active ulcerative colitis. There were no differences in compliance or stiffness between the patients with active ulcerative colitis and normal subjects, but the hypersensitivity was abolished after muscle relaxation. This finding demonstrated that the smooth muscles were tonically contracted in the inflamed rectum, resulting in a decreased rectal circumference.

Recently Yang et al [2009] demonstrated that Oxazolone-induced colitis in BALB/C mice induced increasing wall thickness, opening angles, circumferential and longitudinal wall stiffness in the colon. Furthermore, the mechanical remodelling was associated with the morphometric and histological changes (Figure 9).

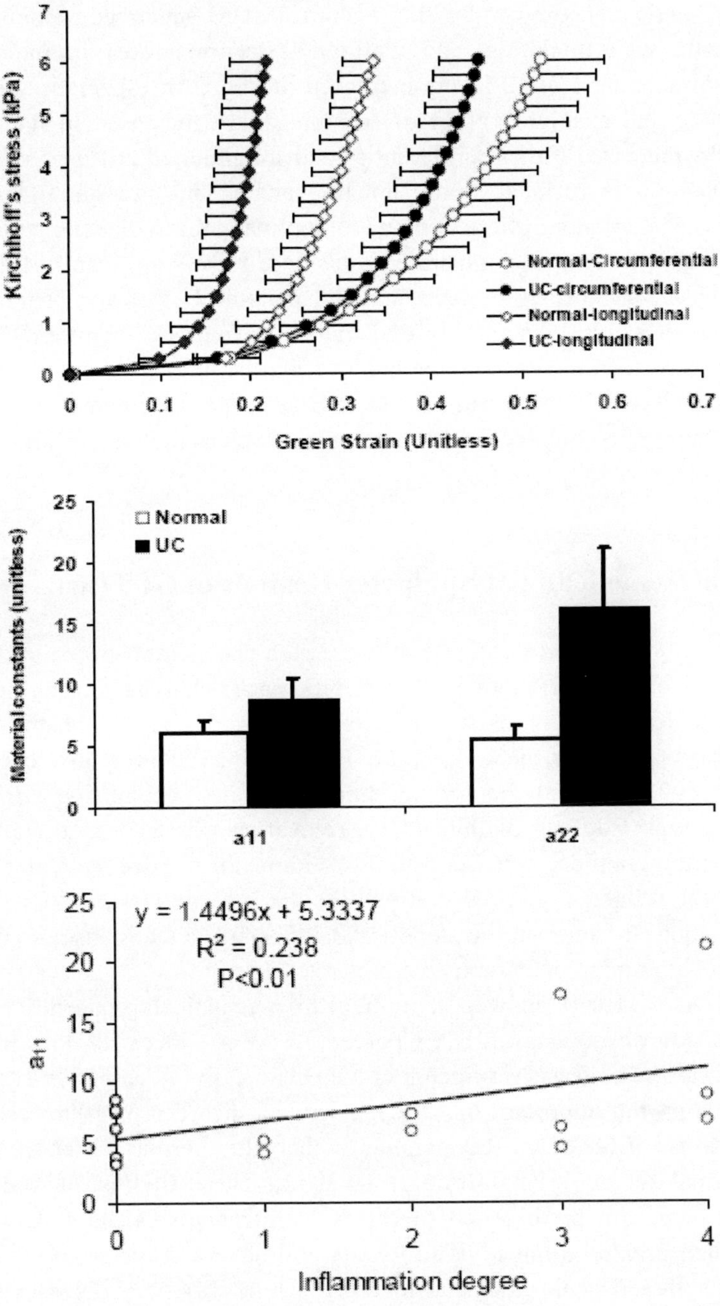

Figure 9. Stress-strain data for both circumferential and longitudinal direction for normal and UC mice (top). The UC mice were stiffer for both directions ($P<0.01$). Similarly, the circumferential and longitudinal constants increased in the UC colon (middle). Association between a_{11} and the degree of inflammation were found (bottom).

Stiffness of the colon wall will lead to increased resistance to normal passage of feces and air in the affected segment, proximal dilatation, and resulting pain and urgency [Drewes et al., 2006].

More recently, Zhao et al [2008b] showed that biomechanical and histomorphometric colon remodelling occurred in STZ-Induced diabetic rats. The blood glucose level increased 3-4-fold in the diabetic rats compared with the controls. DM generated pronounced increasing weight per length, wall thickness, and wall cross-sectional area in the diabetic colon. Histologically, the thickness of all layers increased during DM, especially the mucosa layer. The opening angle and absolute values of residual strain increased in the diabetic group. Furthermore, DM increased the circumferential and longitudinal stiffness of the colon wall. The observed changes in residual strain, opening angle, and stress-strain relation may be contributing factors to colonic dysfunction and abdominal pain in diabetic patients.

Park et al [2008] recently demonstrated that there were significant differences in static compliance between the normal subjects and the irritable bowel syndrome (IBS) patients. However, Drewes et al [2001] demonstrated that no differences between IBS patients and normal subjects were seen in strain and tension of the rectum and sigmoid colon. Therefore, they thought that visceral hypersensitivity in IBS seems to be related to alterations in the nervous system rather than biomechanical parameters such as the tension and strain of the gut wall.

Biomechanical Remodeling of Sphincter Regions in GI Tract

Less UES flexibility in the swallows of older men was demonstrated by Logemann et al [2000]. The mechanical malfunction of the UES and pharynx during mastication and swallowing were found in patients with Parkinson's disease [Johnston et al., 1995]. Manometric abnormalities of the esophageal body and LES have also been documented. Williams et al [2002] studied the biomechanics of failed deglutitive UES relaxation in neurogenic dysphagia. Etiology of failed UES relaxation was 56% medullary disease, 33% Parkinson's or extrapyramidal disease; and 11% idiopathic. Adequacy of UES opening is influenced by the degree of preservation of the pharyngeal swallow response and hyolaryngeal traction. In contrast, the stenotic UES displays a static loss of compliance, only apparent once the limit of sphincter expansion is reached.

Biancani P et al [1984] showed that in vitro mechanical properties of the LES are affected by induction of esophagitis. Sixty percent of the patients with documented increased esophageal exposure to acid gastric juice had a mechanically incompetent LES [Zaninotto et al., 1988], indicating the important of LES mechanical disorder in gastro-esophageal reflux disease (GERD). The EGJ of GERD patients with hiatus hernia was more distensible and shorter than normal subjects [Pandolfino et al, 2002]. These findings partially explain why hiatus hernia patients are predisposed to reflux by mechanisms other than transient LES relaxations, sustain greater volumes of refluxate, and have a reduced ability to discriminate gas from liquid reflux. Achalasia is considered a primary motility disorder confined to the esophagus. The LES in achalasia is frequently hypertonic and manifests absent or incomplete relaxation in response to deglution. On the other hand, the LES and the proximal stomach act physiologically as a functional unit whereby relaxation of the LES during deglution is associated with receptive relaxation of the proximal stomach. Mearin et al [1995] showed that the proximal stomach relaxation was impaired in patients with achalasia. Furthermore, Mearin et al [2000] investigated the mechanical properties of the GEJ comprising intraluminal pressure (measured by manometry), resistance to flow and compliance (measured by resistometry) in health, achalasia, and scleroderma. They found that GEJ resistance to flow is significantly increased in achalasia and decreased in scleroderma while GEJ compliance is diminished in achalasia and normal in scleroderma. Therefore, achalasia is a disease characterized by increased GEJ resistance and rigidity. McMahon et al [2007a] used FLIP to measure the cross-sectional areas in EGJ and showed that in one achalasia patient (unsuccessfully treated with dilations), the CSA never rose above the minimal measurable value despite the pressure increasing to 50 cmH_2O. In another achalasia patient (successfully treated with dilations), the pressure only reached 15 cmH_2O despite opening to a CSA of 250 mm^2. Therefore, The FLIP can be used practically to evaluate difficult cases of EGJ dysfunction and may provide a role in evaluating patients before and after therapies for diseases affecting the EGJ such as achalasia and GERD. Severe erosive reflux disease and Barrett's esophagus are associated with significantly greater loss of the mechanical antireflux barrier as reflected in the presence of hiatal hernia and LES measurements [Lord et al., 2008].

Diabetes mellitus has been associated with a variety of gastrointestinal motor disturbances. Mearin et al [1986] showed that pyloric dysmotility was existed in some patients with diabetes. The cause of sphincter distortion in rectal prolapse is unknown but may be a response to increased mechanical stress placed on the sphincter from the prolapse or an abnormal response by the sphincter complex to the prolapse [Dvorkin et al, 2004].

Impact of GI Biomechanical Remodeling

It is well known that mechanosensation is importance for GI function. The mechanosensitive nerve endings exist extensively in the GI tract where they serve a critical role in homeostasis. The mechanosensitive afferents in the intrinsic and extrinsic pathways were described as low-, wide-dynamic- or high-threshold tension-receptors. Therefore, the GI tract structure as well as the stress and strain distribution in the wall is important for the GI sensory and motor function. The GI wall structure or deformation changes in the disease will alter the relative positions of the mechanosensitive afferents (zero setting of the mechanosensitive afferents). The biomechanical remodelling in the disease such as alterations of residual strain, stress distribution and wall stiffness will alter the tension and stress distribution of the mechanosensitive afferents. Consequently, the perception and motility of the GI tract will change as well. Therefore, the morphological changes and biomechanical remodelling of GI tract due to diseases is likely to affect the function of mechanosensitive afferents in the GI wall and further affect the motor and sensory function.

Biomechanical Modeling of the GI Tract

Gastroenterology research has traditionally been based on experimental approaches rather than on mathematical modelling. Most of the previous modelling efforts in the biological area were in the cardiac and lung field but other areas are in development now. However, in the past five to ten years several groups have independently started to model the GI tract. The use of numerical models and, in particular, of finite element models has been extensively studied in the field of soft tissues mechanics because of the potential they offer in the analysis of the mechanical behaviour of morphologically complex structures, with high structural hierarchy and constituents with non-linear behaviour [Natalia et al., 2006a, 2006b]. The effectiveness of numerical models depends on reliable reconstructions of the morphometry of the anatomical site under investigation and the specific loading and boundary conditions, as well as on the definition of constitutive models capable of describing the mechanical response of the single tissues. The large morphological complexity of the GI tract and the variability in the different parts of the tract are well-known. The complexity increases in the characteristic folds of the connection regions [Liao et al., 2006, 2007; Yang et al., 2007b].

The methods and current development on the biomechanical properties studies on the normal and diseased GI tissues have been described in the above sections. Hereby, the morphometric related modelling establishment on the GI tract will be briefly introduced. According to the reconstruction methods of GI modelling, the establishment of the GI models can be divided into the geometrically based models, the anatomically based models and the theoretical analysis based models.

Geometry Based Models

The geometric modelling is the fundamental part of the GI modelling analysis. Due to the complex geometry of the GI organs, three dimensional (3-D) models of the GI tract are most important. The complex geometry of GI tract can be obtained by advanced imaging techniques.

Medical image based 3-D models of in vivo GI organs have characterized the oesophagus, stomach, small intestine, sigmoid colon, oesophageal gastric junction and rectum based on cross-sectional imaging using ultrasonography, computed tomography (CT), Functional Luminal Imaging Probe (FLIP) or magnetic resonance imaging (MRI) [Frokjaer et al., 2005, 2007a; Li et al., 1994: Jeays et al., 2007; Liao et al., 2004b; McMahon et al., 2007b; Nicosia et al., 2001]. The initial 3-D GI models were reconstructed on the basis of images from ultrasonography, computed tomography (CT) or MRI. With the identified inner and outer wall boundaries of each cross sectional 2-D image, a data cloud of a 3-D model was then obtained by using a segmentation technique. With the development of the medical devices such as IP, it is now possible to record the mechanical parameters as the luminal pressure simultaneously as the cross-sectional medical images. Therefore, the in vivo mechanical behaviour of the organs can be computed on the basis of the reconstructed GI morphometric models and the recorded mechanical parameter. A reconstructed sigmoid-colon model and the corresponding tension and stress distribution on the modelling is illustrated in figure 10. The detailed description about the in vivo GI modelling analysis can be found in studies of Liao et al [2004b] and Frokjaer et al [2005, 2007a].

The 3-D model reconstruction and the tension and stress indicated in figure 10 were only based on the medical images, omitting the information about the tissue material properties and the tissue structures. Therefore, the methods described in this section can be further extended clinically for an individual patient based computational GI tract model analysis.

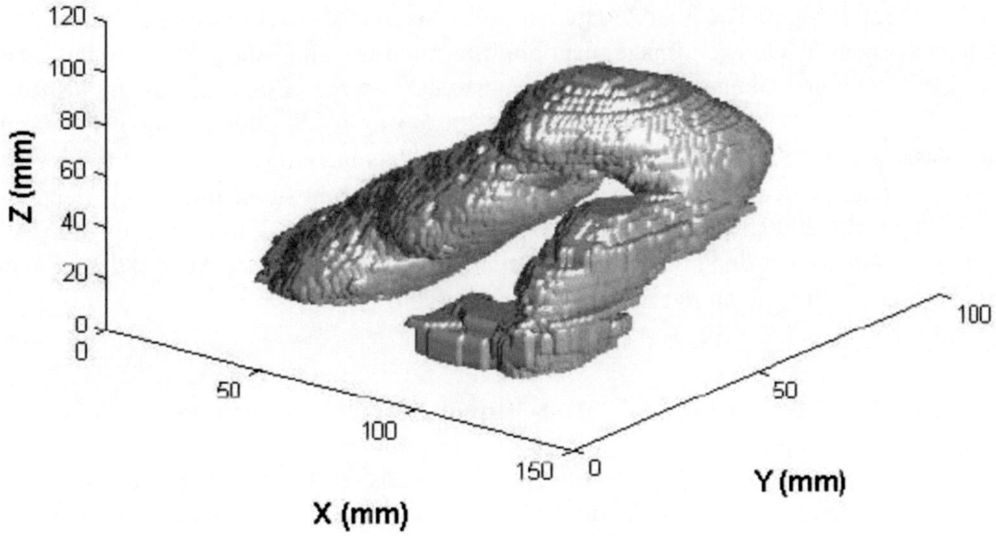

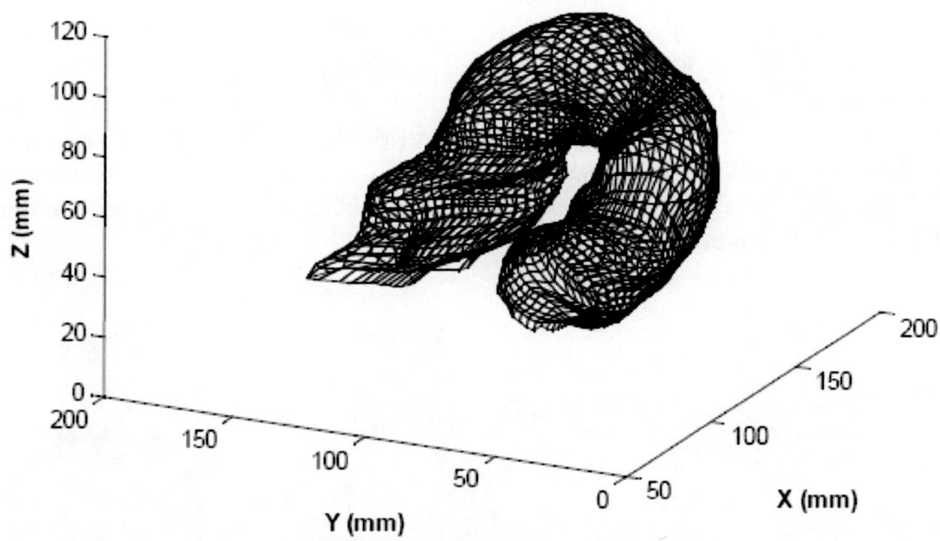

Figure 10. A sigmoid colon solid model generated directly on the basis of the 3-D data cloud (top) and a smoothed surface model generated from the pictures re-sliced along the centre line of the solid model (bottom). The black lines on the top panel are the slices along the centre line of the solid model.

Anatomy Based GI Modelling

For modelling analysis on the in vivo image based models, only the tension or stress was computed on the basis of three dimensional surface geometry by using the Laplace's equation and the wall thickness. The tissue structure and mechanical properties of the GI wall were therefore not taken into account. To aid in understanding of the relationship between the structure and function of the GI tract in health and disease states, anatomically based finite element models are needed. The anatomically based visualization GI model is now commercially available, however the numerical calculation based GI anatomical model existed currently are most built by Andrew Pullan's group and all models were built from the

Visible Human Project. The Visible Human project provided a set of cross sectional images of a human cadaver. On each image, data points around the interesting organ boundary were created and then the geometry models were constructed on the basis of the distinguished data clouds [Buist et al., 2004, 2006; Cheng et al., 2007; Lin et al., 2006a, 2006b; Pullan et a., 2004; Yassi et al., 2004]. The anatomically based models have now been used to investigate normal and pathological electrical activity of the stomach and small intestine [Cheng et al., 2007; Lin et al., 2006a, 2006b; Pullan et a., 2004], the muscle functions on the gastro-oesophageal junction during swallow [Yassi et al., 2004] and the blood flow in the mesenteric arterial system of the human intestine [Buist et al., 2004, 2006].

Theoretical Analysis Based GI Modelling

The morphological complexity of the GI organs makes it difficult to build the anatomically based finite element models. Hence some numerical models of the GI organs on the basis of the theoretical analysis were built by simplifying the complex GI morphometry as a regular geometry such as the circular cylinder for the oesophagus [Brasseur et al., 2007; Li et al., 2006; Liao et al., 2003a; Nicosia et al., 2001; Yang et al., 2006] , sphere for the stomach pouch [Gao et al., 2008] and some regular tubes for describing the antrodudenal junction [Dillard et al., 2007] and esophago–gastric junction [McMahon wt al., 2007a]. Based on the simplified model, most of the biomechanical features such as the tissue structure, tissue properties and bolus flow thus can be expressed mathematically and the mechanical function of the GI tract can be simulated.

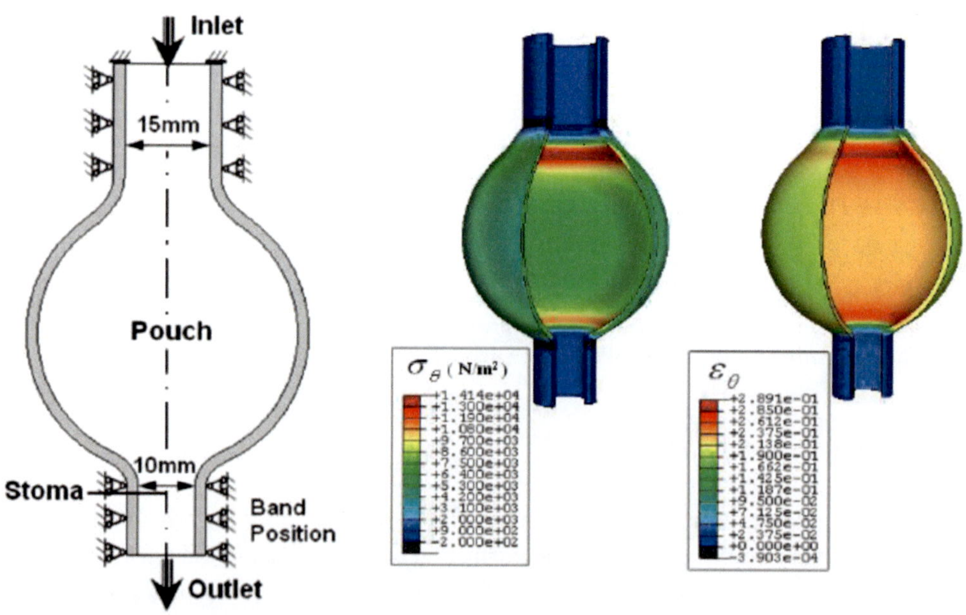

Figure 11. The 3-D pouch finite element model (left) and stress (middle) and strain (right) distribution in a pouch model before emptying.

The simplified GI tract models are now existed for describing the muscle function [Liao et al., 2004b; Nicosia., 2001; Brasseur et al., 2007], the food transportation [Dillard et al.; 2007; Gao et al., 2008; Pal., et al., 2007; Yang et al., 2007c] and blood flow in the GI tract on healthy and diseased situations. A simplified pouch model for describing the obesity patient gastric emptying is illustrated in figure 11 as an example. It is indicated that the pouch emptying curves calculated on the basis of the simplified model agrees well with the clinical recorded results [Gao et al., 2008]

CONCLUSION

The GI tract is functionally subjected to dimensional changes. Hence, biomechanical properties are of particularly importance. Due to the complexity of geometry, anatomy and structure, the biomechanical properties of GI tract are anisotropic and vary along the GI tract. Therefore, analysis of the GI biomechanical properties must take the complex geometry and the anisotropic biomechanical properties into account. The relationship between stress and strain is desirable for a more thorough understanding of the mechanical properties of the GI tract. The stress-strain relationship is better to be determined in reference to the zero-stress state. The biomechanical properties will remodel due to normal growth, surgical invention, systemic diseases, local disorders and etc. Understanding the biomechanical properties of GI tact and its remodeling due to different situations can help us to better understand physiology and pathophysiology of GI tract. Based on the obtained morphometry and biomechanical properties of the GI tract, the normal and disease remodeled GI tract can be modeled mathematically. With the development of the imaging techniques, new analytic methods and medical devices such as IP, the in vivo mechanical behaviour of the organs can be computed on the basis of the reconstructed GI models. The present studies on the GI modeling can be integrated in order to analyze complex structures for understanding biomechanical properties and visceral perception in other visceral organs and further to be integrated as a global GI model.

ACKNOWLEDGMENTS

NIH grant 1RO1DK072616-01A2 and Karen Elise Jensens Foundation supported this work.

REFERENCES

Abell TL, Werkman RF, Familoni BO, Baggous W, Massie D, Vera S. Biliary, pancreatic, and sphincter of Oddi electrical and mechanical signals recorded during ERCP. *Dig. Dis. Sci*. 1998; 43: 540-546.

Adachi T, Mori C, Sakurai K, Shihara N, Tsuda K, Yasuda K. Morphological changes and increased sucrase and isomaltase activity in small intestines of insulin-deficient and type 2 diabetic rats. *Endocr J.* 2003; 50: 271-279.

Arhan P, Devroede G, Danis K, Dornic C, Faverdin C, Persoz B, Pellerin D. Viscoelastic properties of the rectal wall in Hirschsprung's disease. *J. Clin. Invest.* 1978; 62: 82-87.

Assentoft JE, Gregersen H, O'Brien WD Jr. Determination of biomechanical properties in guinea pig esophagus by means of high frequency ultrasound and impedance planimetry. *Dig. Dis. Sci.* 2000; 45: 1260-1266.

Azpiroz F, Malagelada JR. Physiologic variations in canine gastric tone measured by an electronic barostat. *Am. J. Physiol.* 1985; 248: G229-G237.

Azpiroz F, Malagelada JR. Gastric tone measured by an electronic barostat in health and postsurgical gastroparesis. *Gastroenterology.* 1987; 92: 934-943.

Barlow JD, Gregersen H, Thompson DG. Identification of the biomechanical factors associated with the perception of distension in the human esophagus. *Am. J. Physiol Gastrointest Liver Physiol.* 2002; 282: G683-G689.

Berstad A, Hausken T, Gilja OH, Hveem K, Nesje LB, Odegaard S. Ultrasonography of the human stomach. *Scand. J. Gastroenterol.* Suppl 1996; 220: 75-82.

Berstad A, Hausken T, Gilja OH, Nesland A, Odegaard S. Imaging studies in dyspepsia. *Eur. J. Surg.* 1998; Suppl: 42-49.

Bharucha AE, Hubmayr RD, Ferber IJ, Zinsmeister AR. Viscoelastic properties of the human colon. *Am. J. Physiol. Gastrointest Liver Physiol.* 2001; 281: G459-G466.

Bhor VM, Raghuram N, Sivakami S. Oxidative damage and altered antioxidant enzyme activities in the small intestine of streptozotocin-induced diabetic rats. *Int. J. Biochem. Cell Biol.* 2004; 36: 89-97.

Biancani P, Goyal RK, Phillips A, Spiro HM. Mechanics of sphincter action. Studies on the lower esophageal sphincter. *J. Clin. Invest.* 1973; 52: 2973-2978.

Biancani P, Zabinski MP, Kerstein MD, Behar J. Mechanical characteristics of the cat pylorus. *Gastroenterology.* 1980; 78: 301-309.

Biancani P, Barwick K, Selling J, McCallum R. Effects of acute experimental esophagitis on mechanical properties of the lower esophageal sphincter. *Gastroenterology.* 1984; 87: 8-16.

Bouchoucha M, Jais JP, Arhan P, Landais P, Faverdin CL, Pellerin D. Variation of rheological properties of the human rectal wall with distending volume. *Clin. Invest. Med.* 1994; 17:107-114.

Bouchoucha M, Jais JP, Arhan P, Landais P, Pellerin D. Adaptation of the rectal wall to distension in children with constipation. *Clin. Invest. Med.* 1997; 20: 35-40.

BrassEur. J.G, Nicosia MA, Pal A, Miller LS. Function of longitudinal vs circular muscle fibers in esophageal peristalsis, deduced with mathematical modeling. *World J. Gastroenterol.* 2007; 13: 1335-1346.

Bueno L, Fioramonti J, Ruckebusch Y. Rate of flow of digesta and electrical activity of the small intestine in dogs and sheep. *J. Physiol.* 1975: 249: 69-85.

Bueno L, Fioramonti J. Neurohormonal control of intestinal transit. *Reprod. Nutr. Dev.* 1994; 34: 513-525.

Buist ML, Cheng LK, Yassi R, Bradshaw LA, Richards WO, Pullan AJ. An anatomical model of the gastric system for producing bioelectric and biomagnetic fields. *Physiol. Meas.* 2004; 25: 849-861.

Buist ML, Cheng LK, Sanders KM, Pullan AJ. Multiscale modelling of human gastric electric activity: can the electrogastrogram detect functional electrical uncoupling? *Exp. Physiol.* 2006; 91: 383-390.

Charlton M, Ahlman B, Nair KS. The effect of insulin on human small intestinal mucosal protein synthesis. *Gastroenterology*. 2000; 118: 299-306.

Chen P, Zhao J, Nielsen VH, Clausen T, Gregersen H. Intestinal remodelling in mink fed with reduced protein content. *J. Biomech*. 2009; 42: 443-448.

Chen X, Zhao J, Gregersen H. The villi contribute to the mechanics in the guinea pig small intestine. *J. Biomech*. 2008; 41: 806-812.

Cheng LK, Komuro R, Austin TM, Buist ML, Pullan AJ. Anatomically realistic multiscale models of normal and abnormal gastrointestinal electrical activity. *World J. Gastroenterol*. 2007; 13: 1378-1383.

Christensen H, Jørgensen PH, Oxlund H, Laurberg S. Growth hormone increases the mass, the collagenous proteins, and the strength of rat colon. *Scand. J. Gastroenterol*. 1990; 25: 1137-1143.

Christensen H, Andreassen TT, Oxlund H. Age-related alterations in the strength and collagen content of left colon in rats. *Int. J. Colorectal Dis*. 1992a; 7(2): 85-88.

Christensen H, Andreassen TT, Oxlund H. Increased mechanical strength of left colon in old rats treated with growth hormone. *Gerontology*. 1992b; 38: 245-251.

Clapworthy GJ, Viceconti M. Seeing the EuroPhysiome: A roadmap to the Virtual Physiological Human. (Coordination Action #027642, STEP: A strategy for the EuroPhysiome) . Bedfordshire: University of Bedfordshire, 2007.

Cohen S, Green F. The mechanics of esophageal muscle contraction. Evidence of an inotropic effect of gastrin. *J. Clin. Invest*. 1973; 52(8): 2029-2040.

Dall FH, Jørgensen CS, Houe D, Gregersen H, Djurhuus JC. Biomechanical wall properties of the human rectum. A study with impedance planimetry. *Gut*. 1993; 34: 1581-1586.

Dent J. Approaches to driving the evolving understanding of lower oesophageal sphincter mechanical function. *J. Smooth Muscle Res*. 2007; 43: 1-14.

Dillard S, Krishnan S, Udaykumar HS. Mechanics of flow and mixing at antroduodenal junction. *World J. Gastroenterol*. 2007; 13: 1365-1371.

Dobrin PB. Mechanical properties of arterises. *Physiol. Rev*. 1978: 58: 397-460.

Dogan I, Bhargava V, Liu J, Mittal RK. Axial stretch: A novel mechanism of the lower esophageal sphincter relaxation. *Am. J. Physiol Gastrointest Liver Physiol*. 2007; 292: G329-G334.

Distrutti E, Azpiroz F, Soldevilla A, Malagelada JR (2000) Gastric wall tension determines perception of gastric distention. *Gastroenterology*; 118: 641-643.

Dou Y, Gregersen S, Zhao J, Zhuang F, Gregersen H. Effect of re-feeding after starvation on biomechanical properties in rat small intestine. *Med. Eng. Phys*. 2001; 23: 557-566.

Dou Y, Gregersen S, Zhao J, Zhuang F, Gregersen H. Morphometric and biomechanical intestinal remodeling induced by fasting in rats. *Dig. Dis. Sci*. 2002a; 47: 1158-1168.

Dou Y, Lu X, Zhao J, Gregersen H. Morphometric and biomechanical remodelling in the intestine after small bowel resection in the rat. *Neurogastroenterol Motil*. 2002b; 14: 43-53.

Dou Y, Zhao J, Gregersen H. Morphology and stress-strain properties along the small intestine in the rat. *J. Biomech. Eng*. 2003; 125: 266-273.

Dou Y, Fan Y, Zhao J, Gregersen H. Longitudinal residual strain and stress-strain relationship in rat small intestine. *Biomed. Eng. Online*. 2006 Jun 7;5:37.

Drewes AM, Petersen P, Rössel P, Gao C, Hansen JB, Arendt-Nielsen L. Sensitivity and distensibility of the rectum and sigmoid colon in patients with irritable bowel syndrome. *Scand. J. Gastroenterol.* 2001; 36: 827-832.

Drewes AM, Pedersen J, Liu W, Arendt-Nielsen L, Gregersen H. Controlled mechanical distension of the human oesophagus: sensory and biomechanical findings. *Scand. J. Gastroenterol.* 2003a; 38: 27-35.

Drewes AM, Gregersen H. Multimodal pain stimulation of the gastrointestinal tract. *World J. Gastroenterol.* 2006; 12: 2477-2486.

Duch BU, Petersen JA, Vinter-Jensen L, Gregersen H. Elastic properties in the circumferential direction in isolated rat small intestine. *Acta. Physiol. Scand.* 1996,157: 157-163.

Dvorkin LS, Chan CL, Knowles CH, Williams NS, Lunniss PJ, Scott SM. Anal sphincter morphology in patients with full-thickness rectal prolapse. *Dis. Colon. Rectum.* 2004; 47:198-203.

Egorov VI, Schastlivtsev IV, Prut EV, Baranov AO, Turusov RA. Mechanical properties of the human GI tract. *J. Biomech.* 2002; 35: 1417-1425.

Elbrønd H, Tøttrup A, Forman A. Mechanical properties of isolated smooth muscle from rabbit sphincter of Oddi and duodenum. *Scand. J. Gastroenterol.* 1991; 26: 289-294.

Fackler K, Klein L, Hiltner A. Polarizing light microscopy of intestine and its relationship to mechanical behaviour. *J. Microsc.* 1981; 124(Pt 3): 305-311.

Fan Y, Gregersen H, Kassab GS. A two-layered mechanical model of the rat esophagus. Experiment and theory. *Biomed Eng Online.* 2004 Nov 1; 3(1): 40.

Fan Y, Zhao J, Liao D, Gregersen H. The effect of digestion of collagen and elastin on histomorphometry and the zero-stress state in rat esophagus. *Dig. Dis. Sci.* 2005; 50: 1497-1505.

Folwaczny C, Riepl R, Tschop M, Landgraf R. GI involvement in patients with diabetes mellitus: Part I (first of two parts). Epidemiology, pathophysiology, clinical findings. *Z.Gastroenterol.* 1999: 37: 803-815.

Ford MJ, Camilleri M, Wiste JA, Hanson RB. Differences in colonic tone and phasic response to a meal in the transverse and sigmoid human colon. *Gut.* 1995; 37: 264-269.

Frokjaer JB, Liao D, Bergmann A, McMahon BP, Steffensen E, Drewes AM, Gregersen H. Three-dimensional biomechanical properties of the human rectum evaluated with magnetic resonance imaging. *Neurogastroenterol Motil.* 2005; 17: 531-540.

Frokjaer JB, Andersen SD, Lundbye-Christensen S, Funch-Jensen P, Drewes AM, Gregersen H. Sensation and distribution of stress and deformation in the human oesophagus. *Neurogastroenterol Motil.* 2006a; 18: 104-114.

Frokjaer JB, Andersen SD, Drewes AM, Gregersen H. Ultrasound-determined geometric and biomechanical properties of the human duodenum. *Dig. Dis. Sci.* 2006b; 51: 1662-1669.

Frokjaer JB, Liao D, Steffensen E, Dimcevski G, Bergmann A, Drewes AM, Gregersen H. Geometric and mechanosensory properties of the sigmoid colon evaluated with magnetic resonance imaging. *Neurogastroenterol Motil.* 2007a; 19: 253-262.

Frokjaer JB, Andersen SD, Ejskjaer N, Funch-Jensen P, Drewes AM, Gregersen H. Impaired contractility and remodeling of the upper GI tract in diabetes mellitus type-1. *World J. Gastroenterol.* 2007b; 13 (36): 4881-4890.

Fung YC. What are the residual stresses doing in our blood vessels? *Ann. Biomed. Eng.* 1991:19: 237-249.

Fung YC, Liu SQ. Strain distribution in small blood vessels with zero-stress state taken into consideration. *Am. J. Physiol.* 1992: 262: H544-H552.

Fung YC. *Biomechanics. Properties of living tissues*. Berlin, Springer-Verlag,1993.

Gabella G, Blundell D. Gap junctions of the muscles of the small and large intestine. *Cell Tissue Res.* 1981: 219: 469-488.

Gao C, Zhao J, Gregersen H. Histomorphometry and strain distribution in pig duodenum with reference to zero-stress state. *Dig. Dis. Sci.* 2000a; 45: 1500-1508.

Gao C, Gregersen H. Biomechanical and morphological properties in rat large intestine. *J. Biomech.* 2000b; 33:1089-1097.

Gao C, Arendt-Nielsen L, Liu W, Petersen P, Drewes AM, Gregersen H. Sensory and biomechanical responses to ramp-controlled distension of the human duodenum. *Am. J. Physiol Gastrointest Liver Physiol.* 2003; 284: G461-G471.

Gao F, Liao D, Zhao J, Drewes AM, Gregersen H. Numerical analysis of pouch filling and emptying after laparoscopic gastric banding surgery. *Obes. Surg.* 2008; 18: 243-250.

Gilja OH, Lunding J, Hausken T, Gregersen H. Gastric accommodation assessed by ultrasonography. *World J. Gastroenterol.* 2006; 12: 2825-2829.

Gilja OH, Hatlebakk JG, Odegaard S, Berstad A, Viola I, Giertsen C, Hausken T, Gregersen H. Advanced imaging and visualization in GI disorders. *World J. Gastroenterol.* 2007; 13: 1408-1421.

Glavind EB, Forman A, Madsen G, Svane D, Andersson KE, Tøttrup A. Mechanical properties of isolated smooth muscle from human rectum and internal anal sphincter. *Am. J. Physiol.* 1993; 265(4 Pt 1): G792-G798.

Gravesen FH, McMahon BP, Drewes AM, Gregersen H. Measurement of the axial force during primary peristalsis in the oesophagus using a novel electrical impedance technology. *Physiol. Meas.* 2008; 29: 389-399.

Gregersen H, Jensen LS, Djurhuus JC. Changes in oesophageal wall biomechanics after portal vein banding and variceal sclerotherapy measured by a new technique. An experimental study in rabbits. *Gut.* 1988; 29: 1699-1704.

Gregersen H, Orvar K, Christensen J. Biomechanical properties of duodenal wall and duodenal tone during phase I and phase II of the MMC. *Am. J. Physiol.* 1992a; 263(5 Pt 1): G795-G801.

Gregersen H, Giversen IM, Rasmussen LM, Tøttrup A. Biomechanical wall properties and collagen content in the partially obstructed opossum esophagus. *Gastroenterology.* 1992b; 103: 1547-1551.

Gregersen H, Jorgensen CS, Dall FH, Jensen SL Characteristics of spontaneous and evoked motility in the isolated perfused porcine duodenum. *J. Appl. Physiol.* 1992c; 73: 9-19.

Gregersen H, Dall FH, Jorgensen CS, Jensen SL, Ahren B. Effects of noradrenaline and galanin on duodenal motility in the isolated perfused porcine pancreatico-duodenal block. *Regul Pept.* 1992d; 39: 157-167.

Gregersen H, Kassab G. Biomechanics of the GI tract. *Neurogastroenterol. Motil.* 1996; 8: 277-297.

Gregersen H, Kassab G, Pallencaoe E, Lee C, Chien S, Skalak R, Fung YC. Morphometry and strain distribution in guinea pig duodenum with reference to the zero-stress state. *Am. J. Physiol.* 1997; 273(4 Pt 1): G865-G874.

Gregersen H, Lee TC, Chien S, Skalak R, Fung YC. Strain distribution in the layered wall of the esophagus. *J. Biomech. Eng.* 1999; 121: 442-448.

Gregersen H, Kassab GS, Fung YC. The zero-stress state of the GI tract: biomechanical and functional implications. *Dig. Dis. Sci.* 2000; 45: 2271-2281.

Gregersen H, Weis SM, McCulloch AD. Oesophageal morphometry and residual strain in a mouse model of osteogenesis imperfecta. *Neurogastroenterol Motil.* 2001; 13: 457-464.

Gregersen H. *Biomechanics of the Gastrointestinal Tract.* London : Springer-Verlag, 2002a: 1-268.

Gregersen H, Lundby L, Overgaard J. Early and late effects of irradiation on morphometry and residual strain of mouse rectum. *Dig. Dis. Sci.* 2002b; 47: 1472-1479.

Gregersen H, Gilja OH, Hausken T, Heimdal A, Gao C, Matre K, Ødegaard S, Berstad A. Mechanical properties in the human gastric antrum using B-mode ultrasonography and antral distension. *Am. J. Physiol Gastrointest Liver Physiol.* 2002c; 283: G368-G375.

Gregersen H, Lu X, Zhao J. Physiological growth is associated with esophageal morphometric and biomechanical changes in rats. *Neurogastroenterol Motil.* 2004; 16: 403-412.

Gregersen H, Hausken T, Yang J, Ødegaard S, Gilja OH. Mechanosensory properties in the human gastric antrum evaluated using B-mode ultrasonography during volume-controlled antral distension. *Am. J. Physiol Gastrointest Liver Physiol.* 2006b; 290: G876-G882.

Gregersen H, Liao D, Fung YC. Determination of homeostatic elastic moduli in two layers of the esophagus. *J. Biomech. Eng.* 2008; 130: 011005.

Haffner JFW, Stadaas J. Pressure responses to cholinergic and adrenergic agents in the fundus, corpus, and antrum of isolated rabbit stomachs. *Acta. Chir. Scand.* 1972; 138: 713-719.

He CL, Soffer EE, Ferris CD, Walsh RM, Szurszewski JH, Farrugia G. Loss of interstitial cells of cajal and inhibitory innervation in insulin-dependent diabetes. *Gastroenterology.* 2001; 121: 427-434.

Hecker L, Baar K, Dennis RG, Bitar KN. Development of a three-dimensional physiological model of the internal anal sphincter bioengineered in vitro from isolated smooth muscle cells. *Am. J. Physiol Gastrointest Liver Physiol.* 2005; 289: G188-G196.

Helm JF, Dodds WJ, Christensen J, Sarna SK. Control mechanism of spontaneous in vitro contractions of the opossum sphincter of Oddi. *Am. J. Physiol.* 1985; 249(5 Pt 1): G572-G579.

Helm JF, Dodds WJ, Christensen J, Sarna SK. Intramural neural control of opossum sphincter of Oddi. *Am. J. Physiol.* 1989; 257(6 Pt 1): G925-G929.

Hind JA, Nicosia MΛ, Roecker EB, Carnes ML, Robbins J. Comparison of effortful and noneffortful swallows in healthy middle-aged and older adults. *Arch. Phys. Med. Rehabil.* 2001; 82: 1661-1665.

Holland CT, Satchell PM, Farrow BR. Oesophageal compliance in naturally occurring canine megaoesophagus. *Aust. Vet. J.* 1993; 70: 414-420.

Horowitz M, Edelbroek M, Fraser R, Maddox A, Wishart J. Disordered gastric motor function in diabetes melliitus. *Scand. J. Gastroenterol.* 1991; 26: 673-684.

Hunter PJ, Nielsen PM, Bullivant D. The IUPS Physiome Project. International Union of Physiological Sciences. *Novartis Found Symp.* 2002; 247: 207-217.

Hunter PJ, Nielsen PM. A strategy for integrative computational physiology. *Physiology (Bethesda).* 2005; 20: 316-325.

Indireshkumar K, BrassEur. J.G, Faas H, Hebbard GS, Kunz P, Dent J, Feinle C, Li M, Boesiger P, Fried M, Schwizer W. Relative contributions of "pressure pump" and "peristaltic pump" to gastric emptying. *Am. J. Physiol Gastrointest Liver Physiol.* 2000; 278: G604-G616.

Jacob P, Kahrilas PJ, Logemann JA, Shah V, Ha T. Upper esophageal sphincter opening and modulation during swallowing. *Gastroenterology*. 1989; 97: 1469-1478.

Jeays AD, Lawford PV, Gillott R, Spencer P, Barber DC, Bardhan KD, Hose DR. Characterisation of the haemodynamics of the superior mesenteric artery. *J. Biomech.* 2007; 40: 1916-1926.

Johnston BT, Li Q, Castell JA, Castell DO. Swallowing and esophageal function in Parkinson's disease. *Am. J. Gastroenterol.* 1995; 90: 1741-1746.

Jorgensen CS, Dall FH, Storkholm J, Jensen SL, Gregersen H. Elastic properties of the isolated perfused porcine duodenum. *Dig. Dis.*. 1991: 9: 401-407.

Jorgensen CS, Dall FH, Jensen SL, Gregersen H. A new combined high-frequency ultrasound-impedance planimetry measuring system for the quantification of organ wall biomechanics in vivo. *J. Biomech.* 1995; 28: 863-867.

Juhl CO, Vinter-Jensen L, Djurhuus JC, Gregersen H, Dajani EZ. Biomechanical properties of the oesophagus damaged by endoscopic sclerotherapy. An impedance planimetric study in minipigs. *Scand. J. Gastroenterol.* 1994; 29: 867-873.

Jørgensen CS, Ahrensberg JM, Gregersen H, Flyvberg A. Tension-strain relations and morphometry of rat small intestine in experimental diabetes. *Dig. Dis. Sci.* 2001; 46: 960-967.

Karakida T, Homma S. Compliance changes of the GI tract in streptozotocin-induced diabetic rats. *Jpn J. Physiol.* 1989: 39: 559-570.

Keinke O, Schemann M, Ehrlein HJ. Mechanical factors regulating gastric emptying of viscous nutrient meals in dogs. *Q. J. Exp. Physiol.* 1984; 69: 781-795.

Kahrilas PJ, Logemann JA, Krugler C, Flanagan E. Volitional augmentation of upper esophageal sphincter opening during swallowing. *Am. J. Physiol.* 1991; 260: G450-G456

Karakida T, Homma S. Compliance changes of the GI tract in streptozotocin-induced diabetic rats. *Jpn J. physiol.* 1989; 39: 559-570.

Kissmeyer-Nielsen P, Christensen H, Laurberg S. Decrease in collagenous proteins and mechanical strength of distal colon after diverting colostomy in rats. *Int. J. Colorectal Dis.* 1993; 8: 120-124.

Kumar A, Attaluri A, Hashmi S, Schulze KS, Rao SS. Visceral hypersensitivity and impaired accommodation in refractory diabetic gastroparesis. *Neurogastroenterol Motil.* 2008; 20: 635-642.

Kuriyama H, Mishima K, Suzuki H. Some differences in contractile responses of isolated longitudinal and circular muscle from the guinea-pig stomach. *J. Physiol.* 1975; 251: 317-331.

Li J, Zhao J, Liao D, Gregersen H. Effect of smooth muscle tone on morphometry and residual strain in rat duodenum, jejunum and ileum. *J. Biomech.* 2008; 41: 2667-2672.

Li L, Zhang J, Lu G. [Tension-stress effect on genesis and growth of the external anal sphincter] [Article in Chinese]. *Zhonghua Wai Ke Za Zhi.* 1995; 33: 123-126.

Li M, BrassEur. J.G, Dodds WJ. Analyses of normal and abnormal esophageal transport using computer simulations. *Am. J. Physiol.* 1994; 266: G525-G543.

Liao D, Fan Y, Zeng Y, Gregersen H. Stress distribution in the layered wall of the rat oesophagus. *Med. Eng. Phys.* 2003a; 25: 731-738.

Liao D, Yang J, Zhao J, Zeng Y, Vinter-Jensen L, Gregersen H. The effect of epidermal growth factor on the incremental Young's moduli in the rat small intestine. *Med. Eng. Phys*. 2003b; 25: 413-418.

Liao D, Gregersen H, Hausken T, Gilja OH, Mundt M, Kassab G. Analysis of surface geometry of the human stomach using real-time 3-D ultrasonography in vivo. *Neurogastroenterol Motil*. 2004a; 16: 315-324.

Liao D, Zhao J, Fan Y, Gregersen H. Two-layered quasi-3D finite element model of the oesophagus. *Med. Eng. Phys*. 2004b; 26: 535-543.

Liao D, Zhao J, Gregersen H. Regional surface geometry of the rat stomach based on three-dimensional curvature analysis. *Phys Med. Biol*. 2005; 50: 231-246.

Liao D, Zhao J, Gregersen H. Three-dimensional geometry analysis of the stomach in type II diabetic GK rats. *Diabetes Res. Clin. Pract*. 2006a; 71: 1-13.

Liao D, Cassin J, Zhao J, Gregersen H.The geometric configuration and morphometry of the rabbit oesophagus during luminal pressure loading. *Physiol. Meas*. 2006b; 27: 703-711.

Liao D, Zhao J, Yang J, Gregersen H. The oesophageal zero-stress state and mucosal folding from a GIOME perspective. *World J. Gastroenterol*. 2007; 13: 1347-1351.

Liao D, Lelic D, Gao F, Drewes AM, Gregersen H. Biomechanical functional and sensory modelling of the GI tract. *Philos. Transact. A Math. Phys. Eng. Sci*. 2008; 366: 3281-3299.

Liao DH, Zhao JB, Gregersen H. Gastrointestinal tract modelling in health and disease. *World J. Gastroenterol*. 2009; 15: 169-176.

Lin AS, Buist ML, Cheng LK, Smith NP, Pullan AJ. Computational simulations of the human magneto- and electroenterogram. *Ann. Biomed. Eng*, 2006a; 34: 1322-1331.

Lin AS, Buist ML, Smith NP, Pullan AJ. Modelling slow wave activity in the small intestine. *J. Theor. Biol*. 2006b; 242: 356-362.

Loening-Baucke V, Metcalf AM, Shirazi S. Anorectal manometry in active and quiescent ulcerative colitis. *American Journal of Gastroenterology*. 1989; 84: 892-897.

Logemann JA, Pauloski BR, Rademaker AW, Colangelo LA, Kahrilas PJ, Smith CH. Temporal and biomechanical characteristics of oropharyngeal swallow in younger and older men. *J. Speech Lang Hear Res*. 2000; 43: 1264-1274.

Lord RV, Demeester SR, Peters JH, Hagen JA, Elyssnia D, Sheth CT, Demeester TR. iatal Hernia, Lower Esophageal Sphincter Incompetence, and Effectiveness of Nissen Fundoplication in the Spectrum of Gastroesophageal Reflux Disease. *J. Gastrointest Surg*. 2008 Dec 3. [Epub ahead of print].

Lu X, Gregersen H. Regional distribution of axial strain and circumferential residual strain in the layered rabbit oesophagus. *J. Biomech*. 2001; 34: 225-233.

Lu X, Zhao J, Gregersen H. Small intestinal morphometric and biomechanical changes during physiological growth in rats. *J. Biomech*. 2005; 38: 417-426.

Malagelada JR, Rees WD, Mazzotta LJ, Go VL.Gastric motor abnormalities in diabetic and postvagotomy gastroparesis: effect of metoclopramide and bethanechol. *Gastroenterology*. 1980; 78: 286-293.

Mandrek K, Kreis S. Regional differentiation of gastric and of pyloric smooth muscle in the pig: mechanical responses to acetylcholine, histamine, substance P, noradrenaline and adrenaline. *J. Auton. Pharmacol*. 1992; 12: 37-49.

Massey BT, Simuncak C, LeCapitaine-Dana NJ, Pudur S. Transient lower esophageal sphincter relaxations do not result from passive opening of the cardia by gastric distention. *Gastroenterology*. 2006; 130: 89-95.

Matsufuji H, Yokoyama J. Neural control of the internal anal sphincter motility. *J. Smooth Muscle Res.* 2003; 39: 11-20.

McMahon BP, Frokjaer JB, Kunwald P, Liao D, Funch-Jensen P, Drewes AM, Gregersen H. The functional lumen imaging probe (FLIP) for evaluation of the esophagogastric junction. *Am. J. Physiol Gastrointest Liver Physiol.* 2007a; 292: G377-G384.

McMahon BP, Odie KD, Moloney KW, Gregersen H. Computation of flow through the oesophagogastric junction.*World J. Gastroenterol.* 2007b; 13: 1360-1364.

Mearin F, Camilleri M, Malagelada JR. Pyloric dysfunction in diabetics with recurrent nausea and vomiting. *Gastroenterology*. 1986; 90: 1919-1925.

Mearin F, Papo M, Malagelada JR. Impaired gastric relaxation in patients with achalasia. *Gut.* 1995; 36: 363-368.

Mearin F, Fonollosa V, Vilardell M, Malagelada JR. Mechanical properties of the gastro-esophageal junction in health, achalasia, and scleroderma. *Scand. J. Gastroenterol.* 2000; 35: 705-710.

Milenov K, Golenhofen K. Contractile responses of longitudinal and circular smooth muscle of the canine stomach to prosglanding E and F2 alpha. *Prostaglandings Leukotrienes Med.* 1982; 8: 287-300.

Miller LS, Dai Q, Sweitzer BA, Thangada V, Kim JK, Thomas B, Parkman H, Soliman AM. Evaluation of the upper esophageal sphincter (UES) using simultaneous high-resolution endoluminal sonography (HRES) and manometry. *Dig. Dis. Sci.* 2004; 49: 703-709.

Moragas G, Azpiroz F, Pavia J, Malagelada JR (1993) Relations among intragastric pressure, postcibal perception, and gastric emptying. *Am. J. Physiol.* 264: G1112-G1117.

Natalia A, Carniel EL, Pavan PG, Dario P, Izzo I, Menciassi A. Evaluation of the constitutive parameters for the numerical analysis of non linear mechanical response of soft tissues. *Proceedings of Computer Methods in Biomechanics and Biomedical Engineering* 2006a, Antibes, France.

Natalia A, Carniel EL, Pavan PG, Dario P, Izzo I, Menciassi A. Hyperelastic models for the analysis of soft tissue mechanics: definition of constitutive parameters. *IEEE/RAS-EMBS International Conference on Biomedical Robotics and Biomechatronics.* 2006b (pp. 1-8).

Nicosia MA, BrassEur. J.G, Liu JB, Miller LS. Local longitudinal muscle shortening of the human esophagus from high-frequency ultrasonography. *Am. J. Physiol. Gastrointest. Liver Physiol.* 2001; 281: G1022-G1033.

Noda T, Iwakiri R, Fujimoto K, Yoshida T, Utsumi H, Sakata H, Hisatomi A, Aw TY. Suppression of apoptosis is responsible for increased thickness of intestinal mucosa in streptozotocin-induced diabetic rats. *Metabolism.* 2001; 50: 259-264.

Noriyuki O, Sengupta JN, Gebhart GF. Mechanosensitive properties of gastric vagal afferent Fibers in the rat. *J. Neurophysiol.* 1999; 82: 2210–2220.

Orvar KB, Gregersen H, Christensen J. Biomechanical characteristics of the human esophagus. *Dig. Dis. Sci.* 1993; 38: 197-205.

Pal A, BrassEur. J.G, Abrahamsson B. A stomach road or "Magenstrasse" for gastric emptying. *J. Biomech.* 2007; 40: 1202-1210.

Pandolfino JE, Shi G, Curry J, Joehl RJ, BrassEur. J.G, Kahrilas PJ. Esophagogastric junction distensibility: a factor contributing to sphincter incompetence. *Am. J. Physiol Gastrointest Liver Physiol.* 2002; 282: G1052-G1058.

Park JH, Baek YH, Park DI, Kim HJ, Cho YK, Sohn CI, Jeon WK, Kim BI, Rhee PL. Analysis of rectal dynamic and static compliances in patients with irritable bowel syndrome. *Int. J. Colorectal Dis.* 2008; 23: 659-664.

Penninckx F, Lestar B, Kerremans R. The internal anal sphincter: mechanisms of control and its role in maintaining anal continence. *Baillieres Clin Gastroenterol.* 1992; 6: 193-214.

Patel RS, Rao SS. Biomechanical and sensory parameters of the human esophagus at four levels. *Am. J. Physiol.* 1998; 275(2 Pt 1): G187-G191.

Petersen P, Gao C, Rössel P, Qvist P, Arendt-Nielsen L, Gregersen H, Drewes AM. Sensory and biomechanical responses to distension of the normal human rectum and sigmoid colon. *Digestion.* 2001; 64: 191-199.

Pedersen J, Gao C, Egekvist H, Bjerring P, Arendt-Nielsen L, Gregersen H, Drewes AM. Pain and biomechanical responses to distention of the duodenum in patients with systemic sclerosis. *Gastroenterology.* 2003; 124: 1230-1239.

Pullan A, Cheng L, Yassi R, Buist M. Modelling gastrointestinal bioelectric activity. *Prog. Biophys. Mol. Biol.* 2004; 85: 523-550.

Qiao Y, Pan E, Chakravarthula SS, Han F, Liang J, Gudlavalleti S. Measurement of mechanical properties of rectal wall. *J. Mater Sci. Mater Med.* 2005; 16: 183-188.

Rao SS, Read NW, Davison PA, Bannister JJ, Holdsworth CD. Anorectal sensitivity and responses to rectal distention in patients with ulcerative colitis. *Gastroenterology.* 1987; 93, 1270-1275.

Rao SS, Hayek B, Summers RW. Impedance planimetry: an integrated approach for assessing sensory, active, and passive biomechanical properties of the human esophagus. *Am. J. Gastroenterol.* 1995; 90: 431-438.

Rao SS, Gregersen H, Hayek B, Summers RW, Christensen J. Unexplained chest pain: the hypersensitive, hyperreactive, and poorly compliant esophagus. *Ann Intern Med.* 1996; 124: 950-958.

Rao SS, Mudipalli RS, Mujica VR, Patel RS, Zimmerman B. Effects of gender and age on esophageal biomechanical properties and sensation. *Am. J. Gastroenterol.* 2003; 98: 1688-1695.

Rao GN, Drew PJ, Monson JR, Duthie GS. Incremental elastic modulus--a challenge to compliance. *Int. J. Colorectal Dis.* 1997; 12: 33-36.

Rayner CK, Verhagen MAMT, Hebbard GS, DiMatteo AC, Doran SM, Horowitz M. Proximal gastric compliance and perception of distension in type 1 diabetes mellitus: Effect of hyperglycemia. *Am. J. Gastroenterology.* 2000; 95: 1175-1183.

Reddy H, Staahl C, Arendt-Nielsen L, Gregersen H, Drewes AM, Funch-Jensen P. Sensory and biomechanical properties of the esophagus in non-erosive reflux disease. *Scand. J. Gastroenterol.* 2007; 42: 432-440.

Remes-Troche JM, Maher J, Mudipalli R, Rao SS. Altered esophageal sensory-motor function in patients with persistent symptoms after Nissen fundoplication. *Am. J. Surg.* 2007; 193: 200-205.

Rothstein RD. GI motility disorders in diabetes mellitus. *Am. J. Gastroenterology.* 1990; 85: 782-785.

Salet GA, Samsom M, Roelofs JM, van Berge Henegouwen GP, Smout AJ, Akkermans LM. Responses to gastric distension in functional dyspepsia. *Gut.* 1998; 42: 823-829.

Schulze-Delrieu K, Shirazi SS. Pressure and length adaptations in isolated cat stomach. *Am. J. Physiol* 1987; 252: G92-G99.

Schulze-Delrieu K. Intrinsic differences in the filling response of the guinea pig duodenum and ileum. *J. Lab. Clin. Med.* 1991: 117: 44-50.

Schulze-Delrieu K, Herman RJ, Shirazi SS, Brown BP. Contractions move contents by changing the configuration of the isolated cat stomach. *Am. J. Physiol* 1998; 274: G359-G369.

Schwizer WA, Steingotter M, Fox T, Zur T, Thumshirn M, Bosiger P, Fried M. Non-invasive measurement of gastric accommodation in humans. *Gut.* 2002; 51(Suppel): i59-i62.

Secondulfo M, Iafusco D, Carratu R, deMagistris L, Sapone A, Generoso M, Mezzogiomo A, Sasso FC, Carteni M, De Rosa R, Prisco F, Esposito V. Ultrastructural mucosal alterations and increased intestinal permeability in non-celiac, type I diabetic patients. *Dig. Liver Dis*. 2004; 36: 35-45.

Shafik A. A concept of the anatomy of the anal sphincter mechanism and the physiology of defecation. *Dis. Colon. Rectum.* 1987; 30: 970-982.

Shafik A. Effect of duodenal distension on the pyloric sphincter and antrum and the gastric corpus: duodenopyloric reflex. *World J. Surg*. 1998; 22: 1061-1064.

Shirazi S, Schulze-Delrieu K, Brown CK. Duodenal resistance to the emptying of various solutions from the isolated cat stomach. *J. Lab. Clin. Med.* 1988; 111: 654-660.

Siegle ML, Ehrlein HJ. Interdigestive contractile patterns of the ileum in dogs. *Am. J. Physiol.* 1987: 253: G452-G460.

Siegle ML, Ehrlein HJ. Neurotensin changes the motor pattern in canine ileum from propulsive to segmenting. *Dig. Dis. Sci.* 1989: 34: 1521-1527.

Siegle ML, Buhner S, Schemann M, Schmid HR, Ehrlein HJ. Propagation velocities and frequencies of contractions along canine small intestine. *Am. J. Physiol.* 1990: 258: G738-G744.

Smith JB, Zhao JB, Dou YL, Gregersen H. Time-dependent viscoelastic properties along rat small intestine. *World J. Gastroenterol.* 2005; 11: 4974-4878.

Storkholm JH, Villadsen GE, Jensen SL, Gregersen H. Passive elastic wall properties in isolated guinea pig small intestine. *Dig. Dis. Sci.* 1995: 40: 976-982.

Storkholm JH, Villadsen GE, Jensen SL, Gregersen H. Mechanical properties and collagen content differ between isolated guinea pig duodenum, jejunum, and distal ileum. *Dig. Dis. Sci.* 1998; 43: 2034-2041.

Storkholm JH, Zhao J, Villadsen GE, Hager H, Jensen SL, Gregersen H. Biomechanical remodeling of the chronically obstructed Guinea pig small intestine. *Dig. Dis. Sci.* 2007; 52: 336-346.

Storkholm JH, Zhao J, Villadsen GE, Gregersen H. Spontaneous and bolus-induced motility in the chronically obstructed guinea-pig small intestine in vitro. *Dig. Dis. Sci.* 2008; 53: 413-420.

Sun WM, Hebbard GS, Malbert CH, Jones KL, Doran S, Horowitz M, Dent J. Spatial patterns of fasting and fed antropyloric pressure waves in humans. *J. Physiol.* 1997; 503: 455-462.

Szurszewsky JH. A 100-year perspective on gastrointestinal motility. *Am. J. Physiol.* 1998; 274: G447-G453.

Takeda T, Kassab G, Liu J, Puckett JL, Mittal RR, Mittal RK. A novel ultrasound technique to study the biomechanics of the human esophagus in vivo. *Am. J. Physiol. Gastrointest Liver Physiol.* 2002; 282: G785-G793.

Takeda T, Kassab G, Liu J, Nabae T, Mittal RK. Effect of atropine on the biomechanical properties of the oesophageal wall in humans. *J. Physiol.* 2003; 547(Pt 2): 621-628.

Tormo MA, Martinez IM, Romero dT, Gil-Exojo I, Campillo JE. Morphological and enzymatic changes of the small intestine in an n0-STZ diabetes rat model. *Exp. Clin. Endocrinol Diabetes*. 2002; 110: 119-123.

Vanags I, Petersons A, Ose V, Ozolanta I, Kasyanov V, Laizans J, Vjaters E, Gardovskis J, Vanags A. Biomechanical properties of oesophagus wall under loading. *J. Biomech.* 2003; 36: 1387-1390.

Verne GN, Sninsky CA. Diabetes and the GI tract. *Gastroenterol Clin. North Am.* 1998: 27: 861-874.

Watters DA, Smith AN, Eastwood MA, Anderson KC, Elton RA. Mechanical properties of the rat colon: the effect of age, sex and different conditions of storage. *Q. J. Exp. Physiol.* 1985a; 70: 151-162.

Watters DA, Smith AN, Eastwood MA, Anderson KC, Elton RA, Mugerwa JW. Mechanical properties of the colon: comparison of the features of the African and European colon in vitro. *Gut.* 1985b; 26: 384-392.

Whitehead WE, Delvaux M, Working Team. Standardization of barostat procedures for testing smooth tone and sensory thresholds in the GI tract. *Dig. Dis. Sci.* 1997; 42: 223-241.

Williams RB, Pal A, BrassEur. J.G, Cook IJ. Space-time pressure structure of pharyngo-esophageal segment during swallowing. *Am. J. Physiol. Gastrointest Liver Physiol.* 2001; 28: G1290-G1300.

Williams RB, Wallace KL, Ali GN, Cook IJ. Biomechanics of failed deglutitive upper esophageal sphincter relaxation in neurogenic dysphagia. *Am. J. Physiol. Gastrointest Liver Physiol.* 2002; 283: G16-G26.

Yang J, Zhao JB, Zeng YJ, Gregersen H. Biomechanical properties of ileum after systemic treatment with epithelial growth factor. *World J. Gastroenterol.* 2003a; 9: 2278-2283.

Yang J, Zhao J, Zeng Y, Vinter-Jensen L, Gregersen H. Morphological properties of zero-stress state in rat large intestine during systemic EGF treatment. *Dig. Dis. Sci.* 2003b; 48: 442-448.

Yang J, Liao D, Zhao J, Gregersen H. Shear modulus of elasticity of the esophagus. *Ann. Biomed. Eng.* 2004a; 32: 1223-1230.

Yang J, Zhao J, Zeng Y, Gregersen H. Biomechanical properties of the rat oesophagus in experimental type-1 diabetes. *Neurogastroenterol Motil.* 2004b; 16: 195-203.

Yang J, Zhao J, Liao D, Gregersen H. Biomechanical properties of the layered oesophagus and its remodelling in experimental type-1 diabetes. *J. Biomech.* 2006; 39: 894-904.

Yang J, Zhao J, Nakaguchi T, Gregersen H. Biomechanical changes in oxazolone-induced colitis in BALB/C mice. *J Biomech.* 2009; 42: 811-817.

Yang W, Fung TC, Chian KS, Chong CK. Directional, regional, and layer variations of mechanical properties of esophageal tissue and its interpretation using a structure-based constitutive model. *J. Biomech. Eng.* 2006; 128: 409-418.

Yang W, Fung TC, Chian KS, Chong CK. Three-dimensional finite element model of the two-layered oesophagus, including the effects of residual strains and buckling of mucosa. *Proc. Inst. Mech. Eng* [H]. 2007a; 221: 417-426.

Yang W, Fung TC, Chian KS, Chong CK. Instability of the two-layered thick-walled esophageal model under the external pressure and circular outer boundary condition. *J. Biomech.* 2007b; 40: 481-490.

Yang W, Fung TC, Chian KS, Chong CK. Finite element simulation of food transport through the esophageal body. *World J. Gastroenterol.* 2007c; 13: 1352-1359.

Yassi R, Cheng LK, Al-Ali S, Smith NP, Pullan AJ, Windsor JA. An anatomically based mathematical model of the gastroesophageal junction. *Conf. Proc. IEEE Eng. Med. Biol. Soc.*, 2004; 1: 635-638.

Zaninotto G, DeMeester TR, Schwizer W, Johansson KE, Cheng SC. The lower esophageal sphincter in health and disease. *Am. J. Surg.* 1988; 155: 104-111.

Zhao JB, Sha H, Zhuang FY, Gregersen H. Morphological properties and residual strain along the small intestine in rats. *World J. Gastroenterol.*2002a; 8: 312-317.

Zhao J, Sha H, Zhou S, Tong X, Zhuang FY, Gregersen H. Remodelling of zero-stress state of small intestine in streptozotocin-induced diabetic rats. Effect of gliclazide. *Dig. Liver Dis.* 2002b; 34: 707-716.

Zhao J, Yang J, Vinter-Jensen L, Zhuang F, Gregersen H. The morphometry and biomechanical properties of the rat small intestine after systemic treatment with epidermal growth factor. *Biorheology.* 2002c; 39: 719-733.

Zhao J, Yang J, Gregersen H. Biomechanical and morphometric intestinal remodelling during experimental diabetes in rats. *Diabetologia.* 2003a; 46: 1688-1697.

Zhao J, Liao D, Yang J, Gregersen H. Viscoelastic behavior of small intestine in streptozotocin-induced diabetic rats. *Dig. Dis. Sci.* 2003b; 48: 2271-2277.

Zhao J, Yang J, Vinter-Jensen L, Zhuang F, Gregersen H. Biomechanical properties of esophagus during systemic treatment with epidermal growth factor in rats. *Ann. Biomed. Eng.* 2003c; 31: 700-709.

Zhao J, Liao D, Gregersen H. Tension and stress in the rat and rabbit stomach are location- and direction-dependent. *Neurogastroenterol Motil.* 2005; 17: 388-398.

Zhao J, Frokjaer JB, Drewes AM, Ejskjaer N. Upper GI sensory-motor dysfunction in diabetes mellitus. *World J. Gastroenterol.* 2006; 12: 2846-2857.

Zhao J, Jørgensen CS, Liao D, Gregersen H. Dimensions and circumferential stress-strain relation in the porcine esophagus in vitro determined by combined impedance planimetry and high-frequency ultrasound. *Dig. Dis. Sci.* 2007a; 52: 1338-1344.

Zhao J, Chen X, Yang J, Liao D, Gregersen H. Opening angle and residual strain in a three-layered model of pig oesophagus. *J. Biomech.* 2007b; 40: 3187-3192.

Zhao J, Liao D, Gregersen H. Biomechanical and histomorphometric esophageal remodeling in type 2 diabetic GK rats. *J. Diabetes Complications.* 2007c; 21: 34-40.

Zhao J, Liao D, Chen P, Kunwald P, Gregersen H. Stomach stress and strain depend on location, direction and the layered structure. *J. Biomech.* 2008a; 41: 3441-3447.

Zhao J, Nakaguchi T, Gregersen H. Biomechanical and Histomorphometric Colon Remodelling in STZ-Induced Diabetic Rats. *Dig. Dis. Sci.* 2008b Nov 7. [Epub ahead of print].

Zoubi SA, Mayhew TM, Sparrow RA. The small intestine in experimental diabetes: cellular adaptation in crypts and villi at different longitudinal sites. *Virchows Arch*. 1995: 426: 501-507.

In: Biomechanics: Principles, Trends and Applications
Editor: Jerrod H. Levy, pp. 207-231
ISBN: 978-1-60741-394-3

Chapter 6

ELECTROMYOGRAPHY IN THE 21ST CENTURY: FROM VOLUNTARY SIGNALS TO MOTOR EVOKED POTENTIALS

Petra S. Williams[1,2] and Brian C. Clark[1,3]
[1]Institute for Neuromusculoskeletal Research
[2]School of Physical Therapy
[3]Department of Biomedical Sciences at Ohio University, Ohio, USA

ABSTRACT

The force produced by skeletal muscle is controlled by the electrical signals being sent from motor neurons to muscle fibers. These electrical signals, which are known as action potentials, can be recorded as they travel along the muscle cell membrane and are referred to as an electromyogram (EMG) signal. It has been more than a century since the first recording of a voluntary EMG signal was reported, and today it has become a classic technique for evaluating and recording the activation of skeletal muscles during human movement. In recent years, the advent and development of transcranial magnetic stimulation has re-invigorated EMG research, and it is now possible to safely and painlessly evoke EMG signals directly from the motor cortex of conscious humans. This chapter reviews the recording and measurement issues associated with EMG and its respective applications. Particular attention is paid to its role in understanding the neuromechanics of human movement.

INTRODUCTION

Skeletal muscle fibers produce force and allow for motor acts *via* the excitation-contraction coupling processes (MacIntosh et al., 2006). As such, skeletal muscle fibers are both electrical and mechanical in nature. The electrical activation of muscle fibers is under direct control of the nervous system, specifically the α-motoneurons which serve as the final common pathway to muscle fibers, and integrate inputs from descending pathways, spinal cord interneuron circuits, and peripheral sensory afferents. *Electromyography (EMG)* is the recording and analysis of the electrical activity from the muscle fiber membrane that initiates the mechanical activity of the muscle (Basmajian and DeLuca, 1985). Throughout its long and storied history EMG, as a technique, has been used for a wide-variety of purposes. For example, EMG has been used to study the degree and timing patterns of muscle activation associated with voluntary movement in exercise and ergonomics research, and has been used to assess the physiological function of both the nervous and muscular systems. The former approach commonly involves recording the EMG response to evoked contractions elicited at different spatial locations along the corticospinal pathway which has proven useful for clinical diagnostic applications as well as for developing a better basic understanding of the physiological properties of the neuromuscular system.

The historical underpinnings of EMG date back more than 300-years. During the 17th century, Italian physician Francesco Redi documented the connection between muscles and the generation of electricity when he discovered that specialized muscles of *Torpediniformes* fish (e.g., electric ray fish) were capable of producing electric discharges (Basmajian and DeLuca, 1985). At the end of the 18th century, Italian physicist and physician Luigi Galvani demonstrated that electricity could directly initiate muscle contraction (Galvani, 1791), and by the end of the 19th century French physiologist Etienne-Jules Marey had recorded the electrical activity associated with voluntary muscle contraction and introduced the term 'electromyography' (Figure 1) (Marey, 1876). During the first half of the 20th century, as technology advanced and more electrodes were developed, the ability to detect and the quality of EMG signals gradually improved.

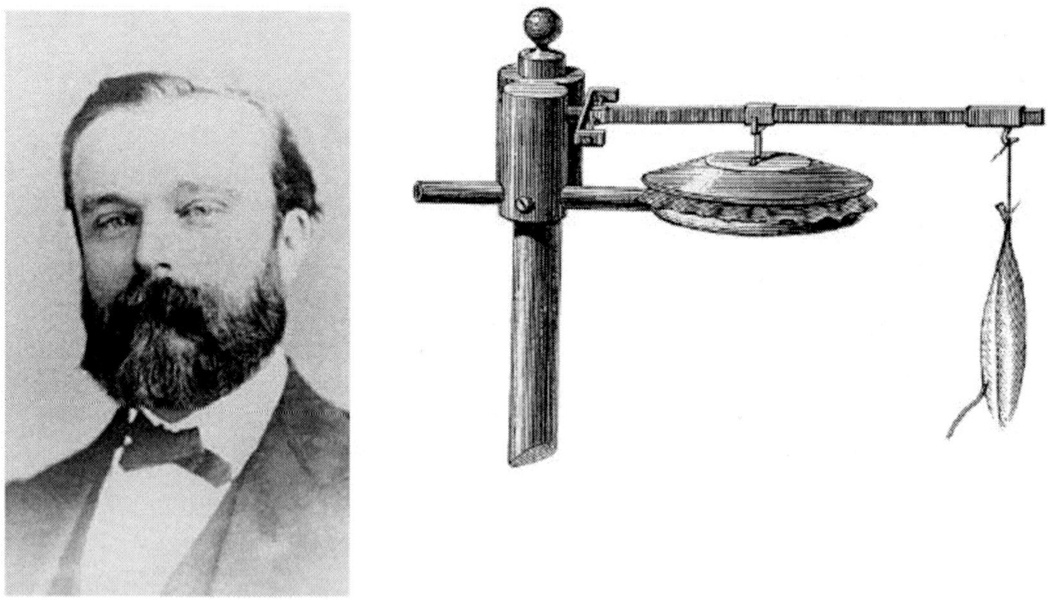

Figure 1. Etienne-Jules Marey around 1850 (left) and his myograph (right) (Marey, 1876). Marey was the first individual to record the electromyogram signal.

During the latter part of the 20th century the utilization and understanding of EMG dramatically improved in association with the development of computers and data recording equipment that allowed for improvements in signal-to-noise ratio, signal recording and processing and computer simulation studies. Today, in the 21st century, EMG continues to be used clinically and innovative approaches using EMG to studying physiology continue to be developed (e.g., using transcranial magnetic brain stimulation to evoke motor potentials at the level of the motor cortex).

Today, the myoelectric signal can be readily and easily detected from within the muscle by an indwelling electrode (intramuscular EMG), such as a needle or fine wire, or non-invasively through the skin overlying the muscle with surface electrodes (surface EMG). Depending upon the type of electrode and task being performed, electrical potentials can be recorded from single muscle fibers, that are commonly referred to as single motor unit recordings because individual muscle fibers are under the direct control of individual motor units, or from the summation of superimposed motor unit action potentials. Regardless of the recording electrodes, the detected EMG signals are generally amplified, filtered and converted from their analog form to a digital signal prior to their analysis which can be both qualitatively and quantitatively analyzed depending on the scientific or clinical reasoning behind the given recordings. However, for virtually all EMG methods, it is essential to recognize that the quantitative and qualitative characteristics of the recorded signal do not faithfully represent the original myoelectric signal (Basmajian and DeLuca, 1985; Farina et al., 2004; Keenan et al., 2005; Keenan et al., 2006a; Keenan et al., 2006b). Accordingly, this chapter will review the biophysical basis of the EMG signal, measurement issues of EMG and signal processing techniques, and discuss current applications of electromyography in the 21st century. A single chapter can not do complete justice to fully understanding the complexities of EMG; rather, in this chapter we aim to provide an overview of the basic fundamental principles of EMG and its many uses. To fully appreciate, utilize, employ and understand electromyography one must understand these basic principles and applications, because as Carlo J. DeLuca, Ph.D., a pioneer of modern EMG, eloquently stated: *'Electromyography is a seductive muse because it provides easy access to physiological processes that cause the muscle to generate force, produce movement, and accomplish the countless functions that allow us to interact with the world around us... To its detriment, electromyography is too easy to use and consequently too easy to abuse' (De Luca, 1997).*

Biophysical Basis, Measurement Issues and EMG Signal Processing

Biophysical Basis of EMG

Excitable cells, such as those of nerve and muscle, have the ability to generate a propagating wave of depolarization (e.g., an action potential) (Mathews, 2003). During muscle activity, both voluntary and evoked (a.k.a. electrically stimulated), there is an electrical potential change in the surface membrane of the skeletal muscle cell that is transmitted as an impulse across the sarcolemma to the interior of the muscle cell *via* a complex system of tubules (Kandel et al., 2000; Mathews, 2003). In muscle tissue, the action potentials are generated at the neuromuscular junctions which are located in the middle of the fibers; thus, action potentials are propagated in both directions towards the end of the fibers (Enoka, 2002; Mathews, 2003). In muscle cells, transmembrane potential remains relatively stable over time based on the relative proportions of sodium, potassium, and chloride in the extracellular and intracellular mediums (Mathews, 2003). When a neural impulse propagates down an α -motoneuron (motor unit action potential) and arrives at the motor endplate a depolarizing wave spreads across the muscle cell that, following an electro-mechanical delay of ~ 25-50 msec (Cavanagh and Komi, 1979; Norman and Komi, 1979), results in force production from muscle fibers. For both voluntary and evoked recordings, the origin of the EMG signal is the electrical activity going across the muscle fiber membrane. Specifically, the EMG signal comprises the extracellular waveform manifestation of the transmembrane voltage reversal process resultant from the potentials generated from the active motor units (Dumitru, 2000). In the case of numerous motor units being activated within the electrode's detection area, the EMG signal represents the sum of these respective motor units (Figure 2) (Basmajian and DeLuca, 1985; Farina et al., 2004; Keenan et al., 2006a).

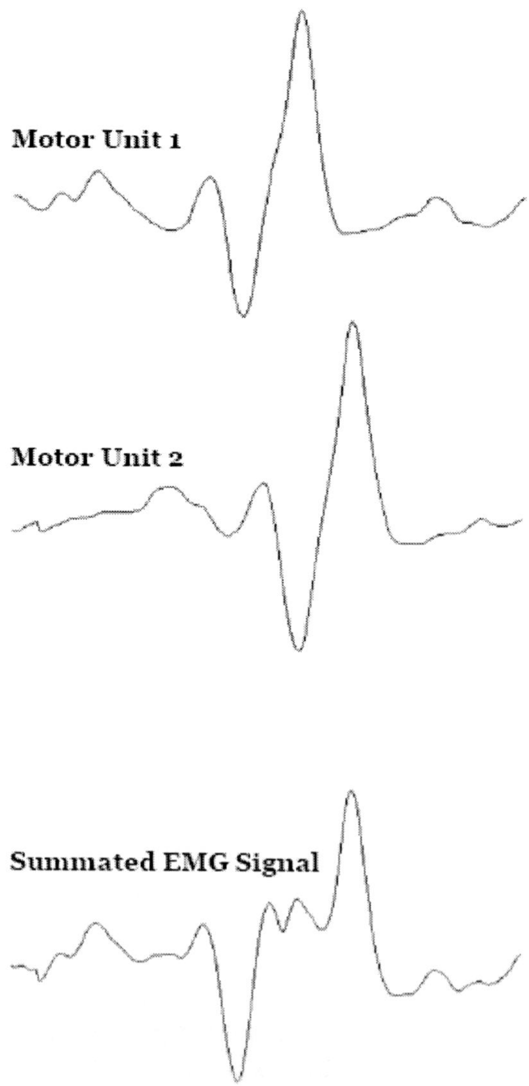

Figure 2. Two simulated motor unit action potentials discharging asynchronously, and the resultant EMG signal representing the sum of these respective motor units.

Recording and Measurement of the EMG Signal

The basic equipment required for modern EMG recordings consists of electrodes that measure the change in voltage conducted through the tissues due to the flux of sodium and potassium ions across the muscle cell membrane associated with propagating action potentials connected with amplifiers, bandwidth filters (commonly 10-500 Hz for surface recordings and 10-1,000 Hz for intramuscular recordings) and analog-to-digital converters or oscilloscopes to display and measure the respective signals. There are a wide variety of electrode choices today with each providing advantages and disadvantages. The appropriate choice of electrodes eventually depends upon what the scientist or clinician is wishing to measure and assess. For example, surface electrodes range in size from very large (30 cm diameter) to small (< 4-mm diameter), while fine wire intramuscular electrodes are extremely small (0.05-mm diameter). The smaller the diameter of the respective electrodes, the smaller the detection area.

Table 1. Factors that influence the surface EMG

Factors That Influence the Surface EMG	
Non-physiological	
Anatomic	Shape of the volume conductor Thickness of the subcutaneous tissue layers Tissue inhomogeneities Distribution of the motor unit territories in the muscle Size of the motor unit territories Distribution and number of fibers in the motor unit territories Length of the fibers Spread of the endplates and tendon junctions within the motor units Spread of the innervation zones and tendon regions among motor units Presence of more than one pennation angle
Detection system	Skin-electrode contact (impedance, noise) Spatial filter for signal detection Interelectrode distance Electrode size and shape Inclination of the detection system relative to muscle fiber orientation Location of the electrodes over the muscle
Physical	Conductivities of the tissues Amount of crosstalk from nearby muscles
Physiological	
Fiber membrane properties	Average muscle fiber conduction velocity Distribution of motor unit conduction velocities Distribution of conduction velocities of the fibers within the motor units Shape of the intracellular action potentials
Motor unit properties	Number of recruited motor units Distribution of motor unit discharge rates Statistics and coefficient of variation for discharge rate Motor unit synchronization
EMG, electromyogram.	

Surface EMG recordings represent the activity of multiple motor units, whereas the intramuscular recordings can represent the activity of multiple motor units, but can also be used to detect single motor unit activity during low-force contractions where a limited number of motor units are activated. With surface EMG recordings the amplitude of the EMG signal is often considered a global measure of motor unit activity. However, because the characteristics of the surface EMG signal depend on many other factors such as the membrane properties of the muscle fibers and the timing of the motor unit action potentials the EMG signal reflects both peripheral (muscle) and central (nervous) properties of the neuromuscular system (Table 1) (Farina et al., 2004).

Regardless of the recording method (surface versus intramuscular) the measurement principles of EMG are similar. EMG activity can be recorded using either a monopolar or a bipolar recording arrangement (Figure 3). Monopolar recordings consist of a single electrode being placed on or in the muscle of interest while a second neutral (reference/ground) electrode being placed at an electrically quiescent site (e.g., bone). Bipolar recordings consist of two electrodes being placed on or in the muscle of interest along with a neutral electrode. The bipolar configuration allows for the determination of the electrical difference between the two recording electrodes which results in detected signals that are not common between the two electrodes to be dramatically attenuated. The bipolar configuration is utilized more as these signals are generally more stable, although monopolar recordings are certainly appropriate under certain conditions when measurement of the absolute magnitude of the voltage is desired.

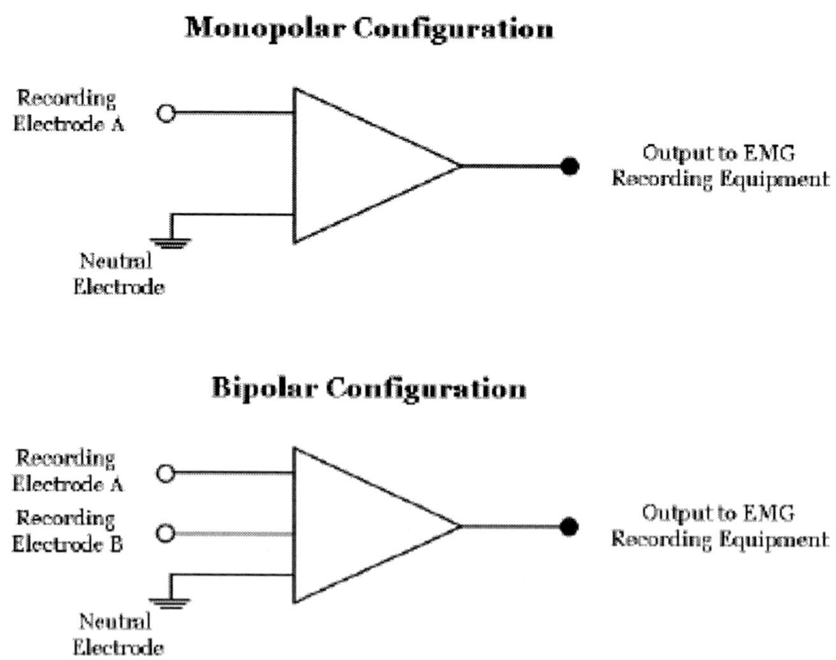

Figure 3. Schematic representation of a monopolar vs. bipolar electrode configuration. Monopolar recordings consist of a single electrode being placed on or in the muscle of interest while a second neutral (reference/ground) electrode being placed at an electrically quiescent site (e.g., bone), whereas bipolar recordings consist of two electrodes being placed on or in the muscle of interest along with a neutral electrode.

The bipolar configuration results in a dramatic reduction in the amplitude of the recorded signal, and the degree of effect varies depending on the interelectrode distance (distance between the two recording electrodes) and their relative location to the neuromuscular junction (Merletti et al., 2001). Specifically, the amplitude will be lower with a smaller interelectrode difference as the signals detected between the two electrodes will be more similar due to the spatial proximity, and if the two electrodes span the neuromuscular junction (innervation zone) the amplitude will be reduced as the electrodes will record symmetrical potentials propagating in each direction.

EMG Signal Processing

In most applications, quantification of the EMG signal is desired. Throughout history there have been a wide number of approaches used, but in general the basic goal of all is to quantify a given component associated with the amplitude or frequency characteristics of the EMG signal. First, consideration will be given to quantifying the 'interference EMG' signal, that is, the EMG signal resultant of the detection of many motor units being asynchronously and concurrently active during a task (the voluntary EMG signal) (Figure 4).

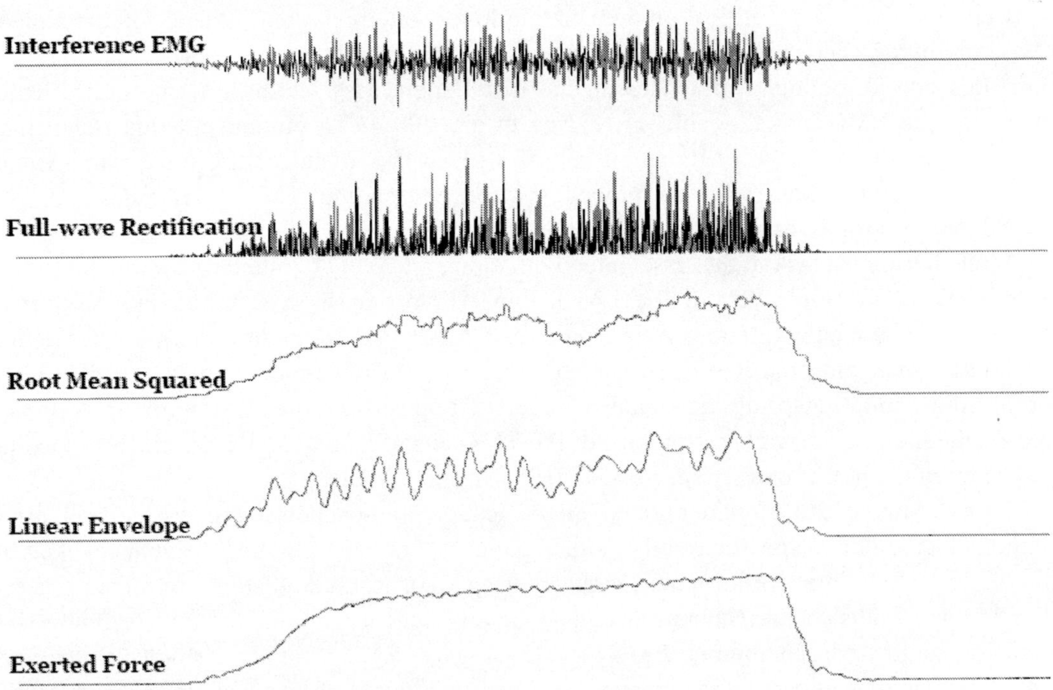

Figure 4. Example of the interference EMG signal recorded with bipolar surface electrodes from the vastus lateralis muscle during a maximal voluntary knee extension task (intereference EMG signal shown in the top trace and the exerted force shown in bottom trace). Because the interference EMG signal varies in both the positive and negative direction the mean of the signal is zero. Thus, to quantify the amplitude of the signal mathematical processing is required. Common approaches involve averaging the full-wave rectified EMG (2nd trace from top), calculating the root mean squared value (3rd trace from top; over a 500-msec epoch in this example) or applying a linear envelope (4th trace from top; using a 10-Hz low-pass Blackman-61dB filter).

Because the interference EMG signal varies in both the positive and negative direction the mean of the signal is zero. Thus, to quantify the amplitude of the signal mathematical processing is required. The following are brief descriptions of common methods for quantifying the amplitude of the interference EMG signal:

- *Average Rectified EMG*: This quantification process involves full-wave rectifying of the EMG signal (converting negative values to positive values), and then taking the mean of the selected time of interest.
- *Root Mean Squared EMG*: This quantification process involves calculating the square root of the mean squared values of the EMG signal at any given time point. It is also sometimes called the quadratic mean, and in cases where the baseline is at zero (as is desirable) this also simply represents the standard deviation. This quantification approach does not require full-wave rectification as the calculation incorporates the squared values of the original interference EMG signal.
- *Linear Envelope EMG*: The linear envelope approach involves passing a low-pass filter through the full-wave rectified signal. As such, it is a type of moving average indicator of EMG amplitude. Low-pass filter frequencies tend to range between 3-50 Hz, with a 10-Hz frequency being relatively common.

Regardless of the quantification method employed, interpretation of the absolute value of EMG amplitudes between different individuals is, in general, not appropriate as there are numerous non-physiologic influences on the EMG signal (e.g., adipose tissue dramatically attenuates the EMG signal; Table 1) (Farina et al., 2004). To circumvent this issue, it is necessary to normalize the EMG amplitude of a given task of interest to a reference value. The most common normalization procedures for interpreting EMG amplitudes between individuals or groups of individuals during exercise or ergonomic tasks is to express the amplitude during the task to that associated with either a maximal voluntary contraction of the given muscle, or better yet, to the EMG signal in response to supramaximal electrical stimulation (commonly referred to as M*max* or CMAP [for compound muscle fiber action potential]). Analyzing the interference EMG signal in the frequency domain can also yield useful information regarding the signals characteristics. Perhaps the most common approach to describe the frequency characteristics of the EMG signal is to quantify the mean or median frequency of the power density spectrum (PDS) (Hagg, 1992).

In addition to the aforementioned quantification approaches, there are several other methods that are more specific to other signal patterns. For example, EMG signals evoked by stimulating along the corticospinal pathway (e.g., cortical stimulation, peripheral nerve stimulation) results in the generation of a relatively synchronous motor unit response that yields a summated, compound EMG signal (Figure 5A). Similarly, single motor unit recordings yield one given action potential of interest (Figure 5B). As such, in these cases it is common to quantify the peak-to-peak amplitude of the EMG signal or the peak of the rectified signal. Additionally, the simple calculation of time or duration is commonly applied to these signals (e.g., to determine the time interval between a single motor unit discharge to then calculate its discharge rate or the variability in motor unit discharge rate).

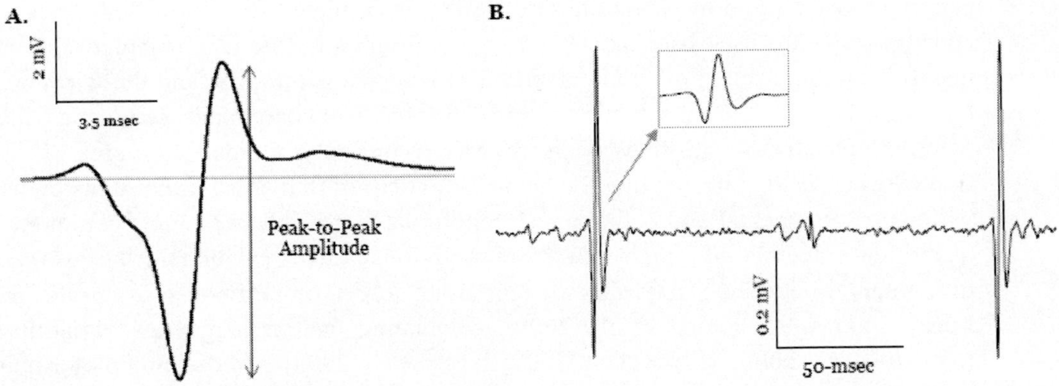

Figure 5. A: Compound surface EMG signal from the soleus muscle in response to a single, supramaximal electrical stimulus to a peripheral nerve (tibial nerve). Evoked potentials are commonly quantified by the simple calculation of their peak to peak amplitude and/or the duration of a given potential. B: Intramusular EMG recordings obtained from the medial head of the biceps brachii. Fine wires were inserted into the muscle and a single motor unit was recorded discharging at a rate of 6 Hz during a low force isometric contraction (inset shows the single unit EMG with greater temporal resolution).

THE MANY FACES OF EMG: APPLICATIONS AND INTERPRETATION

Kinesiological EMG

- *Muscle Activation Patterns:* EMG can be used to determine whether a muscle is simply active (on) or inactive (off). It has been utilized in this fashion by biomechanists for decades to characterize the timing of muscle activation patterns associated with various activities of daily living (e.g., walking), exercises (e.g., cycling), and ergonomic tasks (e.g., lifting). For applications of this nature surface EMG is most commonly used, with the EMG onset typically being defined as the point when the EMG amplitude exceeds a given threshold level that is based on a set amount above baseline levels (e.g., 2-3 x greater than baseline noise for a specified amount of time) (Thomas et al., 2007). It is also relatively common for the onset latency of certain muscles to be calculated relative to those of other muscles using similar criteria. In addition to using EMG to determine whether a muscle is active or not, it can also be used in a variety of other more sophisticated ways to evaluate muscle activation patterns during certain tasks. For example, the relative degree (magnitude) of activation of a given muscle relative to a maximum contraction or other agonistic and antagonistic muscles can be determined (Clark et al., 2005; Clark et al., 2003a; Clark et al., 2003b; Hunter et al., 2004; Hunter and Enoka, 2001), as well as whether EMG bursting is present (transient changes in the interference EMG signal during a constant force voluntary task) (Clark et al., 2005; Kouzaki et al., 2002; Kouzaki et al., 2003).
- *Single Motor Unit Behavior*: Using concentric needle and/or fine-wire EMG it is possible to observe and record activity of single motor units especially during lower force (< 20% of maximal strength) contractions (Figure 5B). Fine-wire EMG is most

commonly performed by threading two sterilized, insulated fine-wires with hooked ends that have exposed bare wire (0.05 mm) through a needle (25-27 gauge) that is inserted into a muscle. Following insertion, the needle is removed and the wires are left embedded in the muscle tissue with the small exposed ends serving as the recording electrodes until they are removed. Accordingly, the detection area of the fine wires is small, and studying the behavior pattern of individual motor units under various environmental and pathological conditions is possible. The most commonly quantified outcomes of single motor unit activity involves calculating the average time interval between a single motor unit firing to determine its discharge rate (de Luca et al., 1996; Farina et al., 2009), calculating the discharge rate variability (Kornatz et al., 2005; Moritz et al., 2005; Tracy et al., 2005), and defining motor unit recruitment thresholds (Farina et al., 2009; Riley et al., 2008). In recent years investigations on single motor unit behavior have provided valuable insight on motor unit adaptations associated with motor control and aging (Enoka et al., 2003).

- *Muscle Fiber Membrane Properties:* The compound muscle fiber action potential (CMAP) (Figure 5A), recorded with either surface or intramuscular electrodes, is of interest in studying changes in muscle membrane properties. In general, CMAP's are elicited by delivering a single, supramaximal electrical stimulation pulse (0.5-1 msec in duration) to a peripheral nerve and recording the EMG response (Clark et al., 2007; Clark et al., 2006a; Russ and Kent-Braun, 2003). The CMAP represents the summated electrical activity (arising from motor unit action potentials) resultant of the synchronous depolarization of the muscle fiber innervated by the depolarized nerve. Thus, the CMAP waveform is determined by the effectiveness of temporal and spatial summation which is affected by various factors. Changes in the CMAP's amplitude is influenced by various factors such as the muscle fiber conduction velocity, number of activated motor units, intracellular, while the CMAP duration is mostly notably affected by temporal summation r action potential characteristics (Keenan et al., 2006b).
- *Spinal Reflexes and Motoneuron Excitability:* The EMG activity associated with electrical stimulation of peripheral nerves has also been utilized to investigate *in vivo* excitability of the spinal reflexes. The most common approach to investigating spinal reflex function has involved measuring the H-reflex (Zehr, 2002). The H-reflex is evoked by electrically stimulating the peripheral nerve which elicits action potentials in the sensory Ia afferents that propagate to the spinal cord where they give rise to excitatory postsynaptic potentials and activate α-motoneuron axons (Figure 6). As such, the amplitude of the EMG reflex response provides a global measure of spinal excitability as it can be modified by a number of factors such as presynaptic inhibition, the amount of Ia neurotransmitter released, and the excitability of the α-motoneurons. The H-reflex (along with the stretch reflex which is similar to the H-reflex except for its induction thru rapid stretch of the muscle spindle fibers) have been and will continue to be powerful investigative tools that are centrally dependent upon EMG (Aagaard et al., 2002; Clark et al., 2006b; Shinohara et al., 2005). Additionally, recent methodological developments indicate that magnetic and electrical stimulation at the level of the cervicomedullary junction can evoke single descending volleys that can be detected with EMG recordings (Taylor, 2006). These cervicomedullary evoked potentials activate α-motorneuron axons primarily through

a monosynaptic connection, and have been suggested to more directly assess α-motorneuron excitability *in vivo* without the confounds inherent in the H-reflex (Martin et al., 2006a; Martin et al., 2006b; Taylor, 2006; Taylor and Gandevia, 2004; Zijdewind et al., 2006).

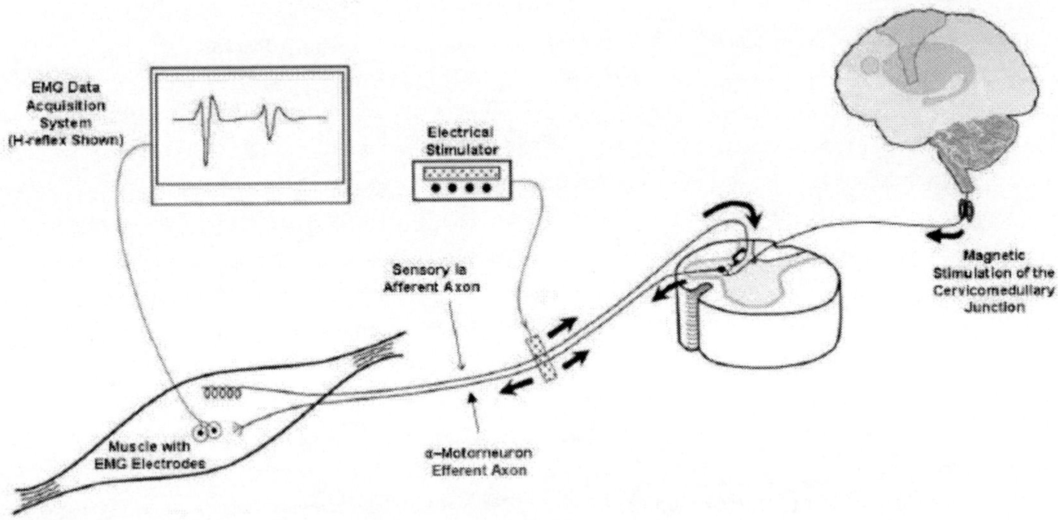

Figure 6. Neuropathway of the Hoffmann (H) reflex response and that of cervicomedullary junction stimulation. The H-reflex is evoked by electrically stimulating the peripheral nerve which elicits action potentials in the sensory Ia afferents that propagate to the spinal cord where they give rise to excitatory postsynaptic potentials and activate alpha-motoneuron axons. As such, this reflex response provides a global measure of spinal excitability as it can be modified by a number of factors such as presynaptic inhibition, the amount of Ia neurotransmitter released, and the excitability of the α-motorneurons. Conversely, magnetic stimulation at the level of the cervicomedullary junction evokes single descending volleys which activate alpha-motorneuron axons primarily through a monosynaptic connection, and can be used to more directly assess α-motorneuron excitability *in vivo*. Figure modified from Aagaard et al., J Appl Physiol, 92: 2309-2318, 2002. Used with permission.

- *Motor Cortical Function and Excitability:* Transcranial magnetic stimulation (TMS) and transcranial electrical stimulation (TES) can be used to activate the human motor cortex and assess the integrity of the central motor pathways (Kobayashi and Pascual-Leone, 2003; Reis et al., 2008; Ridding and Rothwell, 2007). While both methods are non-invasive, TMS is also relatively painless and as such it has rapidly increased in popularity and use. TMS is based on the principle of electromagnetic induction, where a pulse of current passing through a coil placed over a person's head creates a rapidly changing magnetic pulse that penetrates the skull and induces a secondary ionic current in the brain (Figure 7A). When the stimulus intensity is of sufficient strength to depolarize a sufficient number of descending neurons a motor action will occur in the stimulated muscles and an evoked EMG response can be detected (commonly referred to as motor evoked potentials [MEP's]). While TMS was introduced in humans more than 20-years ago (Barker et al., 1985), its use to study human cortical physiology has dramatically increased since the turn of the century.

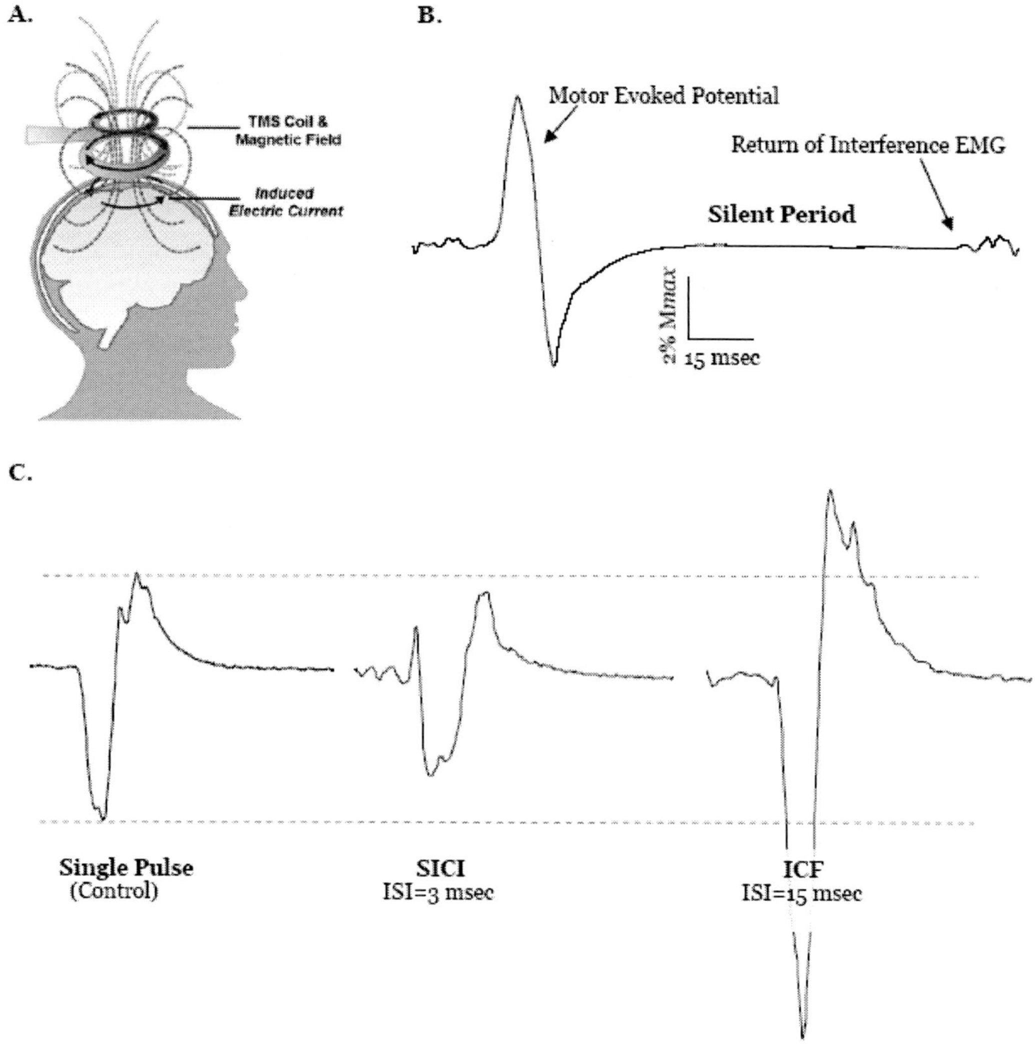

Figure 7. A: Transcranial magnetic stimulation (TMS) induces electrical currents in excitable tissue by electromagnetic induction (here being demonstrated in the brain). B: TMS to the cortex during a muscle contraction produces a motor evoked potential (MEP) followed by electrical quiescence before activity resumes that is indicative of corticospinal inhibition and commonly referred to as the silent period. C: The paired-pulse TMS technique involves coupling a conditioning stimulus with a test stimulus at different intervals. Specifically, the intensity of a conditioning pulse is set below motor threshold (the intensity where an MEP is not elicited), and the test pulse is set to a suprathreshold level. At short interstimulus intervals (e.g., 3-msec) the conditioning stimulus inhibits the MEP, whereas at longer interstimulus intervals (e.g., 15-msec) it facilitates the MEP. Paired pulses with inter-stimulus intervals (ISI's) between 1–5 milliseconds results in short-interval intracortical inhibition (SICI) and it provides a means of studying the activity of GABAA inhibitory circuits within the primary motor cortex (decreases the MEP amplitude in comparison to a control pulse), whereas ISI's between 10-25 milliseconds results in intracortical facilitation (ICF) and it allows for the study of intracortical facilitation that is controlled by GABAA and NMDA receptors.

- From an EMG methodological perspective MEP's are generally recorded in the same fashion as other EMG signals, and to the untrained eye appear similar to any evoked potential (i.e., a compound EMG potential). However, because the EMG signal is being elicited at the level of the brain the physiological underpinnings are more complicated than those of other responses. For example, the CMAP characteristics are primarily ascribed to those of the muscle cell membrane, whereas MEP characteristics are influenced by the muscle cell membrane properties along with cortical and spinal excitability. Because there are many applications and outcomes possible with brain stimulation, a complete discussion is beyond the scope of this chapter, and readers are referred to excellent reviews on this topic area for more detailed information (Kobayashi and Pascual-Leone, 2003; Reis et al., 2008; Ridding and Rothwell, 2007). However, we will briefly provide an overview description of some of the more common applications of TMS that incorporates the evaluation of and quantification of MEP's recorded using EMG.

Motor threshold refers to the lowest TMS intensity necessary to evoke MEP's in the target muscle when single-pulse stimuli are applied to the motor cortex. Motor threshold is generally defined as the lowest intensity required to elicit MEP's of a set threshold level (e.g., > 50 μV peak-to-peak amplitude in at least 50% of trials at rest). Changes in MT can reflect changes at a variety of levels (i.e., the neural membrane, axonal electronic properties, the structure and number of excitatory projections onto the primary motor cortex, or upregulation of receptors in this region) and hence represents a global assessment of the membrane excitability of pyramidal neurons (Maeda and Pascual-Leone, 2003; Ziemann, 2004). The MEP amplitude is also a commonly quantified variable. When TMS is applied to the motor cortex at an intensity above MT high-frequency indirect waves (I waves) are elicited in the corticospinal tract (Di Lazzaro et al., 2004). These waves are modifiable by many mechanisms including neurotransmitters (i.e., glutatmate, GABA), modulators of neurotransmission (i.e., acetylcholine, norepinephrine, and dopamine), and interneurones contacted by corticospinal tract cells which all function to influence the amplitude of the MEP (Figure 7B and 7C) (Maeda and Pascual-Leone, 2003; Ziemann, 2004). The silent period is a duration of electrical quiescence following an MEP when TMS is performed during a voluntary muscle contraction (Figure 7B) (Damron et al., 2008). There are several mechanisms thought to contribute to the silent period, with spinal inhibitory mechanisms thought to be active in the early part and the latter part being specifically cortical in its origin and most likely mediated by GABAergic and dopaminergic cortical inhibitory mechanisms (Brasil-Neto et al., 1992; Fuhr et al., 1991; Inghilleri et al., 1993; Ziemann et al., 1993).

In addition to the aforementioned parameters evoked with a single TMS pulse applied to the motor cortex, coupling two pulses (paired-pulse TMS) can also be used to modify evoked MEP's. Here, a conditioning stimulus is delivered in combination with a test stimulus at different intervals. Specifically, the intensity of a conditioning pulse can be set below motor threshold, and the test pulse is set to a suprathreshold level to study intracortical inhibitory and facilitatory processes. Paired pulses with inter-stimulus intervals (ISI's) between 1–5 milliseconds results in short-interval intracortical inhibition and it provides a means of studying the activity of $GABA_A$ inhibitory circuits within the primary motor cortex (decreases the MEP amplitude in comparison to a control pulse), whereas ISI's between 10-25 milliseconds results in intracortical facilitation and it allows for the study of intracortical facilitation that is controlled by $GABA_A$ and NMDA receptors (Figure 7C) (Kobayashi and Pascual-Leone, 2003; Reis et al., 2008). Further, longer interstimulus intervals (50-200 milliseconds) with suprathreshold conditioning and test stimuli also results in inhibition (long interval intracortical inhibition) (Kobayashi and Pascual-Leone, 2003; Reis et al., 2008). While the underlying mechanisms of long interval intracortical inhibition are not fully understood, it is thought that it is mediated within the primary motor cortex rather than subcortical structures (Nakamura et al., 1997). Pharmacological studies indicate that long interval intracortical inhibition is mediated by $GABA_B$ receptors (McDonnell et al., 2006; Werhahn et al., 1999), and is likely to be mechanistically linked to the silent period.

- Mapping cortical function and reorganization: Since 1991, TMS evoked motor responses have also been used to map brain functions in a direct stimulus/evoked response manner previously only possible during invasive surgery when the surface of the brain was exposed (Cohen et al., 1991; Penfield and Boldrey, 1937; Sohn and Hallett, 2004). During cortical mapping, a grid is placed on the scalp (e.g., a swim cap with a grid pattern) and the MEP amplitudes evoked at numerous sites are determined and the values are plotted to create a 3-dimensional representation between spatial location (x and y axis') and MEP amplitude (z-axis) (Figure 8) (Thickbroom and Mastagliam, 2002). These cortical maps provide three pieces of information: the total *area* on the scalp from which MEP's for the target muscle were recorded, the *"hot spot"* for a muscle (the location where the largest MEP is observed), and the amplitude weighted *center of gravity (COG)* (Wolf et al., 2005). The COG corresponds to the center of the TMS map or the scalp location/topography where the most neurons can be activated for a muscle or a movement which may or may not be equivalent to the hot spot (Butler and Wolf, 2007; Thickbroom and Mastagliam, 2002). Shifts in the location of COG (medial lateral or anterior posterior directions) are commonly suggested to demonstrate cortical reorganization or plasticity in response to injury, spontaneous recovery, or due to rehabilitation intervention (Butler and Wolf, 2007; Curra et al., 2002).

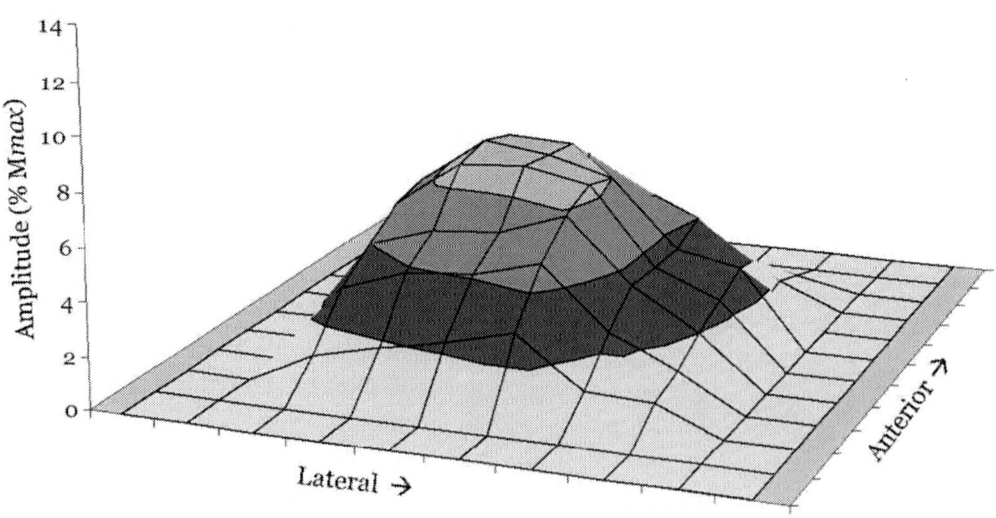

Figure 8. 3-D contour map of cortical representation of skeletal muscle plotting the motor evoked potential amplitude relative to the spatial location of the TMS coil. The *x–y* grid represents the surface of the contralateral scalp, marked into 0.5 cm squares. The z-axis represents the peak to peak EMG amplitude (expressed relative to M*max*) evoked at each point on the scalp for 10 stimuli.

These cortical maps, while insightful, need to be interpreted cautiously. Although the stimulation protocol is similar to the principles used by Penfield, it is important to recognize that the maps created using this technique do not compare in precision to maps created using intracortical microstimulation (Butler and Wolf, 2007; Thickbroom and Mastagliam, 2002). Animal studies have demonstrated that individual corticospinal neurons innervate several motor neuron pools and thus different muscles and corticospinal neurons that innervate a particular muscle are distributed among other corticospinal neurons projecting to different muscle combinations (Nudo, 2006; Rossini and Dal Forno, 2004). This mosaic somatotopy of the cortex and the overlapping spinal cord projections in combination with the lack of stimulus precision with TMS means that multiple muscles will respond to a single TMS pulse delivered at one point on the scalp matrix (Thickbroom and Mastagliam, 2002). The maps usefulness can be further confounded by electrode placement that permits cross talk, or signals evoked at the same time from other muscles, to interfere with the specificity and quality of the recorded MEP (Wolf et al., 2005).

- *Muscle Fatigue*: Skeletal muscle fatigue has fascinated physiologists for more than a century (Mosso, 1904). While muscle fatigue is associated with a variety of changes in physiological properties- many of these properties are electrophysiological in nature (or at a minimum manifest themselves with acute adjustments in electrophysiological properties) (DeLuca, 1985; Hagg, 1992; Merletti et al., 2001). For example, during a sustained, submaximal fatigue task a gradual increase in the amplitude of the interference EMG signal is observed that is primarily due to an increase in the motor units recruited as fatigue progresses (Figure 9A) (Basmaijian and DeLuca, 1985). Additionally, during a sustained contraction the depolarization and propagation of muscle fiber action potentials are modified. These modifications produce time-dependent changes in the surface EMG signal, which result in a shift of the power density spectrum to the lower frequencies (spectral compression) (Figure 9B) (Hagg, 1992). Spectral compression during a fatiguing submaximal contraction has been attributed to a number of underlying physiological factors. One of the most popular hypotheses states that the decrease in muscle fiber conduction velocity seen with fatigue influences the power density spectrum, resulting in spectral compression (Eberstein and Beattie, 1985; Lindstrom et al., 1977; Lindstrom et al., 1970; Merletti et al., 1984). This is most likely due to an accumulation of metabolites (i.e., hydrogen ion and extracellular potassium) (Bigland-Ritchie et al., 1981; Juel, 1988; Tesch et al., 1983), reducing intracellular pH (Brody et al., 1991) and, thus, decreasing sarcolemma excitability. However, this explanation appears to be incomplete, as a disassociation between median frequency and conduction velocity is observed during ischemia and different types of muscular contractions (Masuda et al., 1999; Zwarts et al., 1987). In addition to the changes in the interference EMG signal that are of interest for the study of muscle fatigue, many of the evoked EMG responses described previously are altered. For example, there is evidence that fatigue decreases the CMAP (Fuglevand et al., 1993) and prolongs the silent period (Gandevia and Taylor, 2006; Taylor et al., 2006). As such, EMG is a valuable tool to assess electrophysiological manifestations of fatigue.

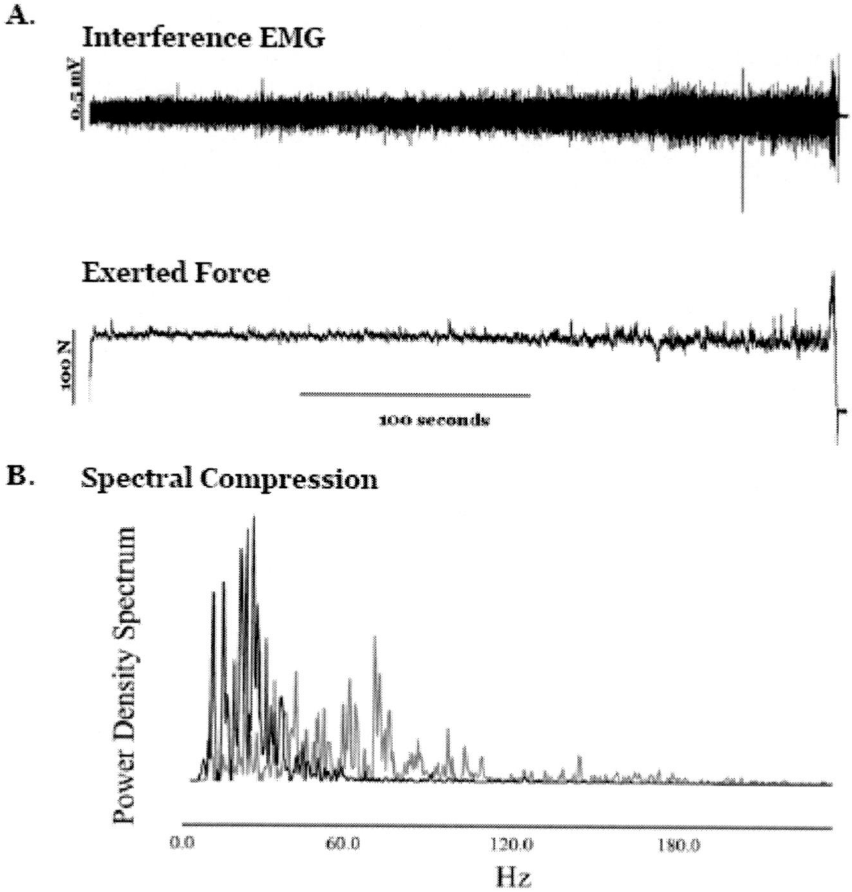

Figure 9. Changes in surface EMG characteristics during a sustained, submaximal contraction. A: Note the gradual increase in the amplitude of the interference EMG signal (top trace) as the force is continually maintained to task failure at 25% of maximum strength (bottom trace of panel A). B: The power density spectrum shifts towards the lower frequencies in association with fatigue. At the start of the contraction the median frequency was 88 Hz (gray trace) and by the end of the contraction this was reduced to 38 Hz (black trace).

Clinical Applications

EMG began to be widely used for clinical applications in the 1950's, and today it is still commonly utilized in many fields of medicine including neurology, neurosurgery, physical therapy, and physical medicine and rehabilitation. In the clinical context, patients with symptoms suggestive of neuromuscular pathology are frequently referred to a neurologist or a clinical neurophysiologist/electromyographer who conducts a variety of tests using EMG. The term "EMG" is commonly used by physicians and healthcare professions to refer to these specific tests; however, as stated earlier- the term EMG should be used to simply refer to the recording of electrical activity from the muscle fiber membrane. These tests frequently involve both surface and intramuscular recordings from resting muscles, during voluntary contraction as well as when electrically evoked and the information (EMG characteristics and responses) can provide insight into the pathologic mechanism, severity, and anatomical location of certain diseases (Daube and Rubin, 2009). Additionally, for the patient with a known neurologic diagnosis, EMG is frequently used to monitor progression, to predict long term potential for functional recovery, to evaluate response to therapy and to elucidate underlying mechanisms of recovery (Gilchrist and Sachs, 2004; Patikas et al., 2005; Thomas and Noga, 2003; van Kuijk et al., 2005). Depending upon the techniques used and the understanding of normal neuromuscular physiology, the source of change in the EMG signal can be localized to the muscle, neuromuscular junction, peripheral nerve, spinal nerve root, spinal cord, or supraspinal structure based upon signature alterations to the EMG signal (Daube and Rubin, 2009; Sella, 2003). Accordingly, it is imperative that clinicians also be familiar with the technical and non-neuromuscular physiological influences on the EMG signal in order to be certain that the differences identified in the recorded EMG signal are truly due to an underlying pathological process and not due to an artifact of the recording and analysis procedures (Daube and Rubin, 2009; Swash, 2002). Below we will briefly describe some of the most common electrodiagnostic EMG tests as well as several other EMG applications that are clinically relevant.

- *Nerve Conduction Studies*: One of the most common clinical uses of EMG is to study the conduction properties of the peripheral neuromuscular system (Katirji, 2002). These are frequently conducted to diagnose conduction disorders such as radiculopathies and neuropathies. Conduction of sensory, mixed and motor nerves are all commonly assessed, and in general utilize surface EMG recordings and peripheral nerve stimulation where the latency of evoked EMG signals (e.g., M-waves or reflex waves) are evaluated or stimulation is applied at two peripheral nerve locations and the temporal separation of the two actions potentials relative to the distance between the two stimulation sites is determined. However, it should also be noted that from a clinical perspective measurements of interest associated with nerve conduction studies typically extend beyond simple velocities and latencies and EMG signal amplitudes and durations are also of interest (Katirji, 2002).
- *Needle EMG*: While we feel the term 'needle EMG' is inappropriate and too non-descriptive to truly be associated with a particular diagnostic test- it is nonetheless commonly used to describe a relatively subjective examination where EMG is recorded and evaluated for spontaneous and insertional activity, as well as the general appearance and characteristics of single motor unit action potentials during low level voluntary contractions (Katirji, 2002). In normal, healthy muscle there is little or no EMG activity under resting conditions; however, in certain disorders (e.g., ALS, polymyositis) spontaneous activity at rest and abnormal action potentials during contraction may be observed (Daube and Rubin, 2009; Echternach, 1994; Katirji, 2002). Additionally, this clinical test can be suggestive of denervation which is commonly associated with increased EMG fibrillations, positive sharp waves, and giant motor unit action potentials (Daube and Rubin, 2009). While the needle EMG recordings can provide valuable clinical insight, it should be noted that one of the limitations of these examinations is its dependence of the skill of the examiner as many of the outcomes are qualitative in nature.
- *Motor Unit Number Estimation*: Quantifying the number of motor axons innervating a muscle (or muscle group) is of clinical importance for diagnosing and monitoring

the progression of a number of neurological diseases (e.g., amyotrophic lateral sclerosis, spinal muscle atrophy). However, counting individual motor units (as would be done in cadaveric studies) is not possible *in vivo*, but through the utilization of EMG recordings estimates can be attained. This EMG technique, referred to as motor unit number estimation (MUNE) is calculated based on a simple ratio of the CMAP divided by the average surface motor unit potential (Bromberg, 2007). The 'surface motor unit potential' is quantified by stimulating the peripheral nerve at different intensities and evaluating the response variability of different groups of axons. There are a number of technical issues and limitations associated with the methods of the MUNE technique, and for a further discussion of these the reader is referred to Bromberg (Bromberg, 2007).

- *Biofeedback*: Another clinical use of EMG is to provide biofeedback, generally based on the degree of surface signal amplitude, as an indicator muscle activity. The use of EMG biofeedback has been used for a wide variety of different applications mainly aimed at 'teaching' people how to exert voluntary control of their muscles (both to increase and decrease the muscle activation depending on the rationale for the EMG biofeedback). For example, it has been used in urinary incontinence programs to provide feedback to patients regarding the activation of their pelvic floor muscles (Dannecker et al., 2005). Additionally, it has been used in ergonomic applications to provide feedback to patients to reduce the overall EMG activity of their trapezius muscles during occupational computer work (Holtermann et al., 2008).
- *Changes in muscle activation patterns associated with movement dysfunction:* As described above in the section *Muscle Activation Patterns,* clinicians use both surface and needle electrodes to detect integrated EMG signals to detect changes in activation patterns of multiple muscles during functional activities such as gait and shoulder elevation (Jaggi et al., 2009; Winter, 1987). The relative onset/offset timing, duration, and sequences of muscle EMG activity can be compared to data generated by biomechanists to identify the muscular sources for the observed kinematic changes during motion which could be caused by musculoskeletal injury or pain (Cram and Kasman, 1998; De Luca, 1997; Thomas et al., 2007).
- *Other Clinical Applications:* There are numerous other clinically relevant applications using EMG that have yet to gain widespread use. For example, clinical applications of transcranial magnetic stimulation have begun to gain attention in recent years for a variety of possibilities, including its use to monitor injury to central motor pathways during surgery by examining changes in MEP's (Calancie and Molano, 2008), as well diagnosing diseases such as ALS, myelopathies, and multiple sclerosis through the evaluation of the evoked EMG responses (e.g., silent period, central motor conduction time) (Chen et al., 2008). TMS methods are frequently used to predict the potential for functional recovery early after hemispheric stroke, especially in the upper extremity and hand (Chen and Winstein, 2009; Curra et al., 2002). Even when the patient presents with significant clinical deficits, the presence and amplitude of an MEP in the acute phase has a high positive predictive value for restoration of function whereas absence of or prolonged MEP latency and conduction times predict poor return of hand motor ability (Butler and Wolf, 2007). Additionally, EMG has been, and will continue to be evaluated as a potential marker

for other diseases and disorders as illustrated by the recent work in evaluating EMG amplitude of the submentalis muscle in sleep behavior disorders (Feri et al., 2008). Thus, it seems that the continual growth and expansion of clinical applications of EMG will continue to be developed in the 21st century.

PERILS OF EMG AND CHAPTER SUMMARY

EMG has been used to expand our knowledge base regarding human movement and neuromuscular disorders for many years. This chapter provides an overview of the fundamentals for a basic understanding and appreciation of electromyography and its many applications. However, we must caution that there are many pitfalls and perils associated with these applications that must be understood for appropriate research design and data interpretation (Basmaijian and DeLuca, 1985; Farina, 2006; Farina et al., 2004). For example, as stated earlier, the absolute amplitude of EMG signals is of trivial consequence when comparisons are being made between individuals, muscles and even over time (Lehman and McGill, 1999). Even normalizing the EMG to a maximal voluntary contraction has limitations due to the assumption that the maximal voluntary contraction is not associated with any failure in central activation. One approach to circumventing this issue would be to normalize the EMG data to an electrically stimulated maximal contraction, but this requires that a peripheral nerve be accessible (as direct muscle stimulation will result in too much artifact to allow for recording stable CMAP's) which is not feasible in many muscles. Other perils of using EMG to evaluate muscle activation patterns are the difficulty in dealing with movement/wire artifact, issues related to cross-talk (signals detected from nearby muscles), as well as issues surrounding interpreting EMG during dynamic tasks where the detection area of underlying muscles changes in relation to surface electrodes located on the skin as the muscle shorten and lengthens during movement (Farina, 2006). Further, day-to-day variation in electrode placement, temperature and state (state-dependency) can confound longitudinal studies. However, despite its limitations, when used appropriately EMG can provide a window into the state of the human neuromuscular system.

ABOUT THE AUTHORS

Petra Williams is an Assistant Professor in the School of Physical Therapy at Ohio University, a practicing physical therapist for 15-years, and a board-certified Neurologic Clinical Specialist. Her clinical and research interest include the mechanisms, diagnosis and treatment of movement dysfunction associated with neuromuscular injury. She has received a B.A. in History and Literature of Religions from Northwestern University, and a certificate of Physical Therapy and a M.S. in Advanced Professional Neurologic Practice from The Ohio State University, She is currently pursuing a Ph.D. in Comparative and Exercise Physiology through the Department of Biological Sciences at Ohio University with a focus on neuromuscular adaptation.

Brian Clark is the Director of the Institute for Neuromusculoskeletal Research at Ohio University, where he is an Assistant Professor in the Department of Biomedical Sciences housed in the College of Osteopathic Medicine. He also serves on the Graduate Faculty in the Comparative and Exercise Physiology Program in the Department of Biological Sciences at Ohio University. He has received a B.S. in Biology from Western Carolina and an M.S. and Ph.D. in Exercise Physiology from Syracuse University. His research interests surround the neuromuscular mechanisms that mediate acute adjustments (e.g., muscle fatigue) and chronic adaptations (e.g., cast immobilization, exercise) in response to physical activity and under pathological conditions. He has over 80 publications (40+ articles and 40+ abstracts) on neuromuscular physiology and EMG. His research has been supported by grants from NASA and the American College of Sports Medicine.

REFERENCES

Aagaard, P., E. B. Simonsen, J. L. Andersen, et al. Neural adaptation to resistance training: changes in evoked V-wave and H-reflex responses *J. Appl. Physiol.* 92(6): 2309-18, 2002.

Barker, A. T., R. Jalinous and I. L. Freeston. Non-invasive magnetic stimulation of human motor cortex *Lancet* 1(8437): 1106-7, 1985.

Basmaijian, J. V. and C. L. DeLuca. Muscle Alive- their functions revealed by electromyography. Baltimore, Williams and Wilkins, 1985.

Bigland-Ritchie, B., E. F. Donovan and C. S. Roussos. Conduction velocity and EMG power spectrum changes in fatigue of sustained maximal efforts *J. Appl. Physiol.* 51(5): 1300-5., 1981.

Brasil-Neto, J. P., L. G. Cohen, M. Panizza, et al. Optimal focal transcranial magnetic activation of the human motor cortex: effects of coil orientation, shape of the induced current pulse, and stimulus intensity *J. Clin. Neurophysiol.* 9(1): 132-6, 1992.

Brody, L. R., M. T. Pollock, S. H. Roy, et al. pH-induced effects on median frequency and conduction velocity of the myoelectric signal *J. Appl. Physiol.* 71(5): 1878-85., 1991.

Bromberg, M. B. Updating motor unit number estimation (MUNE) *Clin. Neurophysiol.* 118(1): 1-8, 2007.

Butler, A. J. and S. L. Wolf. Putting the brain on the map: use of transcranial magnetic stimulation to assess and induce cortical plasticity of upper-extremity movement *Phys. Ther.* 87(6): 719-36, 2007.

Calancie, B. and M. R. Molano. Alarm criteria for motor-evoked potentials: what's wrong with the "presence-or-absence" approach? *Spine* 33(4): 406-414, 2008.

Cavanagh, P. R. and P. V. Komi. Electromechanical delay in human skeletal muscle under concentric and eccentric contractions *Eur J. Appl. Physiol. Occup Physiol* 42(3): 159-63, 1979.

Chen, R., D. Cros, A. Curra, et al. The clinical diagnostic utility of transcranial magnetic stimulation: report of an IFCN committee *Clin. Neurophysiol.* 119(3): 504-532, 2008.

Chen, S. Y. and C. J. Winstein. A systematic review of voluntary arm recovery in hemiparetic stroke: critical predictors for meaningful outcomes using the international classification of functioning, disability, and health *J. Neurol. Phys. Ther.* 33(1): 2-13, 2009.

Clark, B. C., S. R. Collier, T. M. Manini, et al. Sex differences in muscle fatigability and activation patterns of the human quadriceps femoris *Eur. J. Appl. Physiol.* 94(1-2): 196-206, 2005.

Clark, B. C., S. B. Cook and L. L. Ploutz-Snyder. Reliability of techniques to assess human neuromuscular function in vivo *J. Electromyogr. Kinesiol.* 17(1): 90-101, 2007.

Clark, B. C., B. Fernhall and L. L. Ploutz-Snyder. Adaptations in human neuromuscular function following prolonged unweighting: I. Skeletal muscle contractile properties and applied ischemia efficacy *J. Appl. Physiol.* 101(1): 256-63, 2006a.

Clark, B. C., T. M. Manini, S. J. Bolanowski, et al. Adaptations in human neuromuscular function following prolonged unweighting: II. Neurological properties and motor imagery efficacy *J. Appl. Physiol.* 101(1): 264-72, 2006b.

Clark, B. C., T. M. Manini and L. L. Ploutz-Snyder. Derecruitment of the lumbar musculature with fatiguing trunk extension exercise *Spine* 28(3): 282-7, 2003a.

Clark, B. C., T. M. Manini, D. J. The, et al. Gender differences in skeletal muscle fatigability are related to contraction type and EMG spectral compression *J. Appl. Physiol.* 94(6): 2263-72, 2003b.

Cohen, L. G., S. Bandinelli, T. W. Findley, et al. Motor reorganization after upper limb amputation in man. A study with focal magnetic stimulation *Brain* 114 (Pt 1B): 615-27, 1991.

Cram, J. R. and G. S. Kasman. Introduction to Surface Electromyography. Gaithersburg, MD, Aspen Publishers, 1998.

Curra, A., N. Modugno, M. Inghilleri, et al. Transcranial magnetic stimulation techniques in clinical investigation *Neurology* 59(12): 1851-9, 2002.

Damron, L. A., R. L. Hoffman, D. J. Dearth, et al. Quantification of the corticospinal silent period evoked via transcranial magnetic brain stimulation *J. Neurosci. Methods* 173: 121-128, 2008.

Dannecker, C., V. Wolf, R. Raab, et al. EMG-biofeedback assisted pelvic floor muscle training is an effective therapy of stress urinary or mixed incontinence: a 7 year experience with 390 patients *Arch. Gynecol. Obstet.* 273: 93-97, 2005.

Daube, J. R. and D. I. Rubin. Needle electromyography *Muscle Nerve* 39(2): 244-70, 2009.

De Luca, C. J. The use of surface electromyography in biomechanics *Journal of Applied Biomechanics* 13: 135-163, 1997.

De Luca, C. J., P. J. Foley and Z. Erim. Motor unit control properties in constant-force isometric contractions *J. Neurophysiol.* 76(3): 1503-16, 1996.

DeLuca, C. J. Myoelectric manifestations of localized muscular fatigue in humans *Crit. Rev. Biomed .Eng.* (11): 251-279, 1985.

Di Lazzaro, V., A. Oliviero, F. Pilato, et al. The physiological basis of transcranial motor cortex stimulation in conscious humans *Clin. Neurophysiol.* 115(2): 255-66, 2004.

Dumitru, D. Physiological basis of potentials recorded in electromyography *Muscle and Nerve* 23: 1667-1685, 2000.

Eberstein, A. and B. Beattie. Simultaneous measurement of muscle conduction velocity and EMG power spectrum changes during fatigue *Muscle Nerve* 8(9): 768-73., 1985.

Echternach, J. L. Introduction to Electromyography and Nerve Conduction Testing. Thorofare, NJ, SLACK Incorporated, 1994.

Enoka, R. M. Neuromechanics of human movement. Champaign, Human Kinetics, 2002.

Enoka, R. M., E. A. Christou, S. K. Hunter, et al. Mechanisms that contribute to differences in motor performance between young and old adults *J. Electromyogr. Kinesiol.* 13: 1-12, 2003.

Farina, D. Interpretation of the surface electromyogram in dynamic contractions *Exerc. Sport Sci. Rev.* 34(3): 121-7, 2006.

Farina, D., A. Holobar, M. Gazzoni, et al. Adjustments differ among low-threshold motor units during intermittent, isometric contractions *J. Neurophysiol.* 101(1): 350-9, 2009.

Farina, D., R. Merletti and R. M. Enoka. The extraction of neural strategies from the surface EMG *J. Appl. Physiol.* 96(4): 1486-95, 2004.

Feri, R., C. Franceshini, M. Zucconi, et al. Searching for a marker of REM sleep behavior disorder: submentalis muscle EMG amplitude analysis during sleep in patients with narcolepsy/cataplexy *Sleep* 31(10): 1409-1417, 2008.

Fuglevand, A. J., K. M. Zackowski, K. A. Huey, et al. Impairment of neuromuscular propagation during human fatiguing contractions at submaximal forces *J. Physiol.* 460: 549-572, 1993.

Fuhr, P., R. Agostino and M. Hallett. Spinal motor neuron excitability during the silent period after cortical stimulation *Electroencephalogr Clin. Neurophysiol.* 81(4): 257-62, 1991.

Galvani, L. De viribus electricitatis in motu musculari: Commentarius. Bologna, Tip. Istituto delle Scienze, 1791.

Gandevia, S. C. and J. L. Taylor. Supraspinal fatigue: the effects of caffeine on human muscle performance *J. Appl. Physiol.* 100(6): 1749-50, 2006.

Gilchrist, J. M. and G. M. Sachs. Electrodiagnostic studies in the management and prognosis of neuromuscular disorders *Muscle Nerve* 29(2): 165-90, 2004.

Hagg, G. M. Interpretation of EMG spectral alterations and alteration indexes at sustained contraction *J. Appl. Physiol.* 73(4): 1211-7, 1992.

Holtermann, K., K. Sogaard, H. Christensen, et al. The influence of biofeedback training on trapezius activity and rest during occupational computer work: a randomized controll trial *Eur. J. Appl. Physiol.* 104: 983-989, 2008.

Hunter, S. K., A. Critchlow, I. Shin, et al. Fatigability of the elbow flexor muscles for a sustained submaximal contraction is similiar in men and women matched for strength *J. Appl. Physiol.* 96: 195-202, 2004.

Hunter, S. K. and R. M. Enoka. Sex differences in the fatigability of arm muscles depends on absolute force during isometric contractions *J. Appl. Physiol.* 91(6): 2686-94., 2001.

Inghilleri, M., A. Berardelli, G. Cruccu, et al. Silent period evoked by transcranial stimulation of the human cortex and cervicomedullary junction *J. Physiol.* 466: 521-34, 1993.

Jaggi, A., A. A. Malone, J. Cowan, et al. Prospective blinded comparison of surface versus wire electromyographic analysis of muscle recruitment in shoulder instability *Physiother Res. Int.* 14(1): 17-29, 2009.

Juel, C. Muscle action potential propagation velocity changes during activity *Muscle Nerve* 11(7): 714-9., 1988.

Kandel, E. R., J. H. Schwartz and T. M. Jessel. Principles of Neural Science. New York, Mc-Graw Hill, 2000.

Katirji, B. The clinical electromyography examination. An overview *Neurol. Clin.* 20(2): 291-303, v, 2002.

Keenan, K. G., D. Farina, K. S. Maluf, et al. Influence of amplitude cancellation on the simulated surface electromyogram *J. Appl. Physiol.* 98(1): 120-31, 2005.

Keenan, K. G., D. Farina, R. Merletti, et al. Amplitude cancellation reduces the size of motor unit potentials averaged from the surface EMG *J. Appl. Physiol.* 100(6): 1928-37, 2006a.

Keenan, K. G., D. Farina, R. Merletti, et al. Influence of motor unit properties on the size of the simulated evoked surface EMG potential *Exp. Brain Res.* 169(1): 37-49, 2006b.

Kobayashi, M. and A. Pascual-Leone. Transcranial magnetic stimulation in neurology *Lancet Neurol* 2(3): 145-56, 2003.

Kornatz, K. W., E. A. Christou and R. M. Enoka. Practice reduces motor unit discharge variability in a hand muscle and improves manual dexterity in old adults *J. Appl. Physiol.* 98(6): 2072-80, 2005.

Kouzaki, M., M. Shinohara, K. Masani, et al. Alternate muscle activity observed between knee extensor synergists during low-level sustained contractions *J. Appl. Physiol.* 93: 675-684, 2002.

Kouzaki, M., M. Shinohara, K. Masani, et al. Local blood circulation among knee extensor synergists in relation to alternate muscle activity during low-level sustained contraction *J. Appl. Physiol.* 95: 49-56, 2003.

Lehman, G. J. and S. M. McGill. The importance of normalization in the interpretation of surface electromyography: a proof of principle *J. Manipulative Physiol. Ther.* 22(7): 444-6, 1999.

Lindstrom, L., R. Kadefors and I. Petersen. An electromyographic index for localized muscle fatigue *J. Appl. Physiol.* 43(341-353), 1977.

Lindstrom, L., R. Magnusson and I. Petersen. Muscular fatigue and action potential conduction velocity changes studied with frequency analysis of EMG signals *Electromyography* 10(4): 341-56., 1970.

MacIntosh, B. R., P. F. Gardiner and A. J. McComas. Skeletal Muscle: Form and Function. Champaign, IL, Human Kinetics, 2006.

Maeda, F. and A. Pascual-Leone. Transcranial magnetic stimulation: studying motor neurophysiology of psychiatric disorders *Psychopharmacology (Berl)* 168(4): 359-76, 2003.

Marey, E. J. Théorie du myographe qui inscrit le raccourcissement du muscle, 1876.

Martin, P. G., S. C. Gandevia and J. L. Taylor. Output of human motoneuron pools to corticospinal inputs during voluntary contractions *J. Neurophysiol.* 95(6): 3512-8, 2006a.

Martin, P. G., J. L. Smith, J. E. Butler, et al. Fatigue-sensitive afferents inhibit extensor but not flexor motoneurons in humans *J. Neurosci.* 26(18): 4796-802, 2006b.

Masuda, K., T. Masuda, T. Sadoyama, et al. Changes in surface EMG parameters during static and dynamic fatiguing contractions *J. Electromyogr. Kinesiol.* 9(1): 39-46., 1999.

Mathews, G. G. Cellular Physiology of Nerve and Muscle. Malden, MA, Blackwell Publishing Company, 2003.

McDonnell, M. N., Y. Orekhov and U. Ziemann. The role of GABA(B) receptors in intracortical inhibition in the human motor cortex *Exp. Brain Res.* 173(1): 86-93, 2006.

Merletti, R., A. Rainoldi and D. Farina. Surface electromyography for noninvasive characterization of muscle *Exerc. Sport Sci. Rev.* 29(1): 20-5, 2001.

Merletti, R., M. A. Sabbahi and C. J. De Luca. Median frequency of the myoelectric signal. Effects of muscle ischemia and cooling *Eur. J. Appl. Physiol. Occup Physiol* 52(3): 258-65, 1984.

Moritz, C. T., B. K. Barry, M. A. Pascoe, et al. Discharge rate variability influences the variation in force fluctuations across the working range of a hand muscle *J. Neurophysiol.* 93(5): 2449-59, 2005.

Mosso, A. Fatigue. London, Swan Sonnenschein, 1904.

Nakamura, H., H. Kitagawa, Y. Kawaguchi, et al. Intracortical facilitation and inhibition after transcranial magnetic stimulation in conscious humans *J. Physiol.* 498 (Pt 3): 817-23, 1997.

Norman, R. W. and P. V. Komi. Electromechanical delay in skeletal muscle under normal movement conditions *Acta. Physiol. Scand.* 106(3): 241-8, 1979.

Nudo, R. J. Plasticity *NeuroRx* 3(10): 420-427, 2006.

Patikas, D., S. Wolf and L. Doderlein. Electromyographic evaluation of the sound and involved side during gait of spastic hemiplegic children with cerebral palsy *Eur. J. Neurol.* 12(9): 691-9, 2005.

Penfield, W. and E. Boldrey. Somatic motor and sensory representation in cerebral cortex of man as studied by electrical stimulation *Brain* 60: 389-443, 1937.

Petersen, N. T., H. S. Pyndt and J. B. Nielsen. Investigating human motor control by transcranial magnetic stimulation *Exp. Brain Res.* 152(1): 1-16, 2003.

Reis, J., O. B. Swayne, Y. Vandermeeren, et al. Contribution of transcranial magnetic stimulation to the understanding of cortical mechanisms involved in motor control *J. Physiol.* 586(2): 325-51, 2008.

Ridding, M. C. and J. C. Rothwell. Is there a future for therapeutic use of transcranial magnetic stimulation? *Nat. Rev. Neurosci.* 8(7): 559-67, 2007.

Riley, Z. A., A. H. Maerz, J. C. Litsey, et al. Motor unit recruitment in human biceps brachii during sustained voluntary contractions *J. Physiol.* 586(8): 2183-93, 2008.

Rossini, P. M. and G. Dal Forno. Integrated technology for evaluation of brain function and neural plasticity *Phys. Med. Rehabil. Clin. N. Am.* 15(1): 263-306, 2004.

Russ, D. W. and J. A. Kent-Braun. Sex differences in human skeletal muscle fatigue are eliminated under ischemic conditions *J. Appl. Physiol.* 94: 2414-2422, 2003.

Sella, G. E. Neuropathology considerations: clinical and SEMG/biofeedback applications *Appl Psychophysiol Biofeedback* 28(2): 93-105, 2003.

Shinohara, M., C. T. Moritz, M. A. Pascoe, et al. Prolonged muscle vibration increases stretch reflex amplitude, motor unit discharge rate, and force fluctuations in a hand muscle *J. Appl. Physiol.* 99(5): 1835-42, 2005.

Sohn, Y. H. and M. Hallett. Motor evoked potentials *Phys. Med. Rehabil. Clin. N. Am.* 15(1): 117-31, vii, 2004.

Swash, M. What does the neurologist expect from clinical neurophysiology? *Muscle Nerve* Suppl 11: S134-8, 2002.

Taylor, J. L. Stimulation at the cervicomedullary junction in human subjects *J. Electromyogr. Kinesiol.* 16(3): 215-23, 2006.

Taylor, J. L. and S. C. Gandevia. Noninvasive stimulation of the human corticospinal tract *J. Appl. Physiol.* 96(4): 1496-503, 2004.

Taylor, J. L., G. Todd and S. C. Gandevia. Evidence for a supraspinal contribution to human muscle fatigue *Clin. Exp. Pharmacol. Physiol.* 33(4): 400-5, 2006.

Tesch, P. A., P. V. Komi, I. Jacobs, et al. Influence of lactate accumulation of EMG frequency spectrum during repeated concentric contractions *Acta. Physiol. Scand.* 119(1): 61-7., 1983.

Thickbroom, G. W. and F. L. Mastagliam (2002). Mapping Studies. Handbook of Transcranial Magnetic Stimulation. A. Pascual-Leone. London, Arnold Publishers.

Thomas, C. K. and B. R. Noga. Physiological methods to measure motor function in humans and animals with spinal cord injury *J. Rehabil. Res. Dev.* 40(4 Suppl 1): 25-33, 2003.

Thomas, J. S., C. R. France, D. Sha, et al. The effect of chronic low back pain on trunk muscle activations in target reaching movements with various loads *Spine* 32(26): E801-8, 2007.

Tracy, B. L., K. S. Maluf, J. L. Stephenson, et al. Variability of motor unit discharge and force fluctuations across a range of muscle forces in older adults *Muscle Nerve* 32(4): 533-40, 2005.

van Kuijk, A. A., J. W. Pasman, A. C. Geurts, et al. How salient is the silent period? The role of the silent period in the prognosis of upper extremity motor recovery after severe stroke *J. Clin. Neurophysiol.* 22(1): 10-24, 2005.

Werhahn, K. J., E. Kunesch, S. Noachtar, et al. Differential effects on motorcortical inhibition induced by blockade of GABA uptake in humans *J. Physiol.* 517 (Pt 2): 591-7, 1999.

Winter, D. A. The biomechanics of motor control and human gait. Ontario, Canada, University of Waterloo Press, 1987.

Wolf, S. L., A. J. Butler, J. L. Alberts, et al. Contemporary linkages between EMG, kinetics and stroke rehabilitation *J. Electromyogr. Kinesiol.* 15(3): 229-39, 2005.

Zehr, P. E. Considerations for use of the Hoffmann reflex in exercise studies *Eur. J. Appl. Physiol.* 86(6): 455-68, 2002.

Ziemann, U. TMS and drugs *Clin. Neurophysiol.* 115(8): 1717-29, 2004.

Ziemann, U., J. Netz, A. Szelenyi, et al. Spinal and supraspinal mechanisms contribute to the silent period in the contracting soleus muscle after transcranial magnetic stimulation of human motor cortex *Neurosci. Lett.* 156(1-2): 167-71, 1993.

Zijdewind, I., J. E. Butler, S. C. Gandevia, et al. The origin of activity in the biceps brachii muscle during voluntary contractions of the contralateral elbow flexor muscles *Exp. Brain Res.* 175(3): 526-35, 2006.

Zwarts, M. J., T. W. Van Weerden and H. T. Haenen. Relationship between average muscle fibre conduction velocity and EMG power spectra during isometric contraction, recovery and applied ischemia *Eur. J. Appl. Physiol. Occup. Physiol.* 56(2): 212-6, 1987.

In: Biomechanics: Principles, Trends and Applications
Editor: Jerrod H. Levy, pp. 233-249
ISBN: 978-1-60741-394-3

Chapter 7

BIOMECHANICS IN CHILDREN WITH CEREBRAL PALSY

Jessie Chen[1] and Dinah Reilly[2]
[1]Department of Physiology and Neuroscience,
New York University School of Medicine, New York, NY, USA
[2]Department of Occupational and Physical Therapy,
Boise School District, Boise, ID, USA

ABSTRACT

Children with cerebral palsy (CP) lack the higher-level motor skills present in age-matched typically developing (TD) children. The development of postural control is critical to the acquisition of increasingly complex motor skills as well as to the production of coordinated motor behavior, such as locomotion. This chapter examines recent developments in understanding the abnormal postural control in children with CP and assessments of the effectiveness of rehabilitation techniques using biomechanics measurements. We show that the delayed and impaired development of postural control in children with CP is not only due to the immaturity of central nervous system but also abnormal postural alignment and muscle force production.

Children with CP often have difficulty maintaining stability when facing unexpected threat to balance. We present studies comparing reactive balance control in children with spastic diplegic cerebral palsy (SDCP) and TD children using support surface perturbation and show that a number of factors contribute to poor balance control in children with SDCP. 1) There was a temporal disorganization of joint torque activation. 2) There was a spatial disorganization of the joint torque profiles. 3) Children with SDCP also showed slower speed to reach peak torque value. In addition we show that when TD children were asked to mimic crouched stance as that seen in children with SDCP, they exhibited abnormal postural control as well, indicating that musculoskeletal constraints are also contributors to the atypical postural muscle response patterns seen in children with SDCP. These findings suggest that the neuromuscular response patterns of some children with SDCP may be appropriate strategies for their musculoskeletal constraints secondary to deficits in the neural system.

In this chapter we also discuss the culmination of our findings in relation to clinical applications in the management of musculoskeletal impairments to improve postural control in children with SDCP, and the significance of using biomechanical measures to show a direct relationship between the impairments of the musculoskeletal system and reactive postural control as well as possible coping strategies used by children with SDCP. We examine the current biomechanical research used to ascertain the effectiveness of two therapeutic interventions purported to affect the musculoskeletal system for the improvement of function in children with SDCP, ankle foot orthoses (AFO) and strength training.

Finally, we examine the gaps in current clinical research when assessing the effectiveness of interventions to reduce the musculoskeletal impairments constraining static, reactive, and dynamic balance control in children with SDCP.

INTRODUCTION

Children with motor system impairments, such as those seen in cerebral palsy (CP), lack the higher-level motor skills present in age-matched healthy children. It is generally understood that cerebral palsy is a category of chronic non-progressive disorders of movement and posture of early onset (Styer-Acevedo, 1994). Research in the last 20 years has shown that delayed and impaired development of postural control in children with CP is attributed to a number of factors, including an immature central nervous system, the pathophysiology of the neuromuscular system, an impaired postural alignment, and muscle weakness. Postural control is a fundamental motor function involved in nearly every motor task. The development of postural stability is critical for the acquisition of increasingly complex motor skills as well as for the production of coordinated motor behavior seen in the developing child, i.e., walking, stair climbing, jumping, and skipping (Roncesvalles et al, 2001). There has been a rich body of literature studying the alterations of the neural system for postural control in children with CP; however, few studies have examined the biomechanics of the musculoskeletal system on the control of posture.

Most of our knowledge on the development of postural control comes from research using the support surface perturbation paradigm, in which an unexpected perturbation is given and the resulting behaviors are documented as well as the associated muscle activity profiles captured by electromyography (EMG) (Nashner, 1976; Nashner, 1977; Horak and Nashner, 1986; Nashner and McCollum, 1985). Three basic coping strategies have been identified based on the behavioral responses to varying magnitudes of perturbation: ankle, hip and stepping strategies. It was suggested that the forms of automatic postural response elicited following a translation of the support surface became more consistent and efficient with an increase in chronological age (Sveistrup and Woollacott, 1996). In an adult, muscle responses are activated in a distal-to-proximal sequence; i.e., ankle muscles activated prior to knee and hip muscles.

Children with cerebral palsy exhibit exaggerated dynamic stretch reflex responses, a lack of reflex suppression on repetitive stimulation, and abnormal radiation of activity to functionally related muscles (Nashner et al, 1983; Burtner et al, 1998; Woollacott et al, 1998). EMG recordings of muscle activity associated with postural responses to balance threat showed an increase in antagonist recruitment and decreased trunk activation when compared to typically developing children at the same level of walking experience. In addition to high level of agonist and antagonist muscle co-activation, children with spastic cerebral palsy exhibited temporal reversals among the muscles responding to a loss of balance.

Not only do children with CP show abnormal neural responses indicated by altered EMG patterns, they also exhibit abnormality in biomechanical measures. Children with spastic cerebral palsy frequently show restricted range of motion at the ankle, knee, and hip secondary to abnormal muscle tone and disordered movement patterns. The habitual crouched postural pattern during stance results in subsequent shortening of the hamstring muscles, which in turn ensures the continued use of the habitual crouched posture (Shumway-Cook and Woollacott, 2001).

From a biomechanical point of view, in the stance position, the human body can be treated as an inverted pendulum. In its simplest version, the body is considered a single rigid segment with all the external forces acting on the center of mass located near the waist level. The task of postural control is to keep the body's center of mass (COM) within the base of support (BOS), defined by the individual's foot length and stance width (Winter et al, 1990). Throughout development, children face a series of morphological changes to their bodies, including an increase in height and weight; but more importantly, a lowering of the center of mass due to changes in relative mass of the different body parts. Therefore, the seemingly simple task of postural and balance control becomes more complex for growing children as the difficulty of adapting to these changes increases. Researchers using traditional theories of developing postural control, such as reflex-hierarchical theory, find it difficult to uncover a universal solution that governs successful postural control when dealing with changes in so many dimensions.

In the past several decades motor control scientists have been faced with the immense task of discovering how the brain calculates and coordinates a large number of kinetic parameters (forces produced by individual muscle groups) in order to produce appropriate kinematic outcomes (position, velocity and acceleration of individual joints, body segments, and whole body movements) for postural control when facing an external balance threat. Winter, a biomechanist and a researcher in motor control, recently put forth a revolutionary concept using biomechanical models (Winter, 1995) in an effort to simplify this complex issue by reducing the large number of independent parameters into an interaction between two parameters. He proposed that the difference between the center of pressure (COP) (the net pressure force on the body acting through a single point in the BOS) and the center of gravity (COG) (the vertical projection of the center of mass on the ground) is proportional to the acceleration of body sway, and thus is recognized by the balance control system as the "error" signal. Therefore, the greater the COG-COP separation, the faster the acceleration of the COG, and the greater the demand on the postural control system. In response to the balance threat, synergistic muscle responses are triggered automatically in peripheral effectors to keep the COG within BOS. When a corrective movement of the center of pressure to the extremes of the base of support (the edges of the feet) is not adequate to reverse the acceleration of the center of gravity, a step is necessary, or falling occurs.

For children to maintain an upright, quiet stance, the nervous system must detect the deviation of COG away from the vicinity of COP, and then generate sufficient torques around the ankle, knee, and hip joints to counterbalance the tipping force. In dynamic balance, i.e., when walking or climbing stairs, this signal is not only used to keep balance, but used to provide forward momentum as well. The difference between the COG and COP, although a concept borrowed from the field of biomechanics, is considered a sensitive indicator by motor control researchers in speculating the capability of the nervous system to control balance.

It is our understanding that in order to examine the development of balance control, multiple levels of analyses are required. Kinematic data record the motor outcomes. EMG recordings provide insight into the response from the central nervous system, because there is a one-to-one correlation between output from spinal motor neurons and muscle contractions. These two analyses alone, however, are not sufficient to represent the internal and external forces acting on the joints or the whole body to explain the behavioral outcome. As the singular measure of motor responses, the surface EMG is plagued by at least three significant limitations (Bothner, 1998). 1) It can only record from the more superficial muscle groups involved in the task, therefore, an assumption that these are the only muscles involved in recovery of balance is an oversimplification of the total muscle function. 2) No direct correlation has been found between EMG activity and force produced during postural responses, particularly in a reactive balance control task such as postural recovery from a perturbation involving a number of joints. Linear relations have only been observed in single joint movements under unique circumstances such as isometric or isokinetic contractions. 3) The direct relationship between EMG signals and behavior under dynamic circumstances (such as walking and running) remains inconclusive.

Kinetic analyses, on the other hand, have been proposed as a necessary and comprehensive approach to uncover control variables and strategies (Winter, 1995). Kinetic analysis estimates the overall muscle forces around a joint responsible for an observed movement, yielding variables that describe the causes of a motion. Analysis of joint torque, in particular, provides a description and a kinetic account not available through kinematics and EMG analysis alone. In this chapter, we will show results from studies of postural control in children with CP using joint torque analyses and the resultant control of posture by looking at the relationship between COG (kinematic parameter) and COP (kinetic parameter).

Purpose of This Chapter

Research on postural control in children with cerebral palsy has examined four types of postural tasks: 1) steady state postural control, in which the child simply maintains quiet stance balance for a specific period of time; 2) reactive postural control, in which the child reacts to an unexpected threat to balance, like a slip or trip; 3) proactive postural control, in which the child integrates postural responses into voluntary tasks, and 4) dynamic postural control, in which the relationship between the COG and COP continuously changes (locomotion, climbing stairs, stepping over obstacles). Based upon our investigation, it is clear that previous research lacks information on the kinetic changes and musculoskeletal constraints associated with the development of reactive balance control in children with cerebral palsy.

Our group has attempted to fill the gaps by examining the differences in kinetic profiles of the lower extremity joints during reactive postural control in children with different types of CP (diplegic, hemiplegic and ataxic) and typically developing children (Chen and Woollacott, 2007).

Our primary purpose of this section is to examine how torque productions around lower extremity joints of children with spastic diplegic cerebral palsy (SDCP) differ from those of typically developing children during reactive balance control, and the resultant control of balance indicated by the COG-COP relationship. We will describe the differences in balance control in response to a single support surface perturbation in children with spastic diplegic cerebral palsy compared to typically developing children of chronologically younger ages (TDY) and age-matched older children (TDO). The study will investigate the following issues in balance control: 1) the ability to promptly produce joint torque in response to balance threat (torque onset latency); 2) the ability to produce large enough magnitude of torque to counterbalance support-surface perturbation and the relative contributions from the ankle, knee and hip joints in the recovery of balance (maximum torque); and 3) the speed to reach peak torque value in response to balance threats (rate of torque production). See Figure 1 for definition of each variable. In conjunction with our study, we will also discuss other studies that have contributed to the knowledge of impairments in the musculoskeletal system and reactive postural control in children with SDCP.

Secondly, we will discuss the culmination of these findings 1) in relation to clinical applications in the management of musculoskeletal impairments to improve postural control in children with SDCP, and 2) the significance of using biomechanical measures to show a direct relationship between the impairments of the musculoskeletal system and reactive postural control as well as possible coping strategies used by children with SDCP to recover balance. We will examine the current biomechanical research used to ascertain the effectiveness of two therapeutic interventions purported to affect the musculoskeletal system for the improvement of function in children with SDCP: ankle foot orthoses (AFO) and strength training.

Finally, we will examine the gaps in current clinical research when assessing the effectiveness of these interventions to reduce the musculoskeletal impairments constraining static, dynamic, and reactive balance control in children with SDCP.

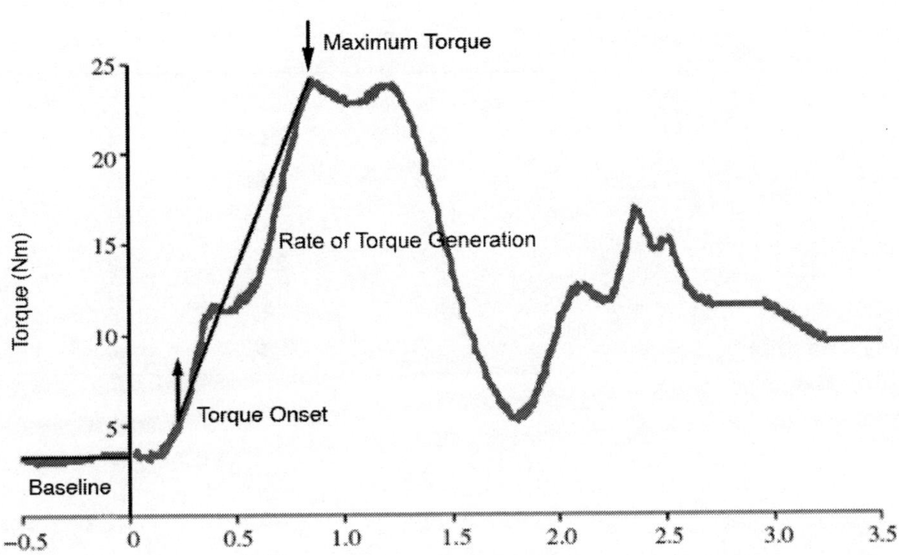

Figure 1. Schematic for deriving torque baseline, onset latency, maximum torque and rate of torque generation. Platform movement onset is plotted at 0 in the x-axis.

Reactive Balance Control in Children with SDCP

We recruited four children with mild spastic diplegic cerebral palsy (SDCP) aged from 8 to 13 years old for this study. All subjects acquired the disorder prenatally, perinatally or shortly after birth. Eight typically developing children with no known musculoskeletal or neurological deficits also participated in the study as control subjects. We formed two subgroups of the healthy children according to their chronological age: younger typically developing children (TDY), 4-7 years old, and older typically developing children (TDO), 8-13 years old, that was age-matched to the children with SDCP. The demographic information of all participants is listed in Table 1.

The participants stood quietly on a movable dual-forceplate system with arms crossed in front of the chest. The width of foot placement was self-selected and was approximately laterally symmetrical with one foot on each forceplate. Perturbations to balance were caused by simultaneous backward movements of the two forceplates at constant amplitude of 10 cm and at varying velocities ranging from 5–85 cm/s.

Table 1. Demographic Information of participants

Subject ID	Gender	GMFCS	Age (yr)	Weight (kg)	Height (m)
SDCP					
1	M	III	11.6	22.4	1.24
2	M	III	8.5	21.8	1.22
3	M	II	12.9	45.5	1.60
4	F	II	13.0	27.3	1.36
M			*11.50*	*32.24*	*1.36*
SD			*2.10*	*13.12*	*0.12*
TDY					
1	M		7.1	23.6	1.23
2	F		5.2	19.5	1.11
3	F		4.9	14.1	1.04
4	F		6.3	20.2	1.14
M			*5.88*	*19.35*	*1.23*
SD			*1.01*	*3.93*	*0.08*
TDO					
1	F		8.3	23.5	1.21
2	F		12.9	34.6	1.41
3	M		11.9	26.1	1.36
4	M		12.5	45.5	1.52
M			*11.40*	*32.43*	*1.38*
SD			*2.11*	*9.92*	*0.13*

We instructed participants not to step unless it was absolutely necessary. Trials were delivered in three blocks. The perturbation velocity increased by block, but was randomized within each block (e.g., the platform velocity ranged from 5–25 cm/s in the first block, from 30–55 cm/s in the second block, and from 60–85 cm/s in the last block). We gave catch trials with no perturbation and with a forward perturbation of 10 cm @ 5 cm/s between blocks to eliminate anticipatory responses. Catch trials were not included in further analysis.

We obtained kinetics data from the custom-built platform system (Neuroscience Technical Service Group, University of Oregon, Eugene) consisting of two hydraulically driven movable forceplates with three axes. Three-dimensional ground reaction forces were measured by each of the plates independently. For every loaded condition (with participant on plate), we collected a corresponding unloaded trial (same perturbation characteristics without participant on plate). We derived net ground reaction forces (ground reaction forces produced by the participant in response to the perturbation, not including the forces driving the plates) by subtracting the respective signal in the unloaded trial from that in the loaded trial. Force data were recorded at 1200 Hz for duration of 5 seconds. Kinematics data were collected using the PEAK Performance Technologies real-time data-acquisition system (Peak Performance, Inc., Centennial, CO). Six infrared cameras were positioned around the participant. We placed reflective markers bilaterally on the following landmarks: (a) the base of the second metatarsophalangeal joints; (b) lateral malleoli; (c) heels; (d) lateral epicondyles of the femur; (e) greater trochanters of the femural head; (f), (g), and (h) anterior superior iliac spines (ASISs) of the pelvis, sacrum, and shoulders, respectively. Reflective wands were attached on the skin overlying midthighs and midshanks. We placed one reflective marker on the upper-right corner of the forceplate to monitor the movement of the plates. Before each experimental session, we took the following anthropometric measures of the participant for future inverse dynamics analysis: (a) body weight, (b) body height, (c) thigh length, (d) midthigh circumference, (e) calf length, (f) calf circumference, (g) knee diameter, (h) foot length, (i) malleolus height, (j) malleolus width, (k) foot breath, and (l) ASIS breadth. Joint torques were calculated using an inverse dynamics model within the PEAK Performance Technologies real-time data-acquisition system. We expressed torque as a reaction torque to all external torques. Counterclockwise torques (view from the right side of the subject) were expressed in positive values and negative values for clockwise torques.

Previous research in postural control suggests that instead of comparing participants with differing capabilities under identical perturbation velocities, it is more important to compare them across conditions that give equivalence of behavioral threat for each individual (Roncesvalles et al, 2001; Chen and Woollacott, 2007). Thus, for this study, we identified a set of behaviorally comparable trials to ensure relative equivalence of balance threat across participants of different sizes, weights and conditions (TDY, TDO and SDCP). As mentioned earlier, the COG-COP difference is recognized by the balance control system as the error signal. It is logical to assume that the threat to the balance control system is equivalent when the sizes of the error signals are similar despite of the different magnitudes of the balance threats. The comparison condition chosen for this study is threshold velocity. It is defined as the fastest velocity at which the subject could utilize his/her own muscle contraction and maintain balance without taking a step.

Due to the uneven force production under the left and right feet for both children with SDCP and TD children, two force plates were used to record the forces from each of the legs independently. To account for the asymmetrical weight distribution between the two feet, the COP used in calculating COG-COP difference was the net COP calculated as follows:

$$COP_{net} = COP_l \frac{GRF_{vl}}{GRF_{vl} + GRF_{vr}} + COP_r \frac{GRF_{vr}}{GRF_{vl} + GRF_{vr}} \quad (1)$$

where

COP_l — COP under the left foot

COP_r — COP under the right foot

GRF_{vl} — vertical ground reaction force under the left foot

GRF_{vr} — vertical ground reaction force under the right foot.

As shown in Figure 2B, the threshold velocity was significantly different among the three groups, $F(2, 135) = 353.5$, $p < .001$. The TDO children tolerated the highest velocities, averaging 71.1 ± 0.4 cm/s without taking a step; the TDY children tolerated lower velocities, 51.4 ± 3.0 cm/s; and children with SDCP showed the lowest perturbation tolerance, with average threshold velocities of 7.9 ± 1.1 cm/s. This indicates that the 8-13 year old children with SDCP showed even lower developmental level for recovery of postural control to that of young TD children ranging 4-8 years of age.

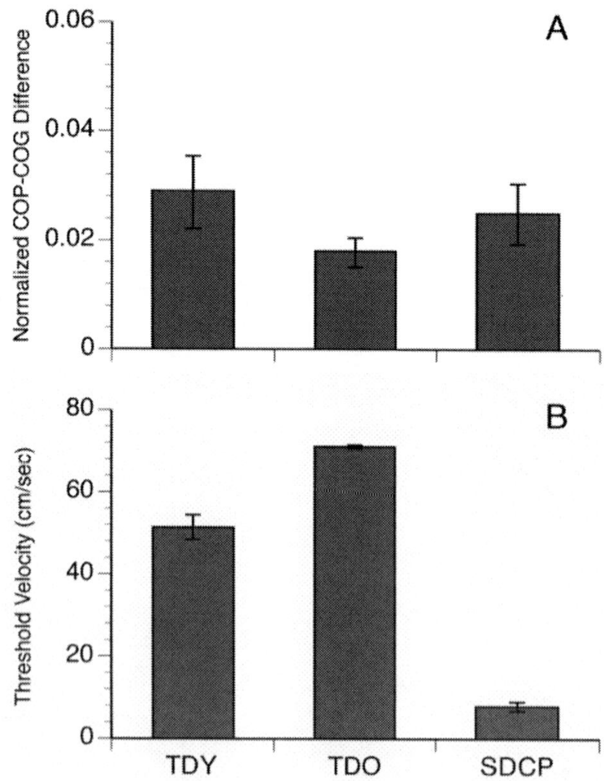

Figure 2. A. Normalized center of gravity (COG) and center of pressure (COP) differences at the threshold velocity. B. Threshold velocity of children in the TDY, TDO and SDCP groups. Error bars represent standard error of the mean.

As predicted by Winter's theory, one-way ANOVA analysis showed that the normalized (by foot length) COG-COP difference in the anterior-posterior direction at the threshold velocity was not significantly different among the three groups, indicating that the relative difficulty of the balance task was equivalent at the threshold velocity, even though the perturbation velocities were different for each subject (Figure 2A). Therefore, the first part of the study focused on the analyses of those trials in which subjects experienced the perturbation at their maximum stability limits.

Torque Onset Latency

One important factor in successfully maintaining stability is activating muscles fast enough to produce torque in response to external balance threats. To do this, we examined the joint torque onset latency, which is defined as the time duration between the onset of platform movement to the time when joint torque exceeded 5 Nm relative to the baseline activity (calculated from first 50 ms of the trial) and maintain the increase for at least 30 ms. When we compared the torque onset latencies within each group, the TD children were found to temporally organize their torque activation pattern; the ankle was activated first, followed by knee, and the hip was activated last (Figure 3A). A one-way ANOVA showed a significant difference of onset latency among three joints in both the TDY group, $F(2, 33) = 14.5, p < .001$, and the TDO group, $F(2, 48) = 113.3, p < .001$. On the contrary, children with SDCP did not show temporally organized activation of joint torques ($F(2, 28) = 0.1, p = .868$). When we compared the torque onset latencies between groups, we found a significant difference at the ankle joint, $F(2, 42) = 45.8, p < .001$, and the knee joint, $F(2, 37) = 10.7, p < .001$, but not the hip joint, $F(2, 30) = 0.4, p < .413$. Post hoc tests further revealed that at the ankle and knee joints children with SDCP responded significantly slower than those of the TD children; however, there was no statistical difference in the ankle torque onset latency between children in the two TD groups of different ages.

Maximum Torque

The comparison of the maximum torque output (normalized by body weight) produced at each joint to counterbalance the external perturbation at the threshold velocity (Figure 3B) showed that children in both TD groups exhibited spatially organized torque contribution among joints, with the ankle contributing the most and the hip contributing the least. In children with SDCP, unlike the pattern shown in TD children, the hip joint contributed significantly more. In fact the maximum hip torque reached a similar level to that of the ankle. The knee, on the other hand, had a significantly reduced contribution. Post hoc tests confirmed that the ankle, knee and hip joints were significantly different in the TDO children, $F(2, 48) = 87.6, p < .001$; for the TDY children, the pattern is still consistent with an ankle strategy, where the ankle joint produced a significantly larger torque, there's no difference between the knee and hip joint, $F(2, 33) = 11.9, p < .001$. In contrast, we found that the ankle and hip joints contributed equally during balance recovery in children with SDCP; the knee joint produced significantly lower torque, which potentially compromised the stabilizing function of the legs.

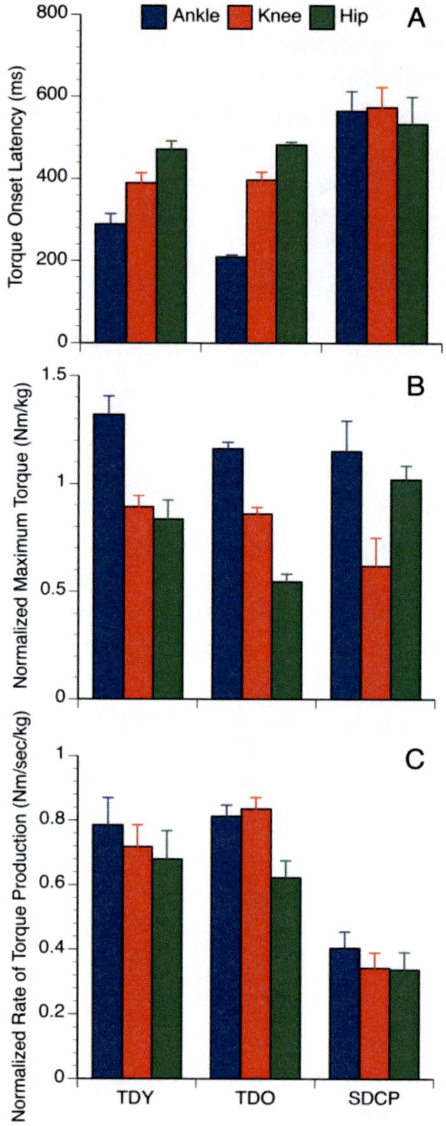

Figure 3. Joint torque onset latencies (A), normalized maximum torques (B), and normalized rate of torque productions (C) at the ankle, knee and hip joints at the threshold velocity.

When comparing maximum torque at each joint between groups of all children, we found no difference between groups at the ankle and knee joints, but the hip torque was significantly lower in the TDO children and there was no difference between the TDY children and the children with SDCP. These findings validate previous results using EMG recording and kinematics measurements indicating that similar to children of chronological younger ages, children with SDCP use more hip strategy during balance recovery (Sveistrup and Woollacott, 1996; Roncesvalles et al, 2001; Shumway-Cook and Woollacott, 2001; Sundermier et al, 2001).

Rate of Peak Torque Production

It has been shown that the rate of torque generation is also critical in balance control (Robinovitch et al, 2002). A larger maximum torque itself isn't sufficient to indicate good balance, if the speed of reaching the peak value is not fast enough to arrest the acceleration of the COM. In this study we calculated the rate of joint torque generation by dividing the maximum joint torque by the duration between the onset of the torque and the time when the torque reached the maximum value, and then normalized using the individual's body weight. The results showed that the ankle, knee and hip joints were turned on equally fast within each group; however, when comparing between groups, it took significantly longer time for children with SDCP to reach peak torque production at all joints compared to the TD children (Figure 3C).

In conclusion, it is clear that although the ability to produce torque at ankle and hip joints are similar in children with SDCP and TD children, maximum torque alone didn't allow them to withstand higher balance threats (see difference in threshold velocity in Figure 2B) comparable to those of the TD children. There are three main factors contributing to their impaired balance control. 1) There was a temporal disorganization of joint activation. The TD children showed clear temporal organization of the torque generation among joints with ankle being activated first and the hip torque activated the last. The SDCP children, on the other hand, activated torques at all joints simultaneously; 2) There was a spatial disorganization of the joint torque profile in children with SDCP. In order to regain balance, children with SDCP produced equal amount of torques at hip and ankle, lacking the distally focused torque contribution pattern (ankle produced the largest maximum torque while hip contributed least) as those found in TD children; and 3) Slow speed to reach peak torque value. It is possible that spasticity during dynamic motion (spasticity of a muscle being more pronounced when the joint is in motion) and co-contraction of opposing muscle groups contribute to the slow rate of torque production, which further impair SDCP children's ability to move COP quickly enough to prevent COG from moving to and beyond the edge of BOS.

An interesting finding in this study is that the normalized maximum joint torques adopted by children with SDCP at the threshold velocity were no less, if no more, than those seen in the TD children (see Figure 3B). Wiley and Damiano (1998) documented that children with spastic cerebral palsy were weaker than age-matched peers in all muscle maximum voluntary contraction tests. The weakness was more pronounced distally. Moreover, this muscle weakness was accompanied by muscle imbalance across joints. Hip flexors and ankle plantar flexors tended to be relatively stronger than their antagonists. Our results in conjunction with Wiley and Damiano's findings suggest that children with SDCP are likely to use joint torques near their maximum capacities when recovering from balance threats.

In addition to our research, other studies have also examined the effects of postural alignment on balance control in children and adults. Burtner et al. (1998) had TD children stand in the crouched posture alignment, typical of the postural alignment seen in many children with SDCP, and sustain unexpected platform perturbations. Using EMGs to record muscle activation patterns in response to recovery of balance, it was shown that 1) ranges of onset latencies for the TD children in the crouched position were more constrained. 2) The onset latencies of gastrocnemius and hamstrings muscles were earlier and the frequency of trunk extensor activation decreased, a trend seen also in children with SDCP. 3) Muscle organization changed in the crouched position. Distal to proximal organization was seen less often; and more proximal to distal organization emerged. 4) TD children in the crouched posture, compared with normal stance, used antagonist muscles more often in response to platform perturbations. These changes in postural muscle response patterns seen when TD children were asked to stand in a crouched stance suggest that musculoskeletal constraints are clear contributors to the atypical postural muscle response patterns seen in children with SDCP. It also suggests that the neuromuscular response patterns of some children with SDCP may be appropriate strategies for their musculoskeletal constraints.

A similar study was done by Potter et al. (1990) in which standing balance was assessed in normal adults while they stood in either 15° or 30° of bilateral knee flexion. The results indicated that the knee flexion posture increased COP excursion in both the medial-lateral and anterior- posterior positions. In addition the amount of this increase was linearly affected by the degree of knee flexion. The greater the degree of the knee flexion angle, the greater the excursion of COP.

Clinical Applications

Our study has added to the knowledge of the effects of musculoskeletal constraints on reactive balance control in children with spastic diplegic cerebral palsy. In addition to muscle weakness (Wiley and Damiano, 1998) and postural alignment (Burtner et al, 1998; Potter et al, 1990), the biomechanical measures of lower extremity kinetics in this study demonstrated that deficits in the temporal and spatial organization of torque production and speed of peak torque production contribute to the impaired reactive postural control in children with SDCP. Therefore, therapeutic interventions to improve musculoskeletal impairments that limit the recovery of balance should not only include techniques and devices to change postural alignment and muscle force production, but interventions to increase or aid in the rate of maximal muscle contraction, especially at the ankle during balance recovery. A study by Engsberg et al. (2006), using 3D gait analysis demonstrated that increasing the strength of the plantar flexors and dorsi flexors in children with SDCP, alone, did not increase torque production at the ankle during gait, as was expected.

When making clinical decisions as to the optimal treatment to improve function in children with cerebral palsy, it's important to distinguish the impairments caused by the primary neural pathophysiology from the impairments caused by coping strategies (Gage, 1991, pg 102). By examining COG-COP difference along with the kinetics of the individual joints and their temporal/spatial organization, we were able to demonstrate the coping strategies children with SDCP use to recover balance. Children with SDCP, lacking the ability to generate torque at the ankle and knee fast enough to recover from perturbations, generate much greater force at the hip joint (hip strategy) to maintain stability and to recover balance (compared to what is seen in TD children). The weakness of our study is that the kinematics at the ankle, knee, and hip joints were not analyzed, because as indicated by previous studies (Burtner et al, 1998) using EMG recordings, it is possible that a crouched posture adopted by children with SDCP might account for the disorganized temporal and spatial torque production at all three joints seen in our study.

AFOs and Strength Training

Next, we will review current research studies using kinematic and kinetic outcomes to examine the effectiveness of clinical interventions commonly used to correct for postural alignment and muscle weakness in children with SDCP: ankle foot orthotics (AFOs) and strength training.

Ankle -foot orthoses (AFO) is one of the most commonly prescribed braces for children with CP for improving gait by optimizing postural alignment and joint angles. Based on the understanding of the biomechanics of the human body, it is a brace worn on the lower leg and foot to support the ankle, hold the foot and ankle in the correct position, and correct foot-drop. To benefit from the use of AFO, clinical decisions about the management of AFO must depend on an understanding of the biomechanics of the foot and ankle during gait, the biomechanics of the orthotic device, and the pathophysiology of gait in children with cerebral palsy (Davids et al, 2007).

There has been one study examining postural control and the effectiveness of orthotics (Burtner et al, 1998), by the same research group demonstrated the contribution of mechanical changes to the altered postural control. In this study the effect of solid and dynamic ankle-foot orthoses (AFOs) on stance balance control was compared. It was found that the use of solid AFOs led to decreased activation of distal musculature as well as decreased recruitment of ankle strategies and increased joint angular velocity at the knee. Therefore, wearing fixed AFOs would not be advantageous for children with SDCP in the recovery of balance since the rate of generating sufficient ankle torque is already compromised. These trends were not found in trials with dynamic AFOs, suggesting that dynamic AFOs are more advantageous for children with spastic diplegic CP when balance control is required during unexpected perturbations in stance. This study is particularly interesting in that it highlighted the importance of using biomechanical-oriented intervention in the rehabilitation of children with cerebral palsy. It should be of great interest to clinicians since it emphasizes the impact of alteration in subtle part of the orthoses design on the efficacy of the treatment.

In the past decade, there has been increasing interest in progressive–resistive strength training protocols as a therapeutic intervention to improve functional mobility in children with cerebral palsy (Mockford and Caulton, 2008). Although there have been many 3D gait analysis (3DGA) studies looking at the effectiveness of strength training on gait, there have been no studies specifically exploring its effects on static, dynamic, or reactive postural control. Lowe et al. (2004) demonstrated a strong relationship between isometric muscle force production and standing balance, but no research studies using the biomechanical measures of kinematics or kinetics have confirmed this finding.

There are five major attributes of a functional gait which are frequently lost in children with cerebral palsy: 1) stability in stance, 2) sufficient foot clearance during swing 3) pre positioning of the foot during terminal swing for initial contact 4) adequate step length, and 5) energy conservation (Gage, 1991, pg 101). Most clinical research studies examining the effectiveness of strength training or the use of AFOS to improve gait in children with SDCP have reported improvements in stride or step length; in the temporal characteristics of cadence and velocity; and in energy conservation, but little in the improvement of postural stability (Figueirdo et al, 2008; Mockford and Caulton, 2008).

Though several studies have explored the effects of the different types of AFOs on gait in children with SDCP (Figueiredo et al, 2008), only a few of these studies have included dynamic postural control as a dependent variable when assessing the efficacy of the AFOs using 3D gait analysis technique (Abel et al, 1998; White et al, 2002). By demonstrating an increase in the percentage of time in single-leg stance versus double-leg stance in the total gait cycle both studies demonstrated an increase in stability with the use of AFOs when compared to the barefoot condition. Although these are important measurements to quantify the increase in stability, without the biomechanical measures of total joint kinetics in relation to stability, the 'how' of the improvement is not clear (Gage, 1991, pg 23).

Gaps in Clinical Research

Postural control is the underlying motor ability for the development of functional mobility and higher-level motor skills (Roncesvalles et al, 2001). In addition, postural control has been shown to change with developmental age and experience in walking in both TD children and children with SDCP (Burtner et al, 1998). Our study examining reactive postural control in children with SDCP demonstrated the importance of studying lower extremity kinetics along with the kinematics and kinetics of whole body movement (COG-COP relationship) to better understand the mechanics of reactive postural control in children with SDCP. However, there have been few clinical research studies specifically examining COG-COP difference as a variable in determining the effectiveness of commonly used interventions of strength training and AFOs. As demonstrated in our study, knowledge of the whole body kinematic and kinetic relationship in the recovery of balance, along with the total joint kinetics has demonstrated the possible coping strategies children with SDCP use to recover balance when muscle force production is impaired.

Only recently has there been a study that examined dynamic postural control in the gait of children with SDCP by specifically analyzing COG-COP separation (Hsue et al, 2009). The separation of COG-COP was measured during an interval of the gait cycle between single stance of one leg (SL1) to single-leg stance of the opposite leg (SL2) in children with SDCP and TD of the same age. They found that the RMS of COG-COP in the medial-lateral (M-L) direction was much greater in children with SDCP compared to the TD children. The authors suggested the children with SDCP either had a lateral body lean due to weak hip abductors, or had dynamic postural instability that required a 'walking by falling" pattern of gait. Further biomechanical measures, i.e., the kinetmatic and kinetic measures of the joints along with EMGs may reveal the cause of the COG-COP separation in the M-L direction. Perhaps the children with SDCP spend a greater percentage of their energy expenditure and generation of joint torgues just to maintain postural stability when walking. Therefore, when examining the increase in stride length that occurred in children with SDCP as a result of strength training or the use of AFOs, using biomechanical measures, the question would be 1) how and if the interventions change the separation of COG-COP and in what direction, 2) how the total joint kinematics and kinetics change in relation to change in dynamic postural control, and finally, 3) whether the interventions assist with the compensation strategy to maintain dynamic postural control, or whether they reduce the musculoskeletal impairments interfering with postural control.

CONCLUSION

Our study, in addition to studies on the effects of postural alignment on postural control in children with SDCP, validate the findings of the study by Lowe et al. (2004), in which it is demonstrated that 54% of the variance in standing postural control (measured by the P-CTSIB) in children with SDCP could be explained by the limited range of motion in hip, knee, and ankle joints, as well as the crouch posture; and to abnormal muscle force production. By analyzing total joint kinetics along with whole body kinematic and kinetics COG-COP, we demonstrated that children with SDCP use a hip strategy as a coping response when recovering from external balance threats due to their inadequacy of producing peak ankle torques fast enough to slow the forward movement of the COM following a perturbation.

Children with spastic diplegia cerebral palsy are a heterogeneous group, each child presenting with a variety of musculoskeletal and neuromuscular impairments. Through 3DGA, analyzing the total joint kinetics along with COG-COP separation to measure postural control, clinicians can better understand the specifics of the coping strategies children with SDCP use to maintain postural control during gait. From this knowledge, clinical interventions can be individualized to each child's specific impairments, developmental age, and level of walking experience. These same biomechanical measures could also be used to examine postural control during other functional activities such as sit-to-stand, crossing obstacles, walking on uneven terrain and climbing stairs.

ACKNOWLEDGMENTS

The authors wish to thank Marjorie H. Woollacott for her guidance during the authors' research at the Motor Control Laboratory of the University of Oregon. The study of reactive balance control in children with SDCP cited in this chapter was supported by the National Institutes of Health Grant 2 R01 NS038714-05 to Marjorie H. Woollacott (PI).

REFERENCES

Abel, M.F., Juhl, G.A., Vaughan, C.L., and Damiano, D.L. (1998). Gait assessment of fixed ankle-foot orthoses in children with spastic diplegia. *Arch. Phys. Med. Rehabil.*, 79(2), 126-33.

Bothner, K. (1998). Postural compensations to a disturbance of balance in human: Intersegmental dynamics. *Diss. Abstr. Internat*, 58, 4746.

Burtner, P., Qualls, C., and Woollacott, M. (1998). Muscle activation characteristics of stance balance control in children with spastic cerebral palsy. *Gait Posture*, 8, 163-174.

Chen, J., and Woollacott, M. (2007). lower extremity kinetics for balance control in children with cerebral palsy. *J. Motor. Behav.*, 39(4), 306-316.

Davids, J.R., Rowan, F., and Davis, R.B. (2007). Indications for orthoses to improve gait in children with cerebral palsy. *J. Am. Acad. Orthop. Surg.*, 15(3), 178-88.

Engsberg, J., Ross, S., and Collins, D. (2006). Increasing ankle strength to improve gait and function in children with cerebral palsy. *Pediatr. Phys. Ther.*, 18, 266-275.

Figueiredo, E.M., Ferreira, G.B., Moreira, R.C.M., Kirkwood, R.N., and Fetters, L. (2008). Efficacy of Ankel-Foot Orthoses on Gait of Children with Cerebral Palsy: Systematic Review of Literature. *Pediatr. Phys. Ther.*, 20, 207-233.

Gage, J.R. Gait Analysis in Cerebral Palsy. London, England: MacKeith Press; 1991.

Horak, F., and Nashner, L. (1986). Central programming of postural movements: Adaptation to altered support surface configurations. *J. Neurophysiol.*, 55, 1369-1381.

Hsue, B.J., Miller, F., and Su, F.C. (2009). The dynamic balance of the children with cerebral palsy and typical developing during gait. Part I: Spatial relationship between COM and COP trajectories. *Gait Posture*, 29(3), 465-70.

Lowes, L.P., Westcott, S.L., Palisano, R.J., Effgen, S.K., and Orlin, M.N. (2004). Muscle Force and Range of Motion as Predictors of Standing Balance in Children with Cerebral Palsy. *Phys. Occup. Ther. Pediatr.*, 24(1-2), 57-77.

Mockford, M., and Caulton, J.M. (2008). Systematic Review of Progressive Strength Training in Children and Adolescents with Cerebral Palsy Who Are Ambulatory. *Pediatr. Phys. Ther.,* 20, 318-333.

Nashner, L. (1976). Adaptation reflexes controlling the human posture. *Exp. Brain Res.*, 26, 59-72.

Nashner, L. (1977). Fixed Patterns of rapid postural responses among leg muscles during stance. *Exp. Brain Res.*, 30, 13-24.

Nashner, L., and McCollum, G. (1985). The organization of human postural movements: A formal basis and experimental synthesis. *Behav. Brain Sci.*, 8, 135-172.

Nashner, L., Shumway-Cook, A., and Marin, O. (1983). Stance posture control in selected groups of children with cerebral palsy: Deficits in sensory organization and muscular coordination. *Exp. Brain Res.*, 49, 393-409.

Potter, P.J., Kirby, R.L., and MacLeod, D.A. (1990). The effects of simulated knee-flexion contractures on standing balance. *Am. J. Phys. Med. Rehabil.*, 69(3), 144-7.

Robinovitch, S., Heller, B., Lui, A., and Cortez, J. (2002). Effect of strength and speed of torque development on balance recovery with the ankle strategy. *J. Neurophysiol.*, 88, 613-620.

Roncesvalles, N., Woollacott, M., and Jensen, J. (2001). Development of lower extremity kinetics for balance control in infants and young children. *J. Mot. Behav.*, 33, 180-192.

Shumway-Cook, A., and Woollacott, M. Motor Control: Theory and practical applications. Second Edition. Baltimore, Maryland: Lippincott Williams and Wilkins. 2001.

Styer-Acevedo, J. Physical therapy for the child with cerebral palsy. In: Tecklin J, editor. Pediatric physical therapy. Philadelphia, Pennsylvania: Lippincott Williams and Wilkins; 1994; 89-134.

Sundermier, L., Woollacott, M., Roncesvalles, N., and Jensen, J. (2001). The development of balance control in children: Comparisons of EMG and kinetics variables and chronological and developmental groupings. *Exp. Brain Res.*, 136, 340-350.

Sveistrup, H., and Woollacott, H. (1996). Longitudinal development of the automatic postural response in infants. *J. Mot. Behav.*, 28, 58-70.

Wiley, E., and Damiano, L. (1998). Lower-extremity strength profiles in spastic cerebral palsy. *Dev. Med. Child Neurol.*, 40, 100-107.

Winter, D. Biomechanics and motor control of human movement. Second Edition. New York, NY: John Wiley and Sons. 1990.

Winter, D. A.B.C. of balance during standing and walking. Waterloo, Ontario: Wiley. 1995.

White, H., Jenkins, J., Neace, W.P., Tylkowski, C., and Walker, J. (2002). Clinically prescribed orthoses demonstrate an increase in velocity of gait in children with cerebral palsy: a retrospective study. *Dev. Med. Child Neurol.*, 44(4), 227-32.

Woollacott, M., Burtner, P., Hensen, J., Jasiewicz, J., Roncesvalles, N., and Sveistrup, H. (1998). Development of postural responses during standing in healthy children and children with spastic diplegia. *Neurosci. Biobeh. Rev.*, 22(4), 583-589.

In: Biomechanics: Principles, Trends and Applications
Editor: Jerrod H. Levy, pp. 251-266
ISBN: 978-1-60741-394-3

Chapter 8

BIOMECHANICAL PROPERTIES OF CORNEA

Sunil Shah and Mohammad Laiquzzaman
Birmingham Heartland and Solihull NHS Trust, Lode Lane, Solihull,
West Midlands Birmingham, UK, B91 2JL Tel 0121 711 2020 Fax 0121 711 4040

ABSTRACT

The knowledge of corneal biomechanical properties of cornea has gained importance in recent years. Investigators have been trying to find easy and practical ways to establish these biomechanical properties but to date have had to rely on corneal thickness measures to give an idea of corneal biomechanics. This review explores what is known about the biomechanical properties of the human cornea.

An overview of corneal thickness measurements, its impact on measurement of intraocular pressure and its importance in various disease states is discussed. The recent advent of the Ocular Response Analyser, an in-vivo measure of ocular hysteresis and corneal resistance factor and the pulse waveform associated with this will be discussed. The importance of this machine with respect to corneal biomechanics will be presented.

INTRODUCTION

The cornea is the transparent outer, fibrous tunic for the eye, which serves a refractive function while maintaining mechanical strength. These properties, i.e. the refractive power, mechanical strength and the shape of the cornea, have been a subject of interest to ophthalmologists for many years. Recently, the importance of the biomechanical properties of the eye have gained prominence due to the increasing popularity of refractive surgery and the knowledge of the effects of biomechanics cornea on glaucoma diagnosis and management.

The cornea consists of five layers: epithelium, Bowman's membrane, stroma, Descemet's membrane and endothelium. The biomechanical properties of the cornea are a function of all of the layers, although some structures within the cornea exert a little effect and indeed, there is no consensus over which layers play a role [1].

Bowman's layer has been suggested to play an important role in the biomechanical stability and rigidity of the cornea [2,3] but other have disputed this suggestion believing that

Bowman's layer does not play a significant role. [4,5] The stroma is believed to be main layer responsible for the corneal biomechanical strength. [6] The stroma constitutes about 90 percent of the corneal thickness, featuring stacked lamellae of collagen fibrils. These collagen fibrils, which are packed in parallel arrays, make up the 300 to 500 lamellae within the stroma. The lamellae extend from limbus to limbus and are oriented at various angles to one another. They are positioned less than 90 degrees to each other in the anterior stroma but nearly orthogonal in the posterior stroma. [7] In the anterior one third of the stroma, the collagen lamellae are narrow and thin, undulating, run in random directions and often branch and interweave in an irregular manner. Studies have postulated that the collagen fibres in the anterior half of the stroma are the ones responsible for maintaining biomechanical stability and the corneal shape and hence any surgical incision of these lamellae can result in a loss of corneal structural stability. [6] However, others have suggested that the collagen fibrils of the central area are responsible for this stability because they are of higher density and densely packed. [8] All the collagen lamellae in the posterior two thirds of the stroma are wider and thicker and tend to be parallel to the corneal surface. The fibrils have a uniform diameter. [9,10] The lamellae are embedded in a proteo-glycan rich matrix compound of glycosaminoglycans. [11] This matrix has been suggested to play a role in the absorbance of mechanical impact and is known to be able to be remodeled.

In endothelial disease, the corneal stroma tends to swell posteriorly more than anteriorly. However, it has been noted that even extreme swelling of the stroma has a minimal effect on the curvature of the cornea. [8,12,13] This is probably attributable to the stromal structure.

To date, the only easy method to measure corneal biomechanical properties has been through the measurement of central corneal thickness (CCT). [14] CCT is used as an important clinical measurement when assessing corneal health and in the diagnosis of corneal disease. Therefore, obtaining an accurate CCT measurement is fundamental prior to planning any surgical intervention of the cornea and assessing the progress of corneal disease.

The Ocular response analyser (ORA) (Name and address of the company)is a new instrument, which measures corneal biomechanical parameters termed corneal hysteresis (CH) and corneal resistance factor (CRF) and also measures corneal biomechanically compensated intraocular pressure (IOPcc). The IOPcc is believed to measure intraocular pressure (IOP) neutralising the influence of corneal biomechanical properties and CH and CRF are believed to be the measurement of visco-elastic properties and elasticity of the cornea respectively.

Central Corneal Thickness (CCT)

CCT in humans has been considered a relatively stable parameter in normal eyes. [15] Doughty and Zaman [16] in their major review after meta-analysis of data from 300 publications reported an average CCT of 534 μm. The range of 'normal' CCT reported in literature is from 430 μm to 600 μm. [17-19] Although CCT has been found to be at its highest in the morning just after awakening, it dehydrates and stabilises after 2 to 3 hours and remains constant throughout the wakening period [20-22].

Measuring CCT is essential to assess the health of the cornea. CCT can be affected by the diseases processes, various medications and contact lens wear. [23-28] This may result in either increased thickness or thinning of the cornea, which in turn will affect the biomechanical properties of the cornea.

MEASUREMENT OF CCT

Various instruments have been used to measure CCT, for example, optical pachymetry [29], non-contact specular microsopy [30], Orbscan II pachymetry [30-34], partial coherence interferometry [35] and videopachymetry [36].

The increased use of CCT in clinical practice could, however, be attributed to the availability of easily usable instrumentation i.e. ultrasonic pachymetry. This instrument provides the gold standard for CCT measurement. [33] Many studies have reported CCT by ultrasound pachymetry [37].

Ultrasonic CCT is determined by passing a known velocity of ultrasound through the corneal tissue. The time difference between echoes of signal values reflected from the anterior and posterior surfaces of the cornea is measured. Typically, high frequency ultrasound between 10 to 20 Mhz is used to measure CCT. This high working frequency does not allow echographic scanning of deeper tissue and allow an accuracy within microns. [38,39] The ultrasound pachymeter requires lower operator skill levels and is portable making it is easy to use. It has the disadvantage of being an invasive method in which the probe touches the anterior corneal surface.

CCT AND INTRAOCULAR PRESSURE

Friedenwald [40] recognized the importance of the biomechanical properties of the eye tunics (ocular rigidity) in the measurement of IOP in his search to develop tonometry. More recent studies have demonstrated the importance of CCT when assessing the true IOP rather than measured IOP. [41-47] These studies essentially used the CCT as a measure of corneal rigidity.

Applanation tonometry (Goldmann Tonometry) has been the gold standard for the measurement of IOP. It has been assumed that for an applanation area of 3.0 mm to 3.5 mm, that the capillary attraction of the tear film on the head of tonometer head equals the force with which the cornea resists flattening. This, however, ignored the effects of surface tension. Goldmann recognized the importance of corneal rigidity and surface tension as opposing force. He thus designed his tonometer tip with a diameter of 3.06 mm, which he assumed would cancel the effect of corneal rigidity and surface tension in an average cornea. Goldmann [48], however, recognised that IOP measurement could be affected by variation in CCT in his seminal paper in 1957.

Over a period of many years, a number of studies have considered that CCT can affect IOP readings. Most of these have expressed an opinion that applanation tonometry readings will be higher in eyes with thicker CCT and eyes with thinner corneas will display lower IOP readings. [49-55] This observation confirms that CCT is an important factor and that it can not be ignored in relation to IOP related diseases and their management.

The relationship between CCT and IOP is, however, not clear. Various factors have been proposed for this relationship. It is postulated that it may be either due to a measurement error in applanation tonometry due to differences in corneal thickness, a physiologic effect of IOP on the corneal tissues, an increase of stromal collagen fibres or corneal stromal rigidity or that all of these factors may be involved [56].

Studies on patients with diabetes, [57] keratoconus, [58,59] corneal dystrophies and corneal degenerations and retinal detachment have also shown a relationship between CCT and IOP [60].

The OHTs study found that ocular hypertensive patients with thicker CCT are at reduced risk of developing glaucoma as the true IOP was lower. [61-63] Bhatt et al [64] and Manni et al [65] found that patients with thinner corneas in glaucoma suspects were found more prone to develop glaucoma i.e. the true IOP was higher. Individual differences in CCT might be responsible for inaccuracies of the tonometric values with underestimation of the true IOP in subjects with thinner CCT, although it is becoming clearer that true IOP is unpredictable with linear correction formulae for CCT and that it is more likely that other biomechanical properties of the cornea are important. It has been suggested that differences in the corneal biomechanics may be an expression of individual structural differences of the ocular tissues with potential consequences on the individual susceptibility to the glaucomatous damage under the same IOP values. It is important that CCT measurement with other corneal biomechanical properties should be integrated into the clinical management of glaucoma and ocular hypertension to better understand the individual risk factors.

It would not be unreasonable to expect that laser ablation of corneal stroma in refractive surgery will affect IOP. Many studies have reported a lower mean IOP after corneal refractive surgery. [66-80] Many hypotheses have been presented to explain the cause of lower IOP readings. It has been postulated that as the collagen fibres are ablated during the refractive surgery with the cut collagen fibres losing their tensile strength. [81] Others have suggested that the general softening of the corneal tissue, following the natural healing process, may be a responsible factor [82] while others suggest that decrease in CCT and change of corneal curvature might be the major cause. [83]

However, without being able to measure these biomechanical properties in vivo, these influences cannot be either predicted or controlled [84].

Corneal Biomechanics

The elasticity of a given tissue or material is described as the relationship between stress and strain. Stress is described as the force per unit cross section applied on a tissue material. Strain measures the stretch of a material and is calculated as a change in the length of the tissue divided by its original length. The elastic modulus, or Youngs Modulus (Y) is the ratio of the change in stress to the associated change in strain. Little deformation occurs with an increase in strain when the elastic modulus is high (Y). The smaller the Y, the more elastic the tissue [85-86].

However, for biological material, Y is not constant. The visco-elastic response consists of an immediate deformation followed by a slow deformation. [86] Hence, the cornea reacts to stress as a typical visco-elastic material [87].

The maintenance of corneal shape is determined by IOP, the elastic properties of the corneal tissue and amount and distribution of the corneal tissue mass. [88] The biomechanical properties of the eye, in the past, has been investigated extensively, but most of these studies have been performed *in-vitro* (either on whole eye or excised corneal tissue). These studies include measuring the tensile strength of strips the corneal tissue isolated by dissection and loaded with weights to cause extension and monitoring changes in the spacing between cellular and extracellular particles under load. [85-96] Recently some studies have reported elasticity of eyes *in vivo* by injecting saline solution in to the anterior chamber [97], by measuring CH and CRF by ORA [98] and by dynamic corneal imaging [99].

Edmund [86] investigated the rigidity of the cornea by measuring the radius of the central corneal curvature, the coefficient of radius variation, the CCT and the coefficient of thickness variation. Friedenwald [40], devised a formula for ocular rigidity where K (an assumed constant scleral rigidity) = dP/dV, (V is intraocular volume and P is the IOP). He assumed that K is constant but later studies have found that K is not constant and that it was found to decrease with IOP in human eyes. Investigators have concluded that the phenomenon of ocular rigidity was complex, and even with very simplifying assumptions, the analysis did not lead to a simple result.

Orssengo and Pye' [100] more recently, determined the modulus of corneal rigidity (elasticity) *in vivo* from corneal dimensions and applanation tonometry.

Although all these above mentioned studies are very interesting and have tried to establish ocular rigidity by different methods. They all involve complicated mathematical calculations [86,88,90,92] and are impractical for clinicians and (also report varying results).

MEASUREMENT OF BIOMECHANICAL PROPERTIES OF THE CORNEA

An instrument that provides *in vivo* measurement values and which is easy to use in a clinical setting, easier for clinicians in terms of data interpretation, sufficiently reliable (accurate) and is reproducible will be ideal for the purpose of measuring biomechanical properties. The first such instrument is the ORA.

The ORA determines the biomechanical properties of the cornea by using a bi-directional applanation process. This is achieved using a force-displacement relationship where the instrument releases a precisely metered air pulse, which strikes the central cornea causing it to move inward. This is followed by flattening of the cornea, which then finally becomes concave. The flattening of the corneal surface is termed 'first applanation' or 'inward applanation'. A millionth of a second after inward applanation, the air pulse is shut down, resulting in a decrease in the pressure applied to the cornea. As the pressure decreases, the cornea returns from the concave state to flatten again (second applanation or outward applanation) and then finally resumes its normal convex shape. The change in the corneal curvature is monitored by an electro-optical collimation detector in the central 3.0 mm of the cornea and the whole process lasts just 20 milliseconds.

A filtered version of the detector signal defines the two applanations as two peaks (inward and outward applanation). The two pressures of internal air supply plenum are determined from the applanation times derived from the detector applanation peaks. The inward and outward applanation pressures are different because of the visco-elastic properties of the cornea. However, theoretical calculations indicate that area density of the cornea is a minor contributing factor to the pressure difference. These are perhaps the result of immediate elastic response and delayed or steady state elastic responses respectively of the corneal tissue. Energy adsorption during the corneal deformation delays the occurrence of the applanation signal peaks resulting in a difference between the applation pressures. This difference between the two peaks is referred as corneal hysteresis (CH). CH is believed to reflect the aggregate effects of thickness, rigidity, hydration and perhaps other factors yet to be determined. Corneal resistance factor (CRF) is believed to be a measure of the elastic property of the cornea [98].

Normal Corneal Hysteresis (CH) and Corneal Resistance Factor (CRF)

The ORA has been reported to give reproducible measurements of these parameters. [101,102] CH and CRF are constant and these parameters do not have a diurnal variation during the day time. [103-105] No correlation has been found between age, gender, corneal steepness, axial length and spherical equivalent refraction do not affect CH and CRF values [106-108].

The mean CH in normal eyes in various publications so far has been reported to be between 10.1 mm Hg and 12. 7 mm Hg (range 4.3 mm Hg to 16.7 mm Hg) and mean CRF values between 10 mm Hg and 11. 7 mm Hg (range 4.0 mm Hg to 17.1 mm Hg). [109-111] There appears to be no agreement on the CH measurement in children, some have found no difference compared to adults while others have found CH and CRF values in children to be slightly higher and others have found lower values [112,113].

As one would expect, there should be a relationship between CCT and CH and CRF, as thicker corneas should be more rigid due to the increased mass of corneal tissue. Studies have reported these parameters to be positively correlated i.e. the higher the CCT the higher CH and CRF. [106,109] However, these reported studies have so far revealed that CH and CRF are independent biomechanical parameters from CCT. In addition, studies have found a moderate to significant dependence of CH and CRF on IOP and CCT [106,114].

ORA Readings in Diseased Eyes

CH and CRF in eyes with glaucoma has been previously investigated. [115-118] Shah et al (2008) reported POAG and normal tension glaucoma eyes (NTG) to have lower mean CH values than OHT eyes [117]. Mean CH in NTG, POAG and OHT eyes was 9.0 +/- 1.9, 9.9 and 10.2 mm Hg; CRF was 9.1, 10.6 and 12.0 mm Hg respectively [117].

Weaker corneas could be screened with ORA parameters, and low CH could be considered a risk factor for underestimation of IOP, however, this does not mean to reduce the effect of CCT on IOP and CCT should continue to be considered a useful parameter [114].

Studies on the keratoconic eyes revealed that CH and CRF show a reduced value compared to normal eyes. CH has been reported to be found between 7.5 mm Hg to 9.6 mm Hg [106,114,119,120].

Several studies have reported on the comparison of CH and CRF in keratoconus eyes with LASIK eyes. [114,120,121] These studies compared CH an CRF of post-surgical eyes with those of keratoconus eyes and found that these parameters in keratoconus eyes were lower than in post-LASIK eyes but Shah and Laiquzzaman [121] found that these biomechanical parameters were very similar when comparing keratoconic and post-refractive surgery eyes. In the study by Shah and Laiquzaman [121] post-refractive surgery mean CH was 9.2 mmHg and CRF 7.6 mmHg. The values for keratoconic eyes were 9.4 mmHg and 7.7 mmHg respectively.

Luce [98] reported that applanation signal morphology in post-refractive surgery eyes indicated similar pattern that of keratoconus and Fuch's dystrophy eyes indicating reduced or low biomechanical properties in these eyes.

Refractive Surgery and Corneal Biomechanical Properties

Corneal refractive surgery results in a reduction of CCT and a change in corneal curvature. Understanding the link between the microstructure of cornea and its biomechanics is fundamental in eyes undergoing laser eye surgery.

There has been a general agreement that individual variation in the biomechanical properties of the cornea all influence treatment outcome and that a method for pre-operative assessment of corneal biomechanical properties of individual eye will may help improve outcome.

To predict refractive surgery outcome, ablation algorithms are derived from the statistical analyses of cumulative databases. These algorithms do not deal with corneal biomechanical measurements and their individual variations, because these formulae have been calculated without an understanding of the physical response of the cornea to structural change from ablation of the corneal stroma. A basic understanding of intra-operative corneal biomechanical changes is necessary to improve predictability, safety and effectiveness in the treatment [122].

Laser refractive surgery involves flap creation and excimer laser ablation, which may affect and alter the biomechanics of the cornea. Many investigators have reported a mean reduction of CH and CRF after laser refractive surgical treatment. Ortiz ct al [120] reported that CH and CRF were significantly reduced in eyes after LASIK. Pepose et al [123] also reported a decreased CH and CRF after laser refractive surgery. They found that CH decreased by 16.2% and CRF decreased by a mean 28.6%. Other studies have found the same results where the CRF values were found to be reduced more than CH. [120,121,123-125] The results of these studies imply that changes in CH and CRF after refractive surgery indicate an alteration in corneal biomechanics correlating with the attempted correction. Chen et al [125] conclude that CRF may be more useful than the CH in assessing biomechanical changes resulting from LASIK.

Refractive surgery carries a risk of inducing iatrogenic keratectasia, although this incidence is low 0.66%. [126] 250 μm has been regarded as arbitrary safe minimum CCT after LASIK. Other corneal surgery protocols have recommend that the total corneal thickness at the end of the procedure should be greater than 400 μm (in order that the residual stroma does not incur ectasia) [127]. According to the suggested protocol, eyes with a total CCT less than 400 μm should be considered at risk for ectasia. [128] Keratoectasia is more common in eyes with residual thin beds and high IOP. [129] Having said this, keratoectasia can also be observed in small corrections. [130,131] These observations imply that there must be differences in individual corneal biomechanical properties.

The Future

Several other instruments, which can be useful in the future to know biomechanical properties of the cornea have also been reported in the literature.

The holographic interferometry technique can determine the biomechanical properties *in vivo*. [132,133] This technique has been reported to measure the elasticity of the cornea. It is a non-destructive optical comparison technique analogous to corneal topography, in that fringes are generated from the surface displacements of the cornea and then recorded. [134] Under stress, the changes in the induced strain are seen as changes in the distribution and spacing of the fringes. The high sensitivity of this method enables measurement of the radial distension of the cornea even for very small IOP changes equal to 0.0375 mm Hg. This technique has practical limitations. The holographic technique generates whole-field data across the cornea. This vast data is manually assessed and then interpreted with simple cross-sectional graphs of displacement. This is impractical in a busy clinical setup.

Another technique reported to measure biomechanical properties is non-contact electronic speckle pattern interferometry (ESPI). This instrument also measures the elasticity of the cornea. ESPI does not require photographic recording of a hologram thus simplifying the measurements and provides rapid real-time measurement of surface displacement. An ESPI has the ability to monitor changes in biological tissue under stress with sub-micron accuracy and to acquire such information rapidly and in a non-destructive manner. The benefits of using the ESPI is that it is non-contact, the measurement can be made of the whole cornea, it has the sensitivity to detect and monitor corneal displacements generated *in vivo* and digital processing allows the real-time measurements and measurements of displacement up to 0.01 μm [122].

Grabner et al [99] reported the measurement of corneal elasticity using dynamic corneal imaging *in vivo* using computer assisted veideokeratography. This is performed using dynamic corneal imaging method by central identation of the cornea to assess the elastic properties of the eye.

The concept of ultrasound elasticity microscope is also a further step towards understanding the elastic properties of cornea. It produces high-quality conventional ultrasonic B-scans over large thicknesses by confocal processing. The ultrasound elasticity microscope can produce high resolution strain images throughout the corneal depth and various corneal layers with different elastic properties appearing as different strains in the images.

Development of these new instruments may potentially lead to an increased understanding of corneal biomechanics.

REFERENCES

[1] Mow CC. A theoretical model of the cornea for use in studies of tonometry. *Bull Math Biophys.* 1968; 30: 437-53.

[2] Leibowitz HM. Corneal Disorders: Clinical diagnosis and management. 1984. Saunders, Philadelphia (p14).

[3] Morishige N, Petroll WM, Nishida T, Kenney MC, Jester JV. Noninvasive corneal stromal collagen imaging using two-photon-generated second-harmonic signals. *J. Cataract. Refract.* Surg. 2006; 32: 1784-91.

[4] Montés-Micó R, Charman WN. Intraocular pressure after excimer laser myopic refractive surgery. Ophthalmic. Physiol. Opt. 2001; 21: 228-35.

[5] Hjortdal JO, Ehlers N. Effect of excimer laser keratectomy on the mechanical performance of the human cornea. *Acta. Ophthalmol.* Scand. 1995; 73: 18-24.

[6] Boote C, Dennis S, Newton RH, Puri H, Meek KM. Colleagen fibrils appear more closely packed in the prepupillary cornea: optical and biomechanical implications. *Invest. Ophthalmol. Vis. Sci.* 2003; 44: 2941-8.

[7] Klyce SD, BeuermanR. Structure and Function of the Cornea In The Cornea(Eds, Kaufman, H., Barron, B. and McDonald, M. B.) 1997 Butterworth-Heinemann, Boston, pp. 3-50.

[8] Müller LJ, Pels E, Vrensen GF. The specific architecture of the anterior stroma accounts for maintenance of corneal curvature. *Br. J. Ophthalmol.* 2001; 85: 437-43.

[9] Pouliquen Y J. 1984 Castroviejo lecture. Fine structure of the corneal stroma. *Cornea.* 1984-85; 3: 168-77.

[10] Komai Y, Ushiki T. The three dimensional organisation of collagen fibrils in the human cornea and sclera. *Invest. Ophthalmol. Vis. Sci.* 1991; 32: 2244-58.

[11] Davison PF, Galbavy EJ. Connective tissue remodeling in corneal and scleral wounds. *Invest. Ophthalmol. Vis. Sci.* 1986; 27: 1478-84.

[12] Erickson P, Comstock TL, Doughty MJ, Cullen AP. The cornea swells in the posterior direction under hydrogel contact lenses *Ophthalmic. Physiol. Opt.* 1999; 19: 475-80.

[13] Moezzi AM, Fonn D, Simpson TL, Sorbara L. Contact lens-induced corneal swelling and surface changes measured with the Orbscan II corneal topographer. 2004. *Optom. Vis. Sci*; 81: 189-93.

[14] Brubaker RF. Tonometry and corneal thickness. Arch Ophthalmol. 1999; 117: 104-5.

[15] Mishima S. Corneal thickness. *Surv. Ophthalmol.* 1968; 13: 57-96.

[16] Doughty MJ, Zaman ML. Human corneal thickness and its impact on intraocular pressure: a review and meta-analysis approach. *Surv. Ophthalmol.* 2000; 44: 367-408.

[17] Alsbirk PH. Corneal thickness. I. Age variation, sex difference and oculometric correlations. *Acta. Ophthalmol.* 1978; 56: 95-104.

[18] Alsbirk PH. Corneal thickness. II. Environmental and genetic factors. *Acta. Ophthalmol.* 1978; 56: 105-13.

[19] Avitabile T, Marano F, Uva MG, Reibaldi A. Evaluation ocentral and peripheral corneal thickness with ultrasound biomicroscopy in normal and keratoconus eyes. *Cornea*. 1997; 16: 639-44.

[20] Manchester PT Jr. Hydration of the cornea. *Trans. Am. Ophthalmol. Soc*. 1970; 68: 425-61.

[21] Kiely PM, Carney LG, Smith G. Diurnal variations of corneal topography and thickness. *Am. J. Optom. Physiol. Opt*. 1982; 59: 976-82.

[22] Harper CL, Boulton ME, Bennett D, Marcyniuk B, Jarvis-Evans JH, Tullo AB, Ridgway AE. Diurnal variations in human corneal thickness. *Br. J. Ophthalmol*. 1996; 80: 1068-72.

[23] Ehlers N, Bramsen T.Central thickness in corneal disorders. *Acta. Ophthalmol*. 1978; 56: 412-6.

[24] Bramsen T, Klauber A, Bjerre P. Central corneal thickness and intraocular tension in patients with acromegaly. *Acta. Ophthalmol*. 1980; 58: 971-4.

[25] Katz B, Melles RB, Schneider JA, Rao NA.Corneal thickness in nephropathic cystinosis. *Br. J. Ophthalmol*. 1989; 73: 665-8.

[26] Yaylali V, Kaufman SC, Thompson HW.Corneal thickness measurements with the Orbscan Topography System and ultrasonic pachymetry. *J. Cataract. Refract. Surg*. 1997; 23: 1345-50.

[27] Høvding G.The central corneal thickness in keratoconjunctivitis sicca. *Acta. Ophthalmol*. 1992; 70: 108-10.

[28] Liu Z, Pflugfelder SC. Corneal thickness is reduced in dry eye. *Cornea*. 1999; 18: 403-7.

[29] Hansen FK. A clinical study of the normal human central corneal thickness. *Acta. Ophthalmol*. 1971; 49: 82-8.

[30] Olsen T, Ehlers N. The thickness of the human cornea as determined by a specular method. *Acta. Ophthalmol*. (Copenh). 1984; 62: 859-71.

[31] Liu Z, Huang AJ, Pflugfelder SC. Evaluation of corneal thickness and topography in normal eyes using the Orbscan corneal topography system. *Br. J. Ophthalmol*. 1999; 83: 774-8.

[32] Kawana K, Tokunaga T, Miyata K, Okamoto F, Kiuchi T, Oshika T. Comparison of corneal thickness measurements using Orbscan II, non-contact specular microscopy, and ultrasonic pachymetry in eyes after laser in situ keratomileusis. *Br. J. Ophthalmol*. 2004; 88: 466-8.

[33] Gherghel D, Hosking SL, Mantry S, Banerjee S, Naroo SA, Shah S. Corneal pachymetry in normal and keratoconic eyes: OrbscanII versus ultrasound. *J. Cataract. Refract. Surg*. 2004; 30:1272-7.

[34] Rainer G, Findl O, Petternel V, Kiss B, Drexler W, Skorpik C, Georgopoulos M, Schmetterer L. Central corneal thickness measurements with partial coherence interferometry, ultrasound, and the Orbscan system. *Ophthalmology*. 2004; 111: 875-9

[35] Hitzenberger CK, Baumgartner A, Drexler W, Fercher AF. Interferometric measurement of corneal thickness with micrometer precision. *Am. J. Ophthalmol*. 1994; 118: 468-76.

[36] Franco S, Almeida JB, Parafita M. Measurement of corneal thickness by videopachymetry: preliminary results. *J. Refract Surg*. 2000; 16: S661-3.

[37] McLaren JW, Bourne WM. A new videopachometer. *Invest. Ophthlamol. Vis. Sci.* 1999; 40: 1593-8.

[38] Ling T, Ho A, Holden BA. Method of evaluating ultrasonic pachometers. *Am. J. Optom Phys Optics*. 1986; 63:462-66.

[39] Sabetti L, Spadea L, Furcese N, Balestrazzi E. Measurement of corneal thickness by ultrasound after photorefractive keratectomy in high myopia. *J. Refract. Corneal. Surg.* 1994; 10(2 Suppl): S211-6.

[40] Friedenwald JS. Contribution to the theory and practice of tonometry. *Am. J. Ophthalmol.* 1937; 20: 985-1024.

[41] Whitacre MM, Stein RA, Hassanein K. The effect of corneal thickness on applanation tonometry. *Am. J. Ophthalmol*. 1993; 115: 592-6.

[42] Herndon LW, Choudhri SA, Cox T, Damji KF, Shields MB, Allingham RR. Central corneal thickness in normal, glaucomatous, and ocular hypertensive eyes. *Arch. Ophthalmol.* 1997; 115: 1137-41.

[43] Foster PJ, Baasanhu J, Alsbirk PH, Munkhbayar D, Uranchimeg D, Johnson GJ. Central corneal thickness and intraocular pressure in a Mongolian population. *Ophthalmology*. 1998; 105: 969-73.

[44] Stodtmeister R. Applanation tonometry and correction according to corneal thickness. *Acta. Ophthalmol.* 1998; 76: 319-24.

[45] Tanaka GH. Corneal pachymetry: a prerequisite for applanation tonometry? *Arch. Ophthalmol.* 1998; 116: 544-5.

[46] Shah S, Chatterjee A, Mathai M, Kelly SP, Kwartz J, Henson D, McLeod D. Relationship between corneal thickness and measured intraocular pressure in a general ophthalmology clinic. *Ophthalmology*. 1999; 106: 2154-60.

[47] Aghaian E, Choe JE, Lin S, Stamper RL. Central corneal thickness of Caucasians, Chinese, Hispanics, Filipinos, African Americans, and Japanese in a glaucoma clinic. *Ophthalmology*. 2004; 111: 2211-9.

[48] Goldmann H, Schmidt T. Applanation tonometry. *Ophthalmologica*. 1957; 134: 221-4.

[49] Tomlinson A, Leighton DA. Ocular dimensions in low tension glaucoma compared with open-angle glaucoma and the normal. *Br. J. Ophthalmol*. 1972; 56: 97-105.

[50] Ehlers N, Hansen FK. Central corneal thickness in low-tension glaucoma. Acta. Ophthalmol.. 1974; 52: 740-6.

[51] Johnson M, Kass MA, Moses RA, Grodzki WJ. Increased corneal thickness simulating elevated intraocular pressure. *Arch. Ophthalmol*. 1978; 96: 664-5.

[52] Swan PG, Brown B, Wlidsoet CF. The influence of sleep and central corneal thickness on intraocular pressure in a case of glaucoma. *Practical Optom*. 1990; 1: 112-5.

[53] Argus WA. Ocular hypertension and central corneal thickness. *Ophthalmology*. 1995; 102: 1810-2.

[54] Herman DC, Hodge DO, Bourne WM. Increased corneal thickness in patients with ocular hypertension. *Arch. Ophthalmol*. 2001; 119: 334-6.

[55] Bhan A, Browning AC, Shah S, Hamilton R, Dave D, Dua HS. Effect of corneal thickness on intraocular pressure measurements with the pneumotonometer, Goldmann applanation tonometer, and Tono-Pen. *Invest. Ophthalmol. Vis. Sci*. 2002; 43: 1389-9.

[56] Wolfs RC, Klaver CC, Vingerling JR, Grobbee DE, Hofman A, de Jong PT. Distribution of central corneal thickness and its association with intraocular pressure: The Rotterdam Study. *Am. J. Ophthalmol*. 1997; 123: 767-72.

[57] Skaff A, Cullen AP, Doughty MJ, Fonn D. Corneal swelling and recovery following wear of thick hydrogel contact lenses in insulin-dependent diabetics. *Ophthalmic. Physiol. Opt.* 1995; 15: 287-97.

[58] Böhm A, Kohlhaas M, Lerche RC, Bischoff B, Richard G. Measuring intraocular pressure in keratoconus. Effect of the changed biomechanics. *Ophthalmologe.* 1997; 94: 771-4.

[59] Patel S, McLaughlin JM. Effects of central corneal thickness on measurement of intra-ocular pressure in keratoconus and post-keratoplasty. *Ophthalmic. Physiol. Opt.* 1999; 19: 236-41.

[60] Winther J. Central corneal thickness in patients with retinal detachment. *Acta. Ophthalmol.* 1982; 60: 945-8.

[61] Brandt JD, Beiser JA, Kass MA, Gordon MO. Central corneal thickness in the Ocular Hypertension Treatment Study (OHTS). *Ophthalmology.* 2001; 108: 1779-8.

[62] Gordon MO, Beiser JA, Brandt JD, Heuer DK, Higginbotham EJ, Johnson CA, Keltner JL, Miller JP, Parrish RK 2nd, Wilson MR, Kass MA. The Ocular Hypertension Treatment Study: baseline factors that predict the onset of primary open-angle glaucoma. *Arch. Ophthalmol.* 2002; 120: 714-20.

[63] Brandt JD. Central corneal thickness, tonometry, and glaucoma risk--a guide for the perplexed. *Can. J. Ophthalmol.* 2007; 42: 562-6.

[64] Bhatt N, Bhojwani R, Morrison A, Kwartz J, Laiquzzaman M, Shah S. A 10-year follow up of ocular hypertensive patients within the Bolton Corneal Thickness Study. Can measured factors predict prognostic outcomes?*Cont. Lens. Anterior. Eye.* 2008; 31: 147-53.

[65] Manni G, Oddone F, Parisi V, Tosto A, Centofanti M. Intraocular pressure and central corneal thickness. *Prog. Brain Res.* 2008; 173: 25-30.

[66] Kohlhaas M, Lercher R-CH, Draeger J, Klemm M, Ehlers N, Hjordtal J. Olsen H, Barraquer C, Barraquer JI, Flicker D, Rivera F, Carriazo C. The influence of central corneal thickness and corneal curvature after corneal refractive surgery. *Eur. J. Implant. Ref. Surg.* 1995; 7: 84-8.

[67] Chatterjee A, Shah S, Bessant DA, Naroo SA, Doyle SJ. Reduction in intraocular pressure after excimer laser photorefractive keratectomy. Correlation with pretreatment myopia. *Ophthalmology.* 1997; 104: 355-9.

[68] Mardelli PG, Piebenga LW, Whitacre MM, Siegmund KD. The effect of excimer laser photorefractive keratectomy on intraocular pressure measurements using the Goldmann applanation tonometer. *Ophthalmology.* 1997; 104: 945-8.

[69] Schipper I, Senn P, Oyo-Szerenyi K, Peter R. Central and peripheral pressure measurements with the Goldmann tonometer and Tono-Pen after photorefractive keratectomy for myopia. *J. Cataract. Refract. Surg.* 2000; 26: 929-33.

[70] Rashad KM, Bahnassy AA. Changes in intraocular pressure after laser in situ keratomileusis. *J. Refract. Surg.* 2001; 17: 420-7.

[71] Oliveira-Soto L, Charman WN. Some possible longer-term ocular changes following excimer laser refractive surgery. *Ophthalmic. Physiol. Opt.* 2002; 22: 274-88.

[72] Zadok D, Raifkup F, Landao D, Frucht-Pery J. Intraocular pressure after LASIK for hyperopia. *Ophthalmology.* 2002; 109: 1659-61.

[73] Kaufmann C, Bachmann LM, Thiel MA. Intraocular pressure measurements using dynamic contour tonometry after laser in situ keratomileusis. *Invest. Ophthalmol. Vis. Sci.* 2003; 44: 3790-4.

[74] Cheng AC, Leung DY, Cheung EY, Fan DS, Law RW, Lam DS. Intraocular pressure measurement in patients with previous LASIK surgery using pressure phosphene tonometer. *Clin. Experiment Ophthalmol.* 2005; 33: 153-7.

[75] Chihara E, Takahashi H, Okazaki K, Park M, Tanito M. The preoperative intraocular pressure level predicts the amount of underestimated intraocular pressure after LASIK for myopia. *Br. J. Ophthalmol.* 2005; 89: 160-4.

[76] Jarade EF, Abi Nader FC, Tabbara KF. Intraocular pressure measurement after hyperopic and myopic LASIK. *J Refract Surg*. 2005; 21: 408-10.

[77] Hjortdal JØ, Møller-Pedersen T, Ivarsen A, Ehlers N. Corneal power, thickness, and stiffness: results of a prospective randomized controlled trial of PRK and LASIK for myopia. *J. Cataract. Refract. Surg*. 2005; 31: 21-9.

[78] Siganos DS, Papastergiou GI, Moedas C. Assessment of the Pascal Dynamic contour tonometer in monitoring intraocular pressure in unoperated eyes and operated eyes after LASIK. *J. Cataract. Refract. Surg*. 2005; 31: 458-459.

[79] Svedberg H, Chen E, Hamberg-Nyström H. Changes in corneal thickness and curvature after different excimer laser photorefractive procedures and their impact on intraocular pressure measurements. *Graefes Arch. Clin. Exp. Ophthalmol.* 2005; 243: 1218-20.

[80] Hsu SY, Chang MS, Lee CJ. Intraocular pressure assessment in both eyes of the same patient after laser in situ keratomileusis. *J. Cataract. Refract. Surg*. 2009; 35: 76-82.

[81] Munger R, Dohadwala AA, Hodge WG, Jackson WB, Mintsioulis G, Damji KF. Changes in measured intraocular pressure after hyperopic photorefractive keratectomy. *J. Cataract. Refract. Surg*. 2001; 27: 1254-62.

[82] Patel S, Aslanides IM. Main causes of reduced intraocular pressure after excimer laser photorefractive keratectomy. *J. Refract. Surg*. 1996; 12: 673-4.

[83] Montés-Micó R, Charman WN. Intraocular pressure after excimer laser myopic refractive surgery. Ophthalmic. *Physiol. Opt*. 2001; 21: 228-35.

[84] Roberts C. Biomechanical customization: the next generation of laser refractive surgery. *J. Cataract. Refract. Surg*. 2005; 31: 2-5.

[85] Battaglioli JL, Kamm RD. Measurements of the compressive properties of scleral tissue *Invest. Ophthalmol. Vis. Sci.* 1984; 25: 59-65.

[86] Edmund C. Corneal elasticity and ocular rigidity in normal and keratoconic eyes. *Acta. Ophthalmol*. 1988; 66: 134-40.

[87] Nash IS, Greene PR, Foster CS. Comparison of mechanical properties of keratoconus and normal corneas. *Exp. Eye Res*. 1982; 35: 413-24.

[88] Edmund C. Assessment of an elastic model in the pathogenesis of keratoconus. *Acta. Ophthalmol*. 1987; 65: 545-5.

[89] Hartstein J, Becker B. Research into the pathogenesis of keratoconus. A new syndrome: low ocular rigidity, contact lenses, and keratoconus. *Arch. Ophthalmol*. 1970; 84: 728-9.

[90] Foster CS, Yamamoto GK Ocular rigidity in keratoconus. *Am. J. Ophthalmol*. 1978; 86: 802-6.

[91] Andreassen TT, Simonsen AH, Oxlund H. Biomechanical properties of keratoconus and normal corneas. *Exp. Eye Res*. 1980; 31: 435-4.

[92] Brooks AMV, Robertson IF, Mahoney A. Ocular rigidity and intraocular pressure in keratoconus. *Aust. J. Ophthalmol.* 1984; 12: 317-24.
[93] Reichel E, Miller D, Blanco E, Mastanduno R. The elastic modulus of central and perilimbal bovine cornea. *Ann. Ophthalmol.* 1989; 21: 205-8.
[94] Vito RP, Carnell PH. Finite element based mechanical models of the cornea for pressure and indenter loading. *Refract. Corneal. Surg.* 1992; 8: 146-51.
[95] Hjortdal JO, Jensen PK. In vitro measurement of corneal strain, thickness, and curvature using digital image processing. *Acta. Ophthalmol. Scand.* 1995; 73: 5-11.
[96] Purslow PP, Karwatowski WSS. Ocular elasticity; Is engineering stiffness a more useful characterization parameter than ocular rigidity? *Ophthalmology.* 1996; 103: 1686-92.
[97] Pallikaris IG, Kymionis GD, Ginis HS, Kounis GA, Tsilimbaris MK. Ocular rigidity in living human eyes. *Invest. Ophthalmol. Vis. Sci.* 2005; 46: 409-14.
[98] Luce DA. Determining in vivo biomechanical properties of the cornea with an ocular response analyzer. *J. Cataract. Refract. Surg.* 2005; 31: 156-62.
[99] Grabner G, Eilmsteiner R, Steindl C, Ruckhofer J, Mattioli R, Husinsky W. Dynamic corneal imaging. *J. Cataract. Refract. Surg.* 2005; 31: 163-74.
[100] Orssengo GJ, Pye DC. Determination of the true intraocular pressure and modulus of elasticity of the human cornea in vivo. *Bull Math Biol.* 1999; 61: 551-72.
[101] Kynigopoulos M, Schlote T, Kotecha A, Tzamalis A, Pajic B, Haefliger I. Repeatability of intraocular pressure and corneal biomechanical properties measurements by the ocular response analyser. *Klin. Monatsbl. Augenheilkd.* 2008; 225: 357-60.
[102] Moreno-Montañés J, Maldonado MJ, García N, Mendiluce L, García-Gómez PJ, Seguí-Gómez M. Reproducibility and clinical relevance of the ocular response analyzer in nonoperated eyes: corneal biomechanical and tonometric implications. *Invest. Ophthalmol. Vis. Sci.* 2008; 49: 968-74.
[103] Laiquzzaman M, Bhojwani R, Cunliffe I, Shah S. Diurnal variation of ocular hysteresis in normal subjects: relevance in clinical context. *Clin. Exp. Ophthalmol.* 2006; 34: 114-8.
[104] González-Méijome JM, Queirós A, Jorge J, Díaz-Rey A, Parafita MA. Intraoffice variability of corneal biomechanical parameters and intraocular pressure (IOP). *Optom. Vis Sci.* 2008; 85: 457-62.
[105] Shen M, Wang J, Qu J, Xu S, Wang X, Fang H, Lu F. Diurnal variation of ocular hysteresis, corneal thickness, and intraocular pressure. *Optom. Vis Sci.* 2008; 85: 1185-92.
[106] Shah S, Laiquzzaman M, Bhojwani R, Mantry S, Cunliffe I. Assessment of the biomechanical properties of the cornea with the ocular response analyzer in normal and keratoconic eyes. *Invest. Ophthalmol. Vis. Sci.* 2007; 48: 3026-31.
[107] Kida T, Liu JH, Weinreb RN. Effects of aging on corneal biomechanical properties and their impact on 24-hour measurement of intraocular pressure. *Am. J. Ophthalmol.* 2008; 146: 567-572.
[108] Kamiya K, Hagishima M, Fujimura F, Shimizu K. Factors affecting corneal hysteresis in normal eyes. *Graefes Arch. Clin. Exp. Ophthalmol.* 2008; 246: 1491-4.
[109] Shah S, Laiquzzaman M, Cunliffe I, Mantry S. The use of the Reichert ocular response analyser to establish the relationship between ocular hysteresis, corneal resistance factor

and central corneal thickness in normal eyes. *Cont. Lens. Anterior Eye*. 2006; 29: 257-62.

[110] Montard R, Kopito R, Touzeau O, Allouch C, Letaief I, Borderie V, Laroche L. Ocular response analyzer: feasibility study and correlation with normal eyes. *J. Fr. Ophtalmol.* 2007; 30: 978-84.

[111] Fontes BM, Ambrósio R Jr, Alonso RS, Jardim D, Velarde GC, Nosé W. Corneal biomechanical metrics in eyes with refraction of -19.00 to +9.00 D in healthy Brazilian patients. *J. Refract. Surg.* 2008; 24: 941-5.

[112] Kirwan C, O'Keefe M, Lanigan B. Corneal hysteresis and intraocular pressure measurement in children using the reichert ocular response analyzer. *Am. J. Ophthalmol.* 2006; 142: 990-2.

[113] Lim L, Gazzard G, Chan YH, Fong A, Kotecha A, Sim EL, Tan D, Tong L, Saw SM. Cornea biomechanical characteristics and their correlates with refractive error in Singaporean children. *Invest. Ophthalmol. Vis. Sci*. 2008; 49: 3852-7.

[114] Touboul D, Roberts C, Kérautret J, Garra C, Maurice-Tison S, Saubusse E, Colin J. Correlations between corneal hysteresis, intraocular pressure, and corneal central pachymetry. *J. Cataract. Refract. Surg*. 2008; 34: 616-22.

[115] Ang GS, Bochmann F, Townend J, Azuara-Blanco A. Corneal biomechanical properties in primary open angle glaucoma and normal tension glaucoma. *J. Glaucoma.* 2008; 17: 259-62.

[116] Schroeder B, Hager A, Kutschan A, Wiegand W. .Measurement of viscoelastic corneal parameters (corneal hysteresis) in patients with primary open angle glaucoma. *Ophthalmologe*. 2008; 105: 916-20.

[117] Shah S, Laiquzzaman M, Mantry S, Cunliffe I. Ocular response analyser to assess hysteresis and corneal resistance factor in low tension, open angle glaucoma and ocular hypertension. *Clin. Experiment. Ophthalmol.* 2008; 36: 508-13.

[118] Sullivan-Mee M, Billingsley SC, Patel AD, Halverson KD, Alldredge BR, Qualls C. Ocular Response Analyzer in subjects with and without glaucoma. *Optom. Vis Sci.* 2008 ; 85: 463-70.

[119] Mollan SP, Wolffsohn JS, Nessim M, Laiquzzaman M, Sivakumar S, Hartley S, Shah S. Accuracy of Goldmann, ocular response analyser, Pascal and TonoPen XL tonometry in keratoconic and normal eyes. *Br. J. Ophthalmol.* 2008; 92: 1661-5.

[120] Ortiz D, Piñero D, Shabayek MH, Arnalich-Montiel F, Alió JL. Corneal biomechanical properties in normal, post-laser in situ keratomileusis, and keratoconic eyes. *J. Cataract. Refract. Surg.* 2007; 33: 1371-5.

[121] Shah S, Laiquzzaman M. Comparison of corneal biomechanics in pre and post-refractive surgery and keratoconic eyes by Ocular Response Analyser. Cont Lens Anterior Eye. (In press).

[122] Jaycock PD, Lobo L, Ibrahim J, Tyrer J, Marshall J. Interferometric technique to measure biomechanical changes in the cornea induced by refractive surgery. *J. Cataract. Refract. Surg.* 2005; 31: 175-84.

[123] Pepose JS, Feigenbaum SK, Qazi MA, Sanderson JP, Roberts CJ. Changes in corneal biomechanics and intraocular pressure following LASIK using static, dynamic, and noncontact tonometry. *Am. J. Ophthalmol.* 2007; 143: 39-47.

[124] Shah S, Laiquzzaman M, Yeung I, Pan X, Roberts. The use of the Ocular Response Analyser to determine corneal hysteresis in eyes before and after excimer laser refractive surgery. *Cont. Lens. Anterior Eye*(In Press).

[125] Chen MC, Lee N, Bourla N, Hamilton DR. Corneal biomechanical measurements before and after laser in situ keratomileusis. *J. Cataract. Refract. Surg.* 2008; 34: 1886-91.

[126] Pallikaris IG, Kymionis GD, Astyrakakis NI. Corneal ectasia induced by laser in situ keratomileusis. *J. Cataract. Refract. Surg.* 2001; 27: 1796-80.

[127] Randleman JB. Post-laser in-situ keratomileusis ectasia: current understanding and future directions. *Curr. Opin. Ophthalmol.* 2006; 17: 406-12.

[128] Vinciguerra P, Munoz MI, Camesasca FI, Grizzi F, Roberts C. Long-term follow-up of ultrathin corneas after surface retreatment with phototherapeutic keratectomy. *J. Cataract. Refract. Surg.* 2005; 31: 82-7.

[129] Baek T, Lee K, Kagaya F, Tomidokoro A, Amano S, Oshika T. Factors affecting the forward shift of posterior corneal surface after laser in situ keratomileusis. *Ophthalmology.* 2001; 108: 317-20.

[130] Ou RJ, Shaw EL, Glasgow BJ. Keratectasia after laser in situ keratomileusis (LASIK): evaluation of the calculated residual stromal bed thickness. *Am. J. Ophthalmol.* 2002; 134: 771-3.

[131] Miyata K, Tokunaga T, Nakahara M, Ohtani S, Nejima R, Kiuchi T, Kaji Y, Oshika T. Residual bed thickness and corneal forward shift after laser in situ keratomileusis. *J. Cataract. Refract. Surg.* 2004; 30: 2251-2.

[132] Calkins JL, Hochheimer BF, Stark WJ. Corneal wound healing: holographic stress-test analysis. *Invest. Ophthalmol. Vis. Sci.* 1981; 21: 322-34.

[133] Kasprzak H, Förster W, von Bally G. Measurement of elastic modulus of the bovine cornea by means of holographic interferometry. Part 1. Method and experiment. *Optom. Vis. Sci.* 1993; 70: 535-44.

[134] Smolek MK. Holographic interferometry of intact and radially incised human eye-bank corneas. *J. Cataract. Refract. Surg.* 1994; 20: 277-86.

In: Biomechanics: Principles, Trends and Applications
Editor: Jerrod H. Levy, pp. 267-280
ISBN: 978-1-60741-394-3

Chapter 9

Some Aspects of the Biomechanics of Skilled Musical Performance

Jessie Chen[1] and George Moore[2]
[1]Department of Physiology and Neuroscience,
New York University School of Medicine, New York, NY, USA
[2]School of Engineering, University of Southern California, Los Angeles, CA, USA

Abstract

This chapter addresses a fundamental aspect of musical performance in string players: how the physical "geography" of the instrument and bow, and the anthropometric dimensions of the player interact to produce the stereotypic motor behavior.

Several factors determine movement, but, unlike most biomechanical tasks, the determining outcome here is acoustic. Both upper extremities are involved in the performance but in very different ways.

The left arm governs the contact position of the fingers on the string, and hence the pitch of a note. The spatial relations of the instrument to the body, the contact point of the finger on the string, the length of the fingerboard, and the dimensions of the arm determine a unique posture, and thus the muscle activation patterns, for the left arm for each individual player. The left arm has quite different postures in relation to the body and the instrument for cello players in comparison to violinists.

The bowing (right) arm draws the bow across the string. Its travel velocity is the principal determinant of loudness, but its distance from the bridge, the contact position along the bow length, bow pressure, and its angle of attack on the string leave a noticeable effect on the tone, or timbre. Bowing movements are essentially determined by the flexion and extension of the elbow, with subtle motions of the shoulder and wrist to keep the bow moving in a straight line perpendicular to the string. Given a fixed spatial relation between the body of the performer and the instrument, the posture of the arm for a given bow/string contact point is uniquely determined.

In addition, the force of gravity plays a role in the control of movement. We show, however, that it affects cellists and violinists in very different ways. Left arm movements of cellists are more affected by gravity than those of violinists; whereas gravity affects the right arm more in violinists.

In this chapter we focus on a series of specialized topics: 1) control of the left arm during shifting movements; 2) fine control of the left arm during corrections of intonation errors; 3) coordination between the upper arm and forearm; and 4) coordination between the left and right arms.

INTRODUCTION

In this chapter we discuss some of the basic biomechanical issues of concern to musicians who play string instruments. We emphasize here the problems specific to cello, with additional commentary for violin.

The instruments in the violin family have four strings, tuned to different pitches by adjusting their tension. Strings on the cello typically have a length of about 70 cm, while on the violin, the strings are much shorter — about 33 cm. The fundamental frequency of a string is determined primarily by its freely vibrating length (and its intrinsic physical characteristics which are not variable), and the pitch to which that string is tuned (Figure 1).

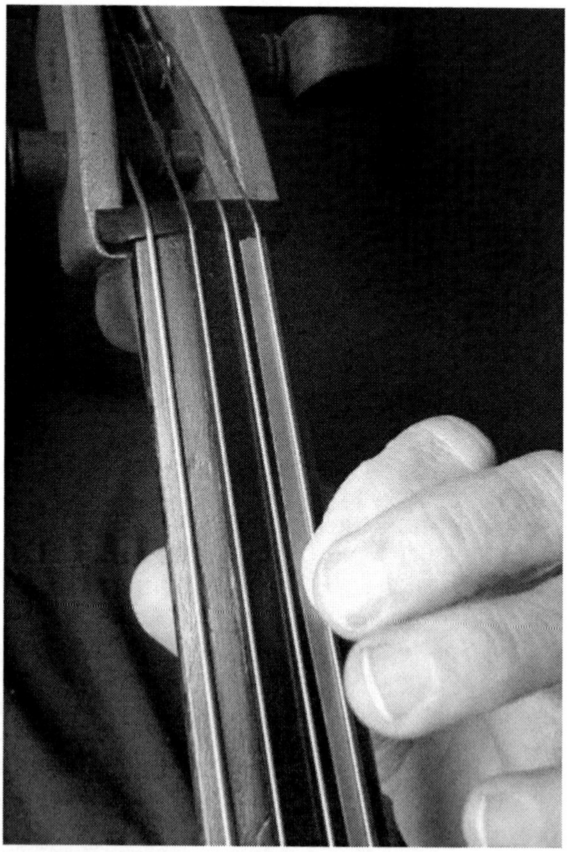

Figure 1. A cellist holds a note using his index finger. The finger placement depresses the string which makes contact with the copper strip affixed to the fingerboard. The string circuit measures the resistance of the string between this point of contact and the bridge.

There are two ways a performer may change from one pitch to another. When the desired note is a short distance away from the current one, the performer may simply switch to another finger placed at a location appropriate to the desired pitch. In the finer aspects of performance, the left (fingering) hand and fingers play a complex and essential role that especially distinguishes beginners from advanced players. But the hand and fingers do not account for the larger movements such as those in the second type of movement, and it is these we shall discuss here. This second type is called a "shift". It is an arm movement that takes the hand along the fingerboard from one location to another. Such movement can require a displacement of the hand of many centimeters in association with movements of the forearm and upper arm, and changes in the posture of the hand. In addition to these movements along a single string, performers often move across strings.

During shifting movements, the finger releases the string, and the arm rotates to bring the hand to the new desired location. If the current and the next notes are played using the same finger on the same string, then that finger essentially moves *along* a straight line trajectory. Much of the motion is achieved by flexion or extension of the elbow joint, with appropriate smaller adjustments of the upper arm (in association with shoulder motion) and the hand. For most purposes the shift must be executed in less than 200 milliseconds.

At the same time, the bow moves *across* the string to produce vibrations, thus generating sound. The velocity of bow movement is the primary determinant of loudness, though other factors play a role as well. The essential fact is that the bow must move along a line perpendicular to the string. Typical bow lengths range from 63-66 cm for the violin to a shorter length of 58-61 cm for the cello. Elbow flexion and extension contribute the largest amount to bow movement, but important contributions to maintaining bow direction and string contact are also due to the shoulder, the forearm, and the (right) hand. Bow velocities may be slow (less than 5 cm/sec) or quite fast (more than 60 cm/sec).

It is important to note that gravity often plays a considerable role in string instrument performance. The violin bow, for example, may sometimes be oriented almost vertically when on the E- string. Owing to the slight tilt at which the violin is held, with its left side elevated, the bow, moving progressively to lower pitch strings (E to A, A to D, D to G) gradually becomes more horizontal (with respect to gravity). On the G-string, gravity ceases to play a major role. For the cello, gravity plays a similar, but lesser, role in bowing movements. In contrast, it contributes a great deal to shifting movements on the cello, far less on the violin. Thus, gravitational forces generally must be included in the force requirements of motion of bowing and shifting, and for maintaining arm position when motion ceases.

Despite the presence of relatively large movements of both arms, considerable precision is also required. In the spatial dimension, fingers of the left hand must make contact with the string and fingerboard with a precision of millimeters for the resulting note to be on pitch. The required precision depends on the position of the note along the string: higher pitches (closer to the bridge) require greater positional precision. For the right arm, the slightest changes in bow velocity or angle make perceptible changes in the tone of notes. Even a slight tremor accompanying stage fright is clearly audible. In the temporal dimension, muscles of different segments in the left arm must be activated in an organized fashion in order to move the hand from one note to another at a precise time to meet the requirements of the music. The right upper arm and forearm must also be coordinated in order to keep the bow moving in a straight trajectory perpendicular to the string at the desired speeds.

Finally, there is also a high demand for coordination between the left and right arms, because the bowing arm is responsible for the tone, the loudness of the note, and special acoustic effects, while the left arm is responsible for positioning the fingers on the precise point on the string for the proper pitch. Whatever the left hand is doing, it is audible if the bow is moving. Sometimes, when the left arm is shifting, it may not be desirable for the tones intermediate between the pre-shift and post-shift notes to be audible. The bow must therefore stop moving between these notes, or greatly reduce its velocity.

Great attention is given by teachers to the ways bowing and shifting movements should be executed, but the *actual* movements also depend on the particular dimensions of the performer. Given the variability of body dimensions between players, movements of arms and hands will be different — even on the same instrument. Movements easily executed by a larger player may be difficult for a smaller, and vice versa. Teachers typically pay little attention to the muscle activation patterns underlying movements, how they should be employed, and for what purpose. The most general guidance is the encouragement of relaxation, but this is often a misguided piece of advice. Another precept is naturalness and optimality, and without question the most skillful performers exhibit this quality. But of what it consists, and how it should be taught, have always been matters of controversy.

In this chapter we focus on a series of specialized topics: 1) control of the left arm during shifting movements; 2) fine control of the left arm: intonation and error correction; 3) coordination between the upper arm and forearm; and 4) coordination between the left and right arms. Our comments are based, in part, on extensive studies of advanced cello students at the University of Oregon School of Music. An electrical circuit adapted to a cello allowed us to track the contact point between the string and the fingerboard resulting from the placement of the finger on that position. We also used solid-state angular-velocity sensors to monitor angular motion of the right and/or left forearms and upper arms. Details of the methodology can be found in our previous publications (Chen et al, 2006; 2008). In addition we monitored the electrical activity (electromyogram-EMG) of various muscle groups of the arm (recorded by electrodes attached to the skin overlying each muscle), which we had observed to be active in controlling the basic movements of interest. Through these simultaneous measurements we were also able to observe the timing relations between the upper arm and forearm, as well as the right and left arms during performance.

Control of the Left Arm During Shifting Movements

Motion during a shifting movement primarily takes place at the elbow joint. On the cello, movement towards a high note (closer to the bridge) requires an extension at the elbow; while flexion is involved in movement toward a low note (away from the bridge). On the violin, where the bridge is closest to the head, the reverse is true. Two primary opposing muscles control elbow rotation (though many more are also activated). Contraction of the biceps flexes the elbow; activation of triceps extends the elbow. The resulting accuracy depends on the precision of the performer's sense of the desired pitch and upon the development of the skill (timing of muscle activation and coordination between the opposing muscles) that enables the performer, after years of training, to move to the required position.

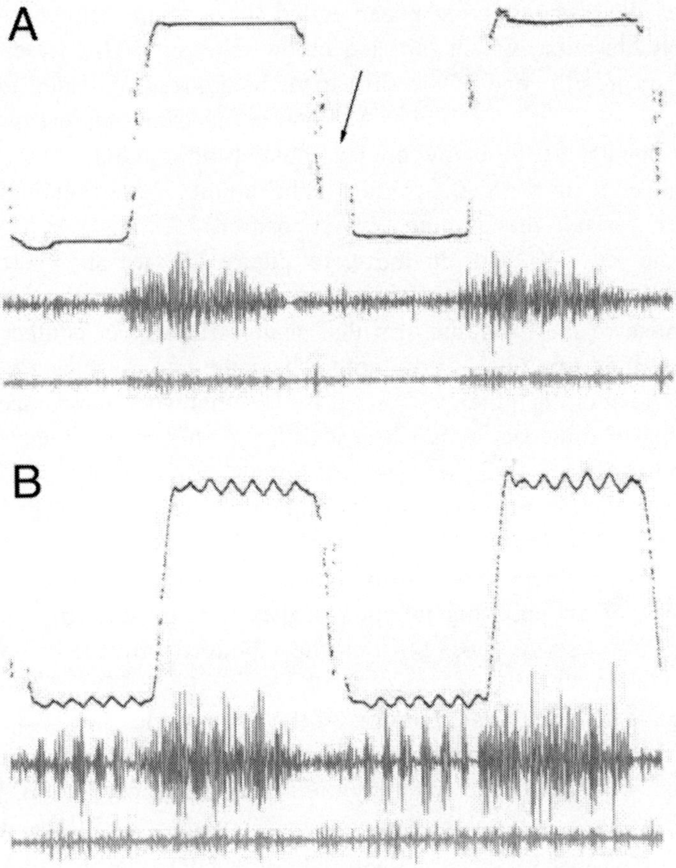

Figure 2. Finger position and associated muscle activities of a cellist performing alternating B-to-A shifts. *A*. Shifts without vibrato. *B*. Shifts with vibrato. Red trace: finger position. Green trace: EMG recording of biceps activity. Cyan trace: EMG recording of triceps activity. The arrow indicates a moment when the finger released the string from contact with the fingerboard.

In Figure 2 we show some of the variables in a relatively simple movement by a cellist during alternate shifts between a higher note (A=440 Hz) and a lower note (B=246.9 Hz) along the cello A-string, a straight-line distance of 26.8 cm. The lower note is closer to the player's shoulder than the higher note. The finger contact position is shown in the red trace. The flat portions of this trace are the periods when the finger stayed in a relatively static position after the performer had completed the movement. Note B is shown at the top of the figure; note A closer to the bottom. The activities of biceps (green) and triceps (cyan) associated with the elbow motion are also shown on the same time scale. To move from A to B requires, among other adjustments, a flexion of the elbow against the force of gravity. Note that biceps became active just prior to the shift from A to B, and that the biceps activity persisted as long as the finger was held at B. This *continued* activity was necessary for countering the action of gravity. When the biceps activity diminished, the finger released the string and the shift to A started (elbow extension).

Triceps activity is even more interesting. For one thing, triceps generated more or less continuous activity along with biceps during the flexion phase that moved the finger towards B and during the holding period at B. These nominally antagonist muscles were *co-*

contracting. When the co-contraction phase ended (prior to the shift to A), triceps fired a brief, higher-amplitude burst, which initiated elbow extension. Gravity carried the arm the rest of the way. Towards the end of the shift to A, biceps was activated to *stop* the motion, and then helped to *maintain* the position at A. Note, in this example, that both muscles served jointly to initiate changes of the elbow angle, to assist the braking of that same action, and then also to stabilize it at a given position. (The thumb, acting against the neck of the instrument, can also assist in maintaining a given position).

But these same muscles do even more. In Figure 2B, we show the same performer shifting between the same two notes. This time, however, the cellist is using *vibrato*, an oscillatory movement of the hand and arm that changes the finger contact position slightly, thereby modulating the basic pitch of the note. The same two muscles are shown again, but now their activation pattern is quite different. It is clear that when the contact position is at A, biceps fires a string of impulses in synchrony with the change in contact position. That is, biceps, in addition to its previous functions, is impulsively changing the elbow angle and, consequently, finger contact position. Given the huge increase in biceps activity when the finger is at B, we cannot be certain of its periodic activity in relation to the vibrato. The same is true for triceps, though there appear to be bursts of impulses in triceps *out of phase* with biceps, an indication of the opposing function of these two muscles in vibrato production in cellists. (Violinists also use vibrato, but often the vibrato is produced primarily using wrist flexion and extension.)

In addition to the different firing patterns of the two muscles, we also note that the same muscle is activated to a different degree at each note position. This is more apparent in biceps (Figure 2B). When producing approximately the same size of oscillation, biceps fired more intensely when the finger was on the lower note, B (where the biceps was activated at a shorter length). This is due to the so-called "length-tension relation" of the muscle, which requires greater or lesser neural activation at different muscle lengths.

Fine Control of The Left Arm: Intonation and Error Correction

In the course of ordinary musical performance, mistakes occur. Non-fixed-pitch instruments, such as string instruments, are especially challenging for pitch accuracy due to the lack of visual and tactile cues of note position. Carl Flesch (1923), the famous violinist and teacher, once claimed that

> …so-called "playing in tune" is nothing else but an extremely rapid and skillful correction of the pitch, which was not accurately "hit" the first time.

Pitch errors will often be corrected immediately after the mistake is perceived. Figures 3A and 3B, illustrate two distinctively different error-correction styles. The performer, a cellist, was asked to shift alternately between two notes on the A string, B (246.9 Hz) and D (293.7 Hz), without vibrato, a movement distance of 26.8 cm. Here we see the final approach to D, and the initial part of the return to B. In Figure 3A is shown a continuous positional adjustment toward the target pitch.

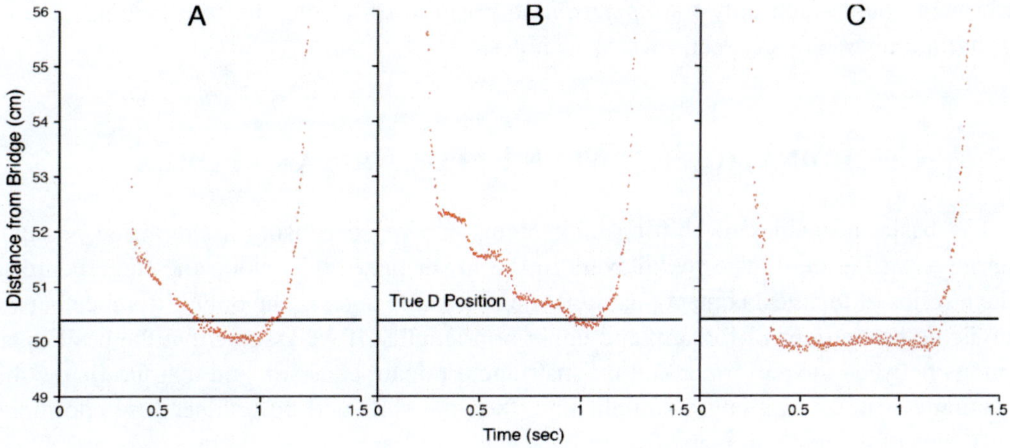

Figure 3. Three short excerpts showing the contact position of the first finger as one subject shifted to note D (293.7 Hz) from note B (246.9 Hz), a total distance of 9.7 cm. *A*. The end of a shift showing a preponderance of slides toward the final position, which was actually sharp, although the pitch might be acceptable to the performer. *B*. The end of a shift where the finger was 2 cm short of the intended point. A series of discrete steps brought the finger closer to the destination. *C*. A shift from B to D without using the bow. The entire movement was quite different, and the initial landing point was scarcely **modified, though it would be "sharp".**

The performer initially landed flat, about 1.5 cm from the true D position, and then slid the finger toward the intended pitch. At 700 ms the finger stopped moving, and this contact position (though about 0.5 cm sharper than the true D position) was held for about 200 ms before the finger began to return to B. This type of continuous change of contact position made up about half the cases where error corrections were observed. This example also **reflects our general finding that a performer's target pitch (intonation)** in the course of a single two-minute trial was variable. Pitch errors on the violin require similar corrections but over considerably smaller distances.

In contrast to the continuous correction, Figure 3B shows that after the initial landing, about 2 cm from the intended note, there was a delay of about 125 msec and a rapid jump in contact position about 1 cm closer to the final point. After a similar delay there was another jump closer to D and this contact position was held for about 200 msec, after which there was a slide to final point. This was the highest pitch within the note, and was held for only 100 msec or so before the required shift returned the finger to note B. Each example in this figure lasted but a second.

The pattern shown in Figure 3C is quite different from the other two. Here, the contact position was held without significant adjustment after the initial landing, even when the landing position was quite far from the correct note position (> 0.5 cm). The only difference between this panel and the other two is that the performer in this trial was not allowed to use the bow: thus no acoustic feedback was available.

In fact, when we asked a group of skilled cellists to shift alternately between note B and note D without using vibrato, at the rate of about one note per second for two minutes, with and without using the bow, we found that all of them had far fewer positional adjustments in the no-bow condition (Chen et al, 2008). This implies that it is the perceived pitch error provided by acoustic feedback that triggers positional adjustments. Without acoustic feedback performers could reach only the *approximate* position of a note, and the landing positions were far more variable between successive notes.

Coordination between Upper Arm and Forearm

The basic mechanics of shifting movements can be seen using a simplified geometric diagram as in Figure 4. The two players in the lower panel shift along the fingerboard and make a series of identical contact positions (i.e., the same notes). The only difference between them is the proportions of forearm and upper arm lengths. If we assume that the basic spatial relations between the performer and the instrument remain constant, and that the finger shifts on a single string, i.e., along a straight line, then the shift will be produced by coordinated movements of the upper arm and forearm.

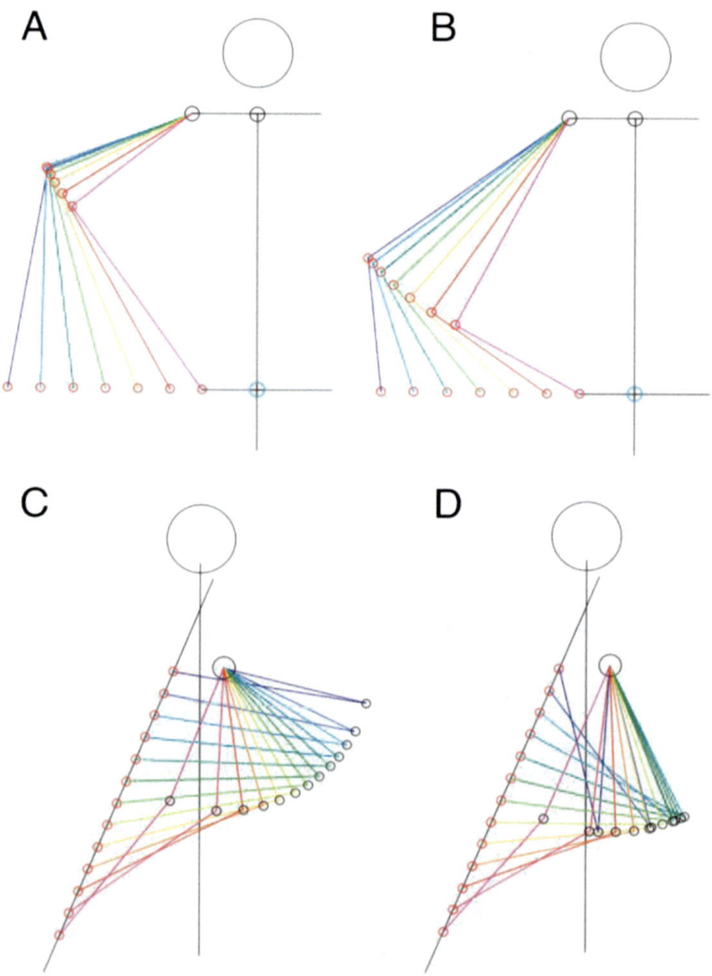

Figure 4. Stick figures of simulated arm postures of players with different upper arm and forearm lengths. The player on the *left* has a shorter upper arm and longer forearm than the player on the *right*, though their total arm lengths are the same. *Top*. Bowing movements. *Bottom*. Shifting movements. Arm sequences are coded from cold to warm colors for postures from low notes to high notes.

The player on the left has a shorter upper arm and longer forearm than the other player on the right (the total arm lengths are the same). Clearly, arm postures are quite different when the players move from the low pitch notes to high pitch notes, though the notes are the same. For the player on the right, the elbow joint first flexes (the upper arm moves backwards in the saggital plane) on successive notes and then extends for the later notes. For the player on the left, all joint angles change monotonically.

It is important to point out that because the elbow joint only permits flexion and extension, the upper arm and forearm always lie in a plane. The spatial orientation of that plane may change with rotations at the shoulder joint. The arm postures shown in Figure 4, lower panel, suggest that the plane in which the arm motion lies is almost parallel to the player's sagittal plane. But with inward medial rotation of the shoulder joint, the arm could lie in the plane of the fingerboard. Our view of the arm would change but not the fundamental relationship between the arm and the string. The shoulder joint rotation would take place along the axis of the string. Indeed, as the player attempts to move on the strings toward his right side (D, G and C), the arm naturally elevates. But the planar position of the upper arm and forearm would remain relatively constant for equal distances along the string (as shown in Figure 4, lower panel). Shifts on the violin are less complex, primarily involving forearm flexion, but the basic posture of the arm in relation to the instrument is more complex.

Let us now consider the task of producing vibrato at each note position. Vibrato is produced by making small oscillating changes in the finger contact point over a total distance of just a few millimeters. The finger never leaves the fingerboard. There are two different ways of moving the contact point. On the cello, one contribution to the contact position change involves alternating elbow flexions and extensions, a few degrees at most. A second contribution can come from oscillating rotations of the forearm (alternating pronations and supinations), again with the contact position being changed by the rotation without the finger leaving the fingerboard. Most cellists employ a combination of these two.

But the choice of vibrato motions is constrained by the geometry of the arm and the instrument. In Figure 4C, a low note on the cello fingerboard (closer to the shoulder), elbow flexions and extensions would move the finger perpendicular to the string and be ineffective for vibrato purposes. On the other hand, forearm rotations would move the finger along the axis of the string. For some players, rotations are the only way to produce vibrato at these locations. Conversely, at the further reaches along the fingerboard, rotations would move perpendicular to the string and only flexions and extensions would support vibrato. At intermediate positions either method would suffice depending on the preferences of the player, though biomechanical factors may play an important role in the player's choice.

In Figure 5, we show activity in the bow arm of a cellist illustrating a special style of bowing — Martelè — that is impulsive in character. The cellist was producing a series of notes at the same pitch; the bow direction changed with each note. Motion involving the bow drawn to the right of the body, away from the instrument, is a so-called "down-bow". Motion toward the fingerboard is called an "up-bow". In both cases the bow motion was essentially perpendicular to the string. The top trace (blue) shows the alternating angular velocity of the right forearm. Negative velocity impulses are down-bows; upward impulses are up-bows. The bottom trace (green) shows the basic bow contact position with the string: rapid movement brought the bow to a new position, where it stopped.

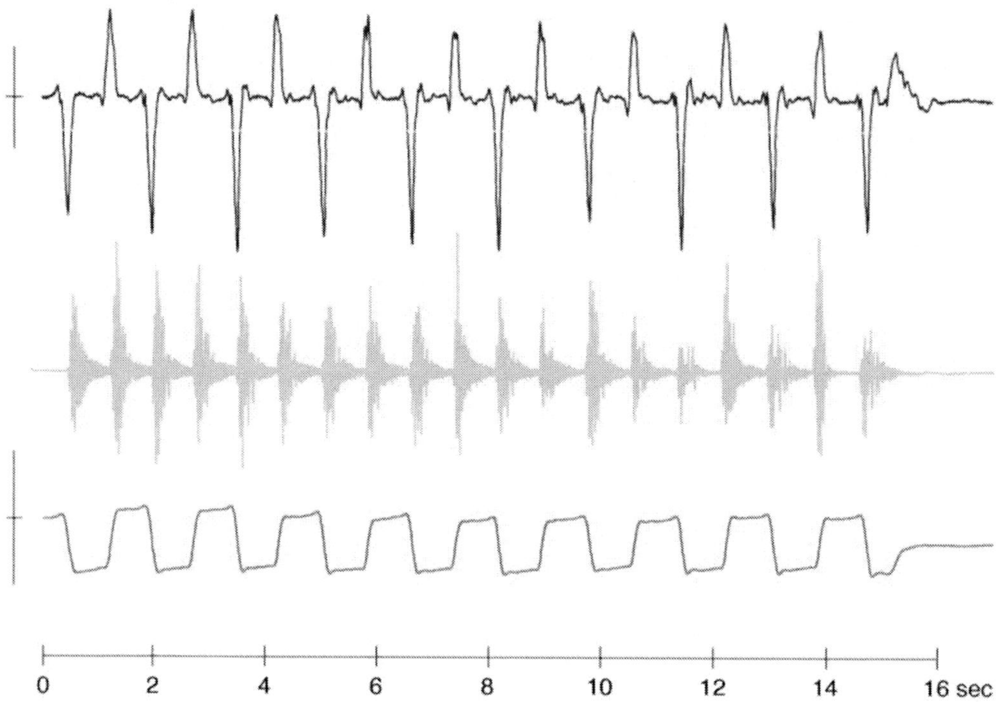

Figure 5. Activity in the bow arm and corresponding acoustic signal during performance of a special style of bowing, "Martelè", that is impulsive in character. *Top*. Angular velocity of the right elbow. Negative impulses are down-bows; upward impulses are up-bows. Scale: +/- 150 deg/sec. *Middle*. Acoustic signal sampled at 360/sec. *Bottom*. Estimated elbow angle. Scale: +/- 5 degrees relative to starting angle.

Regarding the acoustic signal (yellow), each note began with a high amplitude and, after that brief burst, decayed to silence (depending on the physical properties of the instrument). That acoustic pattern is the desired effect of this bowing style, where each note is "attacked".

We also note that the angular velocity patterns of the bow strokes were not anti-symmetric, which is to say that an up-bow motion was not the reverse of the down-bow. What other subtle differences in overall bow control permitted the similarity in the impulsive acoustic responses is a matter of some complexity.

This style of cello bowing — rapid movement from one position of the bow to another, followed by a relatively stationary position — can be analyzed using a simplified geometric representation of the player, the instrument, and the bow. If we assume that the spatial relation between the player and the instrument remains constant, and that the bow moves perpendicular to the string, then Figure 4, top panel, gives a suitable representation. Here we show the same players as those in the lower panel using the same instrument and bow, and the same string contact position. The differences in posture are evident and give rise to rather different overall movements as the notes are played (Figures 4A and 4B). The point here is that the actual bow motion and the resulting acoustic pattern are the same, but the movement by which the performer realizes the notes is different.

It is important to note, once again, that the upper arm and forearm define a plane. In Figure 4, top panel, the plane appears to lie along the plane of the fingerboard. But if the player moves from one string to another it is necessary to raise or lower the elbow to a slightly different spatial plane. Nevertheless the biomechanical constraints remain the same: in changing strings the bow and arm rotate around the axis of the string/fingerboard.

We can also see how the cellist might perform a special modification of this bow motion, namely a sequence of rapid low-amplitude oscillations about a fixed string contact point: a bowing style called "tremolo". Normally, this movement is executed by alternating elbow flexions and extensions. But clearly the required movement cannot be performed when the hand is closer to the string, for here flexion-extension movements would not keep the bow perpendicular to the string. The anatomical conditions for the player in Figure 4A make it relatively simple for tremolo to be produced by the three hand positions furthest from the string. But for the player in Figure 4B tremolo would *only* be possible for the furthest position shown. It is clear that tremolo can best be performed when the bow is near the tip because this arm posture reduces the degrees of freedom the player has to control down to one joint — the elbow — while satisfying the requirement of the bow being perpendicular to the string.

COORDINATION BETWEEN RIGHT AND LEFT ARMS

When playing a string instrument, it is essential to have precise coordination between the right and left arms. In Figure 6, we show a cellist shifting alternately between note B and note D. In Figure 6A, the blue trace represents the angular velocity of the right (bowing) forearm. When the trace crosses the axis, the bow was stopped near the tip (following a down-bow) or near the frog (following an up-bow). The red trace shows the contact position of the finger on the string as the player shifted quickly and alternately between note B (on the A string furthest from the bridge) and D, a short distance away (9.7 cm closer to the bridge). The zero-crossings of bow velocity coincided with the mid-position between notes. The maximum bow velocities occurred while the new note position was held. The green trace is a measure of bow position, obtained by integrating the velocity signal. The peaks of bow position corresponded to the moments of transition between notes.

In Figure 6B, we plot the instantaneous value of finger contact position (left) versus instantaneous value of right elbow angular velocity. The relationship is straightforward; the loop runs clockwise. In Figure 6C, we plot, for comparison, the linear velocities of the wrist (horizontal axis) and elbow of the bowing (right) arm during shifts. Their relationships are far more complex, and bowing motions often involve complex motions of the wrist such as ulnar flexion and extension that occur several times in each cycle. The lower half of the basic "figure-8" loop runs clockwise, the upper half counterclockwise.

Figure 7 shows a similar situation in which a player shifted alternately between two notes, B and A (26.8 cm in distant), on the A string. In red is shown the finger contact position on the string. For reference, the true positions of notes B and A are also shown. From this, it can be seen that both notes were slightly too close to the bridge (i.e., would sound sharp to a perceptive listener). In blue is shown the left forearm angular velocity, part of the motion moving the finger between the two notes. The integral of the right forearm velocity is shown in green, an indicator of bow position. The bow motion was roughly sinusoidal.

Changes in bow direction occurred at the peaks (zero angular velocity of the right elbow). The yellow trace shows the acoustic signal recorded simultaneously.

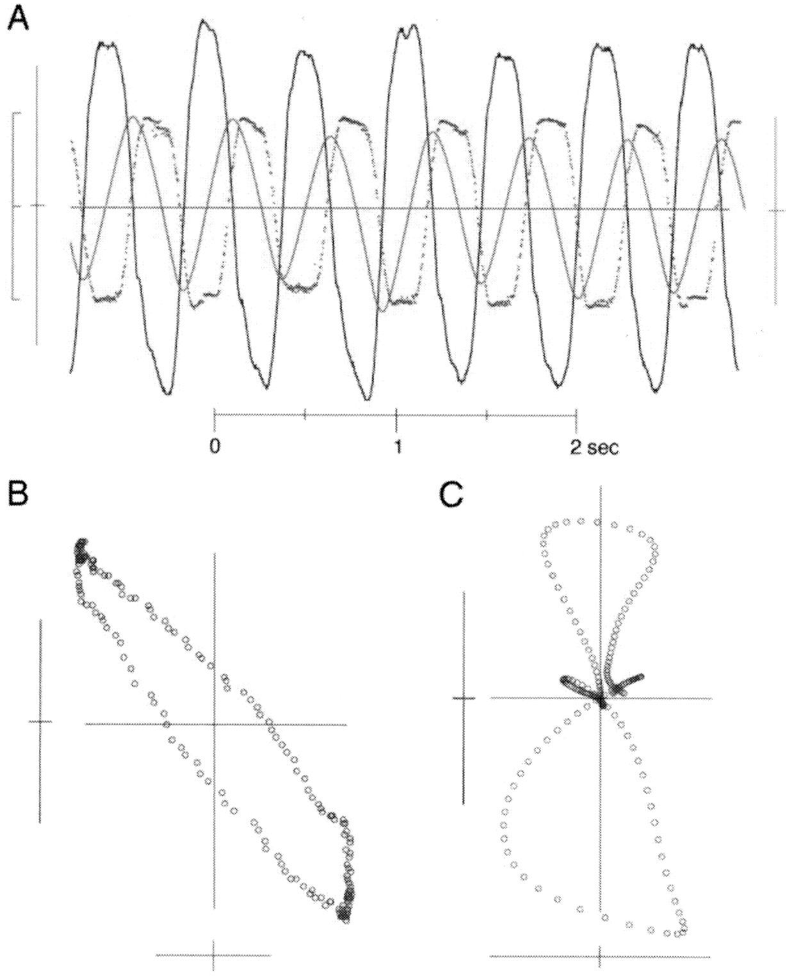

Figure 6. Right and left arm motions during a shifting trial. *A*. Bow and fingering movements during shifts between notes B and D. One bow per note. Red, contact position of the finger on the A string. Top peaks, note B; lower peaks, note D. Scale: +/- 5 cm from average contact position. Blue, right forearm angular velocity. Scale: +/- 100 deg/sec. Green, estimated right forearm angle. Scale: +/- 1 deg from the average angle. Positive impulses are down-bow (elbow extension). *B*. Loop diagram for left arm finger contact position (horizontal axis) versus right forearm angular velocity (data from A). Vertical scale: +/- 100 deg/sec. Horizontal scale: +/- 2.5 cm. *C*. A loop diagram for the right wrist velocity (horizontal axis) versus right elbow velocity when a cellist shifted between notes B and A. Horizontal scale: +/- 100 cm/sec. Vertical scale: +/- 100 cm/sec. Positive values correspond to movements lateral to the mean position of the arm in the horizontal plane. The lower half of the basic "figure-8" loop runs clockwise, the upper half counterclockwise. Data derived from Peak Motion Analysis System (Peak Performance, Inc., Centennial, CO).

The acoustic traces for each note were quite distinct owing to the fact that they were aliased by the sampling rate. It can be seen that the acoustic volume was quite attenuated during the bow changes (arrows), a moment corresponding to a rapid shift in finger position. The diminished bow velocity at the peaks makes the shift inaudible.

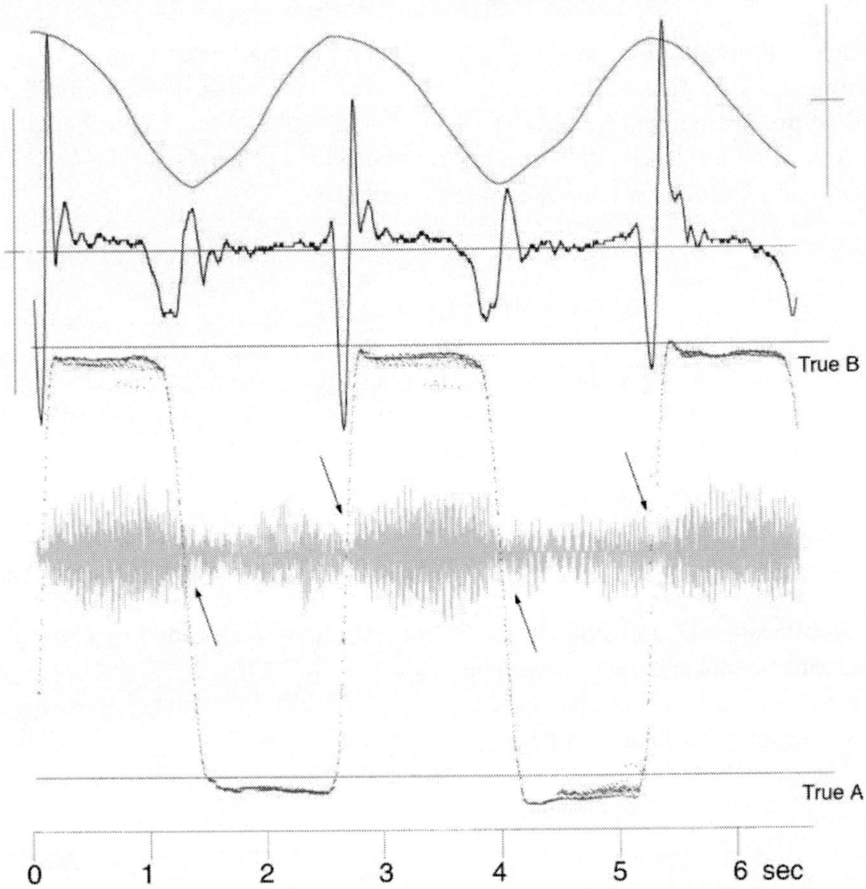

Figure 7. A cellist shifts between notes B and A on the A string. Shift distance: 26.8 cm. Red, finger contact position. Blue, left forearm angular velocity (flexion: negative). Scale: +/- 100 deg/sec. Green, estimated right forearm angle, an indicator of bow position. Scale: +/- 3 degrees. Yellow, acoustic signal (in arbitrary units), sampled at 360/sec.

CONCLUSION

Music pedagogy is not a science. Neither teachers, students or advanced performers can be consciously aware of the details of the extraordinary complexity of movement that is the basis for skilled musical performance.

Here, we have attempted to show just a few examples of the details of movement complexity and coordination. It is unlikely that complete physical descriptions of performance are even possible. If a musician were to repeat a performance, the details would certainly not be the same, because, among other skills, musicians can invent new movements, new interpretations, and new performances.

Mindful of this, the experienced teacher will encourage careful listening, observation, practice, and ever-deeper insights into the music. The teacher should only hope and believe that, somehow, with guidance the gifted student will find a way to produce the desired sound, even without knowing exactly how it is done.

But teachers also need to be aware that the student must also master a unique relationship to an instrument that is dictated by the anthropometric constraints of the body, the physical dimensions of the instrument, the physics of the instrument and the biophysics of the body; always, however, subject to certain as-yet poorly understood principles of the nervous system. Biomechanics of musical performance is based on all these.

ACKNOWLEDGMENTS

The authors wish to thank Gary Felsing (Motus Bioengineering, CA, USA) for support and assistance in the experimental trials. We also wish to thank Marjorie Woollacott and Steve Pologe for contributions to the studies conducted in University of Oregon.

REFERENCES

Chen, J., Woollacott, M., and Pologe, S. (2006). Accuracy and underlying mechanisms of shifting movements in cellists. *Exp. Brain Res*, 174, 467-476.

Chen, J., Woollacott, M., Pologe, S., and Moore, GP. (2008). Pitch and space maps of skilled cellists: Accuracy, variability and error correction. *Exp. Brain Res*, 188, 493-503.

Flesch, C. Die Kunst des Violinspiels. Band I. Reis und Erler, Berlin. 1923.

In: Biomechanics: Principles, Trends and Applications
Editor: Jerrod H. Levy, pp. 281-294
ISBN: 978-1-60741-394-3

Chapter 10

CONTACT HIP STRESS MEASUREMENTS IN ORTHOPAEDIC CLINICAL PRACTICE

Blaž Mavčič[1], Matej Daniel[2], Vane Antolič[3], Aleš Iglič[4] and Veronika Kralj-Iglič[5]

[1]Department of Orthopaedic Surgery, University Medical Centre Ljubljana, Slovenia
[2]Laboratory of Biomechanics, Faculty of Mechanical Engineering, Czech Technical University Prague, Czech Republic
[3]Department of Orthopaedic Surgery, University Medical Centre Ljubljana, Slovenia
[4]Laboratory of Physics, Faculty of Electrical Engineering, University of Ljubljana, Slovenia
[5]Laboratory of Clinical biophysics, Faculty of Medicine, University of Ljubljana, Slovenia

ABSTRACT

There exist several invasive and noninvasive methods to measure the contact hip stress but due to their complexity only few have so far been tested in clinical trials with large numbers of participating subjects. Consequently, the use of contact hip stress measurements in orthopaedic clinical practice is still in its experimental phase.

Biomechanical studies of human hips based on the analysis of 2-D pelvic radiographs have turned out to be a reasonable compromise between the measurement accuracy and the feasibility in clinical setting. Clinical studies have shown significantly higher values of hip stress in adult dysplastic hips when compared to normal hips. It has been found that the cumulative hip stress independently predicts the WOMAC score after 29-years of follow up in dysplastic hips and does so better than morphological radiographic parameters of hip dysplasia or the resultant hip force alone. The preoperative value of the contact hip stress and the magnitude of its operative correction have been found predictors of the long term success of the Bernese periacetabular osteotomy. Elevated shear stress in femoral neck, but not elevated hip contact stress, has been found to be a risk factor for slipping of the capital femoral epiphysis. A statistically significant correlation between the contact hip stress and the age at the total arthroplasty has been shown in a group of hips with idiopathic hip osteoarthritis.

Through advances in 3-D imaging with MRI and CAT scan, visualisation of the femoral head coverage and pelvic muscle attachment points has improved considerably. However, the need to supplement the morphological hip status with biomechanical analysis remains. The current trend is to combine the kinetic gait measurements of the resultant hip force with 3-D imaging of the hip weight-bearing surface in order to better estimate the contact hip stress for a given activity/body position. The added value of such measurements over 2-D pelvic radiograph analysis has not been established yet in clinical trials.

Introduction

Hip stress depends on the magnitude/direction of the resultant hip force, the size of the weight-bearing surface and the stress distribution across this surface. The interest of orthopaedic surgeons in possible relationship between the contact stress and the pathological cartilage conditions has been particularly large for weight-bearing joints with relatively simple kinematics such as the hip joint. A review of the contact hip stress measurement methods has been published by Brand et al. [1]. While many methods have been developed only the noninvasive methods could be used in clinical practice. The aim of this paper is to review the current trends in the clinical use of methods of contact hip stress assessment.

Noninvasive Determination of the Resultant Hip Force

The resultant hip force can be estimated in static or dynamic conditions. Static biomechanical models estimate the resultant hip force for a given body position by solving the static equations for the equilibria of forces and torques [2-4]. The muscles in the static biomechanical models are assumed to be force generators with fixed coordinates. Interindividual variability of muscle attachment points can be achieved by linear scaling of the pelvic configuration in plain anterior-posterior radiographs. In this setting, the standing anterior-posterior pelvic radiographs are assumed to represent the body position of one-legged stance. One of the static biomechanical models applied to a large number of subjects in clinical studies is part of the HIPSTRESS method [4-7]. In this model the one-legged stance is considered to be the representative position for slow gait as the most frequent activity in everyday life [8]. In the one-legged stance, abductor activity is needed to maintain balance on the load bearing leg. The following radiographic parameters are measured on the anterior-posterior radiographs manually [7]: the interhip distance, the pelvic height, the pelvic width, the coordinates of the insertion point of abductors on the greater trochanter (Figure 1). The three-dimensional reference coordinates of the muscle attachment points are taken from a prototype specimen and they are adjusted by linear scaling with regard to the radiographic pelvic parameters for each individual hip. The solution of the vector equations for the equilibria of forces and torques yields the three components of the resultant hip force and the tensions in the abductor muscles.

Noninvasive estimation of the resultant hip force during dynamic activities in different body positions requires the use of the dynamic biomechanical models.

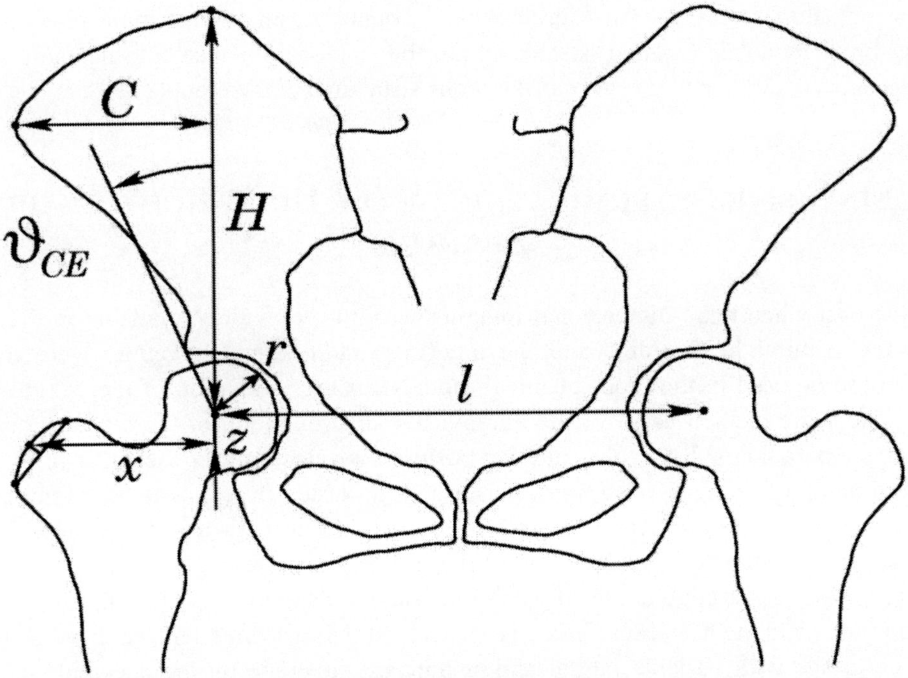

Figure 1. In the HIPSTRESS method of the resultant hip force computation, the following radiographic parameters are measured from the anterior-posterior radiographs manually: the interhip distance l, the pelvic height H, the pelvic width C, the coordinates of the insertion point of abductors on the greater trochanter (x, z). The average of acetabular and femoral head radius r and the Wiberg lateral center-edge angle ϑ_{CE} are used subsequently to estimate the weight-bearing surface and the contact hip stress distribution.

They are based on laboratory measurements with contact force plates, kinematic data of body segment motion and subsequent inverse dynamic analysis of the moving segments [9, 10]. Such approach necessarily includes complex muscle models for several body segments and an appropriate optimization technique to solve the model equations with reduntant muscle forces.

An example of a dynamic biomechanical model based on gait analysis has been developed for use in pre-clinical testing [10]. A computer model of the bones and muscles of the human lower extremities (CT-data, Visible Human, NLM, USA) was scaled to match the anatomies of four THR patients with telemeterized femoral components. Gait analysis data for walking and stair climbing were determined simultaneously with in vivo hip contact forces.

The gait data and the individual musculoskeletal models were then used to calculate the intersegmental resultant joint forces at the ankle, hip and knee joint, as well as the muscle and joint contact forces throughout each gait cycle. The calculated hip joint contact forces were finally validated against the contact forces measured in vivo for both activities.

The latest trend in the noninvasive hip force estimation of individual patients is the polyamid reversed engineering model based on a computed tomography dataset [11]. Pelvis and femur of an individual patient are reproduced in polyamid by selective laser sintering. Hip joint forces can be measured using an experimental setup in which an industrial robot is exerting hip joint forces and moments representing one-legged stance. Hip extensor and abductor actuator forces are measured which counterbalanced the joint moments. The resulting bony model is geometrically accurate, but it does not take into account the joint incongruencies due to the neglected cartilaginous structures in the model [11].

Noninvasive Determination of the Hip Weight-Bearing Surface

In the past when three-dimensional imaging was not yet widely available in the clinical setting, the femoral head radius and the anterior and/or lateral coverage were the main parameters to be used in the biomechanical models for the estimation of the weight-bearing surface. The basic method used for the noninvasive determination of the hip weight-bearing has been plain radiography in the anterior-posterior projection and the false profile view. Many radiographic indices have been described in order to evaluate the femoral head coverage by the acetabulum and to indirectly estimate the weight-bearing surface. The most commonly used parameters include the Wiberg lateral centre-edge angle (CE) [12], the vertical anterior centre-edge angle of Lequesne and de Sèze (VCA) [13], the femoral head extrusion index and the acetabular index [14]. Thus, it was assumed femoral head is a perfect spherical surface with variable lateral and/or anterior coverage by the acetabulum. Soon it became clear that in most physiological conditions the weight-bearing surface is *not* equal in size to the entire articular surface. Variability in some parts of the joint (e.g. lateral coverage) may greatly influence the hip loading while the medial part of the joint bears only small loads [6]. In the HIPSTRESS method [4-7], the weight-bearing surface is assumed to make part of a perfect articular sphere limited on the lateral side by the coverage of the acetabulum. Its medial border depends on the location of the pole of stress distribution. Thus, the size of the weight bearing surface is computed from the average of acetabular and femoral head radius and the Wiberg lateral center-edge angle (Figure 1). The biomechanical role of the horseshoe geometry of the acetabular cartilage has also described using a three-dimensional mathematical model. It has been shown the characteristic horseshoe shape of the articular cartilage in the human acetabulum optimizes the contact stress distribution in the hip joint [15].

Recent advances in the three-dimensional imaging technologies (CAT scan, MRI) have greatly improved the ability to estimate the hip articular surface [16]. There exist contact hip stress studies where the weight-bearing surface was determined with a very precise MRI imaging albeit the resultant hip force magnitude/direction was assumed to be constant for all 82 hips examined [17]. MRI is particularly valuable for the non-invasive estimation of the articular cartilage and the labrum. The disadvantage of classic MRI imaging is that (unlike plain radiography) it cannot be performed with the patient standing and thus the cartilage deformation in different weight-bearing positions cannot be estimated directly. Although the three-dimensional image of the *articular* surface is much more precise it does not suffice to estimate the *weight-bearing* surface directly and again biomechanical models must be used to estimate the importance of different parts of the joint in hip loading. With the development of MRI and hip arthroscopy, some authors have suggested that impingement due to local irregularities near the weight-bearing surface may be more important than the contact hip stress on the weight-bearing surface itself. This has lead to the theory of femoroacetabular impingement, emphasizing the importance of joint incongruencies and the need to identify regions of the femoral head surface with locally increased hip stress values due to the cam/pincer type of impingement [18].

ESTIMATION OF THE CONTACT HIP STRESS IN THE CLINICAL SETTING

Theoretically, any noninvasive method of the resultant hip force determination could be combined with any noninvasive method of the hip weight-bearing surface determination (Figure 2). In the initial research phases the studies mostly focused on the relationship between the biomechanical parameters based on the analysis of individual cases [1]. In such setting it was possible to apply very precise methods, although costly in terms of time and financial resources. However, when the methods were to be applied in clinical studies with larger numbers of patients (over 100) the aspect of timely/financial feasibility turned out to be one of the critical factors. Thus, there exist only few studies of contact hip stress performed in the clinical setting.

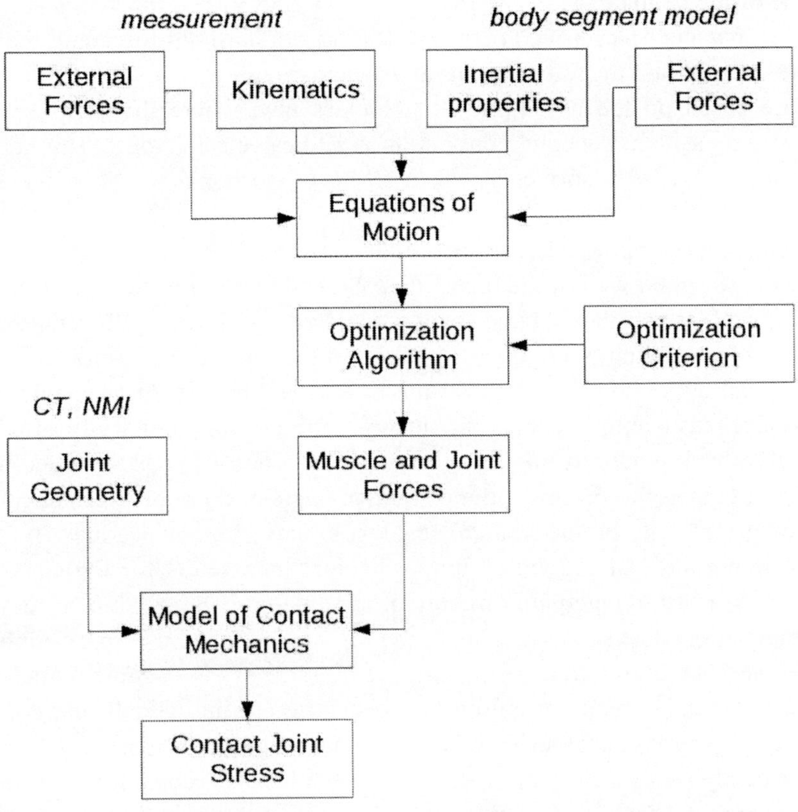

Figure 2. The general algorithm of contact hip stress computation includes the determination of the resultant hip force (combining measurements with the body segment model) and assessment of the weight-bearing surface with morphological imaging of the joint geometry.

As Brand et al. [1] pointed out reports on contact hip stress in the literature may describe different stress distributions. Some authors have estimated spatially averaged values of the contact hip stress (average stress = net force / total weight-bearing surface). Other authors report values of the peak contact stress, i.e. the maximal contact stress value on the weight-bearing surface. Several mathematical approaches to estimate the weight-bearing surface (and consecutively the contact stress distribution) have been proposed [1, 7]. Legal developed a practical method for calculation of the contact hip joint stress for a specific case based on the frontal plane equilibrium force analysis [3]. According to this method, the resultant hip joint force is calculated in static one-legged stance body position assuming one effective abductor muscle with effective attachment point on the greater trochanter and certain inclination towards the horizontal plane. Hadley et al. [19] followed the approach outlined by Legal where the calculation of the hip contact joint stress distribution is restricted to the simplest case of uniform contact stress distribution. One of the few methods applied to more than two hundred patients with different types of hip pathology is the HIPSTRESS method [4-7], developed by the authors of this chapter (Figure 3). The major advantage of the HIPSTRESS method for evaluation of the stress distribution in the hip joint is that it takes into account the non-uniform stress distribution over the weight-bearing surface. This could be of special importance as the gradient of contact stress distribution seems to be even more important than the magnitude of the contact stress [1]. In the HIPSTRESS model the weight-bearing surface is not fixed in advance. The hip geometry affects the resultant hip force and the size and the shape of the weight-bearing surface in a self-consistent manner [6]. These theoretical predictions are based on the assumption of Hooke's law, where the cartilage is described macroscopically as a homogeneous continuum and linear elastic solid. This means that the radial stress in the articular surface is taken to be proportional to the radial strain in the cartilage layer [2].

In most biomechanical models the femoral head is considered to be a perfect sphere. This condition may correspond well to the hips with spherical femoral heads, but dysplastic hips in advanced stages of osteoarthritis have incongruent femoral heads with reduced joint space width, which leads to radiographic overestimation of the femoral head radius and the Wiberg lateral center-edge angle and therefore results in underestimation of the peak contact stress [7]. It is therefore reasonable to expect that incorporation of joint congruity assessment in the biomechanical model would result in even higher values of peak contact hip stress in dysplastic hips and therefore improve the predictive value of these biomechanical parameters. Further, conclusions of biomechanical analyses may be misleading if the relative mathematical importance of individual biomechanical parameters for the contact hip stress computation is assumed to be equally important in explaining contact hip stress variability of the general population. The squared value of the femoral head radius, for example, was found to have direct inverse correlation to the contact hip stress; yet a clinical study proved the femoral head radius had small variability between different individuals and did not account for large contact hip stress differences between normal and dysplastic hips [7].

When critically evaluating the results we should also consider that the statistical significance of biomechanical computations from anterior-posterior radiographs is limited by the data dispersion caused by the error in magnification. It was found that the magnification may vary substantially, but the distribution of the magnification is normal [20].

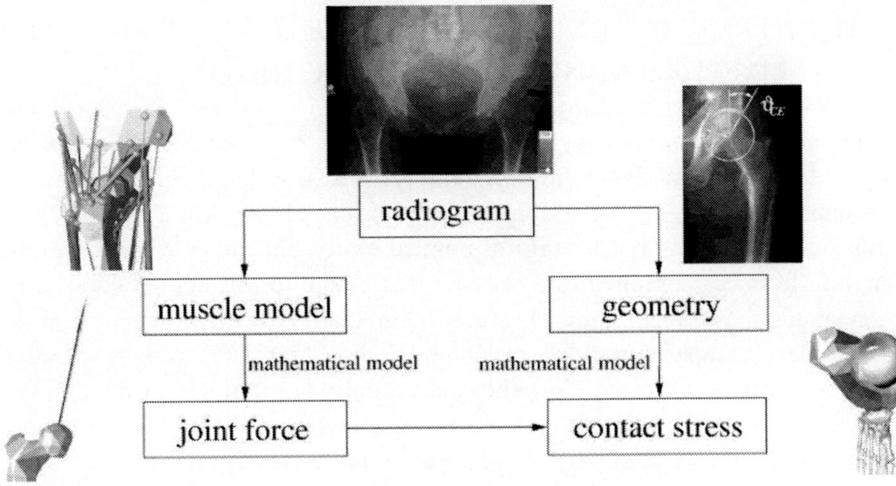

Figure 3. The HIPSTRESS method is based on the general algorithm of contact hip stress computation, as shown in Figure 2: the resultant hip force is computed with a static muscle model of one-legged stance and the joint geometry is assessed by measuring the femoral head radius r and the Wiberg lateral center-edge angle ϑ_{CE}.

Therefore in population studies the magnification should not affect the relative difference between the average or median values but rather increase the data dispersion and decrease the statistical significance of the difference between the considered populations. Also, in determining peak stress, not all the geometrical parameters are equally important as the functional relations between them are nonlinear. It ensues from the mathematical model that stress first depends on the radius of the femoral head, then on the interhip distance, subsequently on the lateral extension of the effective attachment point on the greater trochanter and at last on the pelvic height and width. For example, if the error made in determination of the pelvic height is about 15 per cent, this only yields a 2 per cent error in the peak stress [21].

Three-dimensional imaging techniques have further improved the assessment of the contact hip stress distribution. Hip morphology data can be applied to the finite element models in order to simulate the hip contact stress distribution [22]. Patient-specific, non-linear, contact finite element models of the hip, constructed from computed tomography arthrograms using a custom-built meshing program, were subjected to normal gait cycle loads.

There were significant differences found between the normal control and the asymptomatic hips, and a trend towards significance between the asymptomatic and symptomatic hips of patients afflicted with developmental dysplasia of the hip. The magnitudes of peak cumulative contact pressure differed between apposed articular surfaces. Bone irregularities caused localized pressure elevations and an upward trend between chronic over-pressure exposure and increasing severity of hip dysplasia [22]. However, the method has not been applied to larger numbers of patients in the clinical studies.

Contact Hip Stress in the Assessment of Developmental Hip Dysplasia and Hip Osteoarthritis

Contact hip stress measurements have turned out to be particularly useful in the evaluation of the developmental dysplasia of the hip [1, 7, 23]. With advances in knowledge, it has been established that the developmental dysplasia of the hip (previously called "congenital hip dislocation") is not a uniform clinical entity but rather a broad continuous spectrum ranging from asymptomatic dysplastic acetabula to dislocated hips. Because insufficient acetabular coverage implies the usual hip loads are distributed on a smaller weight-bearing surface compared with normal hips, biomechanical research has focused on estimation of the contact stress in the hip rather than simply morphologic evaluation [24]. In hips with more severe hip dysplasia, several epidemiologic cross-sectional surveys suggested increased incidence of hip osteoarthritis [25, 26]. Furthermore, hip dysplasia was found to be one of the independent risk factors for hip osteoarthritis in addition to age and body mass index [27]. A higher incidence of hip osteoarthritis together with higher average values of contact stress in dysplastic hips have led to the hypothesis that contact hip stress may be one of the key parameters involved in cartilage degeneration [1, 2]. Direct clinical assessment of the predictive value of contact hip stress was reported in two clinical studies of patients with hip dysplasia who were treated with closed reduction and followed up to the average age of 31 years. These authors concluded increased cumulative stress exposure bears higher risk for an unfavorable clinical outcome or osteonecrosis [19, 28]. Although the most severe cases of hip dysplasia are clearly associated with early degeneration, reports on patients with borderline dysplastic hips have been more controversial. In a study with 10-year follow-up of age-matched patients with residual dysplasia without subluxation and normal hips, the authors reported no differences in the reduction of the joint space width or in self-reported hip pain [29]. A recent systematic review found little evidence for a relationship between hip dysplasia and late hip osteoarthritis discovered in patients older than 50 years of age [26]. However, the authors recognized the relationship for the subsequent risk of osteoarthritis in persons diagnosed with dysplasia at a young age compared with the subsequent risk of young patients with osteoarthritis without dysplasia. Some authors have speculated most of the cases of "idiopathic" hip osteoarthritis in fact arise as a result of subtle abnormalities in the anatomic structure of the hip that remained unrecognized during childhood and adolescence and only began to cause clinical symptoms in old age [30].

A clinical study was conducted on the role of the contact hip stress for the development of osteoarthritis in initially asymptomatic human hips, either dysplastic or normal [23]. In the study nonoperatively treated nonsubluxated hips with developmental dysplasia without symptoms at skeletal maturity were identified and compared to adult hips without any disease. Peak contact hip stress was computed with the HIPSTRESS method [4-7] using anterior-posterior pelvic radiographs at skeletal maturity. This method enabled computation of the peak contact hip stress for every individual hip from known values of the body weight, the femoral head radius, the Wiberg center-edge angle, the magnitude of the resultant hip force and the inclination of the resultant hip force with respect to the vertical. The cumulative contact hip stress was determined by multiplying the peak contact hip stress by age at follow-up. The WOMAC scores [31] and radiographic indices of osteoarthritis at a minimum follow-up of 20 years were compared. Dysplastic hips had higher mean peak contact hip stress and higher mean cumulative contact hip stress than normal hips. Mean WOMAC scores and percentage of asymptomatic hips in the study group at the average age of 51 years were equal to the control group at the average age of 68 years. After adjusting for gender and age, the cumulative contact hip stress, Wiberg center-edge angle, body mass index, but not the peak contact hip stress, independently predicted the final WOMAC score in dysplastic hips. Cumulative contact hip stress predicted early hip osteoarthritis better than the Wiberg center-edge angle [23].

The relative maximum hip joint contact stress was found to be higher in healthy women than in healthy men [32]. As women have a higher incidence of the hip osteoarthritis [33], such epidemiological results support the hypothesis that the increased contact hip stress can be one of the risk factors of hip osteoarthritis [32]. This hypothesis was tested by two studies of standard anterior-posterior pelvic radiographs with no or subtle radiological signs of hip osteoarthritis of patients, who underwent hip arthroplasty for primary osteoarthritis years later [34, 35]. In the population of subjects with unilateral osteoarthritis, average peak contact hip stress was significantly higher in hips with arthroplasty than in contralateral hips. In the population of subjects with bilateral osteoarthritis, average peak contact hip stress was significantly higher in hips with earlier arthroplasty than in contralateral hips [34]. Younger age at hip arthroplasty was associated with higher body weight, higher peak contact hip stress normalized to the body weight, higher resultant hip force and larger peak contact hip stress, but not with body mass index [35].

Contact Hip Stress in Preoperative Planning of Orthopaedic Procedures

Osteoarthritis can develop as an idiopathic disease, however, subtle abnormalities could be detected in the hip joint prior to the development of symptoms. Origins of the development of osteoarthritis are ascribed to metabolic resorption of cartilage and/or deformations of anatomical structures. The deviations in the size, shape, mutual proportions or orientation of the acetabulum and/or the femoral head occur frequently. Such deviations are described as hip dysplasia [19]. Although mostly a pediatric problem, hip dysplasia can persist in untreated or unsuccessfully treated cases into adulthood as residual hip dysplasia and may eventually lead to degeneration of the cartilage, presumably due to the pathologically increased stress within the joint. Therefore hip dysplasia represents an important indication for operative procedures that should reduce or redistribute the hip joint stress, thereby stopping or slowing down the pathological processes in the hip cartilage. The hypothesis of secondary osteoarthritis due to decreased femoral head coverage and consecutively higher contact hip stress [28, 36] has led to the invention of surgical procedures of acetabular reorientation in order to prevent the progression of osteoarthritis. Several operative procedures to increase femoral head coverage have been described [37-40] with limited possibilities of correction. Because of such limitations the 'Bernese' periacetabular osteotomy was developed in 1984 and published in 1988 [41]. The procedure has become widely used because it allows optimal correction with minimal exposure and low complication rate, although the operation is technically demanding and its learning curve is long [42]. Ideally this procedure would be indicated in younger adults with concentric hip motion, spherical joint surfaces and no secondary osteoarthritis, but many patients who present with symptomatic hip dysplasia do not meet these criteria [43]. Many attempts have been made to use additional clinical, radiographic or biomechanical factors to carefully select patients that would benefit from the joint preserving surgery [44]. Biomechanical parameters have also been used in the preoperative planning in order to determine the optimal degree of correction in different planes and to achieve the subtle balance between improvement of femoral coverage and restriction of range of motion [5, 16, 17].

In patients who underwent the Bernese periacetabular osteotomy there have been many retrospective analyses published, but only two of them have so far included the pre- and postoperative computations of the contact hip stress [45, 46]. Both studies used the HIPSTRESS method applied to 170 hips altogether. The periacetabular osteotomy was shown to improve the lateral and anterior coverage of the femoral head and accordingly to reduce the normalized peak contact stress in all studied hips. Biomechanical results were consistent with previous studies that have shown contact pressure elevation in dysplastic hips [7, 19] and contact pressure reduction by simulated acetabular reorientation [17]. The postoperative values were in most cases reduced to the level observed in healthy adult hips [47]. Although the Bernese periacetabular osteotomy allowed medialization of the hip joint in addition to improvement of lateral and anterior coverage [48], some studied hips showed lateralization of the center of rotation on the postoperative AP radiographs. Accordingly, the magnitude of the resultant hip force increased in some stratification groups, which is consistent with theoretical predictions of the relationship between the center of rotation and the resultant hip force [5]. These clinical results therefore confirmed theoretical findings [5] that the resultant hip force itself is not a sufficient biomechanical parameter for preoperative planning and that rotational osteotomy can effectively decrease the contact hip stress although the resultant hip force may be slightly increased. The conclusions of this study have shown further reduction of the contact hip stress is only possible through medialization of the femoral head and not through excessive lateral coverage. Accordingly, the procedure was modified with curved periacetabular osteotomy that enables better medialization of the femoral head [46].

Contact hip stress has been tested as a risk factor for slipped capital femoral epiphysis, but with less success [49]. Many hypotheses about the etiology of slippage have been examined, yet the underlying mechanisms have not yet been fully elucidated. Hips contralateral to the slipped ones were compared to age- and gender-matched healthy hips with respect to the shear stress and the contact hip stress. The characteristics of individual hips were incorporated by means of geometrical parameters determined from standard anterior-posterior radiographs. Shear stress was calculated by using a mathematical model where the femoral neck was considered to function as an elastic rod. Contact hip stress was calculated by the HIPSTRESS method [4-7]. Hips contralateral to the slipped ones had higher average shear stress in the femoral neck and more vertically inclined physeal angle in comparison to healthy hips, shear stress in the contralateral hips to the slipped ones remained significantly higher even when normalized to the body weight. However, there was no significant difference in the average contact hip stress [49].

The long-term effect of contact hip stress on the clinical outcome was also studied in the hips operated on by various intertrochanteric osteotomies due to avascular necrosis of the femoral head [50]. The hypothesis stated the hips with a more favorable postoperative distribution of contact hip stress had better clinical outcome. For each hip the peak contact hip stress before/after the operation was determined with the modification of the HIPSTRESS method that took into account the non-weight-bearing necrotic part of the femoral head. The hips were evaluated clinically 9-26 years after the operation and divided into a successful and an unsuccessful group. In the successful group the operation caused an average decrease of the peak hip stress of about 10%, while in the unsuccessful group the operation caused an average increase of the peak hip stress of about 4%, the difference between the respective changes of the peak stress due to the operation being statistically significant [50].

CONCLUSION

The development of noninvasive models to estimate contact hip stress and clinical application of these models have been running hand in hand. Biomechanical studies of human hips based on the analysis of the two-dimensional pelvic radiographs have turned out to be a reasonable compromise between the measurement accuracy and the feasibility in clinical setting. Through wider availability of the three-dimensional imaging techniques, visualization of the femoral head coverage and pelvic muscle attachment points has improved considerably. However, the need to supplement the morphological hip status with biomechanical analysis remains. One of the limitations of the present biomechanical contact hip stress computations is the inability to estimate the patients' physical activity levels throughout their lifetime as such data cannot be reliably acquired retrospectively. The potential confounding effect of differences in physical activity is difficult to estimate because there is no clear evidence lifelong standing, walking, or lifting in the population with normal hips is a risk factor for osteoarthritis. The current trend in biomechanics is to combine the kinetic gait measurements of the resultant hip force with three-dimensional imaging of the hip weight-bearing surface in order to better estimate the contact hip stress for a given activity/body position. Nevertheless, the added value of such measurements over the two-dimensional pelvic radiograph analysis still needs to be established by clinical trials with larger numbers of participating subjects. Eventually, the clinical use of the contact hip stress methods will depend both on their accuracy and feasibility.

REFERENCES

[1] Brand RA, Iglič A, Kralj-Iglič V. Contact stress in the human hip, implications for disease and treatment. *Hip. Int.* 2001;11:117-26.

[2] Brinckmann P, Frobin W, Hierholzer E. Stress on the articular surface of the hip joint in healthy adults and person with idiopathic osteoarthrosis of the hip joint. *J. Biomech.* 1981;14:149-56.

[3] Legal H: Introduction to the Biomechanics of the Hip. In: *Congenital Dysplasia and Dislocation of the Hip.* Ed by Tönnis D. Berlin, Springer-Verlag, 1987: pp 26-57.

[4] Iglič A, Srakar F, Antolič V. Influence of the pelvic shape on the biomechanical status of the hip. *Clin. Biomech.* 1993;8:223-4.

[5] Iglič A, Kralj-Iglič V, Antolič V, Srakar F, Stanič U. Effect of the periacetabular osteotomy on the stress on the human hip joint articular surgace. *IEEE Trans. Rehab. Engr.* 1993;1:207-12.

[6] Ipavec M, Brand R A, Pedersen D R, Mavčič B, Kralj-Iglič V, Iglič A. Mathematical modelling of stress in the hip during gait. *J. Biomech.* 1999;32:1229-35.

[7] Mavčič B, Pompe B, Antolič V, Daniel M, Iglič A, Kralj-Iglič V. Mathematical estimation of stress distribution in normal and dysplastic human hips. *J. Orthop. Res.* 2002;20:1025-30.

[8] McLeish RD, Charnley J. Abduction forces in the one-legged stance. *J. Biomech.* 1970;3:191-209.

[9] Hashimoto N, Ando M, Yayama T, Uchida K, Kobayashi S, Negoro K, Baba H. Dynamic analysis of the resultant force acting on the hip joint during level walking. *Artif. Organs.* 2005;29:387-92.

[10] Heller MO, Bergmann G, Kassi JP, Claes L, Haas NP, Duda GN. Determination of muscle loading at the hip joint for use in pre-clinical testing. *J. Biomech.* 2005;38:1155-63.

[11] Pfeifer R, Hurschler C, Ostermeier S, Windhagen H, Pressel T. In vitro investigation of biomechanical changes of the hip after Salter pelvic osteotomy. *Clin. Biomech.* 2008;23:299-304.

[12] Wiberg G. Studies on dysplastic acetabula and congenital subluxation of the hip joint-with special reference to the complication of osteoarthritis. *Acta. Orthop. Scand.* 1939;83(Suppl 58):1-135.

[13] Lequesne M, de Sèze S. Le faux profil du bassin: Nouvelle incidence radiographique pour l'étude de la hanche: Son utilité dans les dysplasies et les différentes coxopathies. *Rev. Rhum. Mal. Osteoartic.* 1961;28:643-52.

[14] Delaunay S, Dussault RG, Kaplan PA, Alford BA. Radiographic measurements of dysplastic adult hips. *Skeletal. Radiol.* 1997;26:75-81.

[15] Daniel M, Iglič A, Kralj-Iglič V. The shape of acetabular cartilage optimizes hip contact stress distribution. *J. Anat.* 2005;207:85-91.

[16] Mechlenburg I, Nyengaard J R, Romer L, Søballe K. Changes in load-bearing area after Ganz periacetabular osteotomy evaluated by multislice CT scanning and stereology. *Acta. Orthop. Scand.* 2004;75:147-53.

[17] Hipp JA, Sugano N, Millis MB, Murphy SB. Planning acetabular redirection osteotomies based on joint contact pressures. *Clin. Orthop. Relat. Res.* 1999;364:134-43.

[18] Ganz R, Leunig M, Leunig-Ganz K, Harris WH. The etiology of osteoarthritis of the hip: an integrated mechanical concept. *Clin. Orthop. Relat. Res.* 2008;466:264-72.

[19] Hadley NA, Brown TD, Weinstein SL. The effects of contact pressure elevations and aseptic necrosis on the long-term clinical outcome of congenital hip dislocation. *J. Orthop. Res.* 1990;8:504-13.

[20] Vengust R, Drobnič M, Daniel M, Antolič V, Pernus F, Iglič A, Kralj-Iglič V. Role of magnification of standard anterior-posterior radiographs in determination of contact hip joint stress. *Biomed. Eng. Appl. Basis. Comm.* 2000;12:242-4.

[21] Daniel M, Antolič V, Iglič A, Kralj-Iglič V. Determination of contact hip stress from nomograms based on mathematical model. *Med. Eng. Phys.* 2001;23:347-57.

[22] Russell ME, Shivanna KH, Grosland NM, Pedersen DR. Cartilage contact pressure elevations in dysplastic hips: a chronic overload model. *J. Orthop. Surg. Res.* 2006;3:1-6.

[23] Mavčič B, Iglič A, Kralj-Iglič V, Brand RA, Vengust R. Cumulative hip contact stress predicts osteoarthritis in DDH. *Clin. Orthop. Relat. Res.* 2008;466:884-91.

[24] Brand RA, Pedersen DR, Davy DT, Kotzar GM, Heiple KG, Goldberg VM. Comparison of hip force calculation and measurements in the same patient. *J. Arthroplasty* 1994;9:45-51.

[25] Jacobsen S, Sonne-Holm S. Hip dysplasia: a significant risk factor for the development of hip OA. A cross-sectional survey. *Rheumatology* 2005;44:211-8.

[26] Lievense AM, Bierma-Zeinstra SMA, Verhagen AP, Verhaar JAN, Koes BW. Influence of hip dysplasia on the development of OA of the hip. *Ann. Rheum. Dis.* 2004;63:621-6.

[27] Jacobsen S, Sonne-Holm S. Increased body mass index is a predisposition for treatment by total hip replacement. *Int. Orthop.* 2005;29:229-34.

[28] Maxian TA, Brown TD, Weinstein SL. Chronic stress tolerance levels for human articular cartilage: two non-uniform contact models applied to long term follow up of CDH. *J. Biomech.* 1995;28:159-66.

[29] Jacobsen S, Sonne-Holm S, Søballe K, Gebuhr P, Lund B. Joint space width in dysplasia of the hip. *J. Bone Joint Surg. Br.* 2005;87:471-7.

[30] Murphy SB, Kijewski PK, Millis MB, Harless A. Acetabular dysplasia in the adolescent and young adult. *Clin. Orthop. Relat. Res.* 1990;261:214-23.

[31] Bellamy N, Buchanan WW, Goldsmith CH, Campbell J, Stitt LW. Validation study of WOMAC: a health status instrument for measuring clinically important patient relevant outcomes to antirheumatic drug therapy in patients with osteoarthritis of the hip or knee. *J. Rheumatol.* 1988;15:1833-40.

[32] Iglič A, Daniel M, Kralj-Iglič V, Antolič V, Jaklič A. Peak hip joint contact stress in male and female populations. *J. Musculoskeletal. Res.* 2001;5:17-21.

[33] Heliovara M, Makela M, Impivaara O, Knekt P, Aromaa A, Sievers K. Association of everweight, trauma and workload with coxaarthrosis. A health survey of 7217 persons. *Acta. Orthop. Scand.* 1993;64:513-8.

[34] Rečnik G, Kralj-Iglič V, Iglič A, Antolič V, Kramberger S, Vengust R. Higher peak contact hip stress predetermines the side of hip involved in idiopathic osteoarthritis. *Clin. Biomech.* 2007;22:1119-24.

[35] Rečnik G, Kralj-Iglič V, Iglič A, Antolič V, Kramberger S, Rigler I, Pompe B, Vengust R The role of obesity, biomechanical constitution of the pelvis and contact joint stress in progression of hip osteoarthritis. *Osteoarthritis Cartilage* 2009 (in print).

[36] Michaeli DA, Murphy SB, Hipp JA. Comparison of predicted and measured contact pressures in normal and dysplastic hips. *Med. Eng. Phys.* 1997;19:180-6.

[37] Salter RB. Innominate osteotomy in the treatment of congenital dislocation and subluxation of the hip. *J. Bone Joint Surg. Br.* 1961;43-B:518-39.

[38] LeCoeur P. Corrections des défauts d'orientation de l'articulation coxo-fémorale par ostéotomie de l'isthme iliaque. *Rev. Chir. Orthop.* 1965;51:211-2.

[39] Chiari K. Medial displacement osteotomy of the pelvis. *Clin. Orthop. Relat. Res.* 1974;98:55-71.

[40] Tönnis D. *Congenital dysplasia and dislocation of the hip in children and adults*. New York, Springer 1987.

[41] Ganz R, Klaue K, Vinh TS, Mast JW. A new periacetabular osteotomy for the treatment of hip dysplasia. Technique and preliminary results. *Clin. Orthop. Relat. Res.* 1988;232:26-36.

[42] Søballe K. Pelvic osteotomy for acetabular dysplasia. *Acta. Orthop. Scand.* 2003;74:117-8.

[43] Murphy S, Deshmukh R. Periacetabular osteotomy: preoperative radiographic predictors of outcome. *Clin. Orthop. Relat. Res.* 2002;405:168-74.

[44] Trousdale RT, Cabanela ME. Lessons learned after more than 250 periacetabular osteotomies. *Acta. Orthop. Scand.* 2003;74:119-26.

[45] Kralj M, Mavčič B, Antolič V, Iglič A, Kralj-Iglič V. The Bernese periacetabular osteotomy: clinical, radiographic and mechanical 7-15-year follow-up of 26 hips. *Acta. Orthop.* 2005;76:833-40.

[46] Teratani T, Naito M, Shiramizu K, Nakamura Y, Moriyama S. Modified pubic osteotomy for medialization of the femoral head in periacetabular osteotomy: a retrospective study of 144 hips. *Acta. Orthop.* 2008;79:474-82.

[47] Mavčič B, Slivnik T, Antolič V, Iglič A, Kralj-Iglič V. Hip contact stress is related to the development of hip pathology with increasing age. *Clin. Biomech.* 2004;19:939-43.

[48] Crockarell J Jr, Trousdale RT, Cabanela ME, Berry DJ. Early experience and results with the periacetabular osteotomy. The Mayo Clinic experience. *Clin. Orthop. Relat. Res.* 1999;363:45-53.

[49] Zupanc O, Krizančič M, Daniel M, Mavčič B, Antolič V, Iglič A, Kralj-Iglič V. Shear stress in epiphyseal growth plate is a risk factor for slipped capital femoral epiphysis. *J. Pediatr. Orthop.* 2008;28:444-51.

[50] Dolinar D, Antolič V, Herman S, Iglič A, Kralj-Iglič V, Pavlovčič V. Influence of contact hip stress on the outcome of surgical treatment of hips affected by avascular necrosis. *Arch. Orthop. Trauma. Surg.* 2003;123:509-13.

In: Biomechanics: Principles, Trends and Applications
Editor: Jerrod H. Levy, pp. 295-303
ISBN: 978-1-60741-394-3

Chapter 11

External Pelvic Fixation during Lumbar Muscle Resistance Exercise

Michael C. McGlaughlin[1], Philip A. Anloague[1] and Brian C. Clark*[2]

[1]Physical Therapy Program, Department of Health and Sport Science, University of Dayton and University of Dayton, Dayton, OH; USA

[2]Institute for Neuromusculoskeletal Research, Department of Biomedical Sciences, Ohio University College of Osteopathic Medicine, Athens, OH, USA

Abstract

Resistance exercise has long been used to promote musculoskeletal health with the application of training regimens for the clinical treatment and prevention of low back pain growing in popularity over the last couple of decades. A variety of exercise modes have been utilized in an attempt to stimulate and promote increases in muscle function of the lumbar extensors. This chapter examines the current state of knowledge regarding the application of external pelvic fixation during trunk extension exercise and its importance on the concomitant increase of functional outcomes such as muscle strength, muscle activation patterns and compensatory muscle growth.

Keywords: lumbar, exercise, muscle, low back pain.

Introduction

It has been suggested that lumbar muscle training aids in increasing pelvic and spinal stability [1], and over the past few decades a plethora of scientific interest has developed regarding the optimal way to effectively strengthen the integrity of the lumbar musculature that surrounds the spine. In 1994, the US Agency for Health Care and Policy Research (now

* Corresponding Author and Reprint Requests to: Brian C. Clark, Ph.D. Dept Biomedical Sciences Ohio University 211 Irvine Hall, Athens, OH 45701, 740-593-2354, Email: clarkb2@ohio.edu

the Agency for Healthcare Research and Policy) concluded that back extensor muscle conditioning exercises were helpful in the management of acute low back pain (LBP) [2]. However, determining the most effective protocol to strengthen the lumbar musculature has proven somewhat difficult, as researchers have reported wide variations on the effectiveness of exercise prescriptions in increasing lumbar muscle mass and strength. While much of this variation is likely due to the between-study differences in the exercise prescription employed (i.e. training frequency, intensity and global versus targeted training), one potential factor involved in effectively increasing lumbar extensor muscle function surrounds the biomechanics of training and the degree to which the muscles are effectively isolated and targeted during resistance training exercises.

Typically, extension of the trunk is a coupled movement initiated by extension of the hips followed by extension of the lumbar spine [26]. This normal lumbo-pelvic rhythm reduces the demands on the lumbar extensor muscles and underlying apophyseal joints and discs, thereby protecting the region against high stress [27]. Delay in lumbar extension shifts the extensor torque demand to the powerful hip extensors (hamstrings and gluteus maximus), at the time when the external flexion torque on the lumbar region is greatest [27]. This phenomenon creates difficulty in accurately quantifying range-of-motion and torque production making it difficult to determine the relative contribution of each joint segment on these measures [28].

Pollack, Graves and others have previously suggested that to strengthen the lumbar muscles external pelvic fixation (EPF) is required whereby the lumbar extensors are activated in an isolated fashion to limit the compound lumbar-pelvic rhythm and the contributing effects of the gluteals and hamstrings which have a much larger cross sectional area (CSA) and larger moment arms (Figure 1) [3-6,9-14].

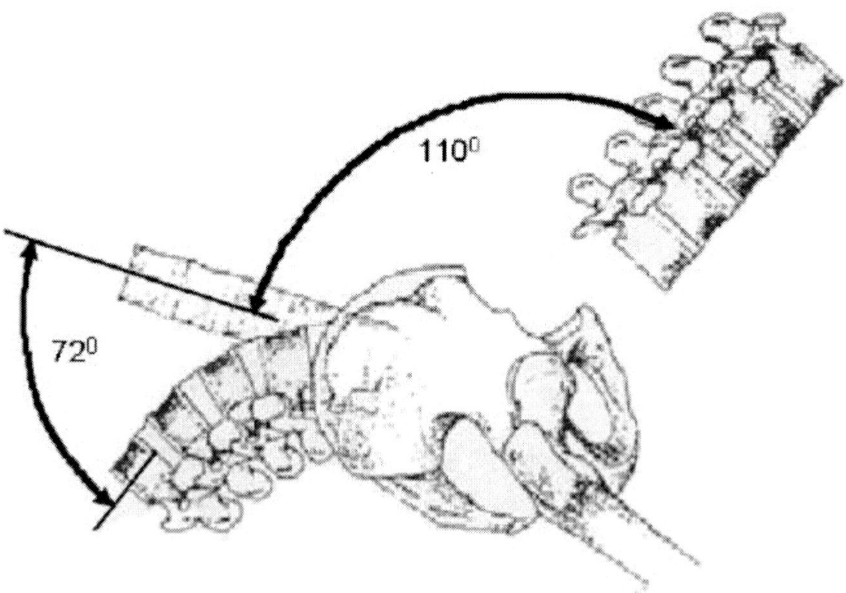

Reprinted with permission from Carpenter and Nelson [6].

Figure 1. During compound trunk extension the contraction of the gluteal and hamstring muscles can produce around 110 degrees of pelvic rotation about the femurs, while simultaneous contraction of the lumbar extensor muscles can produce around 72 degrees of lumbar vertebral movement.

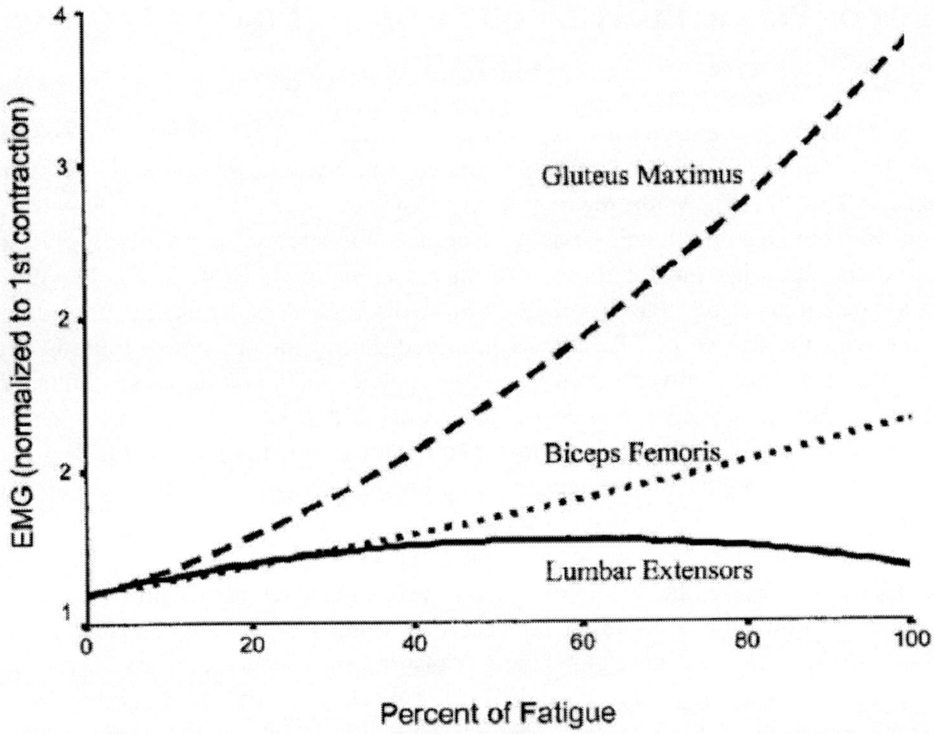

Reprinted with permission from Clark and colleagues [23].

Figure 2. Electromyogram (EMG) activity of the lumbar and hip extensor muscles during fatiguing trunk extension exercise *without* pelvic restriction. Note the plateau and subsequent decline in lumbar muscle EMG amplitude when the subjects reached approximately 55% of their maximal fatigue capacity, and the concomitant increase in hip extensor activity. This observation indicates that during lumbar extension without pelvic restriction the hip extensors accommodate the performance of the task. A similar observation has also been made with an increase in external loading [22, 25].

For example, compound trunk extension involves both contraction of the hip extensor muscles (gluteal and hamstrings) that can result in ~ 110 degrees of pelvic rotation about the femur as well as a contraction of the lumbar extensor musculature that can result in ~ 72 degrees of lumbar vertebral movement (Figure 1) [6].

For the purposes of this chapter, the term *external pelvic fixation (EPF)* has been adopted so that the reader will not be confused with various interchangeable terms, such as neutral pelvis, lumbo-pelvic stabilization, isolated lumbar extension, pelvic restraint, pelvic isolation, or spinal stabilization which have been used when referring to mechanisms performed with internal structures or through external methods [1,4,5,10,11,16,20,21,29,30,31,32].

In this chapter we will discuss the effect that external fixation of pelvic motion has during resistance training on the exercise-induced physiologic responses and muscle performance adaptations. Specifically, we will review the effect of EPF on the exercise training increases in neuromuscular strength and muscle mass as well as its effect on muscle activation patterns.

Effect of Pelvic Fixation on Training-Induced Increases in Strength

When most skeletal muscles are exposed to short-term (i.e. 8-10 weeks) heavy resistance exercise training they tend to increase their maximal voluntary force output (strength) between 20-30% [7, 8]. With regards to the lumbar extensor muscles however, large variations in reported strength gains exist with some authors reporting no increases in strength [9] and others observing greater than 100% increases in strength [4, 5, 10, 11, 30]. While there could be a number of factors explaining these discrepant results one postulated reason is associated with the degree of EPF that is provided during the resistance training protocol. Graves and colleagues investigated whether pelvic stabilization was imperative to strengthening the lumbar extensor muscles and observed that 12-weeks of variable resistance training *with* EPF (1 day/wk) resulted in a significant mean increase of 23.5% in isolated lumbar extension strength (strength testing was conducted with EPF imposed to minimize synergistic contributions of the hip extensors) [4].

Interestingly, subjects who performed the resistance training protocol *without* EPF did not significantly increase their isolated lumbar extension strength despite increasing their training load leading the authors to suggest that trunk extension exercise without EPF restriction primarily exercised the gluteal and hamstring muscles as they are heavily involved in pelvic rotation. Similarly, Mayer et al. did not observe significant increases in strength when trunk extension training *did not include* EPF [9]. In this study, subjects performed Roman chair exercise (a common exercise device found in fitness and rehabilitation settings) which involved progressive resistance back extension exercise on a device that allows one to lift and lower their upper body with variable resistance added by using hand-held metal plates. While subjects performed the training on the Roman chair (which does not limit pelvic rotation) strength testing was performed on a device that does limit the hip extensor contribution to maximal force.

The aforementioned studies all utilized healthy individuals without low back pain. However, a number of authors trained symptomatic patient populations with EPF and found similar results with respect to strength increases, although to a lesser magnitude in comparison to that observed in healthy individuals [3, 12-14].

These patient populations were quite varied including geriatric women [13], a post-surgical population after just 6 weeks recovery from lumbar discectomy [3], and groups with chronic low back pain [12, 14]. When the results of the aforementioned studies performing resistance exercise training *with* EPF are considered collectively, the mean strength increase is on the order of ~ 50%.

This magnitude is exceptional, especially when one considers that the training frequency in many of these studies was as low as one or two times per week. Other evidence of improvements with EPF in the symptomatic population has been documented in the literature, with one dramatic example by Deutsch reporting increases in lumbar extensor strength greater than 300% throughout the range-of-motion with an 8 week rehabilitation training protocol [30].

EFFECT OF PELVIC RESTRICTION ON TRAINING-INDUCED INCREASES IN MUSCLE MASS

A common outcome variable to determine the effectiveness of a resistance training protocol on activating and stimulating a target muscle is the assessment of muscle mass (most commonly cross-sectional area (CSA) or volume). An increased muscle mass is primarily resultant of an increase in the contractile proteins actin and myosin and is in part responsible for the increases in strength described above. Relatively few researchers have investigated CSA changes in the lumbar extensors with EPF exercise training and to our knowledge no one has directly compared exercise training regimens with and without EPF for this outcome variable. Li and colleagues reported that 1 set of resistance exercise at 80% of maximal strength 1 time per week for 12-weeks resulted in a 8% increase in lumbar extensor CSA while three times per week increased CSA 15% [15]. These morphological increases were accompanied with concomitant peak torque improvements of 31-43%. While it is difficult to determine the true effect of EPF on these morphological adaptations without having a comparison group performing the same protocol without restriction, these observations do suggest that lumbar muscle growth can be induced with a relative small degree of muscle overload when training with EPF is performed. More recently, Danneels et al. (2001) reported CSA changes in the multifidus muscle following 10-weeks of exercise training in chronic low back pain patients [16]. They found that dynamic core stability exercises such as leg extension in the four point kneeling position and trunk extension exercise without EPF did not significantly increase multifidus CSA, but when the same exercises were performed with a 5-sec isometric contraction separating the concentric and eccentric contraction phases they observed that multifidus CSA increased ~ 7%. Interestingly, the static component that they observed to be critical for compensatory muscle growth could in effect be construed as 'pelvic fixation,' albeit not external, since the isometric contractions are indeed non-moving and perhaps serving to minimize synergistic contribution from accessory muscles.

MUSCLE ACTIVATION PATTERNS DURING TRUNK EXTENSION EXERCISE

While the efficacy of training paradigms can only directly be determined with a clinical outcome trials study design, one can glean information on the effectiveness of different exercise tasks from acute studies evaluating the muscle activation patterns of synergistic muscles. The most common way to assess muscle activation patterns is through the recording and evaluation of the electromyogram (EMG) signal which is well-documented to have an increased amplitude associated with the recruitment of motor units and increases in discharge rate [17]; although another method to evaluate muscle activity is though the analysis of the transverse relaxation time (T2) obtained from magnetic resonances images, as a slowing in the T2 following exercise is associated with the intensity of skeletal muscle activation [18, 19].

As stated above, it appears that EPF is required to increase voluntary force output from the lumbar extensor muscles [11], thus, one might expect to find greater levels of activation during lumbar extension exercises with EPF than without. A recent report from San Juan and colleagues investigated this potential with asymptomatic subjects, and indeed did report that submaximal lumbar extension exercise (load equal to 50% of maximal strength) with EPF resulted in a 51% greater amplitude of the surface EMG signal recorded from the lumbar paraspinals when compared to the EMG amplitude recorded from trunk extension without EPF (no differences were observed with respect to biceps femoris activity between the two conditions) [20]. While this finding does suggest that EPF is needed for optimal lumbar extensor activation, discrepant results do exist. For instance, Udermann and colleagues reported that pelvic fixation did not influence the muscle activation of the lumbar extensors during submaximal dynamic contractions [21], however, it is possible that the restraint condition imposed in this study did result in some pelvic restriction as they did apply restraint devices, but did not completely tighten them (as opposed to the San Juan paper which did not use any physical restraints to restrict pelvic motion).

Our own data evaluating lumbar muscle activity during resistance exercise protocols has also addressed this question. For example, we have published several studies where we observed that during dynamic trunk extension without EPF, the EMG activity of the hip extensors (gluteus maximum and biceps femoris) increased substantially with loading and muscle fatigue suggesting that these muscles are highly responsible for accommodating the required muscle force production [22-24]. One of these studies reported a decrease in lumbar EMG activity occurred when subjects reached around 55% of their maximal fatigue capacity (the total number of repetitions performed volitionally), and that a concomitant increase in hip extensor activity resulted allowing for the continuation of the exercise task [23]. While this fatigue-induced alteration in lumbar and hip extensor muscle activation patterns has not been investigated under a condition of EPF it is likely that the synergistic contribution from the hip extensors would be minimized and thus the task duration shortened. In a different study we reported that when greater external loads were added to the lumbar extensors *without* EPF, EMG amplitude increased to a greater extent in the larger gluteals and biceps femoris musculature [22]. This finding is corroborated by findings from Mayer and colleagues indicating that MRI signal intensity of the intrinsic lumbar extensor muscles does not increase with an increase in exercise intensity from 50% to 70% of total trunk extension strength [25]. Collectively these findings suggest that when lumbar extension resistance exercise is performed *without* external pelvic fixation the larger hip extensor muscles are more responsible than the lumbar musculature for accommodating an increased load [22, 25].

SUMMARY

Compound lumbar-pelvic rhythm is responsible for trunk extension during normal daily activities when the pelvis is not restrained [4,26,27,28]. Thus, one might argue that since EPF does not occur in activities of daily living that performing more practical exercises that mimic the activities of daily living, functional tasks or even pain management classes is a more effective form of intervention [4, 36]. However, the scientific evidence suggests that this approach primarily targets the larger, more powerful gluteal and hamstring (global) muscles, and that the smaller, less leveraged, (specific) lumbar muscles remain in a relative state of disuse and/or poor neural activation [5,29,31,32]. Conversely, when EPF is employed gains in lumbar muscle strength and function are observed [4,5,6,10,11,12,13,30]. Neural activation also appears to be enhanced as Graves and colleagues noted in their study [4]. They surmised that the relatively large increases in strength, which resulted from only 12 exercise bouts was most likely associated with a large neural component [4]. Similarly, Nelson and others found that 94% of patients who received good or excellent results during a training study were able to maintain their achieved training results even after a 1 year post-discharge [33]. Apparently, the musculoskeletal dysfunction had sufficiently diminished allowing normal neuromotor activity and functional tasks to be performed [34]. Certainly a large portion of the resistance exercise training induced adaptations are a result of neural adaptations, and the role of neuromotor control in both the pathogenesis and rehabilitation of low back pain has long been of interest. A recent meta-analysis investigating the effectiveness of motor control exercises in improving clinical severity of persistent low back pain concluded that motor control exercise training alone is superior to minimal intervention, but that it is not more effective than manual therapy or other forms of exercise [35]. Accordingly, they suggested a multi-modal approach incorporating different exercise prescriptions (e.g., resistance exercise and motor control training) along with manual therapies may be most effective in treating persistent low back pain [35]. Accordingly, the functional relevance of successfully targeting and training the lumbar extensor muscles via EPF (isolation) is suggested as an aspect of this multi-modal approach.

REFERENCES

[1] McGill, S.M. Low back exercises: evidence for improving exercise regimens. *Phys. Ther.* 1998; 78: 754-65.

[2] Bigos, S., O. Bowyer, and G. Braen, *Acute low back problems in adults*, P.H.S. US Department of Health and Human Services, Agency for Health Care Policy and Research, Editor. 1994, Clincal Practice Guideline No. 14, Rockville, MD.

[3] Choi, G., P.P. Raiturker, M.J. Kim, D.J. Chung, Y.S. Chae, and S.H. Lee. The effect of early isolated lumbar extension exercise program for patients with herniated disc undergoing lumbar discectomy. *Neurosurgery*. 2005; 57: 764-72; discussion 764-72.

[4] Graves, J.E., D.C. Webb, M.L. Pollock, et al. Pelvic stabilization during resistance training: its effect on the development of lumbar extension strength. *Arch. Phys. Med. Rehabil.* 1994; 75: 210-5.

[5] Pollock, M.L., S.H. Leggett, J.E. Graves, A. Jones, M. Fulton, and J. Cirulli. Effect of resistance training on lumbar extension strength. *Am. J. Sports Med.* 1989; 17: 624-9.

[6] Carpenter, D.M. and B.W. Nelson. Low back strengthening for the prevention and treatment of low back pain. *Med. Sci. Sports Exerc.* 1999; 31: 18-24.

[7] Fleck, S. and W. Kraemer, *Designing resistance training programs* 1987, Champaign, IL: HUman Kinetics.

[8] O'Shea, P. Effect of selected weight training programs on the development of strength and muscle hypertrophy. *Res. Q.* 1966; 37: 95-102.

[9] Mayer, J.M., B.E. Udermann, J.E. Graves, and L.L. Ploutz-Snyder. Effect of Roman chair exercise training on the development of lumbar extension strength. *J. Strength Cond. Res.* 2003; 17: 356-61.

[10] Carpenter, D.M., J.E. Graves, M.L. Pollock, et al. Effect of 12 and 20 weeks of resistance training on lumbar extension torque production. *Phys. Ther.* 1991; 71: 580-8.

[11] Graves, J.E., M.L. Pollock, D. Foster, et al. Effect of training frequency and specificity on isometric lumbar extension strength. *Spine.* 1990; 15: 504-9.

[12] Risch, S.V., N.K. Norvell, M.L. Pollock, et al. Lumbar strengthening in chronic low back pain patients. Physiologic and psychological benefits. *Spine.* 1993; 18: 232-8.

[13] Holmes, B., S.H. Leggett, and V. Mooney. Comparison of female geriatric lumbar extension strength: asymptomatic versus chronic low back pain patients and their responses to active rehabilitation. . *J. Spinal Disorders.* 1996; 9: 17-22.

[14] Mooney, V., J. UGulick, and M. Perlman. Relationships between myoelectric activity, strength, and MRI of the lumbar extensor muscles in back pain patients and normal subjects. *J. Spinal Disorders.* 1997; 10: 348-356.

[15] Li, Y., L. Ploutz-Snyder, J. Graves, and J. Mayer. Neuromuscular adaptations to lumbar extension strength gains. *Med. Sci. Sports Exerc.* 1998; 30: S207.

[16] Danneels, L.A., G.G. Vanderstraeten, D.C. Cambier, et al. Effects of three different training modalities on the cross sectional area of the lumbar multifidus muscle in patients with chronic low back pain. *Br. J. Sports Med.* 2001; 35: 186-91.

[17] Basmaijian, J.V. and C.L. DeLuca, *Muscle Alive- their functions revealed by electromyography*. 1985, Baltimore: Williams and Wilkins.

[18] Adams, G.R., M.R. Duvoisin, and G.A. Dudley. Magnetic resonance imaging and electromyography as indexes of muscle function. *J. Appl. Physiol.* 1992; 73: 1578-83.

[19] Yue, G., A.L. Alexander, D.H. Laidlaw, A.F. Gmitro, E.C. Unger, and R.M. Enoka. Sensitivity of muscle proton spin-spin relaxation time as an index of muscle activation. *J. Appl. Physiol.* 1994; 77: 84-92.

[20] San Juan, J.G., J.A. Yaggie, S.S. Levy, V. Mooney, B.E. Udermann, and J.M. Mayer. Effects of pelvic stabilization on lumbar muscle activity during dynamic exercise. *J. Strength Cond. Res.* 2005; 19: 903-7.

[21] Udermann, B.E., J.E. Graves, R.G. Donelson, L. Ploutz-Snyder, J.P. Boucher, and J.H. Iriso. Pelvic restraint effect on lumbar gluteal and hamstring muscle electromyographic activation. *Arch. Phys. Med. Rehabil.* 1999; 80: 428-31.

[22] Clark, B.C., T.M. Manini, J.M. Mayer, L.L. Ploutz-Snyder, and J.E. Graves. Electromyographic activity of the lumbar and hip extensors during dynamic trunk extension exercise. *Arch. Phys. Med. Rehabil.* 2002; 83: 1547-52.

[23] Clark, B.C., T.M. Manini, and L.L. Ploutz-Snyder. Derecruitment of the lumbar musculature with fatiguing trunk extension exercise. *Spine.* 2003; 28: 282-7.

[24] Clark, B.C., T.M. Manini, and L.L. Ploutz-Snyder. Fatigue-induced changes in phasic muscle activation patterns during dynamic trunk extension exercise. *Am. J. Phys. Med. Rehabil.* 2007; 86: 373-9.

[25] Mayer, J.M., J.E. Graves, B.C. Clark, M. Formikell, and L.L. Ploutz-Snyder. The use of magnetic resonance imaging to evaluate lumbar muscle activity during trunk extension exercise at varying intensities. *Spine.* 2005; 30: 2556-63.

[26] Nelson JM, Walmsley RP, Stevenson JM: Relative lumbar and pelvic motion during loaded spinal flexion/extension. *Spine*. 1995; 20:199-204.

[27] Neumann, Donald A, *Kinesiology of the musculoskeletal system: foundations for physical rehabilitation*. 2002, St. Louis: Mosby.

[28] Farfan HF. Muscular mechanism of the lumbar spine and the position of power and efficiency. *Ortho. Clinic. N. Am.* 1975; 6:135-44.

[29] Richardson, Carolyn., Hodges, Paul W., Hides, Julie, Therapeutic exercise for lumbopelvic stabilization-a motor control approach for the treatment and prevention of low back pain. 2nd Ed., 2004, Edinburgh: Churchill Livingstone.

[30] Deutsch FE. Isolated lumbar strengthening in the rehabilitation of chronic low back pain. *J. Manip. Physiol. Ther.* 1996; 2:124-33.

[31] Hides JA, Richardson CA, Jull GA. Multifidus muscle recovery is not automatic after resolution of acute, first-episode low back pain. *Spine*. 1996;21:2763-2769.

[32] O'Sullivan PB, Phyty GD, Twomey LT, et al. Evaluation of specific stabilization exercise in the treatment of chronic low back pain with radiological diagnosis of spondylolysis or spondylolisthesis. *Spine*. 1997;22:2959-2967.

[33] Nelson BW, O'Reilly E, Miller M, Hogan M, et al. The clinical effects of intensive, specific exercise on chronic low-back-pain - a controlled-study of 895 consecutive patients with 1-year follow-up. *Orthopedics*. 1995;18(10):971-981.

[34] Mooney V. Research justifies specific spinal training. Retrieved September 10, 2008, from – http://www.biotech.com/db_area/archives/2000/0009sports.41-52.bio.html.

[35] Macedo LG, Maher CG, Latimer J, McAuley JH. Motor control exercise for persistent, nonspecific low back pain: a systematic review. *Physical Therapy*. 2009;89(1):9-25.

[36] Critchely DJ, Ratcliffe J, Noonan S, et al. Effectiveness and cost effectiveness of three types of physiotherapy used to reduce chronic low back pain disability: a pragmatic randomized trial with economic evaluation [with consumer summary]. *Spine*. 2007;32:1474-1481.

In: Biomechanics: Principles, Trends and Applications
Editor: Jerrod H. Levy, pp. 305-313
ISBN: 978-1-60741-394-3

Chapter 12

BONE CELL ADHESION: AN IMPORTANT ASPECT OF CELL BIOMECHANICS IN THE DEVELOPMENT OF SURFACE MODIFICATIONS FOR ORTHOPAEDIC IMPLANTS

Andreas Fritsche[1], Frank Luethen[2], Barbara Nebe[2], Joachim Rychly[2], Ulrich Lembke[3], Carmen Zietz[1], Wolfram Mittelmeier[1] and Rainer Bader*[1]

[1]Department of Orthopaedics, University of Rostock, Medical Faculty, Germany
[2]Department of Cell Biology, University of Rostock, Medical Faculty, Germany
[3]DOT GmbH, Rostock, Germany

ABSTRACT

Most revisions of total joint replacements are due to implant loosening, which is mainly caused by wear particles ("wear disease") and inadequate primary implant stability. The optimised integration of cementless total hip and knee endoprostheses into the bone stock is the most adequate approach to achieve secondary implant stability and to prevent implant loosening. Secondary stability is characterized by bone ingrowth of the implant and decreases the amount of relative implant motion between the implant and bone stock. It has also been suggested that prostheses which are fully occupied by bone cells are less susceptible to infection. The economic impact of implant loosening is immense, hence orthopaedic implant manufactures refine their products continuously.

Many technical developments have improved the survival rate of endoprosthetic implants. Modern materials and surface modifications such as coatings help to reduce wear rates, promote cell ongrowth or prevent infections. The cell adhesion of bone cells onto implant surfaces has not been thoroughly investigated so far. However, different methods to measure cell adhesion have been described. Some workgroups investigate short-term adhesion or proliferation of bone cells on implant materials *in-vitro*, but little is known about the long-term adhesion. Proliferation or short-term adhesion cannot

* Tel.: +49 (0)381 494 9337, Fax: +49 (0)381 494 9308, Email: rainer.bader@med.uni-rostock.de

predict how strong the bonding between bone and implant will be. In most cases, cost intensive animal studies have to be performed in order to gain expressive data. Hence, it is important to assess the bone cell adhesion forces in an adequate experimental setup *in-vitro*.

The exploration of bone cell adhesion on surfaces of orthopaedic implants encourages the development of bio-compatible, bio-active and anti-infectious surfaces. We have developed a test device, based on the spinning disc principle, which allows quantitative measurements of osteoblastic cells on implant surfaces. First results show differences in adhesion forces depending on the substrate. In future assessments different bio-active and anti-infectious surface modifications will be analyzed regarding bone cell adhesion prior to animal studies.

INTRODUCTION

Today, approximately 10 % of all total hip and knee endoprostheses need to be revised after a 10 year period [1,2]. However, implant survival rates of more than 20 years have been reported. The most common reason for implant failure is aseptic loosening mainly caused by particle wear ("wear disease") [3,4]. According to Malchau et al. [1] almost 75 % of all hip endoprostheses revisions are due to aseptic loosening. Other complications such as infections (7 %), dislocations (6 %), fractures (5 %) [1] or implant allergies [5,6] occur less often but can also lead to failure of the implant.

Insufficient primary and secondary implant stability can compromise the survival rate of cementless implants [7]. Primary stability is the initial, post-operative fixation strength with which the endoprostheses is held into place due to a press-fit force exerted by the bone. Secondary stability is characterized by bone ingrowth of the implant and decreases the amount of relative implant motion between the implant and the bone stock. The optimised osseous integration of cementless hip and knee endoprostheses is the most adequate approach to achieve secondary implant stability and to prevent implant loosening [8]. It has also been suggested that endoprostheses which are fully occupied by bone cells are less vulnerable to infection [9].

Besides the risk that patients have to take undergoing surgical treatment, the economic impact of implant revision is immense. About 79,100 total hip and knee arthroplasties were revised in the United States in the year 2005 [10], which amount to treatment expenses and an economic damage of approximately 2.4 billion US-Dollars [11] per year. Infections associated with joint replacement result in 0.36 billion US-Dollars of medical and surgical treatment expenses annually [11].

In total joint replacement much effort has been put into the increase of implant survival rates and hence, the reduction of implant revisions due to aseptic loosening. Since medical implant materials are very demanding regarding biocompatibility and mechanical strength [12,13] the variety of possible materials is highly limited. Therefore, the main emphasis for endoprostheses improvement focuses on surface modifications. Coatings are already successfully applied on implant-tissue interfaces in order to increase wear resistance [14], promote osseous integration or to reduce the release of allergenic ions [6] (eg. Ni^{2+}, Co^{2+}, Cr^{+}). A commonly used surface coating for wear reduction as well as lowering allergenic ion release is the titanium nitride (TiN) coating [15]. Newly developed coatings such as calcium phosphate (CaP) [16] or plasma polymerised allylamine (PPAAm) [17] strive to improve bony ongrowth, while others attempt to reduce infections [18] eg. by integrating copper in titanium dioxide ($TiO2-Cu^{2+}$) [19,20] or biodegradable PDLLA (poly d, l-acid) with antibiotics [21].

In general, cell interaction with the implant interface plays an important role in surface coating development. Cell spreading [22] and proliferation tests are carried out to acquire biocompatibility data of new surface modifications [23]. But proliferation or even short-term adhesion assays [24] cannot predict how strong the bonding between bone and implant will be. Extra cellular matrix (ECM) proteins such as collagens, elastins, proteoglycans, and adhesion proteins such as fibronectins and laminins [25] bind to a biomaterial surface within a few minutes and enable cells to adhere to the implant surface. So far, expressive data of bone cell adhesion on orthopaedic implants can only be obtained in cost intensive animal studies. Hence, it is useful to assess the bone cell adhesion forces in vitro in an adequate experimental setup to predict bone ongrowth and accompany the early stages of implant surface modification development.

Adhesion Assays

Different experimental setups to measure cell adhesion have been proposed. The simplest method is the "wash off" principle (Figure 1a) for short-term adherent cells (less than 60 min) [26]. Non-adherent cells are washed off from the implant surface and can be counted subsequently. This assay has no means of adhesive strength quantification. However, by applying centrifugal forces in a centrifuge (Figure 1b) the adhesion forces can be determined. This low-cost and easy to operate test setup [27,28,29,30] is only applicable to single cells or cell populations (biofilm) that have bonded to the substrate for a short period of time, since the utilised forces are not sufficient enough to detach long-term adherent cells. Nonetheless, bonding strength of long-term, as well as short-term, adherent cells can be determined using micromanipulative assays such as micropipettes (Figure 1c), laser tweezers or atomic-force-microscope (AFM) cantilevers (Figure 1d) [30,31,32,33,34,35,36]. Normal and tangential forces can be applied in order to detach adherent cells from the substrate. Although, results are precise, only single cell measurements are possible and assessment of biofilms is impossible using these techniques and apparatus which are cost-intensive and require specialised operating personnel.

Measurements of short-term as well as long-term adhesive single cells or of cell populations can be achieved using hydrodynamic flow assays. The relatively simple test setup is based on a flow chamber in which a fluid flows over adherent cells on a smooth surface. At a given flow pattern the generated shear stress τ_w (force/area), respectively the resulting shear force, that is exerted on the cells can be calculated. Especially at high flow rates it is required to describe the flow patterns exactly since turbulences can arise and distort the results. Furthermore, the shear force depends on cell size and shape [37,38]. Hydrodynamic flow assays are classified by the type of flow generation or the build-up of the flow chamber. It can be distinguished between linear flow between two parallel plates (Figure 1e), which has been deployed successfully for cell adhesion [39,40], radial flow between two fixed, parallel circular discs (Figure 1f) and flow in the laminar layer on the surface of a rotating disc (spinning disc principle) [26].

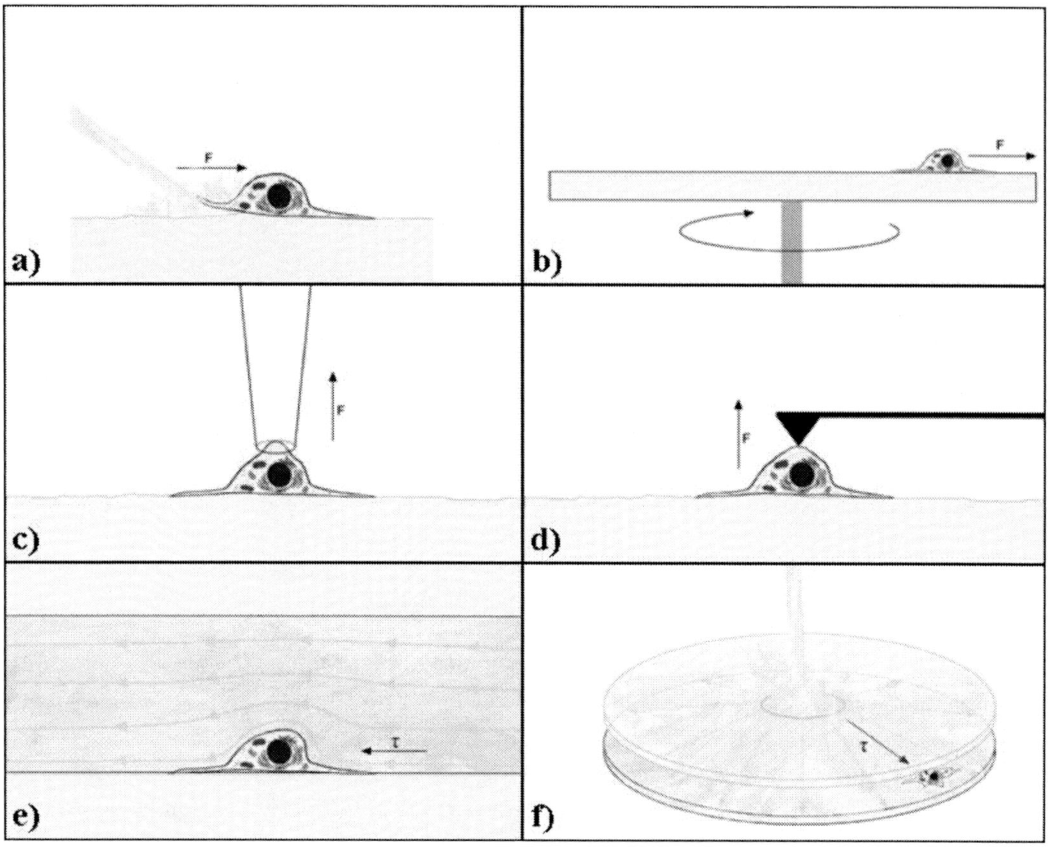

Figure 1. Cell adhesion assays a) "wash of" principle, b) centrifuge, c) micropipette, d) atomic force microscope cantilever, e) linear flow between two parallel plates and f) radial flow between two fixed parallel circular discs.

Approach to Measure Bone Cell Adhesion for Orthopaedic Implant Surfaces

Since little is known about long-term cell adhesion forces of bone cells on implant surfaces relevant for orthopaedic applications, we have developed an experimental adhesion measuring device based on the spinning disc principle [26]. The spinning disc principle describes a rotating disc in a resting fluid where the applied shear stress $\tau_w = r \cdot \omega^{3/2}$ depends on the rotational velocity ω and the radial distance r from the centre of the disc (Figure 2). Furthermore, the shear stress depends on the density ρ and the viscosity η of the fluid used for the experiments. Laminar flow conditions on the surface of the disc even at high rotational velocities are convenient in the attempt to measure cell adhesion. Another advantage is that forces can be applied under constant uniform chemical and biological conditions since the disc can rotate in cell culture medium [30].

Our measuring device consists of a stainless steel (V2A) shaft, a fluid chamber made of polyetheretherketone (PEEK), a glass bottom, an electric motor and a disc shaped test sample (Figure 3).

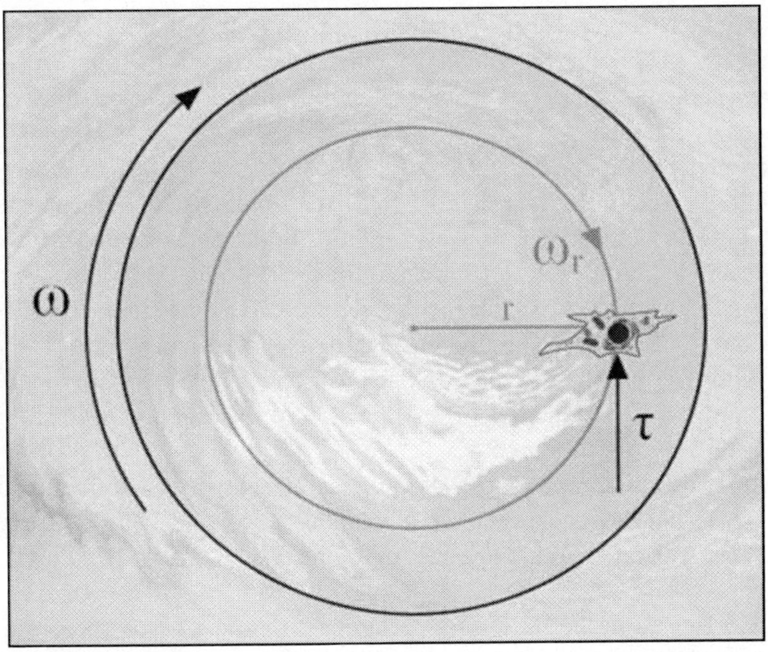

Figure 2. Spinning disc principle.

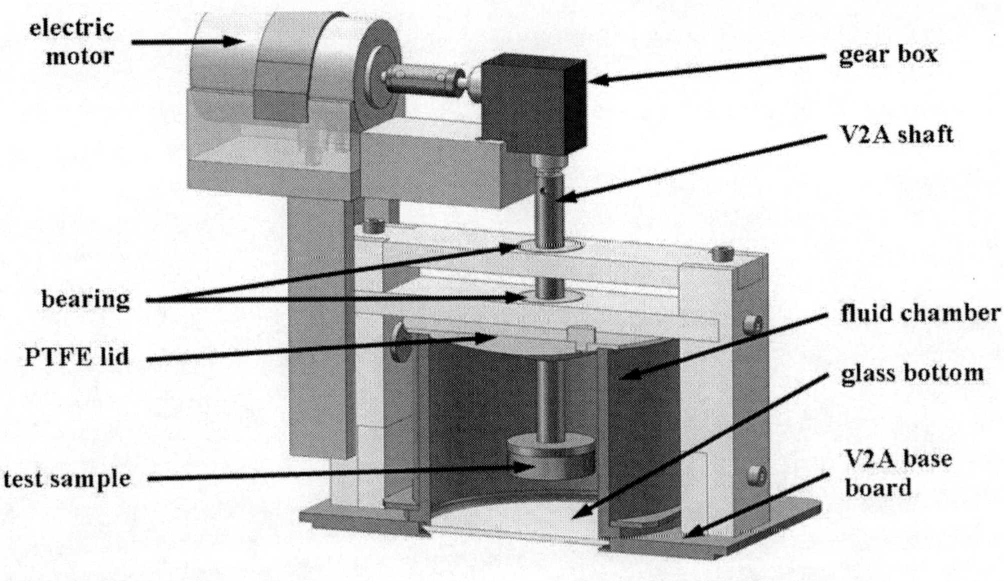

Figure 3. Adhesion measuring device adapted to be used in conjunction with a LSM shown schematically.

The apparatus was designed to be placed on an inverse confocal laser scanning microscope (LSM) so that cells stained with a fluorescent dye can be observed through the glass bottom before and after rotation.

The test setup was built in such a way that the shaft can be moved towards the glass bottom for microscopy and away for disc rotation. It is important to be far away from the bottom of the fluid chamber during rotation in order to avoid interference of the arising convection current [41] with the laminar boundary layer at the disc surface. The test specimens were marked by a laser at different radial positions, so that distinct radial positions could be retraced by microscopy.

First experiments were carried out using MG-63 cells, a human osteosarcoma cell line (CRL-1427, ATCC, Rockville, Maryland, USA), with polished disc shaped test specimens made of titanium alloy (Ti6Al4V) as well as Ti6Al4V samples coated with PPAAm. The bone cells were stained with PKH26 (Sigma-Aldrich, St. Louis, MO, USA) and seeded onto the specimens in 2 µl droplets of Dulbecco's Modified Eagle Medium (DMEM) with 10 % fetal calf serum (FCS, Gibco Invitrogen, Karlsruhe, Germany) and 1 % Gentamicin (Ratiopharm, Ulm, Germany) at defined radial positions. The cells were incubated at 37 °C and 5 % CO_2 and allowed to settle for 20 min before the samples were fully covered by medium.

After a further incubation period of 18 h the test specimens were fixed onto the shaft of the adhesion measuring device and the fluid chamber was filled with DMEM without FCS. Subsequently, the apparatus was mounted onto the LSM (LSM410, Zeiss, Oberkochen, Germany) (Figure 4).

The disc was rotated in the fluid for a period of 3 min at a given speed after moving the sample away from the glass bottom. Imaging of the radial positions was performed before and after rotation at the marked radial positions (Figure 5).

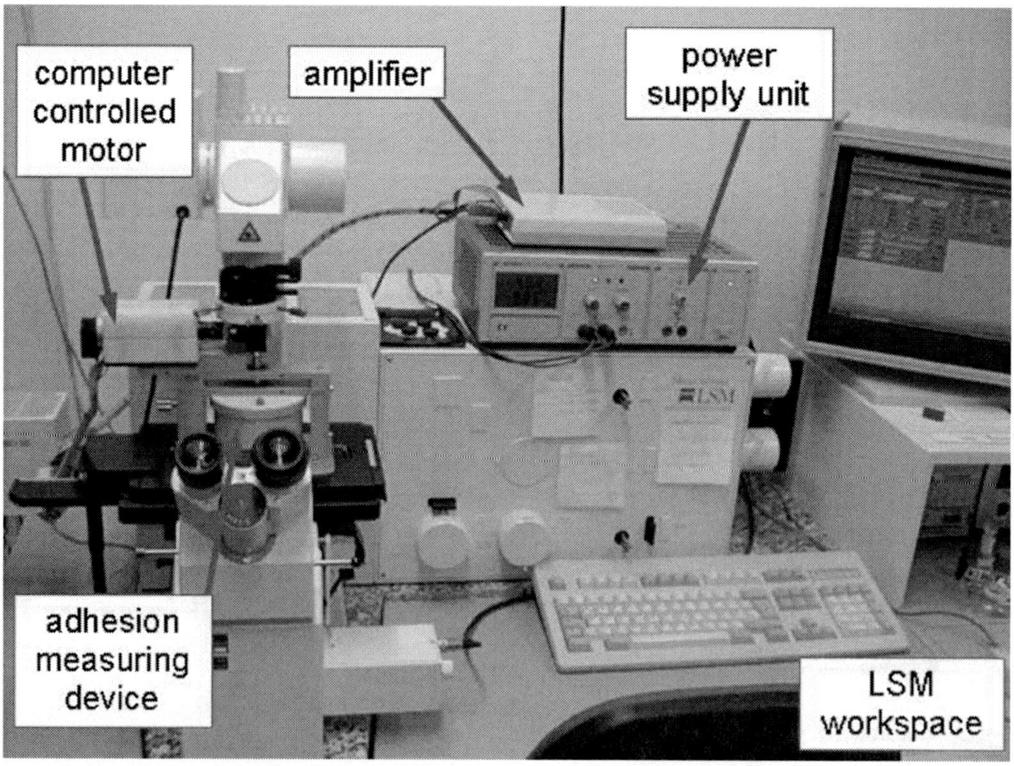

Figure 4. Adhesion measuring device mounted on LSM.

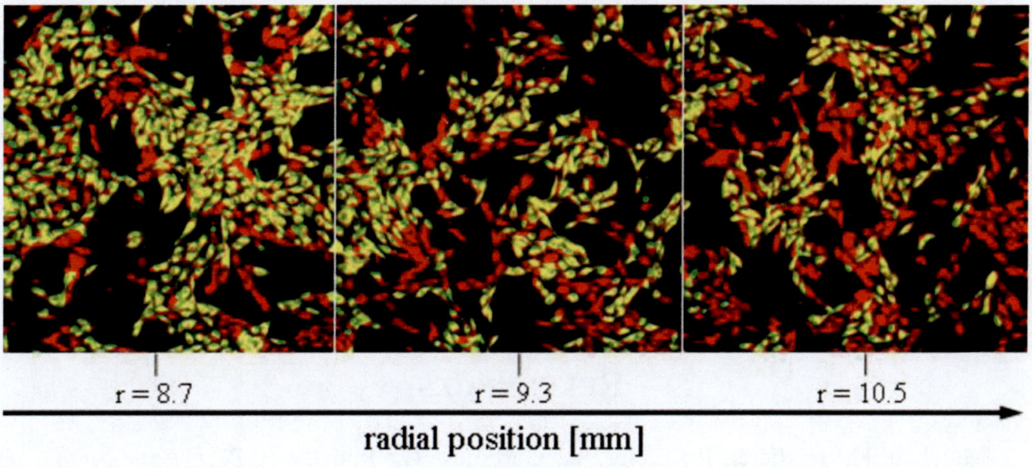

Figure 5. Miscoloured display of MG-63 after rotation showing adherent cells before (red) and after rotation (yellow) at different radial positions.

The resulting shear stress τ_{res} for the spinning disc at the radial position, where 50 % cell detachment was detected, was calculated using equation 1.

$$\tau_{res} = 0.79981 \cdot r\sqrt{\left(\eta\rho\omega^3\right)} \qquad \text{(equation 1)}$$

The obtained results show that the adhesion measurements are reproducible and comparable to bonding strengths found in literature [30,42]. Moreover, a significant increase in cell adhesion strength of MG-63 cells on a PPAAm surface coating compared to a polished Ti6Al4V surface was detected. Animal studies show similar result outcomes. Further exploration of bone cell adhesion on bio-active and anti-infectious orthopaedic implant surfaces is currently in progress.

CONCLUSION

The economic impact of orthopaedic implant revisions due to implant failure is immense which encourages further improvement of endoprostheses survival rates. Because potential biocompatible materials are limited for orthopaedic applications, surface modifications have become the focus of modern implant research. For higher survival rates sufficient osseointegration (i.e. secondary implant stability) is essential. *In-vitro* cell adhesion tests could indicate the potential of newly developed implant surface modifications regarding bone ongrowth in early stages of the development.

To gain expressive test results different cell adhesion assays can be used, each with advantages and disadvantages. We have decided on a rotating disc in a resting fluid (i.e. spinning disc principle). This method is sensitive enough to show differences in cell adhesion between different types of surface modifications. The results are reproducible and comparable to those obtained by other workgroups. However, comparison and interpretation of the results among different workgroups has to be performed carefully, since cell adhesion depends on many influencing factors. Different cell types adhere with different strengths but also the **specimen's material, adhesion time period (i.e. incubation)** or experimental parameters alter the outcome of the results.

Further measurements on different bio-active, anti-infectious and anti-adhesive surfaces and materials are being carried out. These investigations include the analysis of cell adhesion time and different bone cell types (primary osteoblasts, mesenchymal stem cells). Furthermore, the influence of metal ions (e.g. Cu^{2+}) and the significance of specific integrins on cell adhesion by blocking different integrin receptors will be studied extensively.

REFERENCES

[1] Malchau, H.; Herberts, P.; Eisler, T.; Garellick, G.; Söderman, P. *J. Bone Joint Surg. Am.* 2002, *84*, 2-20.

[2] Robertsson, O.; Knutsen, K.; Lewold, S.; Lidgren L. *Acta Orthop. Scand.* 2001, *72*, 503-513.

[3] Zeiler, G.; Distler A. *Biomaterialien* 2000, *1*, 19-24.

[4] Schuh, A.; Thomas, P.; Holzwarth, U.; Zeiler, G. *Orthopäde* 2004, *33*, 727-731.

[5] Holzwarth, U.; Thomas, P.; Kachler, W.; Göske, J.; Schuh, A. *Orthopäde* 2005, *34*, 1046-1051.

[6] Schuh, A.; Thomas, P.; Kachler, W.; Göske, J.; Wagner, L.; Holzwarth, U.; Forst, R. *Orthopäde* 2005, *34*, 327-328.

[7] Doehring, T. C.; Rubash, H. E.; Dore, D. E. *Clin. Orthop. Relat. Res.* 1999, *362*, 230-239.

[8] Widmer, K.-H.; Zurfluh, B.; Morscher, E. W. *Orthopäde* 1997, *26*, 181-189.

[9] Gristina, A. G.; Naylor, P.; Myrvik, Q. *Med. Prog. Technol.* 1988-1989, *14*, 205-224.

[10] Kurtz, S.; Ong, K.; Lau, E.; Mowat, F.; Halpern, M. *J. Bone Joint Surg. Am.* 2007, *89(4)*, 780-785.

[11] Darouiche, R. O. *N. Engl. J. Med.* 2004, *350(14)*, 1422-1429.

[12] Steinau, H.-U.; Schackert, H. K.; Bauer, H. In *Chirurgisches Forum 2007 für experimentelle und klinische Forschung*; Springer: Berlin, (DE); 2007.

[13] Thull, R. *Orthopäde* 2003, *32*, 51-59.

[14] Cook, G.; Costerton, J. W.; Darouiche, R. O. *Int. J. Antimicrob. Agents* 2000, *13(3)*, 169-173.

[15] Neumann, H.-G.; Baumann, A.; Wanke, G.; Hamelynck, K. J.; Morlock, M. *Key Eng. Mater* 2002, *218-220*, 601-604.

[16] Finke, B.; Lüthen, F.; Schröder, K. et al *Biomaterials* 2007, *28*, 4521-4534.

[17] Becker, P.; Neumann, H.G.; Nebe, B. et al *J. Mater Sci. Mater Med.* 2004, *15*, 437-440.

[18] Fritsche, A.; Haenle, M.; Mittelmeier, W. et al *Mat-wiss u Werkstofftech* 2008, *39(9)*, 659-664.

[19] Heidenau, F.; Stenzel, F.; Schmidt, H. et al *Biomaterialien* 2001, *2*, 19-24.

[20] Heidenau, F.; Mittelmeier, W.; Detsch, R. et al *J. Mater. Sci. Mater. Med.* 2005, *16*, 883-888.

[21] Gollwitzer, H.; Thomas, P.; Diehl, P. et al *J. Orthop. Res.* 2005, *23(4)*, 802-809.

[22] Lüthen, F.; Lange, R.; Becker, P.; Rychly, J.; Beck, U.; Nebe, J.G. *Biomaterials* 2005, *26(15)*, 2423-2440.

[23] Nebe, J.G.; Luethen, F.; Lange, R.; Beck, U. *Macromol. Biosci.* 2007, *7(5)*, 567-578.

[24] Nebe, B.; Finke, B.; Lüthen, F.; Bergemann, C.; Schröder, K.; Rychly, J.; Liefeith, K.; Ohl, A. *Biomol. Eng.* 2007, *24(5)*, 447-454.

[25] Stevens, M.M; George, J.H. *Science* 2005, *310*, 1135-1138.

[26] Garcia, A.J.; Huber, F.; Boettiger, D. *J. Biol. Chem.* 1998, *273(18)*, 10988-10993.

[27] Chu, L.; Tempelman, L.A.; Miller, C.; Hammer, D.A. *AIChE Journal* 1994, *40*, 692-703.

[28] Reyes, C.D.; Garcia, A.J. *J. Biomed. Mater Res.* 2003, *67A*, 328-333.

[29] Giacomello, E.; Neumayer, J.; Colombatti, A.; Perris, R. *Biotechniques* 1999, *26*, 756-758.

[30] Garcia, A.J.; Gallant, N.D. *Cell Biochem. Biophys.* 2003, *39*, 61-73.

[31] Evans, E.; Berk, D.; Leung, A.; Mohandas, N. *Biophys.* 1991, *59*, 838-848.

[32] Litvinov, R.I.; Shuman, H.; Bennett, J.S.; Weisel, J.W. *PNAS* 2002, *99*, 7426-7431.

[33] McKeever, P.E. *J. Reticuloendothel Soc.* 1974, *16*, 313-317.

[34] Prechtel, K.; Bausch, A.R.; Marchi-Artzner, V.; Kantlehner, M.; Kessler, H.; Merkel, R., *Phys Rev Lett.* 2002, *89*, 28-101.

[35] Shao, J.Y; Hochmuth, R.M. *Biophys*. 2002, *77*, 587-596.

[36] Tozeren, A.; Sung, K.P.; Chien, S. *Biophys*. 1989, *55*, 479-487.

[37] Hammer, D.A.; Laufenburger, D.A. *Biophys.* 1987, *52*, 475-487.

[38] Xiao, Y.; Truskey, G. A. *Biophys*. 1996, *71*, 2869-2884.

[39] Truskey, G.A.; Pirone, J.S. *J. Biomed. Mater. Res.* 1990, *24*, 1333-1353.

[40] van Kooten, T.G.; Schakenraad, J.M.; van der Mei, H.C.; Busscher, H.J. *J. Biomed. Mater. Res.* 1992, *26*, 725-738.

[41] Piva, M.; Meiburg, E. *Phys. Fluids* 2005, *17,* 63603.

[42] Ma, G. ; Cai, N. ; Tan, S. ; Mhaisaklar, P.S. ; Chian, K.S.; Chan, V.; Liao, K. *J. Biomech.* 2006, *39 (Suppl 1)*, 226.

In: Biomechanics: Principles, Trends and Applications
Editor: Jerrod H. Levy, pp. 315-316
ISBN: 978-1-60741-394-3

Chapter 13

THE DIFFERENCES IN BIOMECHANICAL PATTERNS OF FAST MOTOR LEARNING OF CHILDREN AND ADULTS

***A. Skurvidas*[1]*, A. Zuoza*[1]*, B. Gutnik*[2] *and D. Nash**[2]**

[1]Lithuanian State Academy of Sport and Physical Education, Kaunas, Lithuania

[2]Unitec, Institute of Technology, Auckland, New Zealand

The aim of this work was to establish and compare the patterns expressed on the fast model of motor learning of children and adults executing a fast and accurate task.

The acquisition of a new motor skill follows two distinct stages with continued practice: first, there is an early, fast learning stage in which performance improves rapidly within a single training session; later, there is a slower learning stage within the time period of several sessions of practice (Ungerleider et al., 2002). Motor learning is characterized by a specific set of changes in performance parameters (Flament et al., 1999; Kempf et al., 2001). These changes occur gradually in the course of a learning period. While the decreases or increases in these parameters have been documented in a variety of tasks, it remains to be determined whether the time of fast learning is different for children and adults. Therefore the main aim of this study was to establish if there are differences in reaction time, average and maximal velocity, trajectory, and accuracy as well as the variability of these parameters during motor learning. The tasks involved 5 series with 20 repetitions in each.

For this experiment, we used a standard Analyzer DPA-1 measuring dynamic parameters of movements of different segments of the body. The subjects performed a complicated task. They had to react to a target on the computer screen as quickly as possible and follow the most accurate trajectory towards the target, stopping within it. We measured the specific parameters: reaction time (RT), average velocity (Va), maximal velocity (Vm), path of movement (S), the period of achievement of maximal velocity (Tv) and time from achievement of Vm to reaching the target.

---

* Email: dnash@unitec.ac.nz

We found a significant difference ($P<0.05-0.001$) in RT, Va, Vm and S between children and adults. One-way ANOVA showed a significant effect of time (five series: quick learning) in Vm and S of children ($P<0.01$) and in Va of adults ($P<0.05$). There was a significant difference in the changes of Va, Vm and S between children and adults. It is remarkable that S and Vm decreased significantly after the first series and did not change during the 2nd-5th series while Va increased gradually during all the five series in adults.

This is the first study to show, that the time of fast motor learning in a speed-accuracy task was different between children and adults. We showed that the level of accuracy was highly improved by the children at the expense of quickness of reaction, while adults improved only the average velocity of their performance. It might be considered that the fast motor learning process is expressed more in children than in adults. It is well established that children have different information-processing capabilities compared with adults (Wade 1976).

Besides, most of the indices of variability of performance changed more significantly in children than in adults. The reduction in performance variability was achieved along with increased accuracy in children It might be speculated that children are more prone to change motor variables during the fast learning process than adults. Though the underlying mechanism for the fast learning difference between children and adults has to be elucidated, it might be speculated that feed-forward and error correction mechanisms, as well as the development of internal models of movement, may be more changeable in children than in adults during the fast learning process.

In: Biomechanics: Principles, Trends and Applications
Editor: Jerrod H. Levy, pp. 317-351
ISBN: 978-1-60741-394-3

Chapter 14

APPLYING PRESSURE SENSORS AND SIZE DIFFERENCES IN RUNNING SHOES FIT MEASUREMENT

Y. L. Cheng and Y. L. Hong
Department of Sports Science and Physical Education
The Chinese University of Hong Kong, Hong Kong

INTRODUCTION

Background

Fit is one of the most critical factors affecting footwear comfort (Miller et al., 2000). Blistering, chafing, bunions and pain may be the result of poor fitting shoes. In long run, it may cause the foot skeleton deformity.

In order to find out the proper fitting of footwear, it involves getting to know the size of feet, shoes, and the subjective perception for the shoes selection. Traditional method in measuring the feet size is to measure the length and width of the feet which can be obtained easily by tape measure and devices like Brannock (The Brannock Device Co., Inc., USA). However these are considered to be insufficient for good footwear fitting (Witana et al., 2004). Furthermore, researchers were also encountering problem in quantifying fit as it is rather subjective which was also suggested to be affected by shoe wearing experience such as tightness and looseness of the shoes (Kouchi et al., 2005). Therefore researchers are exploring new method in measuring footwear fit, both objectively and subjectively.

3D technology has been applied to quantify footwear fit. Nacher et al. (2004) used 3D foot digitizing technology to predict fit by getting the foot geometrical characterization data and subjective fit perception from the users. Witana and the colleagues (2004) applied foot scanners in their study to get the 3D foot shape which was then post processed to obtain the 2D outlines, and the dimensional differences between the feet and shoe last were analyzed with the subjective fit perception. The dimensional differences plot allowed the shoe lasts designer to determine the critical position for fit and unfit. One limitation for the 3D technology is that feet and last alignment has to be done manually. The actual position of the feet in the shoes could not be ensured.

Apart from sizing, pressure distribution is likely to be a valid measurement for footwear fit. Dewan et al. (2004) placed pressure sensors between the foot-shoe interfaces for dynamic pressure measurement. The authors suggested that the pressure might have the implication for footwear fitting and they had also identified the feasibility of using pressure sensors for footwear fit. This was the only study that applied pressure sensors for fit measurement. However this study did not relate the subjective perception of fit with the loading and how each particular anatomical position of foot contributed to the overall fitting of the footwear. With the interest to explore this area, Gheorghiu and the colleagues (2004) proposed to investigate the relationship between pressure distribution and subjective fit rating.

The advantage of the pressure sensor application in footwear fit measurement is that it is the real time assessment of the foot – shoe interaction. Unlike 3D technology, manual alignment of feet and last was not required. The loading of the pressure sensors reflects the actual interaction of the feet and the shoes. The investigation of the loading at different foot anatomical position by using pressure sensors and by relating it to subjective fit perception, critical position for fit may be identified. It is also hoped that norm can be built to show how certain parts of the foot and shoe contribute to the overall fit feeling. However, despite the advantages of using pressure sensor, its result can only reflect the shoes tightness but not looseness. It is one of the limitations of pressure sensors application in fit measurement.

Subjective perception of fit of footwear had been measured in many studies (Gheorghiu et al., 2004; Kouchi et al. 2005; Witana, et al. 2004). However none of the questionnaires used above had been proved to be valid and reliable. Only one project had been done in measuring footwear comfort by Mundermann et al. (2002) that tried to validate a footwear comfort questionnaire. They tried to develop a reliable method to assess footwear comfort during running by the Visual Analogue Scales (VASs) (Mundermann et al., 2002). A protocol including a control condition during running was used and the resulting intraclass correlations between comfort ratings for repeated conditions were found to be high. The authors thus concluded that using VASs with a control condition in the protocol provided a reliable measure to assess footwear comfort during running.

Despite the importance of fit in footwear has been announced in a number of researches, only a few studies have been done on relating the shoe-foot size difference to the subjective fit measurement. For the pressure distribution, there is even no concrete finding on the association of the objective pressure measurement of the shoe-foot interface and the subjective fit rating. Therefore, it is worthwhile to explore the feasibility of the application of these objective measurements in measuring the subjective fit rating.

The purpose of this study was to find out the relationship between three variables: subjective fit rating, foot-last size difference and pressure distribution of foot-shoe interface. It is hoped that through the study of the two objective independent variables: foot-last size difference and pressure distribution, variances of subjective fit rating can be more thoroughly explained.

Furthermore, although previous studies have applied fit questionnaire in measuring subjective fit perception, none of the questionnaire has been proven reliable. Therefore another purpose of the present study is to test the reliability of a footwear fit questionnaire that applied VAS scale with the control condition (Mundermunn et al. 2002). The reliable fit questionnaire developed is necessary in quantifying subjective fit perception. Such information will be very useful in the area of research, production and sales in footwear industry.

Although footwear fit is of high importance, present devices available in measuring fit are obviously inadequate. Results of present study will help introduce new technology into this area.

The knowledge of different objective measures in predicting subjective fit rating is especially useful for those who are unable to express their subjective feel towards the fit of the footwear e.g. young children, diabetes patients and the elderly.

Operational Definition

Foot length the distance between pternion (the rearmost point of the heel) and the midpoint of the maximum breadth of the ball cross section. The ball cross section is the vertical cross section passing through MT (metatarsale tibiale: the most medially protruding point on the head of the 1st metatarsal bone) and MF (metatarsale fibulare: the most laterally protruding point on the head of the 5th metatarsal bone).

Foot circumference the circumference of the ball cross section. Ball cross section is the vertical cross section passing through MT. Foot breadth the maximum breadth of the ball cross section. Heel breadth the breadth of the heel measured perpendicular to the foot axis at the 16% of foot length from pternion.

Shoe last is the solid form around which a shoe is constructed resulting the inside shape of the shoe (Figure 1). And each shoe last is designed for a particular heel height, toe shape and type of footwear. Visual Analogue Scale (VAS) is a measurement instrument that tries to measure a characteristic or attitude that is believed to range across a continuum of values and cannot easily be directly measured (Gould et al., 2001). VAS that was used in this study is a 150mm horizontal line, anchored by word descriptors that enquire footwear fit perception at each end. The subject will mark on the line the point that they feel which represents their fit perception of their current state.

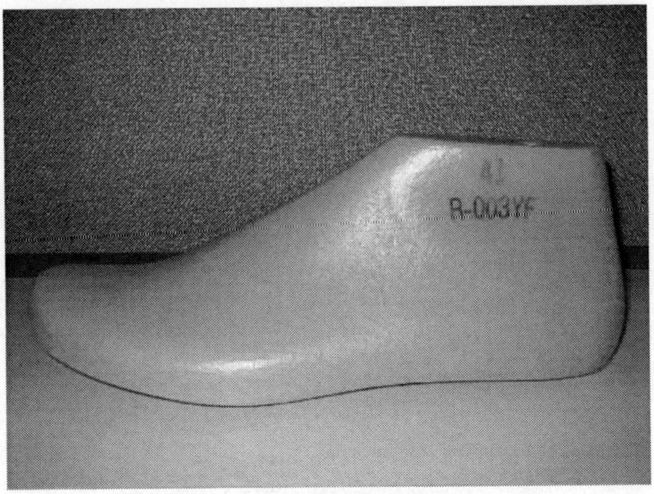

Figure 1. Shoe last.

Unlike the usual VAS used in other studies in which the VAS score was determined by measuring from the left hand end of the line to the point that the subject marks, the measurement starts in the middle in this study. The middle point of the line represents the optimum perception of fit.

REVIEW OF LITERATURE

Problem of Poor Fitting

Fit is one of the most critical factors affecting footwear comfort (Miller et al., 2000). Blistering, chafing, bunions and pain may be the result of poor fitting shoes. In long run, it may be the cause of foot skeleton deformity.

Research regarding footwear fit problem in elderly has been carried out, it was found that among the 65 elderly being investigated, 47 (72%) of them had ill fitting shoes (a discrepancy in length of more than half a British shoe size or more than one British width fitting, 7 mm). It was also found in the study that incorrect shoe length was significantly associated with the presence of ulceration and self reported pain (Burns, Leese and McMurdo, 2002). In addition, ill fitting shoes were also associated with the presence of corns on the toes, hallux valgus deformity and foot pain (Menz and Morris, 2005). Gender was a factor found to be associated with the prevalence of foot problem.

A study conducted by Frey and the colleagues (1993) evaluated the foot size and shoe preference, shoe comfort, shoe size and the presence of foot pain and foot deformities of 356 healthy women between 20 and 60 years of age. In that study, 80% of women reported significant foot pain while wearing shoes. 76% of the women had one or more forefoot deformities, with hallux valgus being the most common (71% of this group had hallux valgus, 50% had hammertoes, 17% had bunionettes, 13% had prominent metatarsal heads, and 4% had miscellaneous deformities). The overall incidence of hallux valgus was 54%.

The same study also showed that 88% percent of the women were wearing shoes that were smaller in width than their feet (average, 1.2 cm smaller). In the women who had no foot pain (20%), the average foot-to-shoe discrepancy was only 0.56 cm. In those women without any deformity (23%), the average foot-to-shoe discrepancy was 0.6 cm.

It is clear from the above study that the majority of women wore shoes that were small for their feet and they would suffer resultant pain and deformities. The patients in the study wore a shoe that was 1.3 to 2.5 cm smaller than the width of the forefoot in average. The patients without pain or deformity wore a shoe that was only smaller than the forefoot by 1.3 cm or less.

Many shoes for females are simply scaled down versions of the same shoe for males. In general, women's feet were in a smaller width to length ratio than male, and the heel is narrower in comparison with the forefoot (Frey, Thompson and Smith, 1995).

Not only gender plays a role in footwear fit, different foot size is also influential. A similar study conducted by Frey and colleagues (1995) found that women with larger feet would have more pain and deformities than the control group in the study. It was suggested that because women with larger feet might have more difficulties in finding shoe with proper fit than those with smaller feet. Women with wider forefeet tended to have wider heels. As foot length increased, both forefoot width and heel width increased but with the heel width increased to a lesser extend. However, as shoe length increased, all the key internal dimensions in the shoes were enlarged in a fixed proportion. Therefore in order for the women with larger feet to find the shoes with heel counter which could grip the heel firmly, smaller shoe should be selected and it might be too snug for the forefoot resulting in deformity and pain.

A study conducted by Nixon and the colleagues (2006) examined the shoes size and width, sex, presence of diabetes, neuropathy and foot wound, type and condition of shoes usually worn in order to find out the relationship of poorly fitting footwear with and without diabetic foot ulceration. 400 US veterans were recruited. Of this population, only 25.5% were wearing appropriately sized shoes. Patients with diabetic foot ulceration were 5.1 times more likely to have poorly fitting shoes than those without a wound. The authors thus suggested that greater accuracy in shoe-foot match could be useful in reducing the risk of foot ulceration.

"Diabetic foot ulcerations are the most common severe complications of diabetes". Ulcers form in patients with diabetes because of a lack of sensation, coupled with repetitive pressure forces (walking). One of the central tenets in reducing the incidence of ulcers is pressure reduction through the use of appropriate shoes and insoles. Poorly fitting shoes may account for a large proportion of diabetic foot ulcers and may also play a role in creating or exacerbating other complications in people without diabetes.

However, the result of a randomized controlled trial conducted by Reiber et al. (2002) seems to contradict the above wisdom. A two year randomized trial of 400 diabetes patients was conducted.

Participants were randomly assigned to receive customized shoes or their usual footwear. It was found that there were as many foot ulcers in specifically designed footwear compared with ordinary footwear. In respond to this finding, Chantelau (2002) commented the ways Reiber et al. (2002) analyzed the results. With a closer look to the result of the study, 9 out of 47 ulcer episodes occurred in well-fitted intervention shoes, compared to 19 out of 37 ulcer episodes occurring in normal shoes and these results were highly statistically significant. However these results were not reported by Reiber et al. (2002). Chantelau (2002) thus elaborated the finding to be supportive of the rationale that fit of footwear was important rather than oppositing it.

Chantelau and Gede (2001) investigated the foot size of the elderly people with and without diabetes mellitus in German. The result showed that the feet of most elderly people were broader than the 'normal' footwear that was currently available on the market. The authors thus suggested that a fixed breadth-to-length ratio for the footwear construction might not be enough to fit all the individual characteristics.

Apart from being one of the factors of foot deformities, fit also appears to govern comfort. If a pair of shoes were tight, the pressure produced between the foot shoe interface may cause pain to the wearer. On the other hand, if a shoe was too loose, the friction caused by the slippery effect would also produce the sense of discomfort. (Luximon, Goonetilleke and Tsui, 2003)

Definition of Fit

However the importance of footwear fit, it is not well defined. It is suggested that "proper shoe fit requires shape or last design with proper toe depth and shape, proper instep depth, proper heel width, and proper curve (flare) of the shoe" (Tremaine and Awad, 1998). However, this definition does not take into the account of individual characteristic of the human feet and it also does not point out the importance of matching between human feet and the shoes. As it was stated 'closeness of match of shoe shape to that of the human foot" was one of the most important factors for comfort of fit (Hawes et al., 1994).

One of the most common methods in measuring the fit of footwear was by finding out the size difference between foot and shoes. Menz and Morris (2005) applied this idea by assessing the footwear and foot dimensions and to find out their length, width and area. Ill fitting footwear was defined as having size smaller than the foot. Similar concept was applied by Nixon et al. (2006), inappropriately sized footwear was defined as a shoe size at least one full size too large or too small using a specific foot measuring device.

Tremaine and Awad (1998), on the other hand, did not regard smaller size of the shoes as unfit. Instead they proposed a range for fit should be equal or no more than 0.25 inch less that fore-foot width. Witana et al. (2004) proposed a linear regression to predict forefoot fit by the summation of the lateral and medial maximum dimensional difference of the forefoot. According to the equation, maximal dimensional difference should be about -8.5mm.

Goonetilleke (2003) summarized a few important factors for footwear fit measurement: dorsal arch height (a measure of foot height), plantar arch height, foot flare, and the angular orientation of the metatarsal; another combination of the foot factors are: length, flare, width and height. Although there are many guidelines available, the ultimate goal of footwear selection is to find out the right match between the foot and the footwear, as Janisse (1992) suggested: "proper fit is achieved when shoe shape is matched to foot shape". The selection of proper fit however is based on the subjective feeling of the person. Thus quantifying the subjective feeling of fit becomes one of the importance research questions in footwear industry.

In order to find out the proper fitting of footwear, it involves getting to know the size of feet, shoes, and the subjective perception for the selection of shoes. Traditional method in measuring the feet size is to measure the length and width of the feet which can be obtained easily by tape measure and other device like Brannock. However these are considered to be insufficient for good footwear fitting (Witana et al., 2004). Furthermore, researchers were also encountering problem in quantifying fit as it is rather subjective and it is suggested to be affected by shoe wearing experience (Kouchi et al., 2005).

Recommendation of Shoe Fit

Determining correct shoe shape and the right size is the primary component in achieving proper shoe fit (Janisse, 1992). Many recommendations are available for footwear selection to ensure fit.

As suggested by American Orthopaedic Foot and Ankle Society[1], a shoe should conform the shape of the forefoot. Apart from shape matching, Fit can be achieved by having proper shoe length that allow extra one half inch in the front. This extra space helps prevent the toes from banging against the shoes. Proper shoe width is also important. Shoes must be wide enough in the front to allow the toes to extend straight ahead and adequate room across the ball of the foot (Janisse, 1992). Metetarsal should be situated in the widest part of the shoe.

Lace up shoes are recommended as laces allow adjustability. The lacing stops the unnecessary movement of the foot (Janisse, 1992).

Sizing

In an attempt to evaluate the relevant dimensions of feet which are necessary for three dimensional foot modeling, Goonetilleke et al. (1997) measured fourteen foot dimensions of 31 male subjects. The result of factor analysis, suggested that length, width, height at midfoot and forefoot, and toe dimensions are needed to be considered for proper footwear fitting and for the foot modeling.

3D technology has been applied to quantify footwear fit. Nacher et al. (2004) used 3D foot digitizing technology to predict fit by getting the foot geometrical characterization data and subjective fit perception from the users. Witana and the colleagues (2004) applied foot scanners in their study to get the 3D foot shape which was then post processed to obtain the 2D outlines, dimensional differences between the feet and shoe last were analyzed with the fit perception. The dimensional differences plot allowed the shoe lasts designer to determine the critical position for fit and unfit. One limitation for the 3D technology, the feet and last alignment has to be done manually that the actual position of the feet could not be ensured.

With an interest to know the 3D foot shape data without using the foot scanner, other technology likes generating 3D foot shape from 2D information was proposed (Luximon et al. 2005). In order to develop the two models suggested by the authors, foot information of 40 participants was used and the models were validated by using another set of 40 Hong Kong male subjects. In the first method, foot height and foot length were used to predict the 3D foot shape by recursive regression equations. In the second method foot profile and foot shape was used to predict the foot shape. Both methods were found to be valid in foot shape prediction and both of them were relatively cheaper than using the foot scanner to determine the foot shape for custom footwear design.

Pressure Distribution

Apart from sizing, pressure seems to be a valid measurement of fit. DeLooze et al. (2003) states that pressure distribution appears to be the objective measure with the most clear association with the subjective ratings. Although this study was done on car seat, this idea should be applicable to footwear as well. Dewan et al. (2004) placed pressure sensors between the foot-shod interfaces for dynamic pressure measurement. The authors suggested

[1] 10 Points for a Proper Shoe fit, *American Orthopaedic Foot and Ankle Society,* http://www.aofas.org/i4a/pages/index.cfm?pageid=3300

that the loading might have the implication for footwear fitting and had also identified the feasibility of using pressure sensors for footwear fit. This was the only study that applied pressure sensors for footwear fit measurement. However this study did not relate the subjective perception of fit with the loading and how each particular anatomical position of the foot contributed to the overall fitting of the footwear. With the interest to explore this area, Gheorghiu and the colleagues (2004) proposed to investigate the relationship between pressure distribution and subjective fit rating.

Subjective Fit

Subjective perception of fit of footwear had been measured in many studies (Gheorghiu et al., 2004; Kouchi et al., 2005; Witana, et al., 2004). However none of the questionnaire used above had been proved to be valid and reliable. Only one project had been done on measuring footwear comfort by Mundermann et al. (2002). They tried to develop a reliable method to assess footwear comfort during running by the Visual Analog Scales (VASs). It will be discussed in greater detail below.

Footwear Comfort

Apart from its influential role in foot health, footwear fit also contributes much in comfort. Comfort is one of the most important factors for the design of footwear. However as comfort is a subjective feeling depending on individual differences and it is difficult to measure directly. Many methods were used to quantify comfort. Borg scale was widely used to evaluate the comfort perception of the subjects. 15-point, 10-point and 5-point scale were used respectively in different studies (Milani et al., 1997, Miller et al., 2000, Jordan and Bartlett, 1995). Ranking was also one of the methods used by investigators (Chen et al, 1994).

Mundermann et al. (2001) used a visual analog scale (VAS) to assess footwear comfort and later, they tried to develop a reliable method to assess footwear comfort during running by the VASs (Mundermann et al., 2002). A protocol including a control condition during running was used and the resulting intraclass correlations between comfort ratings for repeated conditions were found to be high. It was suggested by the authors that a control condition should be used before each test condition in order to have reliable comfort measurement. For long term comfort, assessments with more sessions (four to six) should be used. It was concluded that VASs provide a reliable measure to assess footwear comfort during running when a control condition was included.

Investigators were interested in identifying biomechanical variables to measure the comfort rating. Plantar pressure distribution was one of those under investigation. It was reported that high plantar pressure in the midfoot area and low pressure in the medial forefoot and hallux area were related to better comfort during walking (Chen et al., 1994). In the study, fourteen subjects were first asked to rank the insoles in the order of comfort after walking and running on a treadmill in self paced speed. After then plantar pressure was measured by inserting a pressure distribution measuring device in the shoes. With more comfortable insole, the path of centre of force at the plantar surface showed a movement to the lateral side. It was suggested that measurement of pressure distribution might be useful to detect change of shoe comfort. Similar result was obtained by Jordan and Bartlett (1995) and a conclusion of the use of pressure distribution at foot-shoe interface to be a tool to identify causes of discomfort in footwear.

Fit between the foot and shoe on comfort assessment was also studied. Good fit was reported to be the most critical factor for shoe comfort. The most critical foot dimensions included hallux height, dorsum height and length, forefoot and posterior calcaneus width and toe lengths. If the fitting is good, other factors like skeletal alignment became important. Moreover, shoes comfort was related to the activity performed and so did shoes torsional stiffness and cushioning (Chen et al, 1994).

METHOLOLOGY

Design

The measurement of this study consisted of three parts: Foot scanning, pressure measurement and perceive fit rating of running shoes. Foot scanning was to collect the foot shape data of the subjects and also the shoes lasts' information. Pressure measurement was applied to collect the pressure data between the foot-shoe interfaces. Subjective fit measurement of the running shoes was obtained by fit questionnaire in order to collect the subjective rating of the fit perception of the subjects.

Subject

Fifteen physically healthy male were recruited as subjects. Their foot sizes were screened to be within the range of 25.5-27.5 cm. All subjects were injury free at the moment of study and no injury history to be resulted in abnormal gait. Experiment detail was explained and subjects were required to sign the informed consent before participating in this study).

Instrumentation

Fit Questionnaire. A fit questionnaire composed of 12 questions by using the VAS scale was used. The questions include 1) Overall fit 2) free space in front of toe 3) fit at the side of forefoot 4) upper forefoot fit 5) overall forefoot fit 6) height of dorsal arch 7) height of plantar arch 8) fit at the side of midfoot 9) overall midfoot fit 10) free space at the back of rearfoot 11) fit at the side of the rearfoot 12) overall rearfoot fit.

The questionnaire was first developed in English and then translated into Chinese. Both versions were amended and approved by respected professions respectively. Reliability of the questionnaire was tested by asking the subjects to rate the test shoes again in the second day.

Foot scanner. The InFoot foot scanner (I-Ware Laboratory Co., Ltd) (Figure 2) was used to collect the shoe lasts and subjects' foot shapes data. It is a 3D foot scanner which consists of eight progressive CCD cameras and four semiconductor laser projectors.

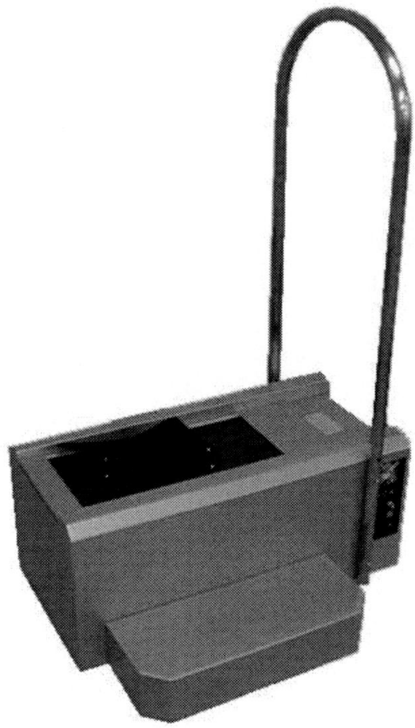

Figure 2. InFoot foot scanner.

Pressure Sensors. 16 flexible pressure sensors (FSA, Canada) (Figure 3) was used to test the shoe-foot interface pressure in the experiment. 12 out of the 16 sensors was attached on the subjects' foot landmarks. Pressure data while standing and walking will be collected with the frequency of 10 Hz.

Running shoes. 5 pairs of running shoes with the size of Eur 41 which were identical in appearance but were made by different shoes lasts were in this study (Figure 4). A pair of running shoes (Gel-Kayano XI, AXIS) was used as the control shoes in this study.

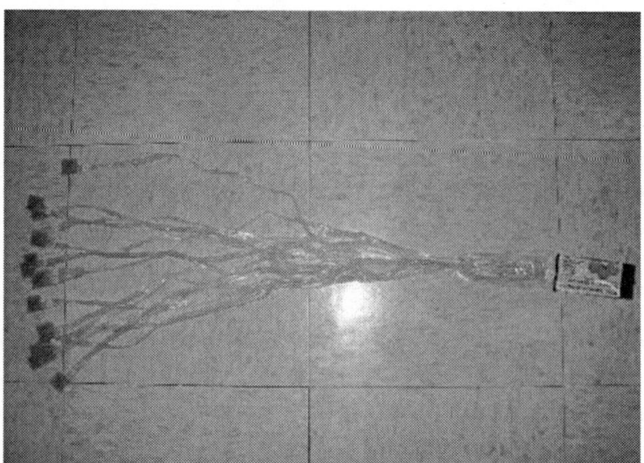

Figure 3. FSA pressure sensors.

Figure 4. Running shoes with identical appearance.

Shoe Lasts. The shoes lasts of the 5 running shoes (Figure 5) were provided by the shoe producing company. They were scanned in order to get the lasts' dimensions: Length, ball girth circumference, breadth, instep circumference and heel breadth.

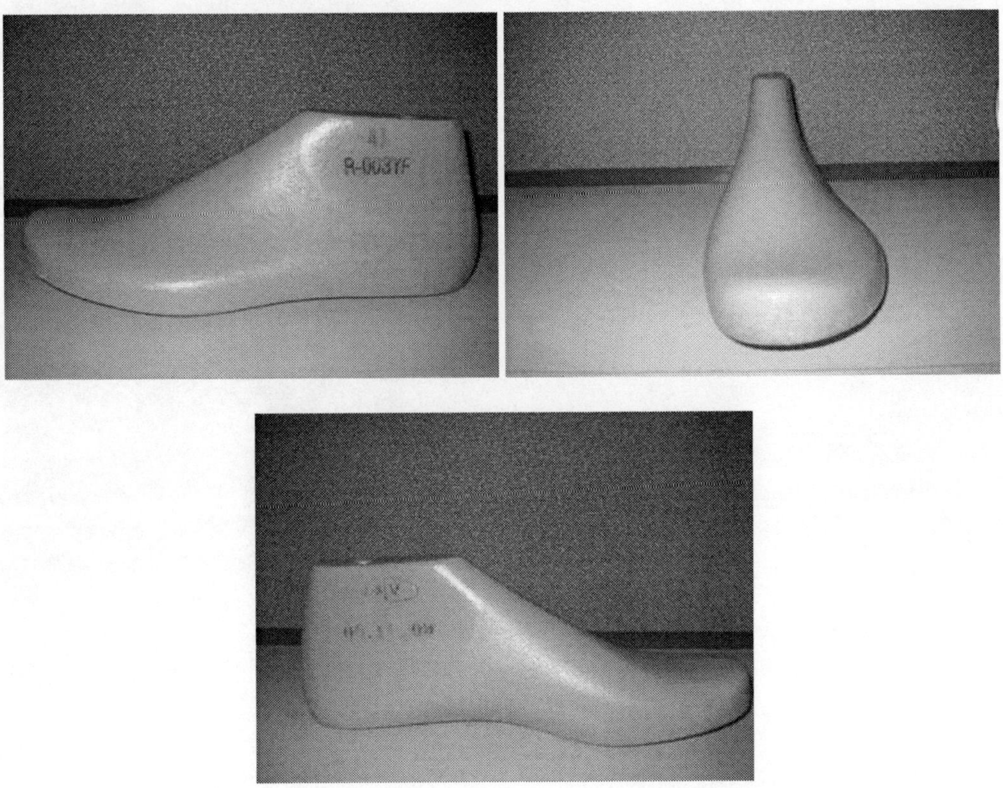

Figure 5. Shoe Lasts: lateral view (left top); frontal view (right top); medial lateral view (left).

Procedure

Subjects came to the Human Biomechanics Laboratory for the experiment. They were required to sign the consent form and their anthropometric data including body height and body weight were measured and recorded. They first completed the foot scanning test and then the subjective fit test and pressure distribution test were completed in the same section.

Foot Scanning. Feet of the subjects were cleaned by alcohol prep pad and twelve markers were attached on the following foot landmark positions: 1) The most medial point of medial malleolus 2) Sphyrion 3) The most lateral point of lateral malleolus 4) sphyrion fibulare 5) Navicular 6) Tuberosity of 5th metatarsalis 7) Metatarsale tibiale 8) metatarsale fibulare 9) toe 1 joint 10) toe 5 joint 11) head of 2nd metatarsal bone 12) tentative junction point (Figure 6).

Foot scanning was done of half weight bearing state. Subjects were required to put one foot in the foot scanner during foot scanning while another foot on a weight. Both feet were scanned(Figure 7 and figure 8).

Landmark positions of the lasts were identified and they were scanned by the foot scanner to get the lasts' dimensions.

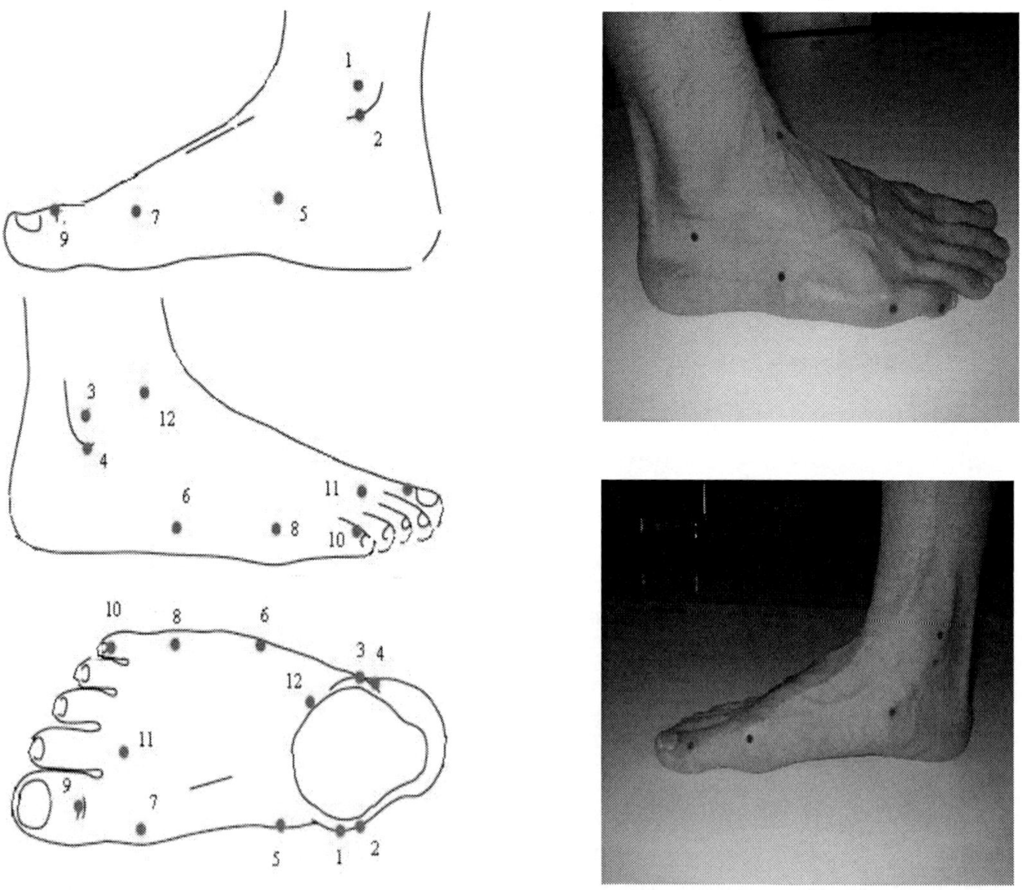

Figure 6. Foot scanning- landmark positions.

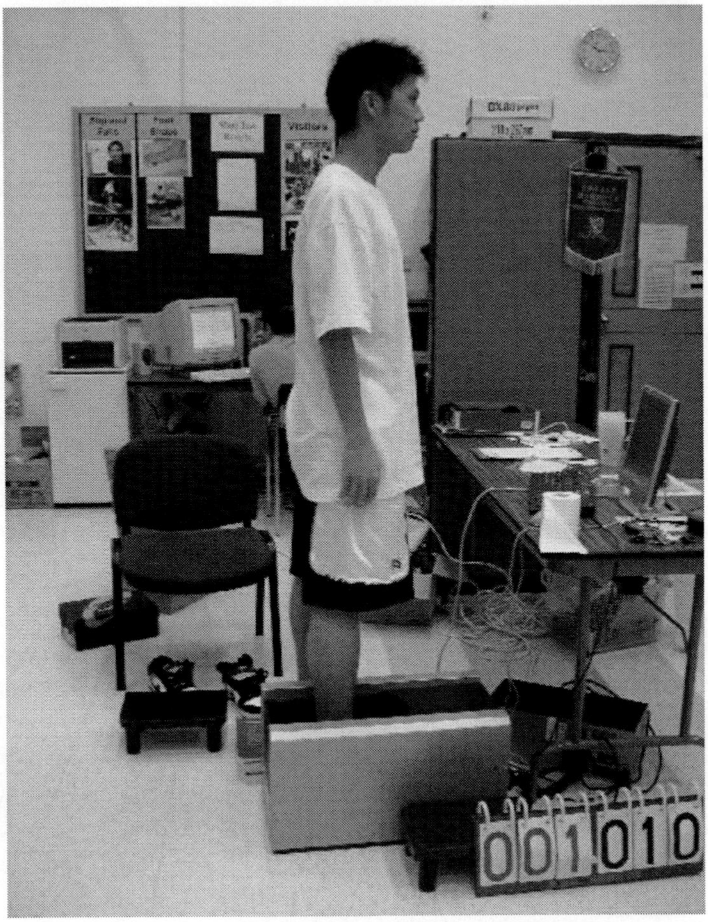

Figure 7. Participants standing posture while foot scanning.

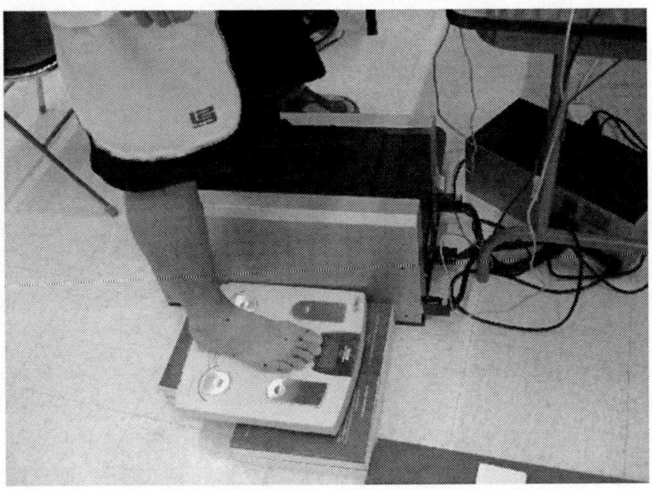

Figure 8. Half weight bearing while foot scanning.

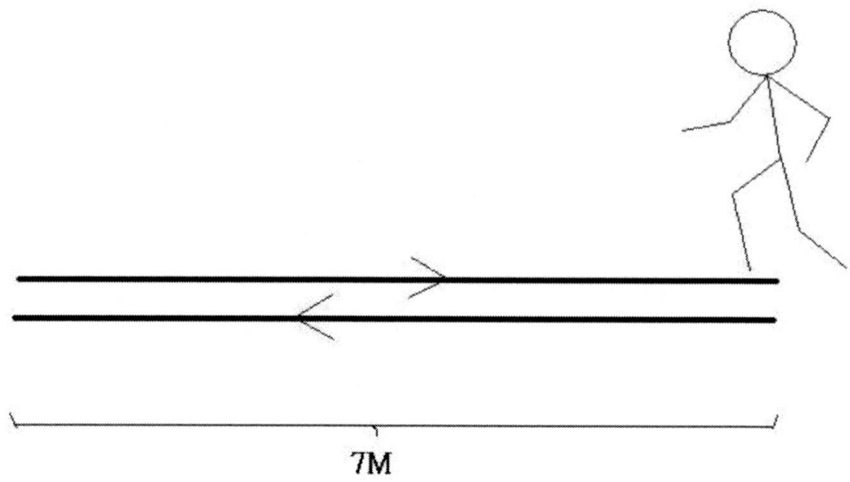

Figure 9. Setting for shoes fit assessment. Participants were asked to walk on a 7 meter long route for 6 times before filling in the questionnaire.

Fit Questionnaire. Subjects were asked to evaluate the tightness or looseness of the running shoes by filling the fit questionnaire which was designed by using a visual analogue scale (VAS). There were totally 12 questions in the questionnaire. Fit at forefoot, mid foot and rear foot position of the footwear were ranked. Detail instructions were given to each subject to make sure every subject understand the questions clearly.

Set-up. Both subjective rating of standing and walking of the running shoes were obtained. For the standing rating, subjects were required to stand for 30 second and then fill in the questionnaire. They were then asked to walk on a 7 meter long route for 6 times before filling in the questionnaire again (Figure 9). Such route was set up to make sure they have completed a minimum walking distance before assessing the shoes' fit. Subjects were allowed to fill in the questionnaire while walking.

Subject would first assess the fit of the control shoes and followed by the assessment of each of the test shoes which were randomized in order. Between each trial of the test shoes, subjects were requested to wear the control shoes and walked on the route for 3 times again in order to make sure that they had the same fit sensation baseline for comparison.

Reliability Test In order to assess the reliability of the questionnaire, subjects were required to come to the laboratory in the other day and to fill in the fit questionnaire again. The sequence of the running shoes were randomized and same protocol as above was used. Reliability for each shoed condition would be obtained to assess the repeatability of the questionnaire.

Fit rating on the first day would be used for data analysis later. The rating got in the second day was used to test the reliability of the questionnaire only.

Pressure measurement. Twelve of the sixteen flexible sensors were attached on the right foot of the subjects on the following location: 1) Tip of toe 1 2) Metatarsal tibiale 3) Tip of toe 2 4) Toe 1 Joint 5) Toe 5 joint 6) Metatarsal fibulare 7) Instep height 8) Navicular 9) Tuberosity of 5th metatarsalis 10) Pternion 11) Medial calcaneous12) Lateral calcaneous (Figure10, figure 11 and figure 12).

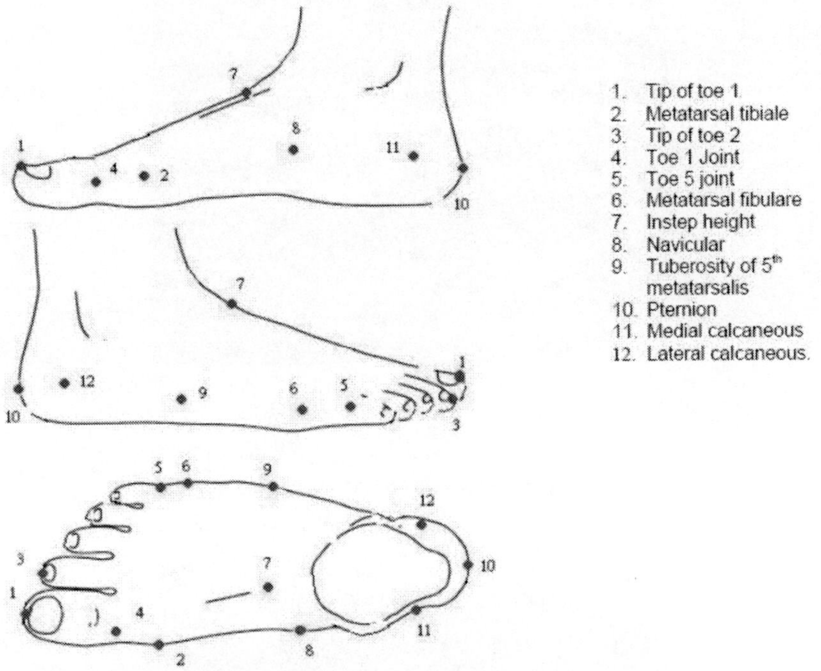

Figure 10. Pressure sensors location.

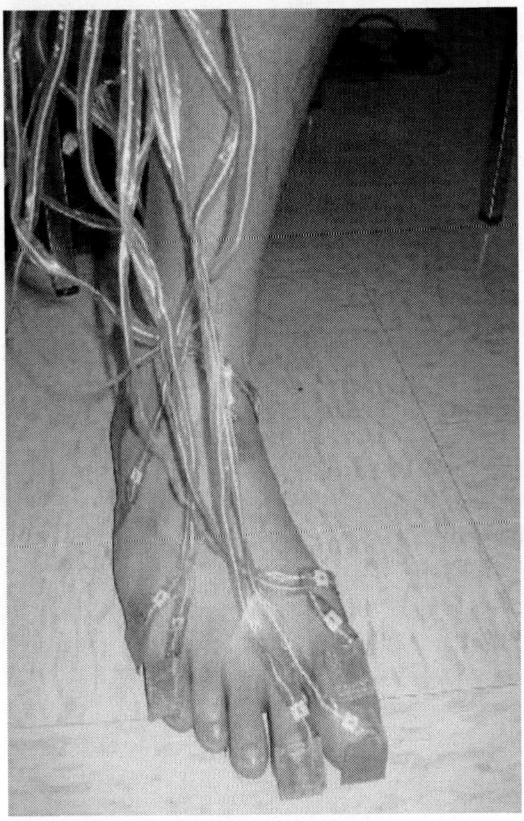

Figure 11. Anterior view of pressure sensors attachment (left).

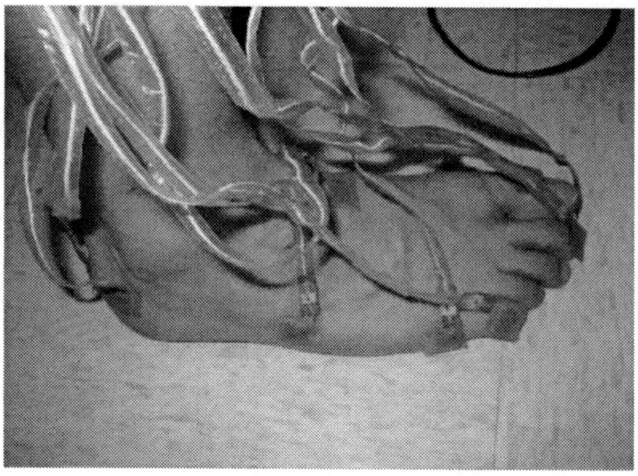

Figure 12. Lateral view of pressure sensors attachment (top right).

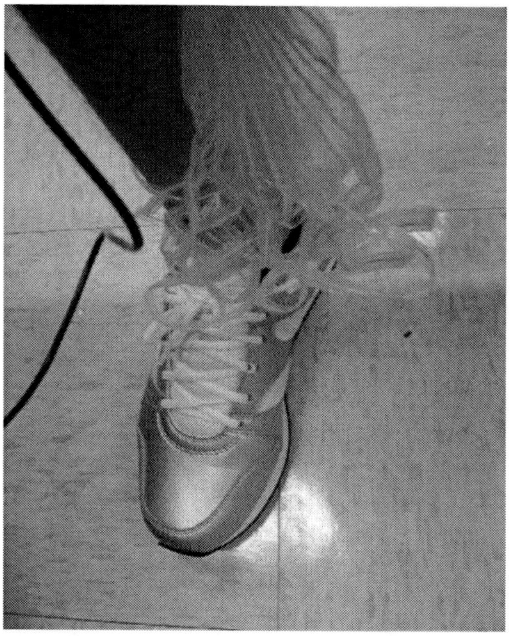

Figure 13. Pressure data collection with the test shoes while standing (right).

Pressure data of three different conditions were collected: 1) socks only when standing 2) test shoes when standing (Figure 13) 3) test shoes when walking. In condition 1, subjects were required to wear the socks and stand with their weights balance at the center. Pressure between the foot shoe interface were collected for 5 seconds. Similar to condition 1, subjects stood steadily with the test shoes on in condition 2. In condition 3, pressure data were collected while the subjects were walking on the route as in the questionnaire test (Figure 14). Participants were asked to perform 3 trails in condition 1 and condition 2; 2 trials in condition 3.

When pressure data collection began, red light on the FSA data collection box flashed. Video were taken to identify the red light and each heel strike during the walking trial. The video was synchronized with pressure sensors.

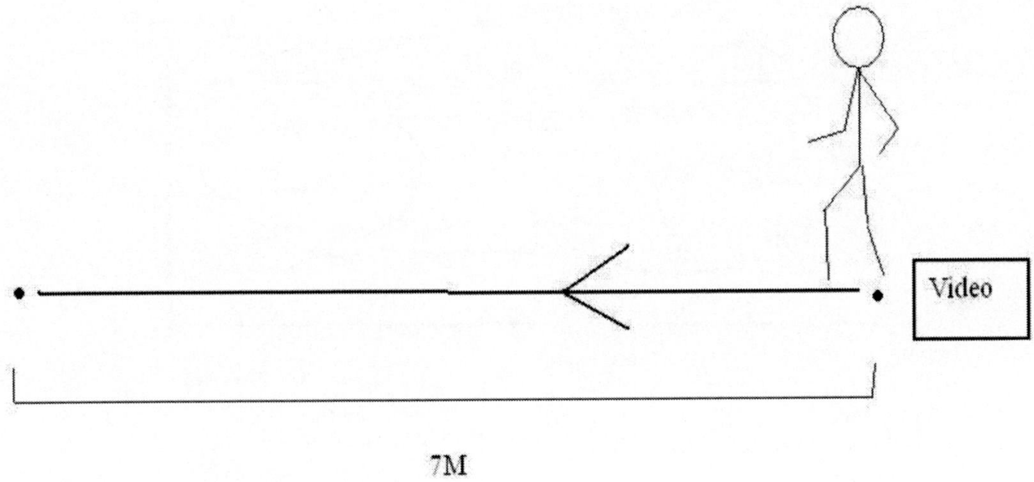

Figure 14. Setting for pressure measurement.

In order to ensure the same shod condition during pressure data collection and subjective fit test, pressure sensors were attached on the subjects' foot before the subjective fit test. Therefore right after fit questionnaire completion, pressure data of the same running shoes condition could be collected. Due to the thinness of the pressure sensors, the interference that would be caused to the subjective fit perception was believed to be very small.

Data Reduction

Foot Scanning. In order to describe the fit of the shoes, size differences between the foot and shoe last were found. Length, Ball girth circumference, foot breadth, instep circumference, heel breadth the foot and last were obtained and five fit indicators were calculated (figure 15) (Kouchi et al., 2005): 1) Foot length allowance (FLA): FL of last – FL of foot; 2) Ball girth circumference allowance (BGCF): BGC of last – BGC of foot 3) Foot breadth allowance (FBA): FB of last – FB of foot; 4) Instep circumference allowance (ICA): IC of last –IC of foot]; 5) Heel breadth allowance (HBA): HB of last – HB of foot.

Fit Questionnaire. For each question, mid point of each line was defined as zero. Distance between the mark and the mid point was measured. When the mark is on the left hand side of the mid point, it was defined as negative while if it was on the right hand side of the point, it was defined as positive.

Data of each question was then input into SPSS version 12 for analysis.

Video. Video taken during the pressure distribution test was captured. The time at which the red light on the FSA collection box flashed was identified and synchronized. Time at each heel strike was recorded.

Pressure Distribution. Pressure data of the sensors were exported to excel file. For the sock standing and shoe standing trials, mean pressure values for each sensor were recorded. As there were 3 trails for each shod conditions (5 pairs of shoe, excluding the sock condition which acted as the baseline value), and there were 16 sensors. The total number of pressure data for each subject was (3 X 5 X16) = 240.

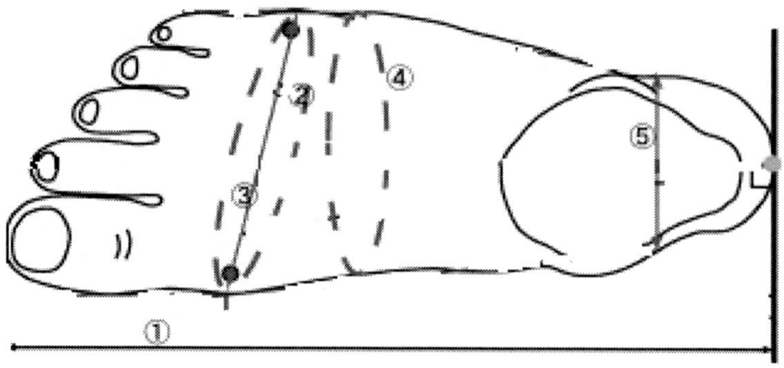

1. Length
2. Ball girth circumference
3. Foot breadth
4. Instep circumference
5. Heel breadth

Figure 15. Foot dimension measurement.

Mean pressures got in sock standing condition was the baseline value. All the other data would subtract the ground value before further analysis.

Data Analysis

SPSS (ver 12.0) was employed to perform all statistical analysis. Pearson Correlation Coefficient and Intraclass correlation coefficient were determined in the reliability test of the fit questionnaire. Stepwise regression was used to evaluate the relationship between the subjective fit perception (dependent variable) and objective measurement (independent variable). Statistical significance was accepted at the 0.05 level of confidence.

RESULT

Subject Detail

15 young male with aged 22.27± 2.05 years, height 175.3± 3.23 cm and weight 68.03± 7.49 kg were recruited in the study (table 1).

Table 1. Summary of descriptive statistic of the subjects

	Mean ± SD	Maximum	Minimum
Age	22.27± 2.05	19	26
Height (cm)	175.3± 3.23	180.2	169.6
Weight (kg)	68.03± 7.49	54.04	70.05

Questionnaire

In order to find out the reliability of the questionnaire, Pearson correlation coefficients of fit ratings measured by the fit questionnaires for different conditions were found. Standing condition and walking condition were combined and analyzed together and were represented as overall condition. The Overall test-retest reliability of each question in the questionnaire ranges from 0.624 to 0.787. All the correlations are significant ($p < 0.01$). The test-retest reliability of each question in standing and walking condition ranges from 0.538 to 0.813 and all the correlations are significant ($p<.01$). Test-retest correlation coefficient of every question for all subjects during standing and walking were shown in table 2 and figure 16.

ICCs of fit rating in different conditions were also found (Table 3). The ICC was 0.841 for overall fit rating in overall condition. The ICC for overall condition is 0.755 – 0.879; standing condition is 0.754 – 0.864; walking is 0.695 – 0.894.

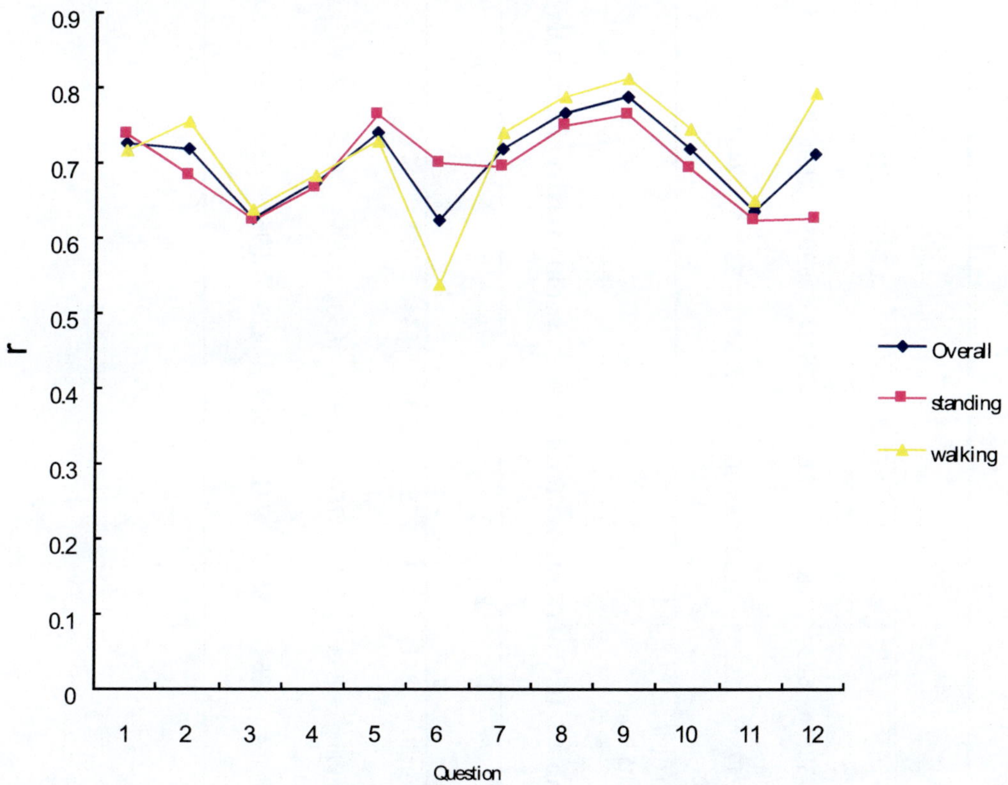

Figure 16. Pearson Correlation coefficient of fit ratings in different conditions (overall, standing, walking).

Table 2. Test-retest Pearson correlation coefficients of fit ratings for different conditions (Overall, standing and walking)

Question		1	2	3	4	5	6	7	8	9	10	11	12
	Overall	.726**	.719**	.626**	.673**	.741**	.624**	.719**	.766**	.787**	.719**	.635**	.712**
r_{tt}	standing	.738**	.683**	.624**	.667**	.764**	.699**	.695**	.751**	.764**	.692**	.625**	.626**
	walking	.717**	.755**	.638**	.683**	.728**	.538**	.741**	.788**	.813**	.745**	.650**	.792**

**. Correlation is significant at the 0.01 level (2-tailed).

*. Correlation is significant at the 0.05 level.

Table 3. ICC of fit ratings for different conditions (Overall, standing and walking)

Question		1	2	3	4	5	6	7	8	9	10	11	12
	Overall	.841	.834	.759	.801	.844	.755	.820	.846	.879	.833	.775	.832
ICC	standing	.848	.810	.754	.797	.859	.799	.809	.845	.864	.812	.764	.767
	walking	.853	.858	.771	.807	.828	.695	.830	.852	.894	.852	.787	.883

Fit Rating

Optimal fitting scored zero, negative values reflect tightness whereas positive values reflect looseness. Apart from the fore foot fit rating for shoe 3, all the other fit rating were ranked as negative which means that the testing shoes were generally ranked as tight by the subject as compared with control shoes(Table 4 and Figure 17). When compared with the other shoes, shoe 3 got all the overall fit, forefoot fit, mid foot fit and rear foot fit rating closest to zero (overall fit = -2.50; forefoot fit = 0.23; mid foot fit = -2.6; rear foot fit = -3.2). As a result, shoe 3 was rated the fittest by the participants among the 5 pairs of shoes.

ANOVA revealed significant difference in the forefoot fit rating. Post Poc Test result showed that there was significant difference between shoe 3 and shoe 4 in forefoot fit rating ($p = 0.05$). The average forefoot fit rating of shoe 4 was -13.53 and that of shoe 3 was 0.23 respectively.

Table 4. Mean score of overall, forefoot, mid foot and rear foot fit rating

Shoe	1	2	3	4	5
Overall fit	-8.07	-8.34	-2.50	-13.34	-8.73
Fore foot fit	-8.4	-9.63	0.23*	-13.53*	-9.97
Mid foot fit	-8.43	-4.97	-2.6	-9.2	-6.93
Rear foot fit	-8.37	-10.93	-3.2	-9.83	-5.40

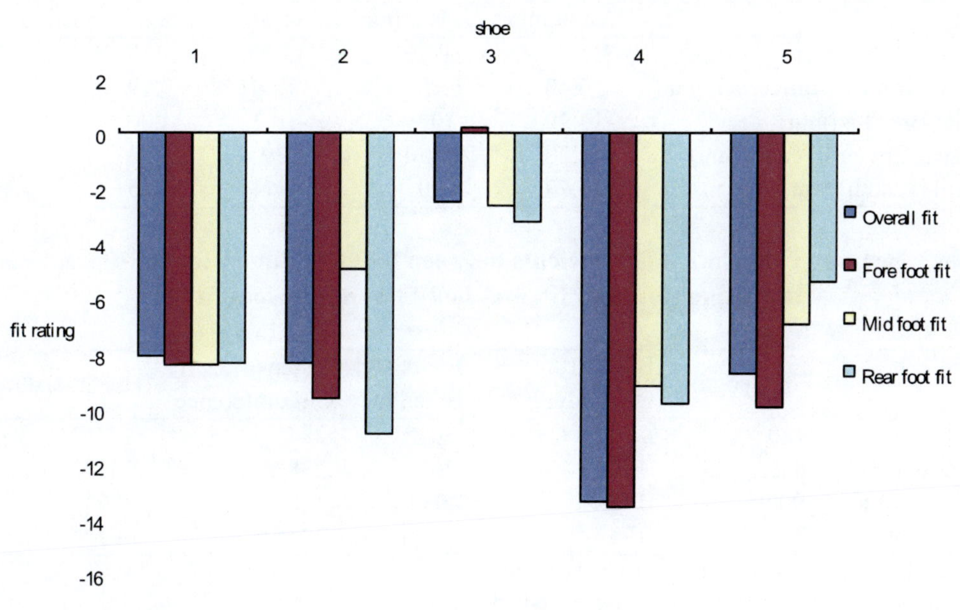

Figure 17. Magnitude of fit ratings of the five testing shoes.

Size Difference

Table 5 **showed the descriptive statistic of the subjects' feet**: foot length (262.08 ± 4.06 mm); ball girth circumference (255.79 ± 8.99 mm); foot breadth (105.14 ± 3.84 mm); instep circumference (105.14 ± 7.76 mm); Heel breadth (67.13 ± 3.48mm).

All the subjects with foot length smaller than the length of the shoe last. For foot breadth size, except in one foot-last combination (subject 15 and shoe last 3), all the subjects have foot breadth larger than that of shoe last. The average size difference between the foot and the shoe lasts were summarized in table 6.

Heel breadth was found to be moderately related to forefoot fit ($r = 0.545$; $p<0.01$) and rear foot fit ($r = 0.467$; $p<0.01$). Low correlations were found between foot length and rear foot fit ($r = 0.254$; $p<0.01$), foot breadth and forefoot fit ($r = 0.203$; $p<0.05$), foot breadth and rear foot fit ($r = 0.370$; $p<0.01$), instep circumference and forefoot fit ($r = 0.2387$; $p<0.01$), instep circumference and rear foot fit ($r = 0.324$; $p<0.01$), heel breadth and mid foot fit ($r = 0.2$; $p<0.05$) (Table 7).

Table 5. Descriptive statistic of subjects' feet

	Foot length	Ball Girth Circumference	Foot Breadth	Instep Circumference	Heel breadth
Mean (mm)	262.08	255.79	105.14	255.60	67.13
SD	4.06	8.99	3.84	7.76	3.48

Table 6. Foot – last size difference

	Minimum	Maximum	Mean	Std. Deviation
Foot length (mm)	3.20	19.70	12.27	4.10
Ball Girth Circumference (mm)	-36.40	5.40	-12.01	9.19
Foot Breadth (mm)	-16.20	.10	-10.22	4.00
Instep Circumference (mm)	-13.20	19.30	1.39	8.34
Heel breadth (mm)	-11.70	3.90	-2.44	3.65

Table 7. Pearson Correlation Coefficients between foot last dimensional difference and fit rating (forefoot fit, mid foot fit and rear foot fit)

r	Foot length	Ball Girth Circumference	Foot Breadth	Instep Circumference	Heel breadth
Overall Fit	.120	.253**	.301**	.284**	.478**
Forefoot Fit	.085	.171*	.203*	.238**	.545**
Mid foot Fit	-0.89	.118	-.032	.120	.200*
Rear foot Fit	.254**	.173*	.37**	.324**	.467**

** Correlation is significant at the 0.01 level (2-tailed).

* Correlation is significant at the 0.05 level (2-tailed).

Pressure Distribution

Pressure at MF was the highest among the 12 foot landmark positions, which was followed by Toe 5 Joint. Average pressures of Shoe 3 were found to be the lowest in the 5 test shoes at 8 landmark positions Toe1, Metarsal Tibiale, Toe 2, Toe 1 Joint, Toe 5 Joint, Pternion, Medial Cancanous, Later Cancanous.(Figure 18).

One way ANOVA was applied to analyze the pressure at different foot landmarks position of the 5 running shoes. No significant difference was found between the test shoes.

Moderate correlation was found between subjective overall fit and Toe 1 joint (-0.422; p<0.01), Toe 5 joint (-0.419; p<0.01) (Table 8).

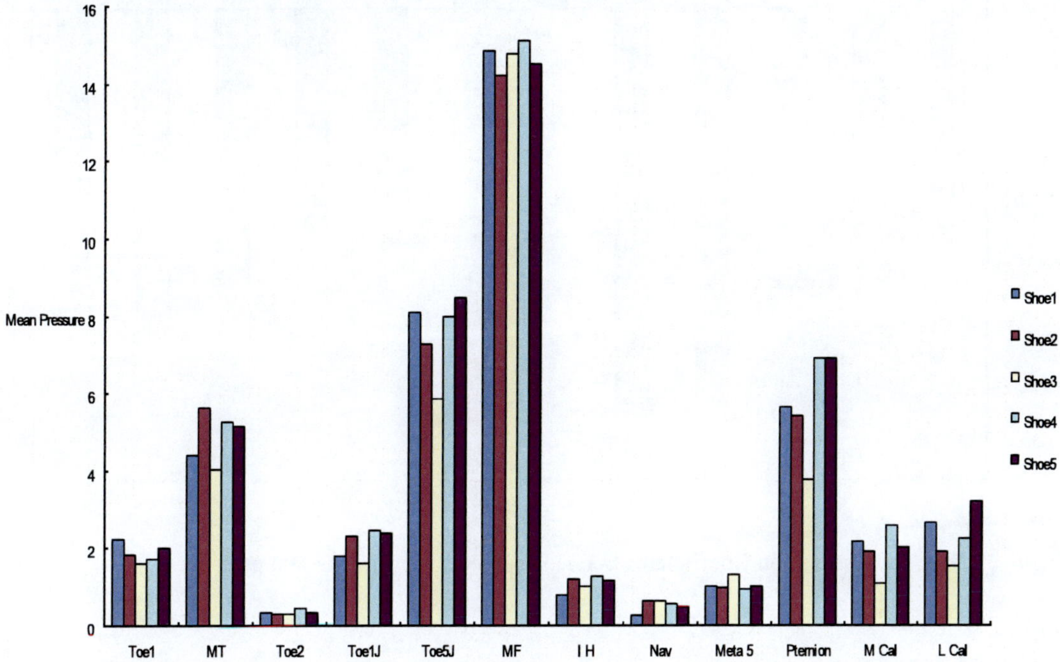

Figure 18. Mean pressure at different foot landmark positions (psi) of the 5 running shoes.

Table 8. Pearson Correlation Coefficients between subjective overall fit and pressure

Foot Landmark	Toe1	MT	Toe2	Toe1J	Toe5J	MF	I H	Nav	Meta 5	Pternion	M Cal	L Cal
r	-.205	-.187	-.11	-.422**	-.419**	.251*	.075	-.201	-.237	-.292**	-.343**	-.264*

Regression

Three regression models were performed by using dimensional differences only, pressure only, dimensional differences and pressure as independent variables to predict the subjective overall fit in standing which is the dependent variable. Stepwise multiple regression was used to find out the model with the greatest R square which indicates that larger variance of subjective fit rating can be explained.

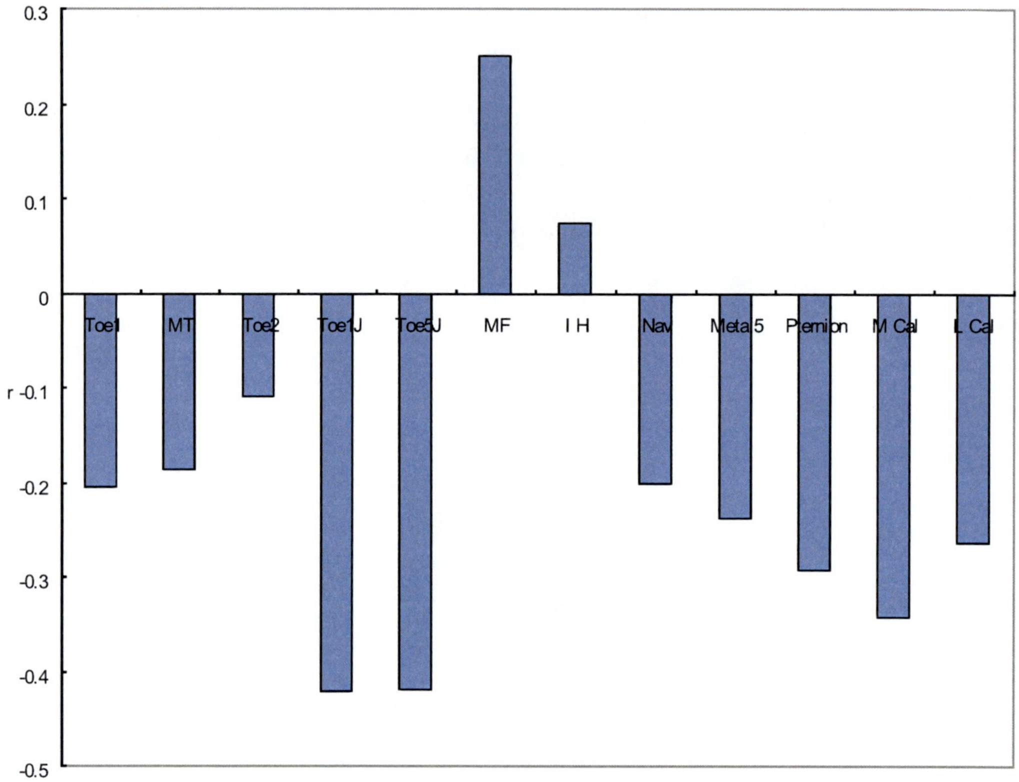

Figure 19. Pearson Correlation Coefficients between subjective overall fit and pressure.

Dimensional differences only:

Table 9. Summary of step-wise regression results by using dimensional difference variables as predictors of subjective fit rating

Model	Variables Entered	R	R Square	Adjusted R Square	Std. Error of the Estimate
1	Heel breadth	.478(a)	.228	.217	13.21577
2	Ball Girth Circumference	.554(b)	.307	.286	12.62078

a. Predictors: (Constant), Heel breadth.

b. Predictors: (Constant), Heel breadth, Ball Girth Circumference.

2 models were run in the stepwise regression for dimensional difference. Heel breadth joint was entered in the first model and the R square was 0.228 (adjusted R square = 0.217). Ball Girth Circumference was added in the second model and the resulting R square increased to 0.307. Summary of the regression was shown in table 9.

Table 10. Coefficients, t values and significant values of step-wise regression results by using dimensional difference variables as predictors of subjective fit rating

Model		Unstandardized Coefficients		Standardized Coefficients	t	Sig.
		B	Std. Error	Beta		
1	(Constant)	-3.040	1.906		-1.595	.115
	Heel breadth	1.938	.435	.478	4.454	.000
2	(Constant)	2.547	2.737		.930	.356
	Heel breadth	2.002	.416	.494	4.810	.000
	Ball Girth Circumference	.451	.165	.280	2.732	.008

Table 10 shows the coefficients, t values and significant values of step-wise regression results by using dimensional difference variables as predictors of subjective fit rating.

Below is the regression equation that uses dimensional difference as predictors of subjective fit rating:

Subjective overall fit = 2.547 + 2.002 (Heel Breadth) + 0.451 (Ball Girth Circumference)

$R^2 = 0.307$ (adjusted $R^2 = 0.286$)

Pressure only:

2 models were run in the stepwise regression for pressure sensors. Toe 1 joint was entered in the first model with the resulting R square to be 0.168. MF was added in the second model and the resulting R square increased to 0.255. Summary of the regression was shown in table 11. Both predictors were found to be significant in the analysis (table 12).

The regression equation for pressure sensors as predictor of subjective overall fit is:

Subjective overall fit = 0.597 (MF) – 2.725 (Toe1 Joint) – 10.285

$R^2 = 0.255$ (adjusted $R^2 = 0.237$)

Table 11. Summary of step-wise regression results by using pressure sensors variables as predictors of subjective fit rating

Model	Variables Entered	R	R Square	Adjusted R Square	Std. Error of the Estimate
1	Toe1J	.422(a)	.178	.168	13.14
2	MF	.505(b)	.255	.237	12.59

a Predictors: (Constant), Toe1J.

b Predictors: (Constant), Toe1J, MF.

Table 12. Coefficients, t values and significant values of step-wise regression results by using pressure sensors variables as predictors of subjective fit rating

Model		Unstandardized Coefficients		Standardized Coefficients	t	Sig.
		B	Std. Error	Beta		
1	(Constant)	-2.043	2.044		-1.000	.321
	Toe1J	-2.620	.625	-.422	-4.189	.000
2	(Constant)	-10.285	3.469		-2.965	.004
	Toe1J	-2.725	.600	-.439	-4.541	.000
	MF	.597	.208	.278	2.878	.005

Dimensional differences and Pressure:

When using both the dimensional differences and the pressure for prediction, a final equation with a higher R square (R^2 = 0.626) was obtained (table 13). Three of the five dimensional differences were entered in the equation: heel breadth, Ball girth circumference and foot length. For the pressure data, pressure of the MF (metatarsal fibulare) and M cal (medial calcaneous) were entered in the equation (table 14).

Table 13. Summary of step-wise regression results by using dimensional difference and pressure sensors as predictors of subjective fit rating

Model	Variables Entered	Variables Removed	R	R Square	Adjusted R Square	Std. Error of the Estimate
1	Toe1J	.	.517(a)	.267	.255	13.11210
2	MF	.	.629(b)	.396	.376	12.00389
3	Heel breadth	.	.670(c)	.449	.421	11.56204
4	Ball Girth Circumference	.	.714(d)	.510	.477	10.98635
5	M_cal	.	.740(e)	.547	.508	10.65697
6	.	Toe1J	.731(f)	.534	.503	10.71514
7	Foot length	.	.791(g)	.626	.594	9.68005

a Predictors: (Constant), Toe1J
b Predictors: (Constant), Toe1J, MF
c Predictors: (Constant), Toe1J, MF, H_breadth
d Predictors: (Constant), Toe1J, MF, H_breadth, BG_Circ
e Predictors: (Constant), Toe1J, MF, H_breadth, BG_Circ, M_cal
f Predictors: (Constant), MF, H_breadth, BG_Circ, M_cal
g Predictors: (Constant), MF, H_breadth, BG_Circ, M_cal, F_length.

The regression equation by using dimensional difference and pressure sensors as predictor of subjective overall fit is:

Subjective overall fit = 22.811 + 0.443 (MF) + 2.62 (Heel Breadth) + 0.680(Ball Girth Circumference) – 2.962 (M Cal) – 1.286 (F Length)

R^2 = 0.626 (adjusted R^2 = 0.594)

Table 14. Coefficients, t values and significant values of step-wise regression results by using dimensional difference and pressure sensors as predictors of subjective fit rating

Model		Unstandardized Coefficients		Standardized Coefficients	t	Sig.
		B	Std. Error	Beta		
1	(Constant)	.500	2.351		.213	.832
	Toe1J	-3.668	.771	-.517	-4.756	.000
2	(Constant)	-10.700	3.781		-2.830	.006
	Toe1J	-3.817	.707	-.538	-5.396	.000
	MF	.777	.216	.359	3.602	.001
3	(Constant)	-7.688	3.853		-1.995	.051
	Toe1J	-3.003	.761	-.423	-3.945	.000
	MF	.641	.215	.296	2.975	.004
	Heel breadth	1.066	.444	.263	2.398	.020
4	(Constant)	.620	4.760		.130	.897
	Toe1J	-2.354	.761	-.332	-3.092	.003
	MF	.481	.213	.222	2.262	.027
	Heel breadth	1.462	.446	.361	3.275	.002
	Ball Girth Circumference	.480	.176	.272	2.730	.008
5	(Constant)	3.241	4.773		.679	.500
	Toe1J	-1.176	.917	-.166	-1.283	.205
	MF	.419	.208	.194	2.009	.049
	Heel breadth	1.759	.454	.434	3.872	.000
	Ball Girth Circumference	.505	.171	.286	2.957	.004
	M_cal	-1.673	.771	-.240	-2.169	.034
6	(Constant)	4.558	4.687		.973	.335
	MF	.346	.202	.160	1.715	.092
	Heel breadth	2.088	.377	.515	5.537	.000
	Ball Girth Circumference	.569	.164	.322	3.463	.001
	M_cal	-2.259	.625	-.324	-3.616	.001
7	(Constant)	22.811	6.422		3.552	.001
	MF	.443	.184	.205	2.406	.019
	Heel breadth	2.620	.369	.646	7.108	.000
	Ball Girth Circumference	.680	.151	.385	4.497	.000
	M_cal	-2.962	.594	-.425	-4.985	.000
	Foot length	-1.286	.340	-.359	-3.781	.000

DISCUSSION

The purpose of this study was to find out the relationship between the three variables: subjective fit rating, foot-last size difference and pressure distribution of foot-shoe interface.

To my knowledge, this is the first study that tried to use both size and pressure distribution as the predictors of subject overall fit of footwear.

Subjective Fit Rating

Reliability of the Fit Questionnaire

Pearson correlation coefficient and ICC were applied to examine the reliability of the fit. It was a usual practice for researchers to use Pearson Correlation to test the reliability of a questionnaire. However, recent research has questioned using correlational methods as a measure of test retest reliability. Correlation is good at comparing deviations from the mean on two measurements, but it is not sensitive to changes in the means of the scores (Vincent, 1999). Therefore ICC was also used in this study to ensure the reliability of the questionnaire. ICC of the questionnaire fell within moderate to high in both standing and walking condition. Similar to the finding of Mundermann and colleagues (2002), the reliability of VAS and a protocol with a control condition to assess footwear fit was good. The satisfactory reliability of the present study allowed the application of the fit questionnaire for fit assessment in the present study.

There were large variations in test retest correlation in fit assessment between individuals (table2). As suggested in Mundermann et al's study (2002), it may be due to the low foot sensitivity of some individuals. Different individual has different sensory threshold to pressure stimuli of the foot. The foot shod contact causes an input signal into the body. The input signal into the system will be modified according to the change in size of the shoes. If the modified input signal is above threshold in some individuals but below the threshold in some individuals, differences in fit perception will be result.

Another reason for that may be due to change of foot sizes at different time of the day or in different day. The foot size of an individual is smaller in the beginning of the day than the end of the day. Therefore it is understandable that the fit perception of the shoe change along with the foot size.

Besides, it is possible that some individuals are initially not able to use VASs in a reliable way (Mundermann et al., 2002). The repeatability was found to improve if they were allowed to fill in the questionnaires in more sessions and it was suggested to use four to six assessments sessions. However, as there were only two sessions included in the present study, it is thus understandable that some individuals might have low repeatability in filling in the VAS questionnaire.

To sum up, the overall satisfactory reliability justifies the use of the questionnaire as the accurate measurement of subjective fit rating in the present study.

Fit Rating

Although significant difference was only found in forefoot fit rating between shoe 3 and shoe 4, shoe 3 was rated the best in average in the overall fit, forefoot fit, mid foot fit and rearfoot fit. The result showed that shoe 3 fitted the subjects best in this study. The negative fit rating of the other test shoes reflected the general sense of tightness of the test shoes felt by the subjects. Shoe 3, which scored less negative then the other test shoes showed that it was rated as less tight by the subjects. When we looked at the shoe last information of shoe 3, it was found to have the largest foot length, ball girth circumference, foot breadth and heel breadth.

When we compared shoe 3 with shoe 4, the instep circumference of shoe 3 is 4.5mm larger. The difference is the largest among the test shoes in term of instep circumference and it accounts for significant difference found in forefoot fit rating between the test shoes.

The difference between last 3 and the other shoe lasts were within 3 mm in all the dimensions but yet, the small differences were proven to have an impact on the subject subjective fit perception. The result shows the sensitivity of the foot to the dimensions changes of the shoes which was within a few millimeters.

Size

Foot Shape

Five foot shape parameters were examined in the study: foot length, ball girth circumference, foot breadth, instep circumference, heel breadth. The average foot length and foot breadth of the subjects in the present study were found to be 262.08mm and 105.14mm respectively. Hawes et al. (1994) found that Japanese and Korean with foot length smaller than 265.7mm had the foot breadth average of 102.5mm. For the North American with the same range of foot length, they had the foot breadth average of 99.1mm. Chantelau and Gede (2002) found that individuals with foot length between 263mm-265mm had the foot breadth between 97mm-101mm. When compared the foot shape of the subjects in present study with those found previously, subjects in this study had wider foot breadth in general.

Dimensional Difference

The average foot shoe-length difference was 12.27mm (3.2-19.7 mm). All the subjects have the foot size smaller than that of the shoes. This result was expected to be due to the selection criteria of the foot size of the participants, those who had foot size larger than that of the last would not have been invited to the study.

The average shoes breadth and foot breadth discrepancy in the present study was -10.22mm. The negative value indicated that breadth of the shoe were smaller than the feet. The dimensional differences here was greater than that found by Menz and Morris (2005) which was -7.27mm for outdoor footwear; but smaller than that found by Frey et al (1993) which was 12mm. According to the recommendation made by Tremaine and Awad (1998), a shoe with proper fit should be equal or no more than 6.25mm (0.25 in) less that the fore-foot width. Therefore the shoes used in the present study might be regarded as too narrow or unfit to the participants. In addition, ball girth circumference and heel breadth also showed a negative discrepancy when compared the foot data with the shoe last.

There is no universal agreement of the definition for footwear fit. Although there are a lot of guidelines available, most of which have not been validated. Menz and Morris (2005) defined incorrect size as at least half a British shoe size too small or too large. Nixon and the colleagues (2006), on the other hand, defined inappropriately sized footwear as a shoe size at least one full size too large or too small by using a foot size measuring device. However, a proper fit of footwear should also put subjective feel of fit into consideration. Therefore it is not possible to assess footwear fit only by looking at the dimensional difference between the shoes and the foot. One would expect to know the function of the shoes, the activity they perform before making the conclusion.

Correlation with Subjective Fit

No substantial relationships were found between size difference at a particular region and subjective overall fit because size differences at different foot regions contribute together to the overall fit rating. In addition, overall fit perception is a subjective measure that involves individual preference of the shoe tightness and looseness (Kouchi et al., 2005). That explains the low correlation found between ball girth circumference, foot breadth, instep circumference and the subjective fit rating($r = 0.253$, 0.301 and 0.284; $p < 0.01$).

As a result, it is interesting to find a moderately strong correlation between heel breadth ($r = .478$, $P < .01$) and subjective fit rating, which is greater than that of the other foot regions mentioned above. Au and Goonetilleke (2007) reported the distributions of 186 responses of the importance of fit rating at different foot regions of comfortable and uncomfortable shoes ladies' dress shoes. The majority of fit ratings were in the MPJ region (25.3%) and Toe region (21.0%). Rear-foot region accounted for 16.7% which was relatively less than the above regions. The result of Au's study showed that people would rate the fit in MPJ region (foot breadth) as more important than rear – foot region (heel breadth).

The different results found showed the disagreement between subjective measurement and objective measurement of fit rating. One might perceive MPJ region as more important in influencing the overall fit but in fact, the rear foot region related more to the overall fit rating. This interesting finding calls for further investigation in the topic. One of the reasons suggested here is the rigidity of the rear foot part of the shoes. In the present study, the fore and mid part of the shoes allowed certain adjustment by the shoe laces. The rear part of the shoes, on the other hand is not adjustable. After the adjustment of the shoe lace, one may perceive the fore part of shoe as more fit. At a result, the correlation between the subjective fit rating and the dimensional difference of the fore part of the shoes was reduced.

No significant relationship was found between the dimensional difference of foot – last length and the subjective fit perception. This finding contradicted the wisdom of the contribution of shoe length in footwear fit. The sizes of the subject's foot were controlled within 25.5cm – 27.5cm and the actual foot sizes ranged between 25.51cm-27.06cm. Such screening prevented extreme unfit condition caused by improper foot length thus reduces the sensitivity in detecting the correlation.

In an attempt to find out how much dimensional difference can explain the variance of subjective overall fit, stepwise regression was applied. Heel breadth and ball girth circumference were input in the equation and the resulting R square was found to be small ($R^2 = 0.307$). Only 30% of subjective fit rating can be explained by foot-last dimensional difference which is small for satisfactory prediction.

This result was expected due to the absence of strong correlation between dimensional difference and subjective fit rating. Contradicting the result of a similar study conducted by Witana and colleagues (2002), regression equation with high R square ($R^2 = 0.8247$) was found. One of the reasons accounting for the difference might due to the manual alignment of the foot and shoe lasts. In Witana and colleagues' study (2002), the foot outlines were aligned with each of the respective last outlines using the heel centerline and the measured end distance between shoe and foot. This alignment, however was not done in the present study.

Furthermore, it was suggested that the perception of fit was significantly affected by experience (Kouchi et al., 2005). The personal preference of tightness of the shoes reduced the predictive power of dimensional difference which results in making generalization of the result difficult.

Pressure

Greater pressures were generally added on the lateral side of the foot as it was shown in the larger average pressure applied at MF and Toe 5 Joint (Figure 17). This may be due to the protruding shape of the metatarsal fibula and the toe 5 joint. The contact of those areas with the shoe produced pressure. Thus it is reasonable to see the relatively smaller average pressure in shoe 3 which is, in average larger then the other test shoe. However, despite the observable variations of the pressure between the foot-shod interface, it wasn't great enough to demonstrate significant difference between the shoes.

Most of the significant correlation of subjective fit rating and pressure were found to be negative. The result implies that tightness of the shoes accompanies poor fit sensation of the subject. The correlation is stronger at the position of Toe 1 joint and toe 5 joint. However, opposite result was found at MF, where fit was directly related to pressure. The result showed that sensation of tightness varied from location to location. One may prefer a pair of shoes to be tight at certain position but not another. This result is in coherence with the finding of Au and Goonetilleke (2007).

When examining forces and pressures at the foot-shoe interface using pressure as the measuring device, one might suspect that the presence of the pressure sensors themselves could influence the parameters that were being considered. The high test and retest correlation of the subjective test measurement, however, provided a proof that the presence of pressure sensors may not influence the subjective fit measurement.

Similar to dimensional difference, stepwise regression was applied to examine how much variance of subjective overall fit can be explained by pressure. MF and toe 1 Joint were input in the equation. The resulting R square was found to be 0.237 showing that around 23% of subjective fit rating can be explained by foot-shod interface pressure measurement. The smaller R square of pressure suggests that dimensional difference is more capable in explaining subjective fit than that of pressure.

Pressure measurement, unlike dimensional difference which can provide real time pressure data while footwear fit is being assessed. However, the incapability of pressure sensors in measuring shoe looseness reduces their prediction power of fit rating. Pressure sensors, in one hand is very sensitive in measuring shoe tightness when the foot-shod interface is in contact. But on the other hand, once the foot is not in contact with the foot, the pressure sensors fail to detect the extend of looseness.

Dimensional Difference and Pressure

In order to improve the predicting power of the objective measures to the subjective fit measurement, both dimensional difference and pressure were input in the stepwise regression. The resulting regression equation was found to have a R square of 0.626 showing that 62.6% of subjective fit rating can be explained by using dimensional difference and foot shod interface pressure together. Four of the five dimensional differences were entered in the equation: heel breadth, Ball girth circumference, medial calcaneous and foot length. For the pressure data, pressure of the MF was entered in the equation.

The predicting power of the objective measures have increased twice from using solely dimensional difference (R^2 = 0.307) or solely pressure sensor (R^2 = 0.255) to using both measures together (R^2 = 0.626). Despite only one pressure value was included in the regression equation.

The result of the present study shows that by using 2 objective measures: dimensional difference and pressure sensors at the same time can improve the explained variance of subjective fit rating.

Subjective Fit and Objective Measures

Footwear fit is a controversial topic which is under the interest in the field of sport biomechanics. Researchers have tried to quantify footwear fit by many methods. The biggest challenge we are facing is the difficulty in quantifying footwear fit as 'right fit' or compatibility is generally unknown in many circumstances. As stated in Witana and colleagues (2004) study, foot-shoe fit depends on many factors. Time of day, activity performed will also influence the fit perception. As a result, footwear purchased at some time on one day may not be as fit at another time of the day. Therefore, the role of subjective fit is always downplayed especially by the professional in sports biomechanics. However, due to the important role that footwear fit is playing in footwear comfort (Mundermann et al., 2003), foot health and sports performance and which subjective perception should be one of the determining factors.

However, not everyone has the precision sense to judge the footwear fit e.g. diabetes patients, elderly and children. It is necessary to develop reliable objective measures in order to predict or to estimate the subjective footwear fit.

Previous studies provided a lot of guidelines in footwear selection in order to ensure fit. Some researchers have also tried different objective measures like foot scanning, foot-shoe dimensional difference, pressure sensors to find out footwear fit. More attention has been put in subjective fit rating nowadays. Many studies are now undergoing to explore the feasibility to use different device to explain subjective footwear fit (Dewan et al., 2004; Gheorghiu et al., 2004; Nacher et al., 2004; Witana et al., 2004). Many of which are of preliminary state of exploration and seldom did these studies try using two kinds of objective measures at the same time. Although the subject group in this study was not the target group we are focusing on, it should be treated as the preliminary study for the application of measuring methods. Therefore, result of the present study is very encouraging and it also gives us insight in future footwear fit research direction.

Conclusion

This study examined the relationship of subjective rating and objective measures of footwear fit by using dimensional difference and pressure sensors. Stepwise regression was applied to find out the amount of explained variance that objective measure(s) could account for the subjective fit rating.

Result showed that using dimension difference alone accounted for 30.7% of subjective fit and 25.5% by using solely pressure sensors. When both objective measures were entered in regression, R square increased to 62.6%. This indicated using two objective measures at the same time increased the predictive power of subjective fit rating.

Dimensional difference and pressure sensors were able to compensate the limitations and restrictions of one another which resulted in an increased predicting power of subjective fit rating.

REFERENCES

Au, E. Y. L. and Goonetilleke, R. S. (in press). A qualitative study on the comfort and fit of ladies' dress shoes. *Applied Ergonomics.*

Burns, S.L., Leese, G.P., and McMurdo, M.E.T. (2002). Older people and ill fitting shoes. *Postgraduate Medical Journal*, 78,344-36.

Chantelau, E. (2002). Letter to the editor: shoe-fitting, doesn't it really matter? *Gerontology*, 48, 408.

Chantelau, E. and Gede, A. (2001). Foot dimension of elderly people with and without diabete mellitus – a data basis for shoe design. *Gerontology*, 48,241-244

Chen H., Nigg B.M., and de Konging J. (1994). Relationship between plantar pressure distribution under the foot and insole comfort. *Clinical Biomechanics,* 9(6)335-41.

DeLooze M.P., Kuijt-evers L.F. M., and Dieen J.V. (2003). Sitting comfort and discomfort and relationships with objective measures, *Ergonomics*, 49(10)985-997

Dewan, C., Pearsall, D., and Turcotte, R. (2004). Dynamic pressure measurement about the foot and ankle. *Proceedings of XXIII International Symposium on Biomechanics in Sports, Ottawa, CA,* 505.

Frey, C. and Roberts, N. E. (2002). Problem shoes, problem: what to tell women about footwear-recommendations on diagnosis, treatment, prevention. *Women's health in primary care,* 5(11)682-91.

Frey, C., Thompson, F., and Smith, J., Sanders, M., Horstman, H. (1993). American Orthopaedic foot and ankle society women's shoe survey. *Foot and Ankle*, 14(2) 78-81.

Gheorghiu, C.R., Pearsall, D.J., and Turcotte, R.A. (2004). Quantifying fit in ice hockey skate books. *Proceedings of XXIII International Symposium on Biomechanics in Sports, Ottawa, CA,* 509.

Goonetilleke, R.S. (2003). Designing footwear: back to basics in an effort to design for people. *Proceedings of SEAMEC, Kuching,* 25-31.

Goonetilleke, R.S., Ho, E.C.F., and So, R.H.Y. (1997-98). Foot sizing beyond the 2-D Brannock Method. *Annual Journal of the Institute of Industrial Engineers (Hong Kong).* 28-31.

Gould, D., Kelly, D., Goldstone, L., and Gammon, J. (2001). Examining the validity of pressure ulcer risk assessment scales: developing and using illustrated patient simulations to collect the data. *Journal of Clinical Nursing*, 10, 697-706.

Hawes, M.R., Sovak, D., Miyashita, M., Kang S.J., Yoshihuku, Y., and Tanaka, S. (1994). Ethnic differences in forefoot shape and the determination of shoe comfort. *Ergonomics*, 37(1) 187-96.

Janisse, D.J. (1992). The art and science of fitting shoes. *Foot and Ankle,* 13(5) 257-262

Jordon, C. and Bartlett R. (1995). Pressure distribution and perceived in casual footwear. *Gait and Posture*, 3(4)215-220

Kouchi, M., Mochimaru M., Nogawa, H., and Ujihashi, S. (2005). Morphological fit of running shoes: perception and physical measurements, *Proceedings of the 7th symposium on footwear biomechanics, Cleveand, OH, USA, 38-39*

Luximon, A., Goonetilleke, R. S. and Tsui, K. L., (2003). Footwear Fit Categorization, the Customer Centric Enterprise:Advances in Mass Customization and personalization, edited by Mitchell M. T. and Frank T. P., Springer Verlag, pp 491-499.

Menz, H. B., and Morris, M.E. (2005). Footwear characteristics and foot problems in older people. *Gerontology*, 51, 346-351

Milani T.L., Hennig E.M., and Lafortune M.A. (1997). Perceptual and biomechanical variables for running in identical shoe constructions with varying midsole hardness. *Clinical Biomechanics,* 12(5)294-300

Miller, J.E., Nigg B.M., Liu W., Stefanyshyn D.J., and Nurse M.A. (2000). Influence of foot, leg and shoe characteristics on subject comfort. *Foot and Ankle international,* 21(9)759-67.

Mundermann A., Stefanyshyn D.J., and Nigg B.M. (2001). Relationship between footwear comfort of shoe inserts and anthropometric and sensory factors. *Medicine and Science in Sports and Exercise,* 33(11)1939-45.

Mundermann A., Nigg B.M., Humble R.N., and Stefanyshyn D.J (2003). Foot orthotics affect lower extremity kinematics and kinetics during running. *Clinical Biomechanics,* 18(3) 254-262

Mundermann A., Nigg B.M., Stefanyshyn D.J., and Humble R. N. (2002). Development of a reliable method to assess footwear comfort during running. *Gait and Posture,* 16(1) 38-45

Nácher, B., Alcántara, E. Alemany, S., García-Hernández, J., and Juan, A. (2004). A 3D foot digitizing and its application to footwear fitting. *Proceedings of 3D Modellling, Paris*, France.

Nixon, B.P., Armstrong D.G., Wendell, C., Vazquez J.R., Rabinovich, Z., Kimbriel, H.R. et al.(2006). Do US Veterans wear appropriately sized shoes?: The veterans affairs shoe size selection study. *Journal of the American Podiatric Medical Association.* 96(4) 290-2.

Jordan C., and Bartlett R. (1995). Pressure distribution and perceived comfort in casual footwear. *Gait and Posture,* 3(4)215-20.

Reiber, G.E., Smith, D.G., Wallance, C., Sullivan K., Hayes, S., Vath, C. Maciejewski, M.L., Yu, O., Heagerty, P.J. LeMaster, J. (2002). Effect of the therapeutic footwear on foot reulceration in patients with diabetes, *Journal of American Journal Association,* 287, 2552-2558.

Tremain, M.D. and Awad, E.M. (1998). The foot and ankle source book. Los Angeles: Lowell House.

Vincent, W.J. (1999). *Statistics in kinesiology.* Champaign, IL : Human Kinetics.

Witana, C.P., Feng, J., Goonetilleke, R.S. (2004). Dimensional differences for evaluating the quality of footwear fit. *Ergonomics*, 47(12)1301-17.

In: Biomechanics: Principles, Trends and Applications
Editor: Jerrod H. Levy, pp. 353-385
ISBN: 978-1-60741-394-3

Chapter 15

IMPROVEMENT OF THE INPUT DATA IN BIOMECHANICS: KINEMATIC AND BODY SEGMENT INERTIAL PARAMETERS

Tony Monnet*[1], *Mickaël Begon*[2], *Claude Vallée*[1], *Patrick Lacouture*[1]*
[1]Université de Poitiers, Laboratoire de Mécanique des Solides
UMR CNRS 6610, Bvd Marie et Pierre Curie BP 30179
86962 Futuroscope Cedex, France
[2]Department of Kinesiology,
University of Montreal,
Research Centre of Sainte-Justine Hospital,
Montreal, Canada

Abstract

Usually, biomechanical models used for human motion analysis are oversimplified, especially for clinical analyses (e.g. Helen Hayes model). The calculated net joint forces and torques are sensitive to the input data: segment kinematics and body segment inertial parameters. It is therefore necessary to improve these input data using new methods and models adapted to the population and movement of interest. The general problem is divided into three parts: (i) minimization of soft tissue artefacts, (ii) joint centre location and (iii) identification of the personalized body segment parameters.

1 Introduction

In sports and clinical biomechanics, most of the analyses are based on the joint kinematics and kinetics using force sensors e.g., force platform) and motion analysis systems. Since the optoelectronical motion analysis systems measure only the spatial coordinates of targets put on the skin, several steps — calculations and assumptions — are involved in the dynamical analysis of the movement. In the eighties, three-dimensional (3D) kinematical models

*tony.monnet@univ-poitiers.fr

appeared for gait analysis using a few markers [28, 39] ; these models and their marker sets have been progressively adopted as standard models by motion analysis software (e.g., Vicon Polygon and OrthoTrack from Motion Analysis Corporation) to automate the process from the marker capture to the joint torque calculation and some interactive multimedia reports.

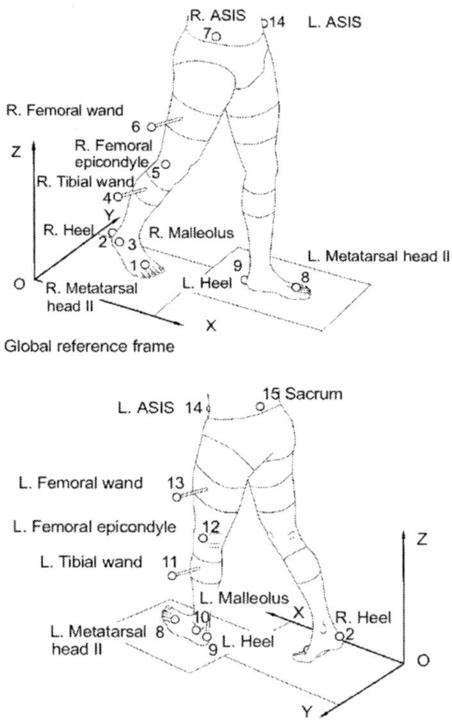

Figure 1: Helen Hayes model in [28].

Helen Hayes model (Figure 1) for example was proposed with a minimal set of markers and assumptions about the lower-limb joint centres. Fifteen markers are needed for capturing the lower-limbs (seven body segments) and then calculating the kinematics and kinetics. This model is still widely used for two main reasons: the markers are easy to track in a 3D space with optoelectronical motion analysis systems [64] and a large database has been obtained from this model. In gait analysis, one of the goals is to discriminate between normal and pathological gaits and to assess changes in dynamics over time. Due to the error magnitude widely quantified by the literature [11, 17, 29, 38], these simplest models are suitable for movie or video game motion capture. To some extent, they can identify intra-subject changes between various conditions and large inter-subject or population differences, since some assumptions (joint centre locations, inertia parameters) can affect differently two subjects or two populations. In applied biomechanics or research fields using human movement measurement tools (e.g., motor control), motion analysis systems combined with analysis software are sometimes used as an "intelligent black box" where the marker placement, height and body mass are the input and the joint angles and torques are the output. However the relationship between marker coordinates and joint torques is not so obvious; it appears indirectly in the motion equation. Using the Newton-Euler formalism proposed by [5], the equations of motion for one segment of a serial chain read:

$$m_1 (\ddot{\mathbf{r}}_{P_1} - \mathbf{g}) + {}^1_0\ddot{R} \left(m_1 \mathbf{c}_1^{(1)} \right) = \mathbf{F}, \tag{1}$$

$$\begin{aligned} &\left[(\ddot{\mathbf{r}}_{P_1} - \mathbf{g}) \otimes m_1 \mathbf{c}_1^{(1)} \right] {}^1_0R^T - {}^1_0R \left[\left(m_1 \mathbf{c}_1^{(1)} \right) \otimes (\ddot{\mathbf{r}}_{P_1} - \mathbf{g}) \right] + \\ &{}^1_0\ddot{R}\, K_{P_1}^{(1)}\, {}^1_0R^T - {}^1_0R\, K_{P_1}^{(1)}\, {}^1_0\ddot{R}^T = \mathcal{M}_{P_1}, \end{aligned} \tag{2}$$

the details of the notations are presented in the second part of this chapter (3.1).

The matrix of rotation (${}^1_0R$) is sensitive to the marker positions. It respects, as much as possible, the anatomical axes of the segment to describe the joint kinematics and kinetics in line with the functional anatomy e.g. flexion-extension, abduction-adduction and mediolateral rotations). Additionally, Equations 1 and 2 show that the joint efforts appear to depend linearly on to the mass (m_1), the first order moments ($m_1\mathbf{c}_1$) and the matrix of inertia (K_{P_1}).

In parallel to the use and abuse of these models for any kind of movements and populations, modeling research and tools have progressed. Nowadays, with the improvement in resolution of the motion analysis systems, increasing the number of markers is no longer a problem. Hundreds of markers can be tracked and identified in real time with a high degree of accuracy; medical imaging devices (e.g. full-body CT scan, DXA) are used to estimate the segment inertial parameters and the human body models become more and more realistic, e.g. introduction of wobbling masses [52] or splitting the foot in subsegments [58].

A first source of uncertainties is the skin movement artefact that affects the skeletal kinematics. Recently, methods and protocols have appeared for estimating the soft tissue artefact or the effect of wobbling masses by placing multiple markers on the body segments (Figure 2a). In their paper, Stagni et al. [62] evidenced that soft tissue artefact is subject- and task-specific, and that critical errors are propagated on knee kinematics. In a recent paper, Monnet et al. [46] proposed a method for determining the most deformed markers and the lesser one on the thigh (Figure 2b). A set of 26 markers was placed on the thigh and the most deformed markers were iteratively eliminated until only three markers remained. First results of this study highlight that the worst markers are located on the medial part of the thigh (1-2, see Figure 2b) or close to joints: the greater trochanter (17, see Figure 2b) and above the patella (10, see Figure 2b). These recent papers have a direct application for the joint centre estimation using functional methods [32]. These methods consist in locating the joint centre from the kinematics of the adjacent segments. Functional methods are not commonly implemented and predictive methods are still used in many research studies and in clinical settings. For human hip joints, predictive approaches have been shown to be accurate to approximately 20 mm when compared with X-ray measurements [10, 50]. The main purpose of the functional methods is to improve this accuracy whatever the marker placement and the subject morphology.

The second source of uncertainties concerns the body segment inertial parameters. They are commonly estimated using regression equations of proportional anthropometrical models determined from specific populations e.g. Bulgarian [51], Caucasian [68]). Recent papers in this field aim at proposing new and more accurate approaches for estimating

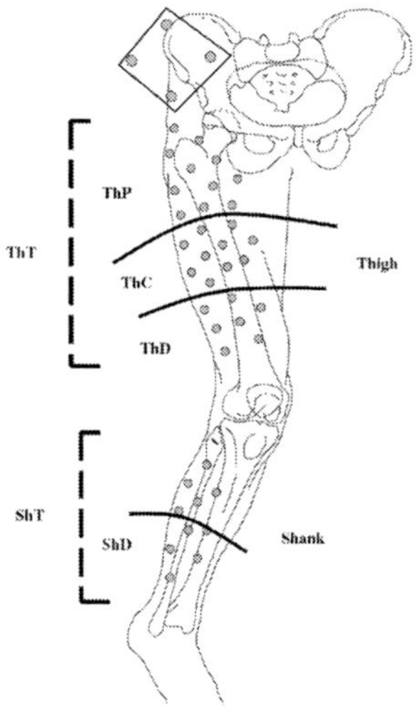

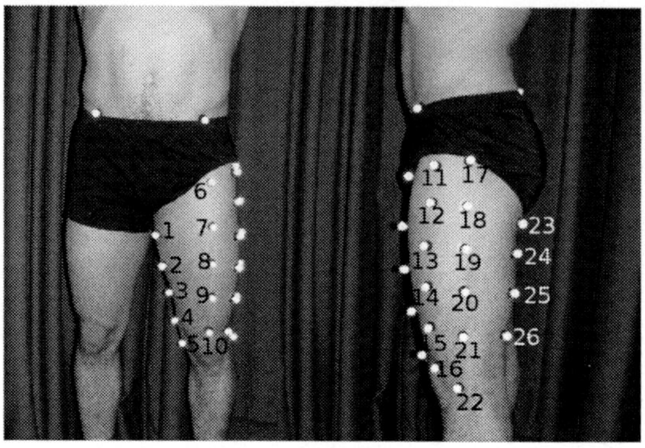

[b] Determination of the most deformed markers [46].

[a] Quantification of soft tissue artefact [62].

Figure 2: Redundant set of markers for [a] the quantification of soft tissue artefact or [b] the determination of the most deformed markers.

subject-specific body segment inertial parameters. Most of the times, they consist in the use of medical imaging [4, 42, 67] and cannot be used easily and widely for sport or clinical analysis.

This chapter aims at describing recent techniques for optimizing the segmental kinematics and identifying subject specific inertia parameters. Both parts are introduced by a short state of art to further explain the limitation of a standard protocol and present a range of solutions proposed in the literature. This chapter is concluded by a few recommendations.

2 Kinematics

To analyse the 3D kinematics of the skeleton, local systems of coordinates should be associated with each segment. Local systems of coordinates are defined using anatomical landmark and joint centre locations [19, 20] to interpret the joint kinematics according to the functional anatomy. Most of the time, anatomical landmark positions are located by placing reflective markers on the skin above bony prominences. These markers are called anatomical markers while markers put elsewhere on the segment are called technical markers. Another approach has been proposed for locating anatomical landmarks because anatomical markers close to the joints are highly sensitive to soft tissue artefacts [18, 20] (e.g. anatomical marker corresponding to the greater trochanter). This alternative method is called the Calibrated Anatomical System Technique (CAST, [17, 19]): anatomical landmarks are pointed using a wand composed of at least two markers with a known position relative to the tip. The location of the anatomical landmark, i.e. the wand tip, may be determined in the laboratory-fixed systems of coordinates and then in a corresponding local system of coordinates using the technical markers. These technical markers can be put all over the segment, specially far from joints and large soft tissues areas. Consequently their kinematics are less affected by skin movement artefact [20].

Concerning the joint centres, they are located through *predictive* or *functional* methods. The glenohumeral and the hip joint centre locations are far from the skin surface and thus difficult to locate from bony landmarks. However they are of many interests because these locations are required to define humeral and femoral local systems of coordinates. The predictive approach is based on regression equations: of the scapula geometry for estimating the glenohumeral joint location, of the pelvis geometry for the hip joint location (Figure 3). Equations of regression can be found in Bell et al. [10] for the hip joint centre and in Meskers et al. [45] for the glenohumeral joint centre. This approach is mainly affected by

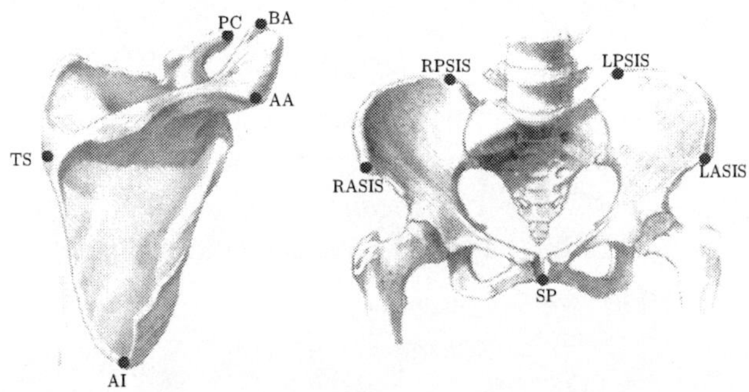

Figure 3: Anatomical landmark positions for locating the glenohumeral joint centre (AA: Angulus Acromialis, AC: AcromioClavicular joint, AI: Angulus Inferior, PC: Processus Coracoideus, TS: Trigonum Spinae) and the the hip joint centre (ASIS: Anterior Superior Iliac Spine, PSIS: Posterior Superior Iliac Spine, SP: pubis symphysis) from regression equations.

errors on anatomical landmark calibration and by the regression uncertainties that are specific to a given population [27]. Furthermore reflective markers need to be stuck on specific bony landmarks with accuracy. Hence for pathological subjects with skeletal deformations, predictive approaches are not suitable. Among the functional methods (e.g. **SCoRE** [32], SARA[33], Helical Axis[69]), none of these has definitively shown a superiority. Several parameters affect the location like the type of the movement (i.e. flexion/extension, abduction/adduction, circumduction) and the nature of movement (velocity and range of motion) [7]. The work of Begon et al. [7] is based on experimental and simulated data. The simulation is constructed from systematic and random noise model to reproduce skin movement artefact that showed a cyclic behaviour with the largest deviations observed near the maximal flexion or abduction. The nature of the movements offers a compromise between a variety of postures and the effect of skin movement. The variety of postures obtained by using different types of movement is always positive in terms of accuracy (Figure 4).

Since in each study participants perform different movements, the comparison and ranking of functional methods remain difficult. Of course, the precision and accuracy of these approaches depend mainly on the quality of the kinematical data. A Recent paper haa

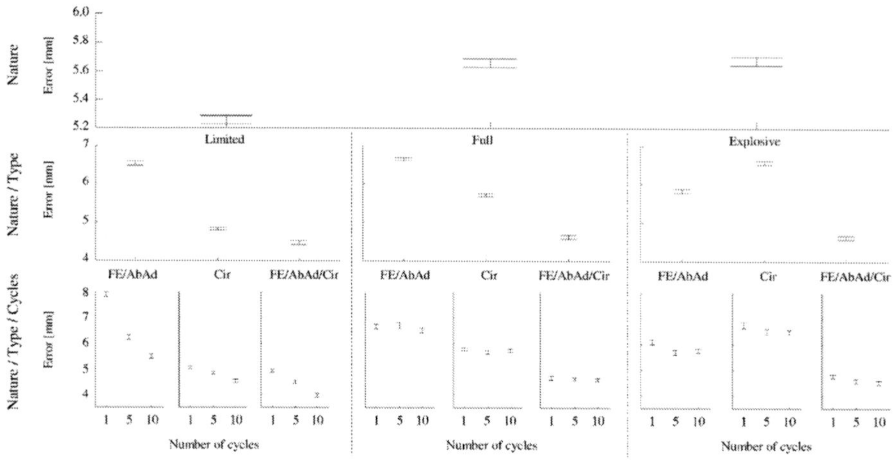

Figure 4: Graphical representation of the multiple comparison procedure of the nested ANOVA. The standard error of the estimated means was calculated using Tukey's post hoc test. The nature of the movement, the nature-type interactions and the nature-type-cycles interactions were tested successively. Two means were considered to be significantly different when their intervals did not overlap, and were considered as not differing significantly when their intervals overlapped ($p < 0.01$) [7].

shown that the Symmetrical centre of Rotation Estimation (**SCoRE**) algorithm is better than the Helical Axis method which is recommended by the ISB [48] because **SCoRE** algorithm is not affected by slow velocities. Therefore its precision can be improved using minimization method for determining the kinematics [8]. Much research has been done and is still going on to improve the accuracy of functional algorithms that determine optimal joint centre location. However, there is a lack of research in whether the obtained locations can be used accurately and effectively when analysing subsequent human movement, specially where the noise from soft tissue deformation is substantially different from the trials used to determine the optimal location. A few studies have been focused on this part of the biomechanical analysis [57, 8].

Soft tissue artefact can also be reduced using optimization methods. Human kinetics calculation is often based on multibody dynamics assuming pin joints without translation. However with at least three markers, each body segment can be considered independently of the proximal one and will have three degrees of freedom in rotation and three degrees of freedom in translation. This results in apparent joint dislocation. To reduce this problem and the skin movement artefacts, Lu and O'Connor [44] proposed a global optimization procedure with a chain model, i.e. the degrees of freedom at each joint and the segment lengths are previously defined from the joint centre locations. The positions of the reflective markers are calculated in the local frames of the corresponding segments and a global optimization solves a multi-target inverse kinematics problem for the body configuration. This method has been successfully applied to computer simulated movements of the lower-limbs [44] and the upper-limbs [59]. Thereafter Begon et al.[9] have shown the interest of this method for movement with unsolvable simulation and for occlusions. Global optimization provides a great opportunity to design optimal marker sets to minimize skin movement artefact, because less than three markers are needed on each body segment and the noisy markers can be removed. This global optimization method can also be improved by a Kalman smoothing filtering [30]. For an accurate analysis of the joint kinematics, the joints cannot be modelled as hinges or ball and sockets. In this case, the global optimization method becomes less interesting compared with a local approach.

The marker placement can follow the ISB recommendations on the definition of joint coordinate systems or be placed over joints. The segment orientation is based on anatomical landmarks indirectly measured. In this chapter, we prefer to associate the CAST protocol with the functional method for locating joint centre. The main advantage is to minimize soft tissue artefact by putting technical markers far from joints. Furthermore, several studies have highlighted that using more than three markers per segment is a good way for minimizing noise and soft tissue artefacts [8, 23, 70]: the segment kinematics is then calculated using minimization methods. However, these methods are not widely used in biomechanics specially in clinical analysis where two markers per segment can be used while the third one is a virtual marker and comes from assumption about joint centre locations.

The first section (2) of this chapter deals with the effect of using a redundant marker set ($n > 3$) for locating joint centre. The functional method is presented (2.1.1) and then the optimization method (2.1.2) for calculating rotation and translation from several markers is detailed. Examples obtained for the thigh and the upper-limb kinematics highlight the benefit of such an approach (2.2 and 2.3).

2.1 Method

2.1.1 Functional method for locating joint centre

The method for locating joint centres will be applied to various joints; the first example concerns the hip joint centre. The positions of hip joint centre with respect to (*wrt*) the local reference frame of the pelvis B_P and *wrt* the local reference frame of the thigh B_T are constant and denoted $\mathbf{u}$ and $\mathbf{v}$, respectively (Figure 5). The matrix R_P (R_T) is the rotation between B_P (B_T) and the global reference frame B_0 and is calculated from the coordinates of the six markers P_i (T_i) using the optimization procedure described in the following section. The vector $\mathbf{p}_P$ ($\mathbf{p}_T$) is the translation from B_0 to B_P (B_T). For each frame f, we have the equality:

$$\underbrace{\begin{bmatrix} {}^fR_P & -{}^fR_T \end{bmatrix}}_{3\times 6} \underbrace{\begin{bmatrix} \mathbf{u} \\ \mathbf{v} \end{bmatrix}}_{6\times 1} = \underbrace{\begin{bmatrix} {}^f\mathbf{p}_T - {}^f\mathbf{p}_P \end{bmatrix}}_{3\times 1}. \tag{3}$$

The vectors $\mathbf{v}$ and $\mathbf{u}$ can be found with at least two different configurations. But more configurations would be desirable because of noise. The vectors $\mathbf{u}$ and $\mathbf{v}$ are therefore estimated in a least square sense:

$$\begin{bmatrix} \mathbf{u} \\ \mathbf{v} \end{bmatrix} = \left(B^T B\right)^{-1} B^T C, \tag{4}$$

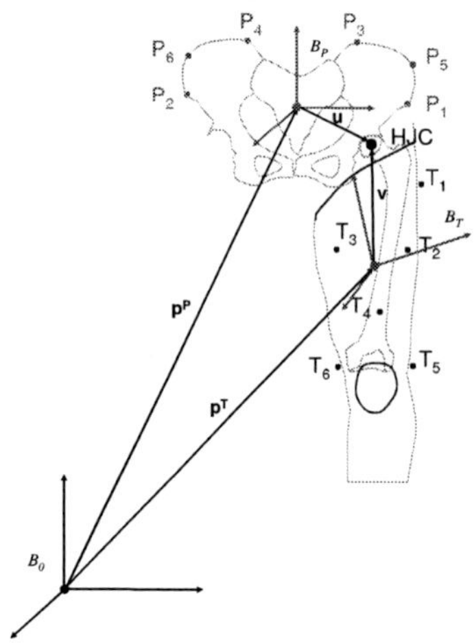

Figure 5: Markers positions and vectors used in the **SCoRE** method. Vector **u** is the position of hip joint centre *wrt* B_P and vector **v** is the position of hip joint centre *wrt* B_T.

with

$$B = \underbrace{\begin{bmatrix} {}^1R_P & -{}^1R_T \\ \vdots & \vdots \\ {}^nR_P & -{}^nR_T \end{bmatrix}}_{3n\times 6}, \tag{5}$$

and

$$C = \underbrace{\begin{bmatrix} {}^1\mathbf{p}_T - {}^1\mathbf{p}_P \\ \vdots \\ {}^n\mathbf{p}_T - {}^n\mathbf{p}_P \end{bmatrix}}_{3n\times 1}. \tag{6}$$

For an accurate estimation of the joint centre location, this equation highlights the impact of the rotation matrix $(R_P, \ R_T)$ and translation vectors $\mathbf{p}_T$, $\mathbf{p}_P$ for each segment. The purpose of the following section is to propose an optimization method for determining these parameters.

2.1.2 Rotation matrix optimization

Usually, rotation matrix of each segment is computed with three markers, that is the minimum. However, in order to limit the uncertainty associated to the motion analysis system and to minimize soft tissue artefact, a few authors have presented optimization methods for calculating rotation matrices with more than three markers [21, 22, 66, 70]. Another method coming from mathematics [14] has the advantage to not using square root or inverse operation. This method is detailed in the following paragraphs.

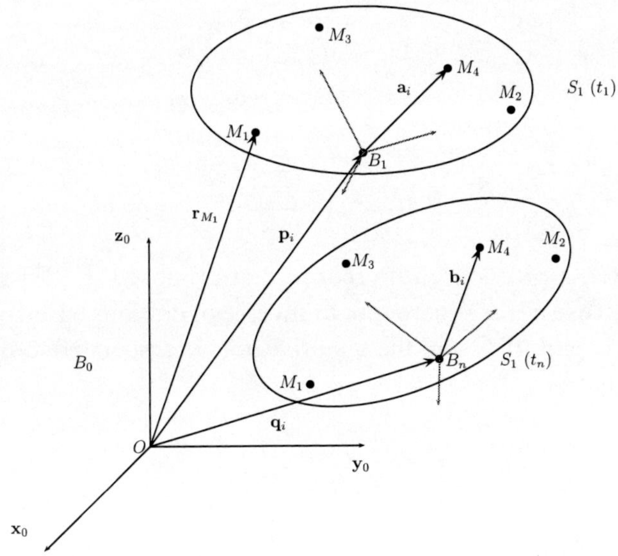

Figure 6: Vector definition for one segment-embedded reference at two times (t_1 and t_n).

All the notations refer to Figure 6. The first step is to calculate the rotation of segment S_1 at time t_1 with respect to B_0 (${}^{t_1}_{0}R$). The rotation matrix ${}^{t_1}_{0}R$ between B_1 and B_0 is:

$$ {}^{t_1}_{0}R = \begin{bmatrix} \mathbf{x} & \mathbf{y} & \mathbf{z} \end{bmatrix}, \tag{7} $$

where

$$ \mathbf{x} = \frac{\mathbf{r}_{M_2}(t_1) - \mathbf{r}_{M_1}(t_1)}{\|\mathbf{r}_{M_2}(t_1) - \mathbf{r}_{M_1}(t_1)\|}, \tag{8} $$

$$ \mathbf{z} = \mathbf{x} \times \frac{\mathbf{r}_{M_3}(t_1) - \mathbf{r}_{M_1}(t_1)}{\|\mathbf{r}_{M_3}(t_1) - \mathbf{r}_{M_1}(t_1)\|}, \tag{9} $$

and

$$ \mathbf{y} = \mathbf{z} \times \mathbf{x}. \tag{10} $$

In the second step, the rotation matrix R is the rotation of the segment S_1 from t_n to t_1: $R = {}^{t_n}_{t_1}R$. The rotation matrix of the segment is such that $\mathbf{p}_i = R\mathbf{q}_i + \mathbf{v}$ and $\mathbf{a}_i = R\mathbf{b}_i$. R can be found by maximizing the following function [23]:

$$ \text{tr}\left[\frac{1}{n}\sum_{i=1}^{n}\left(\mathbf{b}_i\mathbf{a}_i^T\right)R\right] = \text{tr}\left(C^T R\right), \tag{11} $$

(tr represents the trace operator) with

$$ C = \frac{1}{n}\sum_{i=1}^{n}\left(\mathbf{b}_i\mathbf{a}_i^T\right), \tag{12} $$

$$\mathbf{a}_i = \mathbf{r}_{M_i} - \frac{1}{n}\sum_{i=1}^{n}\mathbf{r}_{M_i} \tag{13}$$

and

$$\mathbf{b}_i = \mathbf{r}_{M_i} - \frac{1}{n}\sum_{i=1}^{n}\mathbf{b}_{M_i}. \tag{14}$$

In the present chapter, a procedure different to that proposed in [23] is implemented for calculating R without use of the square root or inverse operations based on the work of [14]. Firstly, the symmetric part of C and the quadrivector $\mathbf{w}$ associated to the skew symmetric part of C are computed as

$$S = \frac{1}{2}(C + C^T) \tag{15}$$

and

$$j(\mathbf{w}) = C - C^T, \tag{16}$$

where $j(\mathbf{w})$ is the skew-symmetric mapping defined by: $j(\mathbf{w})\mathbf{v} = \mathbf{w} \times \mathbf{v}$. To the 3×3 matrix C is associated the 4×4 matrix Q,

$$Q = \begin{bmatrix} 2S - [\mathrm{tr}(S)]\, 1_3 & \mathbf{w} \\ \mathbf{w}^T & \mathrm{tr}(S) \end{bmatrix}. \tag{17}$$

Any rotation can be expressed by the formula:

$$R = \left(m_0^2 - \mathbf{m}.\mathbf{m}\right) 1_3 + 2\mathbf{m}.\mathbf{m}^T + 2m_0 \left[j\left(\mathbf{m}\right)\right], \tag{18}$$

with $\mathbf{m}.\mathbf{m} + m_0^2 = 1$ and $m_0 > 0$ (1_3) is a 3×3 identity matrix. To the rotation R is associated the quadrivector $\mathbf{Y}$ [61]:

$$\mathbf{Y} = \begin{bmatrix} \mathbf{m} \\ m_0 \end{bmatrix}. \tag{19}$$

With the association of $Q$and Y to C and R the expression for tr$(C^T R)$ in Equation 11 becomes:

$$\mathrm{tr}\left(C^T R\right) = \mathbf{Y}.Q\mathbf{Y}. \tag{20}$$

The problem formulated in Equation 11 is now to maximize:

$$\phi(\mathbf{Y}) = \frac{\mathbf{Y}.Q\mathbf{Y}}{\mathbf{Y}.\mathbf{Y}} \tag{21}$$

The conjugate gradient iterative algorithm is then applied [15]:

- Initialise $\mathbf{Y}_0$, $\mathbf{W}_0 = \mathbf{G}_0 = (\mathrm{grad}\phi)\left(\mathbf{Y}_0\right)$
- While $\mathbf{G_i} > \epsilon$

- Search for μ_i maximizing $\phi(\mathbf{Y}_i + \mu \mathbf{W}_i)$
- Compute $\mathbf{Y}_{i+1} = \mathbf{Y}_i + \mu_i \mathbf{W}_i$
- Compute $\mathbf{G}_{i+1} = (\text{grad}\phi)(\mathbf{Y}_{i+1})$
- Compute $\lambda_{i+1} = \frac{\mathbf{G}_{i+1}.(\mathbf{G}_{i+1} - \mathbf{G}_i)}{\mathbf{G}_i.\mathbf{G}_i}$
- Determine the new search direction $\mathbf{W}_{i+1} = \mathbf{G}_{i+1} + \lambda_{i+1}\mathbf{W}_i$

- end

In this algorithm, **G** and μ are calculated by the formulae:

$$G = \frac{2}{\mathbf{Y}.\mathbf{Y}}\left[Q\mathbf{Y} - \phi(\mathbf{Y})\mathbf{Y}\right] \tag{22}$$

and

$$\mu = \frac{-\beta - \sqrt{\Delta}}{2\alpha}. \tag{23}$$

In these formulae,

$$\alpha = (\mathbf{Y}.\mathbf{W})(Q\mathbf{W}.\mathbf{W}) - (Q\mathbf{Y}.\mathbf{W})(\mathbf{W}.\mathbf{W}), \tag{24}$$

$$\beta = (\mathbf{Y}.\mathbf{Y})(Q\mathbf{W}.\mathbf{W}) - (\mathbf{Y}.Q\mathbf{Y})(\mathbf{W}.\mathbf{W}), \tag{25}$$

and

$$\begin{aligned} \Delta &= \left(1 - \frac{(\mathbf{Y}.\mathbf{W})^2}{(\mathbf{Y}.\mathbf{Y})(\mathbf{W}.\mathbf{W})}\right)\beta^2 + 4(\mathbf{Y}.\mathbf{Y})(\mathbf{W}.\mathbf{W}) \\ &\left[Q\mathbf{Y}.\mathbf{W} - \frac{(\mathbf{Y}.\mathbf{W})}{(\mathbf{Y}.\mathbf{Y})(\mathbf{W}.\mathbf{W})}\frac{(\mathbf{Y}.\mathbf{Y})(Q\mathbf{W}.\mathbf{W}) + (\mathbf{W}.\mathbf{W})(Q\mathbf{Y}.\mathbf{Y})}{2}\right]^2. \end{aligned} \tag{26}$$

The rotation matrix ${}^{t_n}_{t_1}R$ is then calculated from the solution (the quadrivector $\mathbf{Y}$) of the gradient algorithm with Equation 18. In a third step, at each instant of time the optimized rotation matrix ${}^{t_n}_{0}R$ between B_n and B_0 is:

$${}^{t_n}_{0}R = {}^{t_1}_{0}R \;\; {}^{t_n}_{t_1}R \tag{27}$$

In current biomechanical analysis, the rotation matrix is calculated with three markers using Equations 7 to 10 at each instant of time and the translation vector is the position in the global reference frame of coordinates of one of these three markers:

$$\mathbf{t} = \mathbf{r}_{M_i}. \tag{28}$$

Whereas in the optimal methods, the translation vector is defined as the mean vector of all the markers fixed on the segment:

$$\mathbf{t} = \frac{1}{n}\sum_{i=1}^{n}\mathbf{r}_{M_i}. \tag{29}$$

To summarize, the conjugate gradient iterative algorithm proposed here calculates the rotation between two orientations of the segment without square root or inverse operations. The orientation of the segment is then recalled in the global reference frame with Equation (27).

2.2 Effects of the rotation matrix on the kinematics

To illustrate the benefit for calculating the rotation matrix from a redundant marker set, a movement was simulated with a noise added to a reference kinematics. The artificial noise included random and continuous components, it was designed to reproduce skin movement artefact plus measurement errors (see [7] for details). The rotation matrix was then calculated from the simulated data (reference) and from noisy data with a number of markers ranging from three to eight. In a paper dealing with the determination of the rotation matrix, Challis [23] proposed to assess the accuracy in the kinematics estimation parameters by calculating the mean absolute relative difference $(\overline{E})$ between the data from the noisy kinematics and those obtained from the reference kinematics. The average of the absolute difference was calculated by the following formula:

$$\overline{E} = \frac{1}{n} \sum_{i=1}^{n} \left\| \frac{VC_i - VE_i}{VC_i} \right\| \tag{30}$$

where n is the number of data sample, VC_i is the real value (obtained from the reference kinematic) and VE_i is the estimated value obtained from the noisy data.

The mean absolute relative difference was calculated on three different sets of kinematical entities: the nine coefficients of the rotation matrix (R), the angular velocity vector (ω) and the nine coefficients of the second time derivative of rotation matrix ($\ddot{R}$). The effect of an increased number of markers for calculating the rotation matrix is not very pronounced when regarding the nine coefficients of the rotation matrix (Figure 7): the mean absolute relative difference is 0.028 for three markers and drops to 0.011 for seven markers. These results as those of Challis [23] show that the accuracy increases with the number of markers and the change in accuracy was most rapid when the number of markers increased from three to four. We can also observe that the number of markers is not always correlated with an improvement of the accuracy because an additional marker can give more noise than extra information. For example from five to six markers, the error drives from 0.0139 to 0.0146. However, this trend disappears with the time derivation. For the angular velocity and for the second time derivative of the rotation matrix, an increased number of markers is always related to a better accuracy.

This simulation study was continued with an experimental approach [46]. A set of 26 markers was methodically fixed on the left thigh of a male participant who was asked to perform several walking trials. The hip joint kinematics were calculated with various combinations of three to six markers while a reference kinematics was calculated with all the 26 markers. The marker locations and the effect of using more than three markers were evaluated for flexion-extension, abduction-adduction and mediolateral rotation angles (Figure 8). From three to six markers the precision increased: from 3.20to 2.20for flexion-extension angle, from 3.18to 2.07for abduction-adduction angle and from 7.33to 3.73for mediolateral hip rotation angle (reference angles were obtained with the 26 markers). Statistical differences (ANOVA, $p<0.05$) were observed between three and four markers, between three and five markers and between three and six markers for all the axes of rotation.

These results highlight the benefit of using more than three markers for calculating the rotation matrix. It has a direct impact on the nine coefficients of the rotation matrix. When hip joint angles are calculated, these results show that the precision is increased

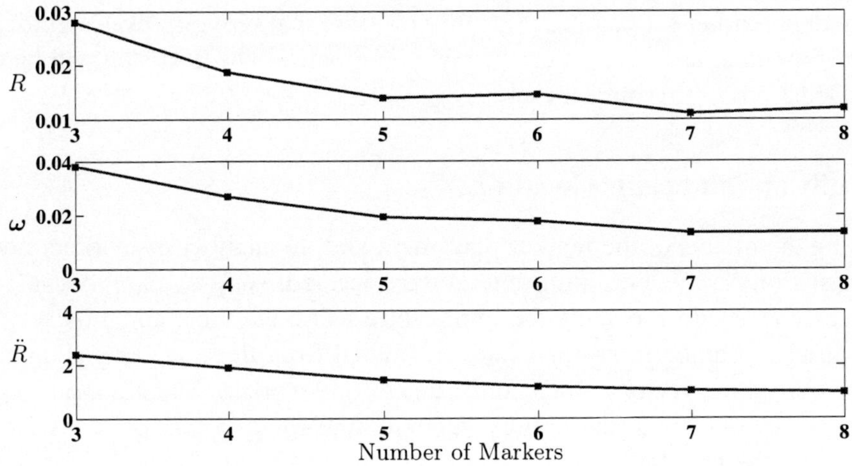

Figure 7: Mean absolute relative difference for the rotation matrix (R), the angular velocity (ω) and the second time derivative of the rotation matrix ($\ddot{R}$) according to the number of markers.

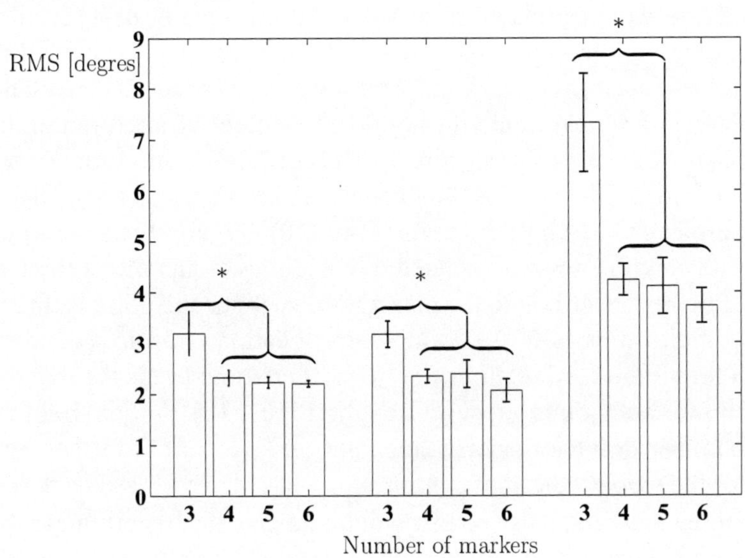

Figure 8: Mean RMS differences of the three angles (flexion-extension, abduction-adduction and mediolateral rotation from left to right) according to the number of markers involved.

with the number of markers. Statistically, the RMS difference is equal for four, five and six markers. For describing the segment kinematics or for locating joint centre, it appears that a redundant marker set composed of four or more markers improves significantly the accuracy and avoid to apply filters on the raw coordinates. The next step will be to define the precise locations of this markers set.

2.3 Results on joint centre location

For assessing the effects of the marker placement and the number of markers involved in the rototranslation parameters, joint centres were located using two approaches: the first approach has consisted in locating the joint centre using the most common way i.e. the rotation matrices of adjacent segments are calculated from three markers (Equations 7 to 10) and the translation vector is the coordinates of one marker. The second approach has located the joint centre with the optimal method: the rotation matrix is calculated with six markers and the translation vector is the centroid of the six markers. Two studies are presented here [8, 57]. The first one is focused on the shoulder and elbow joint centre locations with all the combinations of three markers for both proximal and distal segments. The second study applies the optimal method (rotation matrix and translation vector) for locating the hip joint centre.

In the first study [57], the authors evidenced that the **SCoRE** method (as all the functional methods) located differently the joint centre according to the selected three markers (for both proximal and distal segments) and their order. Results concern the reconstruction of the joint centres during a fast movement with impact, i.e. a punch. In a first step, multiple joint centres were estimated from the set-up movement data for all permutations of three from six markers in both proximal and distal segments. Groups of three markers that were used to define a local reference frame were called triads. Then, all these solutions were reconstructed in a global frame during the movement of interest and these locations were fitted with a 95% sphere of precision. This procedure was performed for the shoulder and elbow joint centres. Most of the time using three markers ($M_{i,j,k}$), the local systems of co-ordinates are calculated by Equations (7) to (10). Figure 9 is a two-dimensional example to illustrate how the noise associated to each marker can affect the local system of co-ordinates differently according to the marker order. Since the noise is three-dimensional and marker-specific, diffences in the joint centre location is up to 44 mm (shoulder) or 56 mm (elbow) for all the permutations of a triad. For some triads the maximum radius is lower than 15 mm for both joints but never nought (Figure 10). Although solidification [26] has reduced random error it has dispersed the joint centre locations further apart, especially along one axis of the elbow. The actual information has also been reduced.

Using only three markers, these results would indicate that determining the joint centre with an accuracy of greater than approximately 20 mm is unlikely. This result confirms our affirmation in the introduction, the quality and accuracy of the kinematics is of highest importance.

It is also pertinent to note that, although the theoretical literature regarding functional methods relates to hinges or ball and sockets joints, the elbow, knee or ankle are neither truly hinge nor ball and socket joints. However, as presented in [57], the elbow is not ideally suited to determine joint centre location. But the **SCoRE** method can be applied to

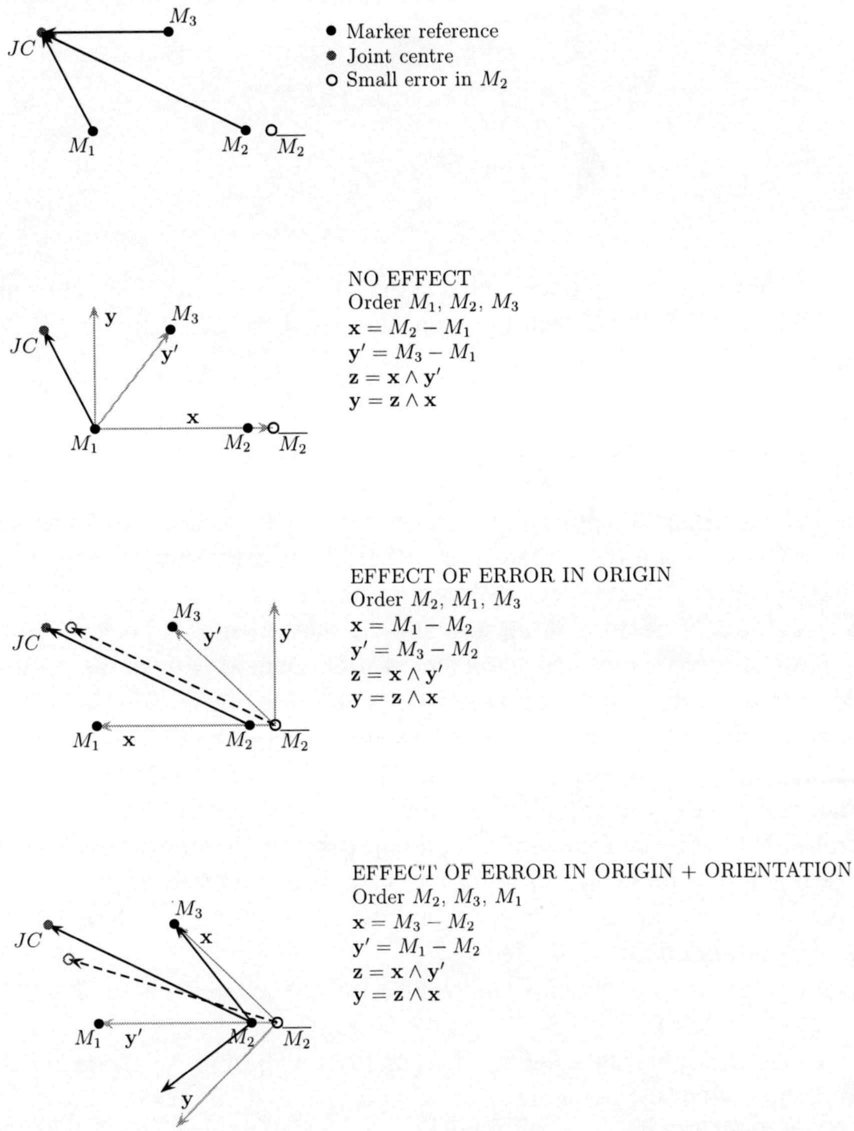

Figure 9: Effect of the noise associated to one marker (M2) on the joint centre position (JC) in a two-dimensional case. When the marker order is M_1, M_2, M_3, the error on M_2 has no influence on the joint centre location. When the marker order is M_2, M_1, M_3, the error on M_2 affects the location of the joint centre but does not affect the segment reference frame. When the marker order is M_2, M_3, M_1, the error on M_2 affects both the rotation and the translation.

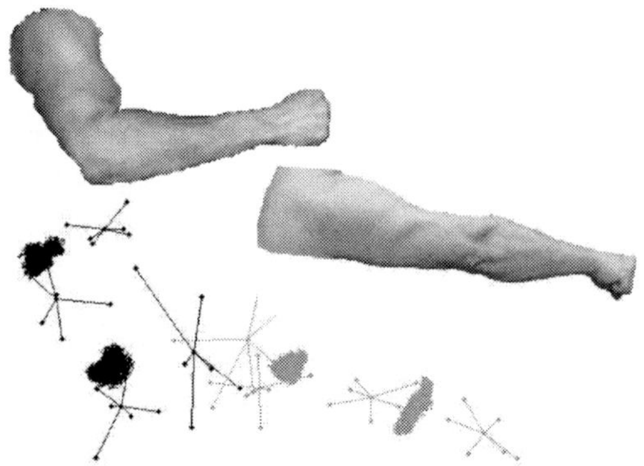

Figure 10: Lateral views of the upper-limb and the 14,400 locations of shoulder and elbow joint centres for the 1st frame (black lines and dots) and the 220th frame (grey lines and dots) of the punch [57].

the elbow joint given that the combined movement from each of the three articulations was not uniaxial.

The second experimental application is related to the hip joint centre location and has assessed the effect of using more than three markers for calculating the rotation matrix. For that, six reflective markers were fixed on the thigh and six on the pelvis (Figure 5). The subject was asked to perform three slow star-arc movements [16] over a large range of motion. Hip joint centre was located using the **SCoRE** method with all the combination of three markers on the distal and proximal segment and using one marker for the translation vector (*Classical* method). From the same data, hip joint centre was located using all the six markers and the centroid vector for the translation vector (*Optimal* method). As the order of the markers used to calculate the rotation matrix affects the orientation of the segment-embedded reference frame [57], for all the permutation of markers in the proximal and distal segment, a joint centre location was calculated. Like in the previous study, a sphere of precision was calculated in static and dynamic conditions (during walking) for the *Classical* and *Optimal* methods.

In static, the locations of the hip joint centres of the spheres varied slightly with the method: 3.7 mm (Table 1).

However the radii of the static spheres ranged from 4.0 mm to 43.5 mm and during the walking trial, the radii of the dynamic spheres increased to 45.6 mm for *Classical* method and to 29.9 mm for *Optimal* method (Table 1). The estimation of an optimal joint centre location is affected by the number of markers and their placement. However the main uncertainty comes from the reconstruction of this optimal location in the motion of interest where only the redundancy of the marker set can provide an accurate location.

2.4 Conclusion

All these results highlight the effects of the numbers of markers involved in the kinematics calculation, i.e. joint centre locations and angles. Increasing the number of markers is important for an accurate estimation of the rototranslation matrix.

The rotation matrix of each segment has an impact on the angles determination and on

Table 1: Location and precision [mm] of the hip joint centre reconstruction for the static and walking trial using classical and optimal methods [8].

		Classical method	Optimal method
3*Static location	x	115.1	113.0
	y	300.5	303.4
	z	897.5	898.5
2*Static radius	min	40.0	4.0
	max	43.5	4.2
2*Dynamic radius	min	31.9	15.5
	max	45.6	29.9

the joint centre locations when using **SCoRE** method. Our results show the inter-marker variability on the kinematics due to soft tissue artefact that cannot be directly measured without invasive approaches. Soft tissue artefact is defined as markers movement relative to the underlying bone. Currently only invasive devices and fluoroscopic techniques can achieve direct measurement of bony segment movement [37]. Error introduced in the joint centre reconstruction came mainly from the lack of redundancy during the reconstruction while the optimal joint centre is calculated for numerous frames. Since joint centre cannot be determined from the motion of interest [54], subjects should perform extra set-up trials which have different skin movement artefacts than those during the movement of interest. This is another source of error; the noise has to be minimized in both movements.

The accuracy of the joint centre location is not only due to the number of markers but also due to the movement characteristics as shown in the results of the first study [57]. Since the joint centre is obtained by a pseudo-inverse procedure, the redundancy of the movement is important while the noise remains small. Once the joint centre locations have been determined, a method of reconstructing them during the movement of interest is normally required. In this second step, the accuracy is not improved by the movement redundancy because the joint centre is reconstructed in each single frame. The studies of Begon et al. [8] and Roosen et al. [57] assess this difficult and blank part of the biomechanical analysis. They recall important principles of human biomechanics analysed using rigid body mechanics. By in vivo examples they have shown the limits of the solid mechanics for biomechanics and have proposed methods for improving the accuracy of the kinematics. Since the human body segments are not rigid, three markers are not enough to calculate accurately the kinematics. The classical approach for calculating the rototranslation matrix using three markers (Equations 7 to 10) holds concurrently error on the origin and on the reference axes. According to the marker placement on the segment and to their order (i.e. $ijk, ikj, jki, jik, kij, kji$ for markers i, j, k), the joint centre location and the joint kinematics will differ. The equality between all the permutations occurs only in one case: the body is perfectly rigid and there is no noise in the marker coordinates measurement. This case is impossible in biomechanics. The hypotheses assessed in [8, 57] are not exhaustive and will be completed in the next future.

3 Body segment parameters

Are the body segment inertial parameters essential in biomechanics? As said in a recent paper [55] the influence of body segment inertial parameters for solving the problem of inverse dynamics is controversial. Some papers reported low importance (< 0.06 Nm.kg^{-1}) of body segment inertial parameters uncertainties [24, 36, 53] while others demonstrated the opposite ($>$ 12%) [1, 40, 60]. The difference between these two findings comes mainly from the movements that each group studied. During walking, low accelerations of lower-limbs and high importance of the reaction force during step minimize effects due to body segment inertial parameters on net moments at the hip, knee or ankle joints. While during other movements [1, 40], specially without contact forces, body segment inertial parameters show non negligible effects on net moments.

Whatever is our opinion and experience on this question, body segment inertial parameters should be estimated for any biomechanical analysis. The researcher can choose between several models or methods: geometric models, proportional models derived from cadavers' studies, in vivo scanning models or identification. Each approach has its advantages and drawbacks (see [13] for an overview). We propose to present and develop the latter method – the identification – for the determination of personalized body segment inertial parameters. It consists in solving a redundant system for the inertial parameters by a numerical computation. This method has been already used by [65] but with some limitations and assumptions (two-dimensional model, symmetric limbs). Originally identification was developed for estimating the segment parameters of robots. Indeed, for robot manipulators, these parameters are not always precisely known [6, 56] and various algorithms of identification exist to determine them [2, 3, 6, 12, 25, 56]. In the classical identification approaches [2, 6], the parameters are estimated from motion data and actuator torques or forces, both measured by internal measurement devices. For human applications, this approach is impossible because net joints forces and moments cannot be measured; they are calculated from kinematics and contact forces using an inverse dynamics procedure. An alternative approach is the external identification. In this method, the external model relates the motion of the multibody system to the reaction forces and torques on a force platform. This technique was initiated by [56] for identifying the segment inertial properties of a three segment robot (PUMA 562).

The method proposed by [56] is based on the recursive Newton-Euler algorithm in which the inertial parameters of each segment are represented by a ten-dimensional vector solved by a least-squared method (pseudo-inverse). The ten inertia parameters of n segments are contained in the $\mathbf{x}$ vector (which is a $10\,n$ dimensional vector) and the following inverse problem is solved for $\mathbf{x}$:

$$A\mathbf{x} = \mathbf{b}, \tag{31}$$

where matrix A contained the kinematics and vector $\mathbf{b}$ groups the external forces and torques together. This method was applied to the human body with degrees of freedom ranging from 6 to 25 [25] but the results were unsatisfactory on account of the high number of degrees of freedom. Furthermore, when the inertia parameters are gathered in a vector, the properties of the inertia matrix (i.e., symmetric and positive definite) are lost. That is why we prefer a matrix form of the motion equations that keeps the properties of the inertia matrix and use them as constraints during the resolution of the inverse problem.

Some authors introduced the matrix formulations for dynamical analysis. [41] [43] and proposed 4×4 and 6×6 matrices, respectively. Fayet et al. [34, 35] have introduced the global inertia tensor for an iterative formulation of the equations of motion in a 4×4 matrix form. However all these methods are recursive formalisms. Since they are applied successively to the different segments, they cannot be used for human applications where the net joint forces and moments are unknown. In answer to this weakness, we propose an original expression of the Newton-Euler equation of motion where all the inertial parameters of a multibody system composed of n segments are grouped together in a single symmetric and positive definite matrix of dimension $(3n + 1) \times (3n + 1)$ [5].

The external identification appears favourable to the determination of the human body segment inertial parameters. The high number of the human body segments could be artificially reduced by soliciting a single joint. In the following example, the human body is modeled as a two-segment system: the right upper-limb and the remainder of the body. The purpose of this section is to detail the matrix formulation of the equation of motion for identifying the inertia parameters of the upper-limb. The following section is organized in three parts: firstly the matrix formulation of the motion is detailed (3.1) and then the subsections 3.2 and 3.3 present the identification process and the experimental protocol.

3.1 An original Newton-Euler formalism for identification

Let S_1 and S_2 be two rigid bodies of mass m_1 and m_2 coupled each other around the joint centre A which is common to both bodies (Figure 11). These bodies move in a Galilean

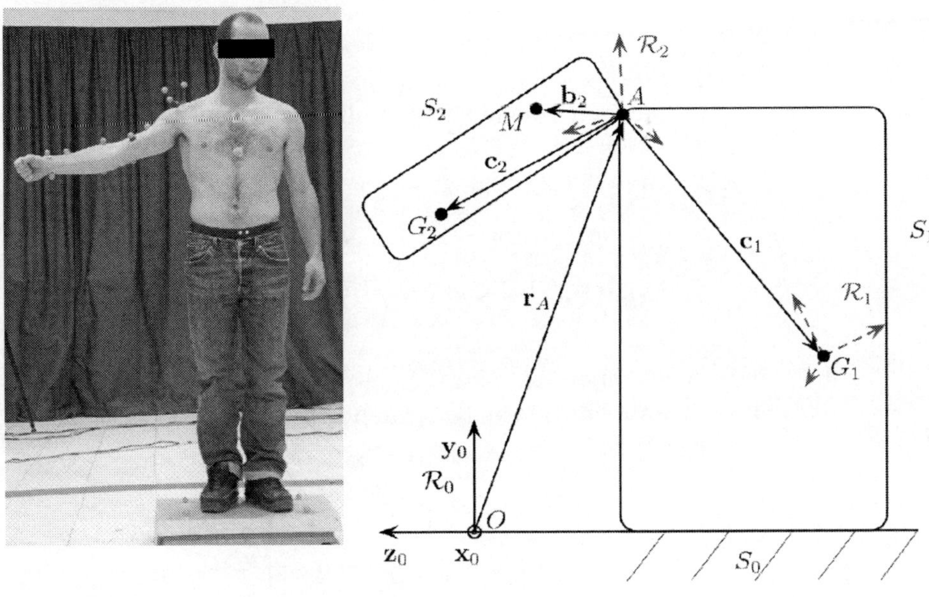

Figure 11: Schema

reference frame with reference frame $\mathcal{R}_0\,(O, \mathbf{x}_0, \mathbf{y}_0, \mathbf{z}_0)$. The mass centre of these bodies are denoted G_1 and G_2 and are expressed with respect to the joint centre by the vectors

$\mathbf{c}_i$ $(i = 1, 2)$. The position of the joint centre A is described by the vector: $\mathbf{r}_A = \overrightarrow{OA}$ (note: the different reference frames are expressed by a superscript in parentheses). The relation between the global coordinates of a vector $\mathbf{c}^{(0)} = \begin{bmatrix} c_{x0} & c_{y0} & c_{z0} \end{bmatrix}^T$ in the absolute reference frame $\mathcal{R}_0$ and its local coordinates $\mathbf{c}^{(i)} = \begin{bmatrix} c_{x_i} & c_{y_i} & c_{z_i} \end{bmatrix}^T$ expressed in the body-fixed reference frame, is described by the rotation matrix $R^{(i)}$: $\mathbf{c}^{(0)} = R^{(i)}\mathbf{c}^{(i)}$ (Figure 11).

The position of any point M of S_2 is expressed with respect to A by the vector $\mathbf{b}_2 = \overrightarrow{AM}$. According to Vallée et al. [63], the position of M with respect to the reference frame $\mathcal{R}_0$ can be written with the rotation matrix $R^{(2)}$:

$$\begin{aligned} \overrightarrow{OM} &= \overrightarrow{OA} + \overrightarrow{AM} \\ &= \mathbf{r}_A^{(0)} + R^{(2)}\mathbf{b}_2^{(2)}. \end{aligned} \tag{32}$$

With this notation, the acceleration of M noted Γ_M is simply:

$$\mathbf{\Gamma}_M = \ddot{\mathbf{r}}_A^{(0)} + \ddot{R}^{(2)}\mathbf{b}_2^{(2)}. \tag{33}$$

So, the dynamic resultant of S_2 is:

$$\chi_2 = \int_{S_2} (\mathbf{\Gamma}_M)\, dm = m_2 \left(\ddot{\mathbf{r}}_A^{(0)} + \ddot{R}^{(2)}\mathbf{c}_2^{(2)} \right), \tag{34}$$

where $m_2\mathbf{c}_2 = \int_{S_2} (\mathbf{b}_2)\, dm$ defines the mass centre. With the same notation, we can express the dynamic resultant of S_1:

$$\chi_1 = \int_{S_1} (\mathbf{\Gamma}_M)\, dm = m_1 \left(\ddot{\mathbf{r}}_A^{(0)} + \ddot{R}^{(1)}\mathbf{c}_1^{(1)} \right). \tag{35}$$

In the inertial frame $\mathcal{R}_0$, the Newton second law is applied successively to S_1 and S_2:

$$\mathbf{F}_1^{(0)} = \chi_2 + m_2\mathbf{g}^{(0)} = m_2 \left(\ddot{\mathbf{r}}_A^{(0)} + \mathbf{g}^{(0)} + \ddot{R}^{(2)}\mathbf{c}_2^{(2)} \right) \tag{36}$$

$$\begin{aligned} \mathbf{F}_0^{(0)} &= \chi_1 + m_1\mathbf{g}^{(0)} + \mathbf{F}_1^{(0)} \\ &= (m_1 + m_2)\left(\ddot{\mathbf{r}}_A^{(0)} + \mathbf{g}^{(0)} \right) + m_1\ddot{R}^{(1)}\mathbf{c}_1^{(1)} + m_2\ddot{R}^{(2)}\mathbf{c}_2^{(2)}, \end{aligned} \tag{37}$$

where $\mathbf{g}^{(0)}$ is the gravity acceleration: $\mathbf{g}^{(0)} = [0 \quad 0 \quad 9.81]^T\, m.s^{-2}$ and $\mathbf{F}_i$ is the reaction force between S_i and S_{i+1}. Equation 37 can be written in matrix form:

$$\begin{bmatrix} \ddot{R}^{(2)} & \ddot{R}^{(1)} & \left(\ddot{\mathbf{r}}_A^{(0)} + \mathbf{g}^{(0)} \right) \end{bmatrix} \begin{bmatrix} m_2\mathbf{c}_2^{(2)} \\ m_1\mathbf{c}_1^{(1)} \\ m_1 + m_2 \end{bmatrix} = \mathbf{F}_0^{(0)}. \tag{38}$$

For the rotation, we choose to express the moment by a 3×3 skew symmetric matrix, using the tensorial product of two vectors defined by:

$$\mathbf{a}(\otimes)\mathbf{b} = \mathbf{a}\mathbf{b}^T. \tag{39}$$

So the dynamic moment of S_2 about A (noted $\delta_A(S_2)$) is:

$$\begin{aligned}\delta_A(S_2) &= \text{skew} \int_{S_2} \left[\left(R^{(2)}\mathbf{b}_2^{(2)}\right) \wedge \mathbf{\Gamma}_M \right] dm \\ &= \int_{S_2} \left[\mathbf{\Gamma}_M \otimes \left(R^{(2)}\mathbf{b}_2^{(2)}\right) - \left(R^{(2)}\mathbf{b}_2^{(2)}\right) \otimes \mathbf{\Gamma}_M \right] dm. \end{aligned} \tag{40}$$

In this equation, skew is the operator which associates to any trivector a 3×3 skew-symmetric matrix:

$$\text{skew} \begin{bmatrix} X \\ Y \\ Z \end{bmatrix} = \begin{bmatrix} 0 & -Z & Y \\ Z & 0 & -X \\ -Y & X & 0 \end{bmatrix}. \tag{41}$$

Using Equation (33), Equation. (40) becomes after simplification:

$$\begin{aligned}\delta_A(S_2) &= \ddot{\mathbf{r}}_A^{(0)} \otimes \left(m_2 R^{(2)}\mathbf{c}_2^{(2)}\right) + \int_{S_2} \left[\left(\ddot{R}^{(2)}\mathbf{b}_2^{(2)}\right) \otimes \left(R^{(2)}\mathbf{b}_2^{(2)}\right) \right] dm - \\ & \left(m_2 R^{(2)}\mathbf{c}_2^{(2)}\right) \otimes \ddot{\mathbf{r}}_A^{(0)} - \int_{S_2} \left[\left(R^{(2)}\mathbf{b}_2^{(2)}\right) \otimes \left(\ddot{R}^{(2)}\mathbf{b}_2^{(2)}\right) \right] dm. \end{aligned} \tag{42}$$

Thanks to the notion of transposition:

$$\ddot{\mathbf{r}}_A^{(0)} \otimes \left(m_2 R^{(2)}\mathbf{c}_2^{(2)}\right) = \left(\ddot{\mathbf{r}}_A^{(0)} \otimes m_2 \mathbf{c}_2^{(2)}\right) R^{(2)T}, \tag{43}$$

we can rewrite Equation (42) as follows:

$$\begin{aligned}\delta_A(S_2) &= \left(\ddot{\mathbf{r}}_A^{(0)} \otimes m_2 \mathbf{c}_2^{(2)}\right) R^{(2)T} + \ddot{R}^{(2)} \left[\int_{S_2} \left(\mathbf{b}_2^{(2)} \otimes \mathbf{b}_2^{(2)}\right) dm \right] R^{(2)T} - \\ & R^{(2)} \left(m_2 \mathbf{c}_2^{(2)} \otimes \ddot{\mathbf{r}}_A^{(0)}\right) - R^{(2)} \left[\int_{S_2} \left(\mathbf{b}_2^{(2)} \otimes \mathbf{b}_2^{(2)}\right) dm \right] \ddot{R}^{(2)T}. \end{aligned} \tag{44}$$

The integral in brackets is the definition of the Poinsot inertia matrix and is noted $K_2^{(2)}$:

$$K_2^{(2)} = \int_{S_2} \left(\mathbf{b}_2^{(2)} \otimes \mathbf{b}_2^{(2)}\right) dm, \tag{45}$$

and represents a form of the inertia tensor of the body at A_2. K_2 is related to the classical definition of the inertia tensor J_2 by the relation:

$$K_2 = \frac{1}{2} \left(\text{tr} J_2\right) I - J_2, \tag{46}$$

where I is the identity matrix (see [63] for details). The dynamic principle is written as:

$$\begin{aligned}\mathcal{M}_A(S_2) &= \left[\left(\ddot{\mathbf{r}}_A^{(0)} + \mathbf{g}^{(0)}\right) \otimes m_2 \mathbf{c}_2^{(2)} \right] R^{(2)T} + \ddot{R}^{(2)} K_2^{(2)} R^{(2)T} - \\ & R^{(2)} \left[\left(m_2 \mathbf{c}_2^{(2)} \otimes \left(\ddot{\mathbf{r}}_A^{(0)} + \mathbf{g}^{(0)}\right)\right) \right] - R^{(2)} K_2^{(2)} \ddot{R}^{(2)T}, \end{aligned} \tag{47}$$

where $\mathcal{M}_A(S_2)$ is the moment exerted by S_1 on S_2 calculated in A and expressed in the reference frame $\mathcal{R}_0$. $\mathcal{M}_A(S_2)$ is the 3×3 skew-symmetric form of the usual moment vector $\mathbf{M}_A(S_2)$:

$$\mathcal{M}_A(S_2) = \text{skew}\left(\mathbf{M}_A(S_2)\right), \tag{48}$$

that is to say:

$$\mathbf{M}_A(S_2) = \begin{bmatrix} M_A(S_2)_x \\ M_A(S_2)_y \\ M_A(S_2)_z \end{bmatrix} \tag{49}$$

$$\mathcal{M}_A(S_2) = \begin{bmatrix} 0 & -M_A(S_2)_z & M_A(S_2)_y \\ M_A(S_2)_z & 0 & -M_A(S_2)_x \\ -M_A(S_2)_y & M_A(S_2)_x & 0 \end{bmatrix}. \tag{50}$$

With the same notations, the dynamic moment of S_1 about A reads:

$$\begin{aligned} \delta_A(S_1) &= \left(\ddot{\mathbf{r}}_A^{(0)} \otimes m_1\mathbf{c}_1^{(1)}\right) R^{(1)T} + \ddot{R}^{(1)} K_1^{(1)} R^{(1)T} - \\ &\quad R^{(1)}\left(m_1\mathbf{c}_1^{(1)} \otimes \ddot{\mathbf{r}}_A^{(0)}\right) - R^{(1)} K_1^{(1)} \ddot{R}^{(1)T}. \end{aligned} \tag{51}$$

Using classical inverse dynamics, the moment at the point A due to S_0 is:

$$\begin{aligned} \mathcal{M}_A(S_0) &= \left[\left(\ddot{\mathbf{r}}_A^{(0)} + \mathbf{g}^{(0)}\right) \otimes m_1\mathbf{c}_1^{(1)}\right] R^{(1)T} + \ddot{R}^{(1)} K_1^{(1)} R^{(1)T} - \\ &\quad R^{(1)}\left[m_1\mathbf{c}^{(1)} \otimes \left(\ddot{\mathbf{r}}_A^{(0)} + \mathbf{g}^{(0)}\right)\right] - R^{(1)} K_1^{(1)} \ddot{R}^{(1)T} + \mathcal{M}_A(S_2) \end{aligned} \tag{52}$$

The equations of translation (1) and rotation (2) can be summarized in matrix form by introducing the 4×7 matrices A and B:

$$A = \begin{bmatrix} \ddot{R}^{(2)} & \ddot{R}^{(1)} & \left(\ddot{\mathbf{r}}_A^{(0)} + \mathbf{g}^{(0)}\right) \\ \mathbf{0}^T & \mathbf{0}^T & 0 \end{bmatrix}, \tag{53}$$

$$B = \begin{bmatrix} R^{(2)} & R^{(1)} & \mathbf{0} \\ \mathbf{0}^T & \mathbf{0}^T & 1 \end{bmatrix}. \tag{54}$$

Let us summarize the inertia parameters of bodies S_1 and S_2 in the 7×7 matrix H:

$$H = \begin{bmatrix} K_2^{(2)} & [0] & m_2\mathbf{c}_2^{(2)} \\ [0] & K_1^{(1)} & m_1\mathbf{c}_1^{(1)} \\ m_2\mathbf{c}_2^{(2)T} & m_1\mathbf{c}_1^{(1)T} & m_1+m_2 \end{bmatrix}. \tag{55}$$

The equations of motion are then written:

$$AHB^T - BHA^T = C = \begin{bmatrix} \mathcal{M}_A(S_0) & \mathbf{F}_0^{(0)} \\ -\mathbf{F}_0^{(0)T} & 0 \end{bmatrix}. \tag{56}$$

Matrices A, B and C are time dependents. Matrix H is constant because all of this parameters are expressed in the body fixed reference frame. So the translational and rotational motions equations are written in matrix form and the inertia parameters of S_1 and S_2 appear therein linearly. The matrix C is easily available with a force platform and matrices A and B are determine with a motion analysis system. So matrix H could be found by a minimization procedure.

3.2 Conjugate gradient algorithm for identifying a symmetric and positive definite matrix

The matrix C is a 4×4 skew-symmetric matrix, matrix C has six independent components. The matrix H is a 7×7 symmetric and positive definite matrix, so matrix H has 28 independent components. Therefore Equation (56) provides six conditions for 28 unknown, but these six conditions hold at each time. This equation must be sampled in at least five different configurations; more configurations would be desirable because of noise. We make n experiments where t_i is the duration of the ith experiment. Using a least squared method, we define the function $J(H)$:

$$J(H) = \frac{1}{4}\sum_{i=1}^{n}\int_0^{t_i} \left\| A_i(t)HB_i(t)^T - B_i(t)HA_i(t)^T - C_i(t)\right\|^2 dt \tag{57}$$

This function J is minimized by the following conjugate gradient algorithm:

1. Initialize H with H_0

   $G_0 = \nabla J(H_0)$

   $D_0 = -G_0 \; ; \; k = 0$

2. iteration k

   $\mu_k = \dfrac{\mathrm{tr}\left(G_k^T D_k\right)}{\mathrm{tr}\left[D_k^T\left(\sum \int BAD_kBA - BB^TD_kA^TA\right)dt\right]}$

   $H_{k+1} = H_k + \mu_k D_k$

3. if $\|G_{k+1}\| < \epsilon$ break, else

   $D_{k+1} = -G_{k+1} + \lambda_k D_k$

   with $\lambda_k = \dfrac{\mathrm{tr}\left[G_{k+1}^T(G_{k+1} - G_k)\right]}{\mathrm{tr}(G_k^T G_k)}$

   $k = k+1$ et return in 2.

Remark: In μ_k, indices, time and duration are omitted.

In this algorithm, the gradient (∇) of the function J is the matrix calculated as follows:

$$\begin{aligned}
\nabla J(H) \quad = - \; & \sum_{i=1}^{n}\int_0^{t_i}\left[\frac{B_i^T(t)A_i(t)HB_i^T(t)A_i(t) + A_i^T(t)B_i(t)H^TA_i^T(t)B_i(t)}{2}\right] dt + \\
& \sum_{i=1}^{n}\int_0^{t_i}\left[\frac{B_i^T(t)B_i(t)HA_i^T(t)A_i(t) + A_i^T(t)A_i(t)H^TB_i^T(t)B_i(t)}{2}\right] dt + \\
& \sum_{i=1}^{n}\int_0^{t_i}\left[\frac{B_i^T(t)C_i(t)A_i(t) + A_i^T(t)C_i^T(t)B_i(t)}{2}\right] dt
\end{aligned} \tag{58}$$

3.3 Application on the upper-limb

Three men participated in this study. They were instructed to perform fifteen cycles of a complex movement of the right upper-limb composed of a flexion, horizontal abduction, adduction and mediolateral rotation. This movement was chosen because the three degrees of freedom of the glenohumeral joint were excited. This condition is necessary for identifying all the inertia parameters of the upper-limb. Subjects were sat on a chair and they were asked to keep their trunk and legs as fixed as possible and to move only the right upper-limb.

A $Saga3^{RT}$ motion analysis system (Biogesta, Valenciennes–France) was used to collect kinematical data with six infrared cameras (50 Hz) inside a calibrated volume of 2 m $\times$ 1 m $\times$ 1 m. Force platform data (Kistler 5233A) were synchronized. Ten markers were fixed on the body: four on the right upper-limb, three on the thorax and three on a rigid tripod fixed to the acromion. The tripod was used to identify the scapula motion for locating the glenohumeral joint centre using the **SCoRE** method. Since only the right upper-limb was moving, only ten inertia parameters were excited: $K_2^{(2)}$, $m_2\mathbf{c}_2$ and $m_1 + m_2$ (see Equation (55)). As observed in the Figure 12, the solution is obtained in ten iterations because the conjugate gradient is a direct method. After the tenth iterations, the cost function is

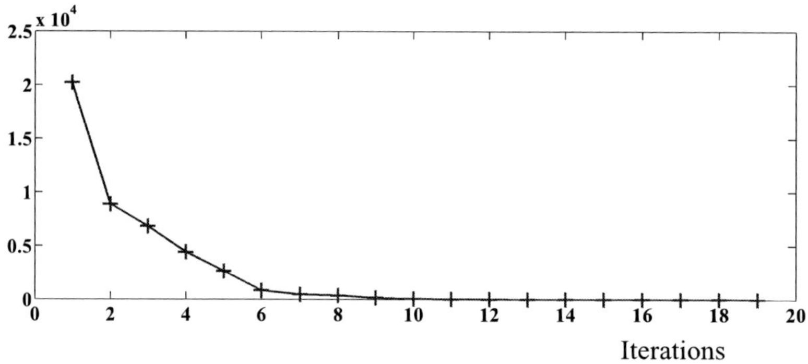

Figure 12: Evolution of the cost function during the iterations.

constant and the solution is obtained.

The inertia matrix identified is the Poinsot inertia matrix:

$$K_2^{(2)} = \int_{S_2} \left(\mathbf{b}_2^{(2)} \mathbf{b}_2^{(2)^T} \right) dm. \tag{59}$$

The usual inertia matrix is obtained from $K_2^{(2)}$ by the formula [63]:

$$J_2^{(2)} = \text{trace}\left(K_2^{(2)} \right) 1_3 - K_2^{(2)} \tag{60}$$

Table 2 presents the inertia matrix and the product of the mass by the centre of mass position ($m_2\mathbf{c}_2$ of the upper-limb obtained with the identification method. These parameters are compared with upper-arm inertial parameters estimated with the proportional model of Zatsiorsky modified by de Leva [31].

For two subjects (S1 and S3), the inertia moment along the longitudinal axis seems to be overestimated compared to the classical body segment inertial parameters. This result

Table 2: Inertia matrix ($kg.m^2$) and first order moment ($m_2\mathbf{c_2}$ in $kg.m$) of the upper-limb for the three subjects obtained with the identification method and with the proportional model of Zatsiorsky modified by de Leva [31].

	de Leva	Identification
S1 (1.80 m, 70 kg)	$\begin{bmatrix} 0.454 & 0 & 0 \\ 0 & 0.008 & 0 \\ 0 & 0 & 0.450 \end{bmatrix}$ $\begin{bmatrix} 0 & 1.0734 & 0 \end{bmatrix}^T$	$\begin{bmatrix} 0.476 & 0.001 & -0.018 \\ 0.001 & 0.040 & 0.035 \\ -0.018 & 0.035 & 0.486 \end{bmatrix}$ $\begin{bmatrix} 0.1123 & 1.2210 & -0.0485 \end{bmatrix}^T$
S2 (1.78 m, 78 kg)	$\begin{bmatrix} 0.495 & 0 & 0 \\ 0 & 0.009 & 0 \\ 0 & 0 & 0.490 \end{bmatrix}$ $\begin{bmatrix} 0 & 1.1954 & 0 \end{bmatrix}^T$	$\begin{bmatrix} 0.504 & 0.021 & -0.001 \\ 0.021 & 0.006 & -0.025 \\ -0.001 & -0.025 & 0.503 \end{bmatrix}$ $\begin{bmatrix} -0.0429 & 0.7603 & 0.0201 \end{bmatrix}^T$
S3 (1.78 m, 75 kg)	$\begin{bmatrix} 0.476 & 0 & 0 \\ 0 & 0.008 & 0 \\ 0 & 0 & 0.472 \end{bmatrix}$ $\begin{bmatrix} 0 & 1.1293 & 0 \end{bmatrix}^T$	$\begin{bmatrix} 0.512 & 0.005 & 0.006 \\ 0.005 & 0.033 & 0.058 \\ 0.006 & 0.058 & 0.528 \end{bmatrix}$ $\begin{bmatrix} 0.1200 & 1.0285 & 0.0787 \end{bmatrix}^T$

could be caused by a small elbow flexion. Unfortunately no marker was fixed to the forearm in order to verify this hypothesis. Identified inertia and first order moment $(m_2\mathbf{c}_2)$ values are realistic values. The main advantage of the identification method is to propose products of inertia and a centre of mass not aligned with the longitudinal axis of the segment. This signified that segment are not homogeneous and not symmetric whereas they are mainly represented in most of the geometric or proportional models. The drawback of this method is to not determine the mass of the segments. Since the centre of mass position and the mass do not appear independently in the Equation 55, only the first order moment is identified and it is not possible to dissociate the mass from the first order moment.

To conclude, the external identification is a recent approach for determining the human body segment inertial parameters. The experimental procedure requires only a motion analysis system and a force platform. The identified values of the matrix of inertia for the upper-limb are realistic values and close to the values estimated by Zatsiorsky's anthropometric model. The difficulty in interpreting results comes from the fact that there is no "gold standard" for the body segment inertial parameters. The values calculated from the model of Zatsiorsky give an idea of the order of magnitude but do not represent actual values. The presented identification method is a good way for the body segment inertial parameters determination without any assumption about the density of the segment or the segment shape.

4 Conclusion

This chapter has presented methods and algorithms for a more accurate estimation of human kinematics and kinetics. The first section focused on the rotation matrix determination and joint centre locations and the second section proposed a new approach for the estimation of the body segment inertial parameters.

As all experimental sciences, biomechanics is dependent on the measurement tool evolution. Nowadays, the high camera resolution and the high sampling rate allow new applications that have not been explored yet. Tracking ten markers on each segment during walking is no longer a problem. Some specific activities (sports for example) may be more difficult to analyse mainly because of marker self-occlusions. It was the purpose of the first section to evidence that using more than three markers on each segment improves significantly the kinematic. This result was already presented in 1995 by Challis [23] but this method is still not commonly used in human movement analysis. An increased number of markers improves the kinematics, especially the time derivative data (angular velocity, linear and angular acceleration) and improves the joint centre location.

Of course, our conclusion is not to put twenty-six markers on each segment as used in [46] for every day analyses because it is too time consuming. This solution accounts on the global or average segment kinematics but studies with a high number of markers will help to determine the best methods and marker placements. The set of experimental studies proposed in this chapter makes recommendation for an accurate human movement analysis:

- combination of flexion-extension, abduction-adduction and circumduction for estimating an optimal joint centre location;

- **SCoRE** is an accurate algorithm for locating joint centre because it is not sensitive to velocity and calculates an average from the optimal locations found in both proximal and distal segments;

- use a redundant marker set combined with a technique of local optimization;

- definition of the origin of the local frame using the marker centroid.

The proposed external identification method is suitable for a personalized estimation of body segment inertial parameters. Moreover, the procedure requires only materials currently used in biomechanical analysis, i.e. motion analysis system and force platform. The procedure proposed in this chapter involved a matrix formulation of the equations of motion for conserving the properties of the inertia matrix. This choice was relevant because all the body segment inertial parameters are grouped together in a symmetric and positive definite matrix. This help to restrict the solution (during the optimization procedure) to the cone of symmetric and positive definite matrix. This study may be continued with a three-segment model, for example by splitting the upper-limb in arm and forearm plus hand. This method needs other specific movements before collecting the movement of interest. As the time available with the subject (especially in clinical analysis) is sometimes short, this could be a limitation of this method. The main advantage of this approach is that no assumption is made about the density of the segment or about the shape of the segment. Indeed, the centre of mass is not located in the longitudinal axis but small components on antero-posterior and medio-lateral axes are evidenced.

5 Perspectives

This was not assessed in this chapter but the use of a redundant set of markers could be an alternative to raw data filtering or smoothing. The choice of the method of derivation or smoothing is never easy because the skin movement artefact has similar frequency to the skeletal kinematics. Derivation increases the noise and the time derivative data must be treated. Optimization method for calculating the rotation matrix is a method for decreasing the skin movement artefact and also the measurement error of the stereophotogrammetric system. In our results [8, 48, 57]) no treatment was applied to the raw data; future work will evaluate if the time derivative data need smoothing when they are calculated through optimized rotation matrix. Future work will use the method proposed in [46] in order to define the best markers set placement for an accurate and reliable determination of the segment kinematics from six markers for example. For that, inter-subjects and inter-movements effects on the markers placement will be measured.

In parallel to the experimental part, the theory on the rotation matrix calculation has to be further explored. In the method proposed, the rotation matrix between two orientations of the segment analysed is calculated by optimization. But, since we need the rotation matrix with respect to the global system of coordinates, we recalled this optimized rotation matrix with the rotation matrix between the orientation of the segment at the first frame with the global system of coordinates. An error in orientation associated with the first frame is then introduced. Future studies may focus on techniques for reducing this systematic error of orientation by using more key frames and calculating average axes.

As skin movement artefact affects differently each marker according to its location, using a redundant set of markers decreases the global skin movement artefact and improve the measurement of the underlying bone kinematics. This hypothesis can be achieved by comparison with invasive or fluoroscopic techniques only. In most applications soft tissue artefact has to be minimized because skeletal kinematics is needed (joint centre location, anatomical landmarks recalling). The inertia parameters of the soft tissue in motion affect the joint dynamics [52]. For specific applications like identification of the body segment parameters [47, 49], movement of the segment including soft tissue motion contribution is needed.

The matrix formulation of the equations of motion presented in this paper may be used for a n segment model. However, as explained in [25], the accuracy of the identified parameters depends on the length of the kinematic chain. They showed that the accuracy of the identified parameter strongly decreases above six to eight degrees of freedom in the kinematic chain. It is clearly inappropriate to use this method for all the body segments considering the high degrees of freedom. Because of the constraints associated to the matrix form of the equations of motion, our method should be more robust and accurate for multibody systems with several degrees of freedom. Although this method could not identify all the body segment parameters simultaneously, we expect that it should be used for each limb. For most of the subject studied, constraints consisting of searching the same body segment inertial properties for the right and left limb may be introduced. This could improve the method in terms of reliability and easiness. The next evolution of this method will focus on specific movements that must be performed for an accurate identification. As the identification is sensitive to the input data, results of the optimization method are different according to the movement realized. Body segment inertial parameters calculated with the identification approach from several movements have to be compared with reference parameters (medical imaging?) in order to select the optimal movements.

Finally, it might be interesting to combine the identification with functional methods to determine with the same setup movements both the inertial parameters and the joint centre locations.

References

[1] Andrews, JG; Mish, SP. Methods for investigating the sensitivity of joint resultants to body segment parameter variations. *Journal of Biomechanics*, 1996 29(5), 651–654.

[2] Antonelli, G; Caccavale, F; Chiacchio, P. Systematic procedure for the identification of dynamic parameters of robot manipulators. *Robotica*, 1999 17(4), 427–435.

[3] Arteaga, MA. Robot control and parameter estimation with only joint position measurements. *Automatica*, 2003 39(1), 67–73.

[4] Arthurs, KL; Andrews, DM. Upper extremity soft and rigid tissue mass prediction using segment anthropometric measures and dxa. *Journal of Biomechanics*, 2009 42(3), 389–394.

[5] Atchonouglo, EK; Vallée, C; Monnet, T; Fortuné, D. Identification of the ten inertia parameters of a rigid body. *Journal of Applied Mathematics and Mechanics*, 2008 72(1), 22–25.

[6] Atkeson, CG; An, CH; Hollerbach, JM. Estimation of inertial parameters of manipulator loads and links. *International Journal of Robotics Research*, 1986 5(3), 101–119.

[7] Begon, M; Monnet, T; Lacouture, P. Effects of movement for estimating the hip joint centre. *Gait & Posture*, 2007 25 (3), 353–359.

[8] Begon, M; Roosen, A; Pain, MTG; Monnet, T. Effect of segment-embedded reference frame definition on hip joint centre location during walking. *Journal of biomechanics*, In Submission.

[9] Begon, M; Wieber, P-B; Yeadon, MR. Kinematics estimation of straddled movements on high bar from a limited number of skin markers using a chain model. *Journal of Biomechanics*, 2008 41(3), 581–586.

[10] Bell, A; Pedersen, D; Brand, R. A comparison of the accuracy of several hip center location prediction methods. *Journal of Biomechanics*, 1990 23, 617–621.

[11] Benedetti, MG; Catani, F; Leardini, A; Pignotti, E; Giannini, S. Data management in gait analysis for clinical applications. *Clinical Biomechanics*, 1998 13(3), 204–215.

[12] Bennis, F; Khalil, W; Gautier, M. Calculation of the base inertial parameters of closed-loops robots. In *Proceedings - IEEE International Conference on Robotics and Automation*, 1992 1, 370–375.

[13] Bjornstrup, J. Estimation of human body segment parameters. 1996. Available from Inter Tech-Report, http://www.cvmt.auc.dk/jorgen/PhD/.

[14] Bouby, C; Fortuné, D; Vallée, C; Pietraszkiewicz, W. Direct determination of the rotation in the polar decomposition of the deformation gradient by maximizing a rayleigh quotient. *ZAMM Zeitschrift fur Angewandte Mathematik und Mechanik*, 2005 85(3), 155–162.

[15] Brousse, P. Optimization in mechanics: problems and methods. *North Holland Editor*, Amsterdam, 1988.

[16] Camomilla, V; Cereatti, A; Vannozzi, G; Cappozzo, A. An optimized protocol for hip joint centre determination using the functional method. *Journal of Biomechanics*, 2006 39(6), 1096–1106.

[17] Cappozzo, A. Gait analysis methodology. *Human Movement Science*, 1984 3, 27–50.

[18] Cappozzo, A. Three-dimensional analysis of human walking: experimental methods and associated artefacts. *Human Movement Science*, 1991 10, 589–602.

[19] Cappozzo, A; Catani, F; Della Croce, U; Leardini, A. Position and orientation in space of bones during movement: anatomical frame definition and determination. *Clinical Biomechanics*, 1995 10, 171–178.

[20] Cappozzo, A; Catani, F; Leardini, F; Benedetti, MG; Della Croce, U. Position and orientation in space of bones during movement: experimental artefacts. *Clinical Biomechanics*, 1996 11(2), 90–100.

[21] Carman, AB; Milburn, PD. Determining rigid body transformation parameters from ill-conditioned spatial marker co-ordinates. *Journal of Biomechanics*, 2006 39(10), 1778–1786.

[22] Challis, JH. An examination of procedures for determining body segment attitude and position from noisy biomechanical data. *Medical Engineering & Physics*, 1995 17(2), 83–90.

[23] Challis, JH. A procedure for determining rigid body transformation parameters. *Journal of Biomechanics*, 1995 28(6), 733–737.

[24] Challis, JH; Kerwin, DG. Quantification of the uncertainties in resultant joint moments computed in a dynamic activity. *Journal of Sports Sciences*, 1996 14(3), 219–231.

[25] Chenut, X; Fisette, P; Samin, J-Cl. Recursive formalism with a minimal dynamic parameterization for the identification and simulation of multibody systems. Application to the human body. *Multibody System Dynamics*, 2002 8(2), 117–140.

[26] Chèze, L; Fregly, BJ; Dimnet, J. A solidification procedure to facilitate kinematic analyses based on video system data. *Journal of Biomechanics*, 1995 28, 879-884

[27] Croce, UD; Cappozzo, A; Kerrigan, C; Lucchetti, L. Bone position and orientation errors: Pelvis and lower limb anatomical landmark identification reliability. *Gait & Posture*, 1997 5, 156-157

[28] Davis, RB; Ounpuu, S; Tyburski, D; Gage, JR. A gait analysis data collection and reduction technique. *Human Movement Science*, 1991 10(5), 575–587.

[29] Davis, RB; Ounpuu, S; De Luca, PA. Joint moments: Evaluation of ground reaction, inertial and segmental weight effects. *Gait & Posture*, 1994 2(1), 58–58.

[30] De Groote, F; De Laet, T; Jonkers, I; De Schutter, J. Kalman smoothing improves the estimation of joint kinematics and kinetics in marker-based human gait analysis. *Journal of Biomechanics*, 2008 41(16), 3390–3398.

[31] de Leva, P. Adjustments to Zatsiorsky-Seluyanov's segment inertia parameters. *Journal of Biomechanics*, 1996 29(9), 1223–1230.

[32] Ehrig, RM; Taylor, WR; Duda, GN; Heller, MO. A survey of formal methods for determining the centre of rotation of ball joints. *Journal of Biomechanics*, 2006 39(5), 2798–2809.

[33] Ehrig, RM; Taylor, WR; Duda, GN; Heller, MO. A survey of formal methods for determining functional joint axes. *Journal of Biomechanics*, 2007 40(10), 2150 – 2157.

[34] Fayet, M; Pfister, F. Analysis of multibody systems with indirect coordinates and global inertia tensors. *European Journal of Mechanics - A/Solids*, 1994 13, 431–457.

[35] Fayet, M; Sandel, A; Maiffredy, L. Utilisation des tenseurs d'inertie globaux dans le cas de systèmes à liaisons complexes et pseudo-paramètres. *17ème congrès français de mécanique*, 2005 Troyes.

[36] Ganley, KJ; Powers, CM. Anthropometric parameters in children: a comparison of values obtained from dual energy x-ray absorptiometry and cadaver-based estimates. *Gait & Posture*, 2004 19(2), 133–140.

[37] Gao, B; Zheng, N. Investigation of soft tissue movement during level walking: Translations and rotations of skin markers. *Journal of Biomechanics*, 2008 41(15), 3189–3195.

[38] Kadaba, M; Ramakrishnan, H; Wooten, M; Gainey, J; Gorton, G; Cochran, G. Repeatability of kinematic, kinetic and electromyographic data in normal adult gait. *Journal of Orthopaedic Research*, 1989 7, 849–860.

[39] Kadaba, MP; Ramakrishnan, HK; Wootten, ME. Measurement of lower extremity kinematics during level walking. *Journal of Orthopaedic Research*, 1990 8(3), 383–392.

[40] Kingma, I; Toussaint, H; De Looze, M; Dieen, J. Segment inertial parameter evaluation in two anthropometric models by application of a dynamic linked segment model. *Journal of Biomechanics*, 1996 29, 693–704.

[41] Kozlowski, K. Computational requirements for a discrete kalman filter in robot dynamics algorithms. *Robotica*, 1993 11, 27–36.

[42] Lee, MK; Le, NS; Fang, AC; Koh, MTH. Measurement of body segment parameters using dual energy x-ray absorptiometry and three-dimensional geometry: An application in gait analysis. *Journal of Biomechanics*, 2009 42(3), 217–222.

[43] Legnani, G; Casolo, F; Righettini, P; Zappa, B. A homogeneous matrix approach to 3D kinematics and dynamics – I. theory. *Mechanism and Machine Theory*, 1996 31(5), 573–587.

[44] Lu, TW; O'Connor, J. Bone position estimation from skin marker co-ordinates using global optimisation with joint constraints. *Journal of Biomechanics*, 1999 32(2), 129–134.

[45] Meskers, CGM; van der Helm, FCT; Rozendaal, LA; Rozing, PM. In vivo estimation of the glenohumeral joint rotation center from scapular bony landmarks by linear regression. *Journal of Biomechanics*, 1997 31(1), 129–134.

[46] Monnet, T; Begon, M; Pain, MTG; Colloud, F. Iterative solidification procedure to determine the optimal marker location. In *10th international symposium on the 3D Analysis of Human Movement. Amsterdam 28-31 October*, 2008.

[47] Monnet, T; Begon, M; Vallée, C. Identification des paramètres inertiels du corps humain. *18ème congrès Franąis de Mécanique*, Grenoble, 2007.

[48] Monnet, T; Desailly, E; Begon, M; Vallée, C; Lacouture, P. Comparison of the **SCoRE** and **HA** methods for locating *in vivo* the glenohumeral joint centre. *Journal of Biomechanics*, 2007 40, 3487–3492.

[49] Monnet, T; Vallée, C; Lacouture, P. Identification of the inertia matrix of the upper limb. *21th congress of the International Society of Biomechanics*, Taipei, 2007.

[50] Neptune, R; Hull, M. Accuracy assessment of methods for determining hip movement in seated cycling. *Journal of Biomechanics*, 1995 28, 423–437.

[51] Nikolova, GS; Toshev, YE. Estimation of male and female body segment parameters of the bulgarian population using a 16-segmental mathematical model. *Journal of Biomechanics*, 2007 40, 3700–3707.

[52] Pain, MTG; Challis, JH. The influence of soft tissue movement on ground reaction forces, joint torques and joint reaction forces in drop landings. *Journal of Biomechanics*, 2006 39(1), 119–124.

[53] Pearsall, DJ; Costigan, PA. The effect of segment parameter error on gait analysis results. *Gait & Posture*, 1999 9(3), 173–183.

[54] Piazza, SJ; Erdemir, A; Okita, N; Cavanagh, PR. Assessment of the functional method of hip joint center location subject to reduced range of hip motion. *Journal of Biomechanics*, 2004 37(3), 349–356.

[55] Rao, G; Amarantini, D; Berton, E; Favier, D. Influence of body segments' parameters estimation models on inverse dynamics solutions during gait. *Journal of Biomechanics*, 2006 39(8), 1531–1536.

[56] Raucent, B; Campion, B; Bastin, G; Samin, J-C; Willems, PY. Identification of the barycentric parameters of robot manipulators from external measurements. *Automatica*, 1992 28(5), 1011–1016.

[57] Roosen, A; Pain, MTG; Begon, M. Limitations of functionally determined joint centres for analysis of athletic human movement: A case study of the upper limb. *Journal of Applied Biomechanics*, In Press.

[58] Rouhani, H; Favre, J; Crevoisier, X; Jolles, BM; Aminian, K. Validation of multisegment models for foot and ankle complex. *Journal of Biomechanics*, 2008 41(Supplement 1), S168–S168.

[59] Roux, E; Bouilland, S; Godillon-Maquinghen, A-P; Bouttens, D. Evaluation of the global optimisation method within the upper limb kinematics analysis. *Journal of Biomechanics*, 2002 35, 1279–1283.

[60] Silva, MPT; Ambrosio, JAC. Sensitivity of the results produced by the inverse dynamic analysis of a human stride to perturbed input data. *Gait & Posture*, 2004 19(1), 35–49.

[61] Souriau, J-M. Géométrie et relativité. *Editions Hermann*, Paris, 1964.

[62] Stagni, R; Fantozzi, S; Cappello, A; Leardini, A. Quantification of soft tissue artefact in motion analysis by combining 3d fluoroscopy and stereophotogrammetry: a study on two subjects. *Clinical Biomechanics*, 2005 20(3), 320 – 329.

[63] Vallée, C; Hamdouni, A; Isnard, F; Fortuné, D. The equations of motion of a rigid body without parametrization of rotations. *Journal of Applied Mathematics and Mechanics*, 1999 63(1), 25–30.

[64] Vaughan, CL; Davis, BL; O'Connor, JC. *Dynamics of Human Gait*. Human Kinetics Publishers, 1992.

[65] Vaughan, CL; Andrews, JG; Hay, JG. Selection of body segment parameters by optimization methods. *Journal of Biomechanical Engineering*, 1982 104(1), 38–44.

[66] Veldpaus, F; Woltring, H; Dortmans, L. A least-square algorithm for the equiform transformation from spatial marker coordinates. *Journal of Biomechanics*, 1988 21, 45–54.

[67] Wicke, J; Dumas, GA; Costigan, PA. A comparison between a new model and current models for estimating trunk segment inertial parameters. *Journal of Biomechanics*, 2009 42(1), 55–60.

[68] Winter, DA. Biomechanics and motor control of human movement. *Wiley interSciences*, second edition, 1990.

[69] Woltring, H. Data processing and error analysis. 1990. In Cappozzo & Berme.

[70] Woltring, H. Representation and calculation of 3D joint movement. *Human movement Sciences*, 1991 10, 603–616.

INDEX

A

B

D

E

F

I

J

K

Q

R

S